AN INTRODUCTION
TO THE
THEORY OF NUMBERS

AN INTRODUCTION
TO THE
THEORY OF NUMBERS

BY

G. H. HARDY

AND

E. M. WRIGHT

Principal and Vice-Chancellor Emeritus of the University of Aberdeen

SIXTH EDITION

OXFORD
UNIVERSITY PRESS

Great Clarendon Street, Oxford OX2 6DP

Oxford University Press is a department of the University of Oxford.
It furthers the University's objective of excellence in research, scholarship,
and education by publishing worldwide in

Oxford New York

Auckland Cape Town Dar es Salaam Hong Kong Karachi
Kuala Lumpur Madrid Melbourne Mexico City Nairobi
New Delhi Shanghai Taipei Toronto

With offices in

Argentina Austria Brazil Chile Czech Republic France Greece
Guatemala Hungary Italy Japan Poland Portugal Singapore
South Korea Switzerland Thailand Turkey Ukraine Vietnam

Oxford is a registered trade mark of Oxford University Press
in the UK and in certain other countries

Published in the United States
by Oxford University Press Inc., New York

First edition 1938
Second edition 1945
Third edition 1954
Fourth edition 1960
Fifth edition 1979
Sixth edition 2008

Reprinted 1983, 1985, 1988 (with corrections), 1989,
1990, 1992, 1993, 1994, 1995, 1996, 1998,
2000 (with additional General Index), 2002, 2003 (twice), 2004, 2005, 2006

Printed in Great Britain
on acid-free paper by
Clays Ltd, St Ives plc

ISBN 978–0–19–921985–8
ISBN 978–0–19–921986–5 (Pbk)

3

FOREWORD BY ANDREW WILES

I had the great good fortune to have a high school mathematics teacher who had studied number theory. At his suggestion I acquired a copy of the fourth edition of Hardy and Wright's marvellous book *An Introduction to the Theory of Numbers*. This, together with Davenport's *The Higher Arithmetic*, became my favourite introductory books in the subject. Scouring the pages of the text for clues about the Fermat problem (I was already obsessed) I learned for the first time about the real breadth of number theory. Only four of the chapters in the middle of the book were about quadratic fields and Diophantine equations, and much of the rest of the material was new to me; Diophantine geometry, round numbers, Dirichlet's theorem, continued fractions, quaternions, reciprocity . . . The list went on and on.

The book became a starting point for ventures into the different branches of the subject. For me the first quest was to find out more about algebraic number theory and Kummer's theory in particular. The more analytic parts did not have the same attraction then and did not really catch my imagination until I had learned some complex analysis. Only then could I appreciate the power of the zeta function. However, the book was always there as a starting point which I could return to whenever I was intrigued by a new piece of theory, sometimes many years later. Part of the success of the book lay in its extensive notes and references which gave navigational hints for the inexperienced mathematician. This part of the book has been updated and extended by Roger Heath-Brown so that a 21st-century-student can profit from more recent discoveries and texts. This is in the style of his wonderful commentary on Titchmarsh's *The Theory of the Riemann Zeta Function*. It will be an invaluable aid to the new reader but it will also be a great pleasure to those who have read the book in their youth, a bit like hearing the life stories of one's erstwhile school friends.

A final chapter has been added giving an account of the theory of elliptic curves. Although this theory is not described in the original editions (except for a brief reference in the notes to §13.6) it has proved to be critical in the study of Diophantine equations and of the Fermat equation in particular. Through the Birch and Swinnerton-Dyer conjecture on the one hand and through the extraordinary link with the Fermat equation on the other it has become a central part of the number theorist's life. It even played a central role in the effective resolution of a famous class number problem of Gauss. All this would have seemed absurdly improbable when the book was written. It is thus an appropriate ending for the new edition to have a lucid exposition of this theory by Joe Silverman. Of course it is only a quick sketch of the theory and the reader will surely be tempted to devote many hours, if not the best part of a lifetime, to unravelling its many mysteries.

A.J.W.

January, 2008

PREFACE TO THE SIXTH EDITION

This sixth edition contains a considerable expansion of the end-of-chapter notes. There have been many exciting developments since these were last revised, which are now described in the notes. It is hoped that these will provide an avenue leading the interested reader towards current research areas. The notes for some chapters were written with the generous help of other authorities. Professor D. Masser updated the material on Chapters 4 and 11, while Professor G.E. Andrews did the same for Chapter 19. A substantial amount of new material was added to the notes for Chapter 21 by Professor T.D. Wooley, and a similar review of the notes for Chapter 24 was undertaken by Professor R. Hans-Gill. We are naturally very grateful to all of them for their assistance.

In addition, we have added a substantial new chapter, dealing with elliptic curves. This subject, which was not mentioned in earlier editions, has come to be such a central topic in the theory of numbers that it was felt to deserve a full treatment. The material is naturally connected with the original chapter on Diophantine Equations.

Finally, we have corrected a significant number of misprints in the fifth edition. A large number of correspondents reported typographical or mathematical errors, and we thank everyone who contributed in this way.

The proposal to produce this new edition originally came from Professors John Maitland Wright and John Coates. We are very grateful for their enthusiastic support.

D.R.H.-B.
J.H.S.

September, 2007

PREFACE TO THE FIFTH EDITION

The main changes in this edition are in the Notes at the end of each chapter. I have sought to provide up-to-date references for the reader who wishes to pursue a particular topic further and to present, both in the Notes and in the text, a reasonably accurate account of the present state of knowledge. For this I have been dependent on the relevant sections of those invaluable publications, the *Zentralblatt* and the *Mathematical Reviews*. But I was also greatly helped by several correspondents who suggested amendments or answered queries. I am especially grateful to Professors J. W. S. Cassels and H. Halberstam, each of whom supplied me at my request with a long and most valuable list of suggestions and references.

There is a new, more transparent proof of Theorem 445 and an account of my changed opinion about Theodorus'method in irrationals. To facilitate the use of this edition for reference purposes, I have, so far as possible, kept the page numbers unchanged. For this reason, I have added a short appendix on recent progress in some aspects of the theory of prime numbers, rather than insert the material in the appropriate places in the text.

E. M. W.

ABERDEEN
October 1978

PREFACE TO THE FIRST EDITION

This book has developed gradually from lectures delivered in a number of universities during the last ten years, and, like many books which have grown out of lectures, it has no very definite plan.

It is not in any sense (as an expert can see by reading the table of contents) a systematic treatise on the theory of numbers. It does not even contain a fully reasoned account of any one side of that many-sided theory, but is an introduction, or a series of introductions, to almost all of these sides in turn. We say something about each of a number of subjects which are not usually combined in a single volume, and about some which are not always regarded as forming part of the theory of numbers at all. Thus chs. XII–XV belong to the 'algebraic' theory of numbers, Chs. XIX–XXI to the 'addictive', and Ch. XXII to the 'analytic' theories; while Chs. III, XI, XXIII, and XXIV deal with matters usually classified under the headings of 'geometry of numbers' or 'Diophantine approximation'. There is plenty of variety in our programme, but very little depth; it is impossible, in 400 pages, to treat any of these many topics at all profoundly.

There are large gaps in the book which will be noticed at once by any expert. The most conspicuous is the omission of any account of the theory of quadratic forms. This theory has been developed more systematically than any other part of the theory of numbers, and there are good discussions of it in easily accessible books. We had to omit something, and this seemed to us the part of the theory where we had the least to add to existing accounts.

We have often allowed out personal interests to decide out programme, and have selected subjects less because of their importance (though most of them are important enough) than because we found them congenial and because other writers have left us something to say. Our first aim has been to write an interesting book, and one unlike other books. We may have succeeded at the price of too much eccentricity, or we may have failed; but we can hardly have failed completely, the subject-matter being so attractive that only extravagant incompetence could make it dull.

The book is written for mathematicians, but it does not demand any great mathematical knowledge or technique. In the first eighteen chapters we assume nothing that is not commonly taught in schools, and any intelligent university student should find them comparatively easy reading. The last six are more difficult, and in them we presuppose a little more, but nothing beyond the content of the simpler university courses.

The title is the same as that of a very well-known book by Professor L. E. Dickson (with which ours has little in common). We proposed at one time to change it to *An introduction to arithmetic*, a more novel and in some ways a more appropriate title; but it was pointed out that this might lead to misunderstandings about the content of the book.

A number of friends have helped us in the preparation of the book. Dr. H. Heilbronn has read all of it both in manuscript and in print, and his criticisms and suggestions have led to many very substantial improvements, the most important of which are acknowledged in the text. Dr. H. S. A. Potter and Dr. S. Wylie have read the proofs and helped us to remove many errors and obscurities. They have also checked most of the references to the literature in the notes at the ends of the chapters. Dr. H. Davenport and Dr. R. Rado have also read parts of the book, and in particular the last chapter, which, after their suggestions and Dr. Heilbronn's, bears very little resemblance to the original draft.

We have borrowed freely from the other books which are catalogued on pp. 417–19 [pp. 596–9 in current 6th edn.], and especially from those of Landau and Perron. To Landau in particular we, in common with all serious students of the theory of numbers, owe a debt which we could hardly overstate.

G. H. H.
E. M. W.

OXFORD
August 1938

REMARKS ON NOTATION

We borrow four symbols from formal logic, viz.

$$\rightarrow, \equiv, \exists, \in .$$

$\rightarrow$ is to be read as 'implies'. Thus

$$l \mid m \rightarrow l \mid n \quad \text{(p. 2)}$$

means ' "l is a divisor of m" implies "l is a divisor of n" ', or, what is the same thing, 'if l divides m then l divides n'; and

$$b \mid a \,.\, c \mid b \rightarrow c \mid a \quad \text{(p. 1)}$$

means 'if b divides a and c divides b then c divides a'.

$\equiv$ is to be read 'is equivalent to'. Thus

$$m \mid ka - ka' \equiv m_1 \mid a - a' \quad \text{(p. 61)}$$

means that the assertions 'm divides $ka—ka'$' and 'm_1 divides $a—a'$' are equivalent; either implies the other.

These two symbols must be distinguished carefully from $\rightarrow$ (tends to) and $\equiv$ (is congruent to). There can hardly be any misunderstanding, since $\rightarrow$ and $\equiv$ are always relations between *propositions*.

$\exists$ is to be read as 'there is an'. Thus

$$\exists l \,.\, 1 < l < m \,.\, l \mid m \quad \text{(p. 2)}$$

means 'there is an l such that (i) $1 < l < m$ and (ii) l divides m'.

$\in$ is the relation of a member of a class to the class. Thus

$$m \in S \,.\, n \in S \rightarrow (m \pm n) \in S \quad \text{(p. 23)}$$

means 'if m and n are members of S then $m + n$ and $m - n$ are members of S'.

A star affixed to the number of a theorem (e.g. Theorem 15*) means that the proof of the theorem is too difficult to be included in the book. It is not affixed to theorems which are not proved but may be proved by arguments similar to those used in the text.

CONTENTS

I

THE SERIES OF PRIMES (1)

1.1. Divisibility of integers. The numbers

$$\ldots, -3, -2, -1, 0, 1, 2, \ldots$$

are called the *rational integers,* or simply the *integers*; the numbers

$$0, 1, 2, 3, \ldots$$

the *non-negative integers*; and the numbers

$$1, 2, 3, \ldots$$

the *positive integers.* The positive integers form the primary subject-matter of arithmetic, but it is often essential to regard them as a subclass of the integers or of some larger class of numbers.

In what follows the letters

$$a, b, \ldots, n, p, \ldots, x, y, \ldots$$

will usually denote integers, which will sometimes, but not always, be subject to further restrictions, such as to be positive or non-negative. We shall often use the word 'number' as meaning 'integer' (or 'positive integer', etc.), when it is clear from the context that we are considering only numbers of this particular class.

An integer a is said to be *divisible* by another integer b, not 0, if there is a third integer c such that

$$a = bc.$$

If a and b are positive, c is necessarily positive. We express the fact that a is divisible by b, or b is a *divisor* of a, by

$$b|a.$$

Thus

$$1|a, \quad a|a;$$

and $b|0$ for every b but 0. We shall also sometimes use

$$b \nmid a$$

to express the contrary of $b|a$. It is plain that

$$b|a \,.\, c|b \rightarrow c|a,$$
$$b|a \rightarrow bc|ac$$

if $c \neq 0$, and

$$c|a \,.\, c|b \rightarrow c|ma + nb$$

for all integral m and n.

1.2. Prime numbers. In this section and until § 2.9 the numbers considered are generally positive integers.† Among the positive integers there is a sub-class of peculiar importance, the class of primes. A number p is said to be *prime* if

(i) $p > 1$,
(ii) p has no positive divisors except 1 and p.

For example, 37 is a prime. It is important to observe that 1 is not reckoned as a prime. In this and the next chapter we reserve the letter p for primes.‡

A number greater than 1 and not prime is called *composite*.

Our first theorem is

THEOREM 1. *Every positive integer, except* 1, *is a product of primes.*

Either n is prime, when there is nothing to prove, or n has divisors between 1 and n. If m is the least of these divisors, m is prime; for otherwise

$$\exists l \,.\, 1 < l < m \,.\, l|m;$$

and

$$l|m \rightarrow l|n,$$

which contradicts the definition of m.

Hence n is prime or divisible by a prime less than n, say p_1, in which case

$$n = p_1 n_1, \quad 1 < n_1 < n.$$

† There are occasional exceptions, as in §§ 1.7, where e^x is the exponential function of analysis.

‡ It would be inconvenient to have to observe this convention rigidly throughout the book, and we often depart from it. In Ch. IX, for example, we use p/q for a typical rational fraction, and p is not usually prime. But p is the 'natural' letter for a prime, and we give it preference when we can conveniently.

Here either n_1 is prime, in which case the proof is completed, or it is divisible by a prime p_2 less than n_1, in which case

$$n = p_1 n_1 = p_1 p_2 n_2, \quad 1 < n_2 < n_1 < n.$$

Repeating the argument, we obtain a sequence of decreasing numbers $n, n_1, \ldots, n_{k-1}, \ldots$, all greater than 1, for each of which the same alternative presents itself. Sooner or later we must accept the first alternative, that n_{k-1} is a prime, say p_k, and then

$$n = p_1 p_2 \ldots p_k. \tag{1.2.1}$$

Thus

$$666 = 2.3.3.37.$$

If $ab = n$, then a and b cannot both exceed $\sqrt{n}$. Hence any composite n is divisible by a prime p which does not exceed $\sqrt{n}$.

The primes in (1.2.1) are not necessarily distinct, nor arranged in any particular order. If we arrange them in increasing order, associate sets of equal primes into single factors, and change the notation appropriately, we obtain

$$n = p_1^{a_1} p_2^{a_2} \ldots p_k^{a_k} \quad (a_1 > 0, a_2 > 0, \ldots, p_1 < p_2 < \ldots). \tag{1.2.2}$$

We then say that n is expressed in *standard form*.

1.3. Statement of the fundamental theorem of arithmetic. There is nothing in the proof of Theorem 1 to show that (1.2.2) is a *unique* expression of n, or, what is the same thing, that (1.2.1) is unique except for possible rearrangement of the factors; but consideration of special cases at once suggests that this is true.

THEOREM 2 (THE FUNDAMENTAL THEOREM OF ARITHMETIC). *The standard form of n is unique; apart from rearrangement of factors, n can be expressed as a product of primes in one way only.*

Theorem 2 is the foundation of systematic arithmetic, but we shall not use it in this chapter, and defer the proof to § 2.10. It is however convenient to prove at once that it is a corollary of the simpler theorem which follows.

THEOREM 3 (EUCLID'S FIRST THEOREM). *If p is prime, and $p \mid ab$, then $p \mid a$ or $p \mid b$.*

We take this theorem for granted for the moment and deduce Theorem 2. The proof of Theorem 2 is then reduced to that of Theorem 3, which is given in § 2.10.

It is an obvious corollary of Theorem 3 that

$$p|abc\dots l \to p|a \text{ or } p|b \text{ or } p|c\dots \text{ or } p|l,$$

and in particular that, if $a, b, \dots, l$ are primes, then p is one of $a, b, \dots, l$. Suppose now that

$$n = p_1^{a_1} p_2^{a_2} \dots p_k^{a_k} = q_1^{b_1} q_2^{b_2} \dots q_j^{b_j},$$

each product being a product of primes in standard form. Then $p_i | q_1^{b_1} \dots q_j^{b_j}$ for every i, so that every p is a q; and similarly every q is a p. Hence $k = j$ and, since both sets are arranged in increasing order, $p_i = q_i$ for every i.

If $a_i > b_i$, and we divide by p_i^{bi}, we obtain

$$p_1^{a_1} \dots p_i^{a_i - b_i} \dots p_k^{a_k} = p_1^{b_1} \dots p_{i-1}^{b_{i-1}} p_{i+1}^{b_{i+1}} \dots p_k^{b_k}.$$

The left-hand side is divisible by p_i, while the right-hand side is not; a contradiction. Similarly $b_i > a_i$ yields a contradiction. It follows that $a_i = b_i$, and this completes the proof of Theorem 2.

It will now be obvious why 1 should not be counted as a prime. If it were, Theorem 2 would be false, since we could insert any number of unit factors.

1.4. The sequence of primes. The first primes are

$$2, 3, 5, 7, 11, 13, 17, 19, 23, 29, 31, 37, 41, 43, 47, 53, \dots.$$

It is easy to construct a table of primes, up to a moderate limit N, by a procedure known as the 'sieve of Eratosthenes'. We have seen that if $n \leqslant N$, and n is not prime, then n must be divisible by a prime not greater than $\sqrt{N}$. We now write down the numbers

$$2, 3, 4, 5, 6, \dots, N$$

and strike out successively

(i) $4, 6, 8, 10, \dots$, i.e. 2^2 and then every even number,

(ii) $9, 15, 21, 27, \dots$, i.e. 3^2 and then every multiple of 3 not yet struck out,

(iii) $25, 35, 55, 65, \ldots$, i.e. 5^2, the square of the next remaining number after 3, and then every multiple of 5 not yet struck out, ...

We continue the process until the next remaining number, after that whose multiples were cancelled last, is greater than $\sqrt{N}$. The numbers which remain are primes. All the present tables of primes have been constructed by modifications of this procedure.

The tables indicate that the series of primes is infinite. They are complete up to 100,000,000; the total number of primes below 10 million is 664,579; and the number between 9,900,000 and 10,000,000 is 6,134. The total number of primes below 1,000,000,000 is 50,847,478; these primes are not known individually. A number of very large primes, mostly of the form $2^p - 1$ (see §2.5), are also known; the largest found so far has just over 6,500 digits.†

These data suggest the theorem

THEOREM 4 (EUCLID'S SECOND THEOREM). *The number of primes is infinite.*

We shall prove this in § 2.1.

The 'average' distribution of the primes is very regular; its density shows a steady but slow decrease. The numbers of primes in the first five blocks of 1,000 numbers are

$$168, 135, 127, 120, 119,$$

and those in the last five blocks of 1,000 below 10,000,000 are

$$62, 58, 67, 64, 53.$$

The last 53 primes are divided into sets of

$$5, 4, 7, 4, 6, 3, 6, 4, 5, 9$$

in the ten hundreds of the thousand.

On the other hand the distribution of the primes in detail is extremely irregular.

In the first place, the tables show at intervals long blocks of composite numbers. Thus the prime 370,261 is followed by 111 composite numbers. It is easy to see that these long blocks must occur. Suppose that

$$2, 3, 5, \ldots, p$$

† See the end of chapter notes.

are the primes up to p. Then all numbers up to p are divisible by one of these primes, and therefore, if

$$2.3.5\ldots p = q,$$

all of the $p-1$ numbers

$$q+2, q+3, q+4, \ldots, q+p$$

are composite. If Theorem 4 is true, then p can be as large as we please; and otherwise all numbers from some point on are composite.

THEOREM 5. *There are blocks of consecutive composite numbers whose length exceeds any given number N.*

On the other hand, the tables indicate the indefinite persistence of prime-pairs, such as 3, 5 or 101, 103, differing by 2. There are 1,224 such pairs $(p, p+2)$ below 100,000, and 8,169 below 1,000,000. The evidence, when examined in detail, appears to justify the conjecture

There are infinitely many prime-pairs $(p, p+2)$.

It is indeed reasonable to conjecture more. The numbers $p, p+2, p+4$ cannot all be prime, since one of them must be divisible by 3; but there is no obvious reason why $p, p+2, p+6$ should not all be prime, and the evidence indicates that such prime-triplets also persist indefinitely. Similarly, it appears that triplets $(p, p+4, p+6)$ persist indefinitely. We are therefore led to the conjecture

There are infinitely many prime-triplets of the types $(p, p+2, p+6)$ *and* $(p, p+4, p+6)$.

Such conjectures, with larger sets of primes, may be multiplied, but their proof or disproof is at present beyond the resources of mathematics.

1.5. Some questions concerning primes. What are the natural questions to ask about a sequence of numbers such as the primes? We have suggested some already, and we now ask some more.

(1) *Is there a simple general formula for the n-th prime* p_n† (a formula, that is to say, by which we can calculate the value of p_n for any given n with less labour than by the use of the sieve of Eratosthenes)? No such formula is known and it is unlikely that such a formula is possible.

† See the end of chapter notes.

On the other hand, it is possible to devise a number of 'formulae' for p_n. Of these, some are no more than curiosities since they define p_n in terms of itself, and no previously unknown p_n can be calculated from them. We give an example in Theorem 419. Others would in theory enable us to calculate p_n, but only at the cost of substantially more labour than does the sieve of Eratosthenes. Others still are essentially equivalent to that sieve. We return to these questions in § 2.7 and in §§ 1, 2 of the Appendix.

Similar remarks apply to another question of the same kind, viz.

(2) *is there a simple general formula for the prime which follows a given prime* (i.e. a recurrence formula such as $p_{n+1} = p_n^2 + 2$)?

Another natural question is

(3) *is there a rule by which, given any prime p, we can find a larger prime q?*

This question of course presupposes that, as stated in Theorem 4, the number of primes is infinite. It would be answered in the affirmative if any simple function $f(n)$ were known which assumed prime values for all integral values of n. Apart from trivial curiosities of the kind already mentioned, no such function is known. The only plausible conjecture concerning the form of such a function was made by Fermat,† and Fermat's conjecture was false.

Our next question is

(4) *how many primes are there less than a given number x?*

This question is a much more profitable one, but it requires careful interpretation. Suppose that, as is usual, we define $\pi(x)$ to be the number of primes which do not exceed x, so that $\pi(1) = 0$, $\pi(2) = 1$, $\pi(20) = 8$. If p_n is the nth prime then $\pi(p_n) = n$, so that $\pi(x)$, as function of x, and p_n, as function of n, are inverse functions. To ask for an exact formula for $\pi(x)$, of any simple type, is therefore practically to repeat question (1).

We must therefore interpret the question differently, and ask '*about* how many primes ...?' Are most numbers primes, or only a small proportion? Is there any simple function $f(x)$ which is 'a good measure' of $\pi(x)$?

We answer these questions in § 1.8 and Ch. XXII.

1.6. Some notations. We shall often use the symbols

$$O,\ o,\ \sim, \tag{1.6.1}$$

† See § 2.5.

and occasionally

(1.6.2) $$\prec, \succ, \asymp .$$

These symbols are defined as follows.

Suppose that n is an integral variable which tends to infinity, and x a continuous variable which tends to infinity or to zero or to some other limiting value; that $\phi(n)$ or $\phi(x)$ is a positive function of n or x; and that $f(n)$ or $f(x)$ is any other function of n or x. Then

(i) $f = O(\phi)$ means that† $|f| < A\phi$,
where A is independent of n or x, for all values of n or x in question;
(ii) $f = o(\phi)$ means that $f/\phi \to 0$;
and
(iii) $f \sim \phi$ means that $f/\phi \to 1$.
Thus

$$10x = O(x), \quad \sin x = O(1), \quad x = O(x^2),$$
$$x = o(x^2), \quad \sin x = o(x), \quad x + 1 \sim x,$$

where $x \to \infty$, and

$$x^2 = O(x), \quad x^2 = o(x), \quad \sin x \sim x, \quad 1 + x \sim 1,$$

when $x \to 0$. It is to be observed that $f = o(\phi)$ implies, and is stronger than, $f = O(\phi)$.

As regards the symbols (1.6.2),

(iv) $f \prec \phi$ means $f/\phi \to 0$, and is equivalent to $f = o(\phi)$;
(v) $f \succ \phi$ means $f/\phi \to \infty$;
(vi) $f \asymp \phi$ means $A\phi < f < A\phi$,

where the two A's (which are naturally not the same) are both positive and independent of n or x. Thus $f \asymp \phi$ asserts that 'f is of the same order of magnitude as ϕ'.

We shall very often use A as in (vi), viz. as an *unspecified positive constant.* Different A's have usually different values, even when they occur in the same formula; and, even when definite values can be assigned to them, these values are irrelevant to the argument.

† $|f|$ denotes, as usually in analysis, the modulus or absolute value of f.

So far we have defined (for example) '$f = O(1)$', but not '$O(1)$' in isolation; and it is convenient to make our notations more elastic. We agree that '$O(\phi)$' denotes *an unspecified f such that $f = O(\phi)$*. We can then write, for example,

$$O(1) + O(1) = O(1) = o(x)$$

when $x \to \infty$, meaning by this 'if $f = O(1)$ and $g = O(1)$ then $f + g = O(1)$ and *a fortiori* $f + g = o(x)$'. Or again we may write

$$\sum_{\nu=1}^{n} O(1) = O(n),$$

meaning by this that the sum of n terms, each numerically less than a constant, is numerically less than a constant multiple of n.

It is to be observed that the relation '=', asserted between O or o symbols, is not usually symmetrical. Thus $o(1) = O(1)$ is always true; but $O(1) = o(1)$ is usually false. We may also observe that $f \sim \phi$ is equivalent to $f = \phi + o(\phi)$ or to

$$f = \phi\{1 + o(1)\}.$$

In these circumstances we say that *f and ϕ are asymptotically equivalent*, or that *f is asymptotic* to ϕ.

There is another phrase which it is convenient to define here. Suppose that P is a possible property of a positive integer, and $P(x)$ the number of numbers less than x which possess the property P. If

$$P(x) \sim x,$$

when $x \to \infty$, i.e. if the number of numbers less than x which do not possess the property is $o(x)$, then we say that *almost all numbers* possess the property. Thus we shall see† that $\pi(x) = o(x)$, so that almost all numbers are composite.

1.7. The logarithmic function. The theory of the distribution of primes demands a knowledge of the properties of the logarithmic function $\log x$. We take the ordinary analytic theory of logarithms and exponentials for granted, but it is important to lay stress on one property of $\log x$.‡

† This follows at once from Theorem 7.

‡ $\log x$ is, of course, the 'Napierian' logarithm of x, to base e. 'Common' logarithms have no mathematical interest.

Since

$$e^x = 1 + x + \cdots + \frac{x^n}{n!} + \frac{x^{n+1}}{(n+1)!} + \cdots,$$

$$x^{-n}e^x > \frac{x}{(n+1)!} \to \infty$$

when $x \to \infty$. Hence e^x tends to infinity more rapidly than any power of x. It follows that $\log x$, the inverse function, *tends to infinity more slowly than any positive power of* x; $\log x \to \infty$, but

(1.7.1) $$\frac{\log x}{x^\delta} \to 0,$$

or $\log x = o(x^\delta)$, for every positive δ. Similarly, $\log\log x$ tends to infinity more slowly than any power of $\log x$.

We may give a numerical illustration of the slowness of the growth of $\log x$. If $x = 10^9 = 1{,}000{,}000{,}000$ then

$$\log x = 20{\cdot}72\ldots.$$

Since $e^3 = 20{\cdot}08\ldots$, $\log\log x$ is a little greater than 3, and $\log\log\log x$ a little greater than 1. If $x = 10^{1{,}000}$, $\log\log\log x$ is a little greater than 2. In spite of this, the 'order of infinity' of $\log\log\log x$ has been made to play a part in the theory of primes.

The function

$$\frac{x}{\log x}$$

is particularly important in the theory of primes. It tends to infinity more slowly than x but, in virtue of (1.7.1), more rapidly than $x^{1-\delta}$, i.e. than any power of x lower than the first; and it is the simplest function which has this property.

1.8. Statement of the prime number theorem. After this preface we can state the theorem which answers question (4) of § 1.5.

THEOREM 6 (THE PRIME NUMBER THEOREM). *The number of primes not exceeding* x *is asymptotic to* $x/\log x$:

$$\pi(x) \sim \frac{x}{\log x}.$$

This theorem is the central theorem in the theory of the distribution of primes. We shall give a proof in Ch. XXII. This proof is not easy but, in the same chapter, we shall give a much simpler proof of the weaker

THEOREM 7 (TCHEBYCHEF'S THEOREM). *The order of magnitude of* $\pi(x)$ *is* $x/\log x$:

$$\pi(x) \asymp \frac{x}{\log x}.$$

It is interesting to compare Theorem 6 with the evidence of the tables. The values of $\pi(x)$ for $x = 10^3$, $x = 10^6$, and $x = 10^9$ are

$$168, \quad 78,498, \quad 50,847,534;$$

and the values of $x/\log x$, to the nearest integer, are

$$145, \quad 72,382, \quad 48,254,942.$$

The ratios are

$$1{\cdot}159\ldots, 1{\cdot}084\ldots, 1{\cdot}053\ldots;$$

and show an approximation, though not a very rapid one, to 1. The excess of the actual over the approximate values can be accounted for by the general theory.

If

$$y = \frac{x}{\log x}$$

then

$$\log y = \log x - \log\log x,$$

and

$$\log\log x = o(\log x),$$

so that

$$\log y \sim \log x, \quad x = y\log x \sim y\log y.$$

The function inverse to $x/\log x$ is therefore asymptotic to $x \log x$.

From this remark we infer that Theorem 6 is equivalent to

THEOREM 8:

$$p_n \sim n \log n.$$

Similarly, Theorem 7 is equivalent to

THEOREM 9:

$$p_n \asymp n \log n.$$

The 664,999th prime is 10,006,721; the reader should compare these figures with Theorem 8.

We arrange what we have to say about primes and their distribution in three chapters. This introductory chapter contains little but definitions and preliminary explanations; we have proved nothing except the easy, though important, Theorem 1. In Ch. II we prove rather more: in particular, Euclid's theorems 3 and 4. The first of these carries with it (as we saw in § 1.3) the 'fundamental theorem' Theorem 2, on which almost all our later work depends; and we give two proofs in §§ 2.10–2.11. We prove Theorem 4 in §§ 2.1, 2.4, and 2.6, using several methods, some of which enable us to develop the theorem a little further. Later, in Ch. XXII, we return to the theory of the distribution of primes, and develop it as far as is possible by elementary methods, proving, amongst other results, Theorem 7 and finally Theorem 6.

NOTES

§ 1.3. Theorem 3 is Euclid vii. 30. Theorem 2 does not seem to have been stated explicitly before Gauss (*D.A.*, § 16). It was, of course, familiar to earlier mathematicians; but Gauss was the first to develop arithmetic as a systematic science. See also § 12.5.

§ 1.4. The best table of factors is D. N. Lehmer's *Factor table for the first ten millions* (Carnegie Institution, Washington 105 (1909)) which gives the smallest factor of all numbers up to 10,017,000 not divisible by 2, 3, 5, or 7. The same author's *List of prime numbers from 1 to* 10,006,721 (Carnegie Institution, Washington 165 (1914)) has been extended up to 10^8 by Baker and Gruenberger (*The first six million prime numbers*, Rand Corp., Microcard Found., Madison 1959). Information about earlier tables will be found in the introduction to Lehmer's two volumes and in Dickson's *History*, i, ch. xiii. Our numbers of primes are less by 1 than Lehmer's because he counts 1 as a prime. Mapes (*Math. Computation* 17 (1963), 184–5) gives a table of $\pi(x)$ for x any multiple of 10 million up to 1,000 million.

A list of tables of primes with descriptive notes is given in D. H. Lehmer's *Guide to tables in the theory of numbers* (Washington, 1941). Large tables of primes are essentially obsolete now, since computers can generate primes afresh with sufficient rapidity for practical purposes.

Theorem 4 is Euclid ix. 20.

For Theorem 5 see Lucas, *Théorie des nombres,* i (1891), 359–61.

Kraitchik [*Sphinx*, 6 (1936), 166 and 8 (1938), 86] lists all primes between $10^{12}-10^4$ and $10^{12}+10^4$; and Jones, Lal, and Blundon (*Math. Comp.* 21 (1967), 103–7) have tabulated all primes in the range 10^k to $10^k+150,000$ for integer k from 8 to 15. The largest known pair of primes $p, p+2$ is

$$2003663613.2^{195000} \pm 1,$$

found by Vautier in 2007. These primes have 58711 decimal digits.

In § 22.20 we give a simple argument leading to a conjectural formula for the number of pairs $(p, p+2)$ below x. This agrees well with the known facts. The method can be used to find many other conjectural theorems concerning pairs, triplets, and larger blocks of primes.

§ 1.5. Our list of questions is modified from that given by Carmichael, *Theory of numbers*, 29. Of course we have not (and cannot) define what we mean by a 'simple formula' in this context. One could more usefully ask about algorithms for computing the nth prime. Clearly there is an algorithm, given by the sieve of Eratosthenes. Thus the interesting question is just how fast such an algorithm might be. A method based on the work of Lagarias and Odlyzko (*J. Algorithms* 8 (1987), 173–91) computes p_n in time $O(n^{3/5})$, (or indeed slightly faster if large amounts of memory are available). For questions (2) and (3) one might similarly ask how fast one can find p_{n+1} given p_n, or more generally, how rapidly one can find any prime greater than a given prime p. At present it appears that the best approach is merely to test each number from p_n onwards for primality. One would conjecture that this process is extremely efficient, in as much as there should be a constant $c > 0$ such that the next prime is found in time $O((\log n)^c)$. We have a very fast test for primality, due to Agrawal, Kayal, and Saxena (*Ann. of Math.* (2) 160 (2004), 781–93), but the best known upper bound on the difference $p_{n+1} - p_n$ is only $O\left(p_n^{0.525}\right)$. (See Baker, Harman, and Pintz, *Proc. London Math. Soc.* (3) 83 (2001), 532–62). Thus at present we can only say that p_{n+1} can be determined, given p_n, in time $O\left(p_n^{\theta}\right)$, for any constant $\theta > 0.525$.

§ 1.7. Littlewood's proof that $\pi(x)$ is sometimes greater than the 'logarithm integral' $\mathrm{Li}(x)$ depends upon the largeness of logloglog x for large x. See Ingham, ch. v, or Landau, *Vorlesungen*, ii. 123–56.

§ 1.8. Theorem 7 was proved by Tchebychef about 1850, and Theorem 6 by Hadamard and de la Vallée Poussin in 1896. See Ingham, 4–5; Landau, *Handbuch*, 3–55; and Ch. XXII, especially the note to §§ 22.14–16.

A better approximation to $\pi(x)$ is provided by the 'logarithmic integral'

$$\mathrm{Li}(x) = \int_2^x \frac{dt}{\log t}.$$

Thus at $x = 10^9$, for example, $\pi(x)$ and $x/\log x$ differ by more than 2,500,000, while $\pi(x)$ and $\mathrm{Li}(x)$ only differ by about 1,700.

II

THE SERIES OF PRIMES (2)

2.1. First proof of Euclid's second theorem. Euclid's own proof of Theorem 4 was as follows.

Let 2, 3, 5,…, p be the aggregate of primes up to p, and let

$$q = 2.3.5\ldots p + 1. \tag{2.1.1}$$

Then q is not divisible by any of the numbers 2, 3, 5,…, p. It is therefore either prime, or divisible by a prime between p and q. In either case there is a prime greater than p, which proves the theorem.

The theorem is equivalent to

$$\pi(x) \to \infty. \tag{2.1.2}$$

2.2. Further deductions from Euclid's argument. If p is the nth prime p_n, and q is defined as in (2.1.1), it is plain that

$$q < p_n^n + 1$$

for $n > 1$,[†] and so that

$$p_{n+1} < p_n^n + 1.$$

This inequality enables us to assign an upper limit to the rate of increase of p_n, and a lower limit to that of $\pi(x)$.

We can, however, obtain better limits as follows. Suppose that

$$p_n < 2^{2^n} \tag{2.2.1}$$

for $n = 1, 2, \ldots, N$. Then Euclid's argument shows that

$$p_{N+1} \leqslant p_1 p_2 \ldots p_N + 1 < 2^{2+4+\cdots+2^N} + 1 < 2^{2^{N+1}}. \tag{2.2.2}$$

Since (2.2.1) is true for $n = 1$, it is true for all n.

† There is equality when

$$n = 1, \quad p = 2, \quad q = 3.$$

Suppose now that $n \geqslant 4$ and

$$e^{e^{n-1}} < x \leqslant e^{e^n}.$$

Then†

$$e^{n-1} > 2^n,\ e^{e^{n-1}} > 2^{2^n};$$

and so

$$\pi(x) \geqslant \pi(e^{e^{n-1}}) \geqslant \pi(2^{2^n}) \geqslant n,$$

by (2.2.1). Since $\log\log x \leqslant n$, we deduce that

$$\pi(x) \geqslant \log\log x$$

for $x > e^{e^3}$; and it is plain that the inequality holds also for $2 \leqslant x \leqslant e^{e^3}$. We have therefore proved

THEOREM 10:

$$\pi(x) \geqslant \log\log x \quad (x \geqslant 2).$$

We have thus gone beyond Theorem 4 and found a lower limit for the order of magnitude of $\pi(x)$. The limit is of course an absurdly weak one, since for $x = 10^9$ it gives $\pi(x) \geqslant 3$, and the actual value of $\pi(x)$ is over 50 million.

2.3. Primes in certain arithmetical progressions. Euclid's argument may be developed in other directions.

THEOREM 11. *There are infinitely many primes of the form* $4n + 3$.

Define q by

$$q = 2^2.3.5\ldots p - 1,$$

instead of by (2.1.1). Then q is of the form $4n+3$, and is not divisible by any of the primes up to p. It cannot be a product of primes $4n+1$ only, since the product of two numbers of this form is of the same form; and therefore it is divisible by a prime $4n+3$, greater than p.

THEOREM 12. *There are infinitely many primes of the form* $6n + 5$.

† This is not true for $n = 3$.

The proof is similar. We define q by

$$q = 2.3.5 \dots p - 1,$$

and observe that any prime number, except 2 or 3, is $6n+1$ or $6n+5$, and that the product of two numbers $6n+1$ is of the same form.

The progression $4n+1$ is more difficult. We must assume the truth of a theorem which we shall prove later (§ 20.3).

THEOREM 13. *If a and b have no common factor, then any odd prime divisor of* $a^2 + b^2$ *is of the form* $4n + 1$.

If we take this for granted, we can prove that there are infinitely many primes $4n+1$. In fact we can prove

THEOREM 14. *There are infinitely many primes of the form* $8n+5$.

We take

$$q = 3^2.5^2.7^2 \dots p^2 + 2^2,$$

a sum of two squares which have no common factor. The square of an odd number $2m+1$ is

$$4m(m + 1) + 1$$

and is $8n+1$, so that q is $8n+5$. Observing that, by Theorem 13, any prime factor of q is $4n+1$, and so $8n+1$ or $8n+5$, and that the product of two numbers $8n+1$ is of the same form, we can complete the proof as before.

All these theorems are particular cases of a famous theorem of Dirichlet.

THEOREM 15* (DIRICHLET'S THEOREM).† *If a is positive and a and b have no common divisor except* 1, *then there are infinitely many primes of the form* $an+b$.

The proof of this theorem is too difficult for insertion in this book. There are simpler proofs when b is 1 or -1.

† An asterisk attached to the number of a theorem indicates that it is not proved anywhere in the book.

2.4. Second proof of Euclid's theorem. Our second proof of Theorem 4, which is due to Pólya, depends upon a property of what are called 'Fermat's numbers'.

Fermat's numbers are defined by

$$F_n = 2^{2^n} + 1,$$

so that

$$F_1 = 5, \quad F_2 = 17, \quad F_3 = 257, \quad F_4 = 65537.$$

They are of great interest in many ways: for example, it was proved by Gauss† that, if F_n is a prime p, then a regular polygon of p sides can be inscribed in a circle by Euclidean methods.

The property of the Fermat numbers which is relevant here is

THEOREM 16. *No two Fermat numbers have a common divisor greater than* 1.

For suppose that F_n and F_{n+k}, where $k > 0$, are two Fermat numbers, and that

$$m|F_n, \quad m|F_{n+k}.$$

If $x = 2^{2^n}$, we have

$$\frac{F_{n+k} - 2}{F_n} = \frac{2^{2^{n+k}} - 1}{2^{2^n} + 1} = \frac{x^{2^k} - 1}{x + 1} = x^{2^k - 1} - x^{2^k - 2} + \cdots - 1,$$

and so $F_n | F_{n+k} - 2$. Hence

$$m|F_{n+k}, \quad m|F_{n+k} - 2;$$

and therefore $m|\,2$. Since F_n is odd, $m = 1$, which proves the theorem.

It follows that each of the numbers $F_1, F_2, \ldots, F_n$ is divisible by an odd prime which does not divide any of the others; and therefore that there are at least n odd primes not exceeding F_n. This proves Euclid's theorem. Also

$$p_{n+1} \leqslant F_n = 2^{2^n} + 1,$$

and it is plain that this inequality, which is a little stronger than (2.2.1), leads to a proof of Theorem 10.

† See § 5.8.

2.5. Fermat's and Mersenne's numbers. The first four Fermat numbers are prime, and Fermat conjectured that all were prime. Euler, however, found in 1732 that

$$F_5 = 2^{2^5} + 1 = 641.6700417$$

is composite. For

$$641 = 5^4 + 2^4 = 5.2^7 + 1$$

divides each of $5^4 . 2^{28} + 2^{32}$ and $5^4 . 2^{28} - 1$ and so divides their difference F_5.

In 1880 Landry proved that

$$F_6 = 2^{2^6} + 1 = 274177.67280421310721.$$

More recent writers have proved that F_n is composite for

$$7 \leqslant n \leqslant 16, n = 18, 19, 21, 23, 36, 38, 39, 55, 63, 73$$

and many larger values of n. No factor is known for F_{14}, but in all the other cases proved to be composite a factor is known.

No prime F_n has been found beyond F_4, so that Fermat's conjecture has not proved a very happy one. It is perhaps more probable that the number of primes F_n is finite.† If this is so, then the number of primes 2^n+1 is finite, since it is easy to prove

THEOREM 17. *If $a \geqslant 2$ and $a^n + 1$ is prime, then a is even and $n = 2^m$.*

For if a is odd then $a^n + 1$ is even; and if n has an odd factor k and $n = kl$, then $a^n + 1$ is divisible by

$$\frac{a^{kl} + 1}{a^l + 1} = a^{(k-1)l} - a^{(k-2)l} + \cdots + 1.$$

† This is what is suggested by considerations of probability. Assuming Theorem 7, one might argue roughly as follows. The probability that a number n is prime is at most

$$\frac{A}{\log n},$$

and therefore the total expectation of Fermat primes is at most

$$A \sum \left\{ \frac{1}{\log(2^{2^n} + 1)} \right\} < A \sum 2^{-n} < A.$$

This argument (apart from its general lack of precision) assumes that there are no special reasons why a Fermat number should be likely to be prime, while Theorems 16 and 17 suggest that there are some.

It is interesting to compare the fate of Fermat's conjecture with that of another famous conjecture, concerning primes of the form $2^n - 1$. We begin with another trivial theorem of much the same type as Theorem 17.

THEOREM 18. *If $n > 1$ and $a^n - 1$ is prime, then $a = 2$ and n is prime.*

For if $a > 2$, then $a - 1 | a^n - 1$; and if $a = 2$ and $n = kl$, then we have $2^k - 1 | 2^n - 1$.

The problem of the primality of $a^n - 1$ is thus reduced to that of the primality of $2^p - 1$. It was asserted by Mersenne in 1644 that $M_p = 2^p - 1$ is prime for

$$p = 2, 3, 5, 7, 13, 17, 19, 31, 67, 127, 257,$$

and composite for the other 44 values of p less than 257. The first mistake in Mersenne's statement was found about 1886,† when Pervusin and Seelhoff discovered that M_{61} is prime. Subsequently four further mistakes were found in Mersenne's statement and it need no longer be taken seriously. In 1876 Lucas found a method for testing whether M_p is prime and used it to prove M_{127} prime. This remained the largest known prime until 1951, when, using different methods, Ferrier found a larger prime (using only a desk calculating machine) and Miller and Wheeler (using the EDSAC 1 electronic computer at Cambridge) found several large primes, of which the largest was

$$180M_{127}^2 + 1,$$

which is larger than Ferrier's. But Lucas's test is particularly suitable for use on a binary digital computer and it has subsequently been applied by a succession of investigators (Lehmer and Robinson, Hurwitz and Selfridge, Riesel, Gillies, Tuckerman and finally Nickel and Noll). As a result it is now known that M_p is prime for

$$\begin{aligned} p = {} & 2, 3, 5, 7, 13, 17, 19, 31, 61, 89, 107, \\ & 127, 521, 607, 1279, 2203, 2281, 3217, \\ & 4253, 4423, 9689, 9941, 11213, 19937, 21701, \end{aligned}$$

and composite for all other $p < 21700$. The largest known prime is thus M_{21701}, a number of 6533 digits.‡

† Euler stated in 1732 that M_{41} and M_{47} are prime, but this was a mistake.

‡ See the end of chapter notes.

We describe Lucas's test in § 15.5 and give the test used by Miller and Wheeler in Theorem 101.

The problem of Mersenne's numbers is connected with that of 'perfect' numbers, which we shall consider in § 16.8.

We return to this subject in § 6.15 and § 15.5.

2.6. Third proof of Euclid's theorem. Suppose that $2, 3, \ldots, p_j$ are the first j primes and let $N(x)$ be the number of n not exceeding x which are not divisible by any prime $p > p_j$. If we express such an n in the form

$$n = n_1^2 m,$$

where m is 'squarefree', i.e. is not divisible by the square of any prime, we have

$$m = 2^{b_1} 3^{b_2} \ldots p_j^{b_j},$$

with every b either 0 or 1. There are just 2^j possible choices of the exponents and so not more than 2^j different values of m. Again, $n_1 \leqslant \sqrt{n} \leqslant \sqrt{x}$ and so there are not more than $\sqrt{x}$ different values of n_1. Hence

$$N(x) \leqslant 2^j \sqrt{x}. \tag{2.6.1}$$

If Theorem 4 is false, so that the number of primes is finite, let the primes be $2, 3, \ldots, p_j$. In this case $N(x) = x$ for every x and so

$$x \leqslant 2^j \sqrt{x}, \quad x \leqslant 2^{2j},$$

which is false for $x \geqslant 2^{2j} + 1$.

We can use this argument to prove two further results.

THEOREM 19. *The series*

$$\sum \frac{1}{p} = \frac{1}{2} + \frac{1}{3} + \frac{1}{5} + \frac{1}{7} + \frac{1}{11} + \cdots \tag{2.6.2}$$

is divergent.

If the series is convergent, we can choose j so that the remainder after j terms is less than $\frac{1}{2}$, i.e.

$$\frac{1}{p_{j+1}} + \frac{1}{p_{j+2}} + \cdots < \frac{1}{2}.$$

The number of $n \leqslant x$ which are divisible by p is at most x/p. Hence $x - N(x)$, the number of $n \leqslant x$ divisible by one or more of $p_{j+1}, p_{j+2}, \ldots$, is not more than

$$\frac{x}{p_{j+1}} + \frac{x}{p_{j+2}} + \cdots < \tfrac{1}{2}x.$$

Hence, by (2.6.1),

$$\tfrac{1}{2}x < N(x) \leqslant 2^j \sqrt{x}, \quad x < 2^{2j+2},$$

which is false for $x \geqslant 2^{2j+2}$. Hence the series diverges.

THEOREM 20:

$$\pi(x) \geqslant \frac{\log x}{2 \log 2} \quad (x \geqslant 1); \quad p_n \leqslant 4^n.$$

We take $j = \pi(x)$, so that $p_{j+1} > x$ and $N(x) = x$. We have

$$x = N(x) \leqslant 2^{\pi(x)} \sqrt{x}, \quad 2^{\pi(x)} \geqslant \sqrt{x},$$

and the first part of Theorem 20 follows on taking logarithms. If we put $x = p_n$, so that $\pi(x) = n$, the second part is immediate.

By Theorem 20, $\pi(10^9) \geqslant 15$; a number, of course, still ridiculously below the mark.

2.7. Further results on formulae for primes. We return for a moment to the questions raised in § 1.5. We may ask for 'a formula for primes' in various senses.

(i) We may ask for a simple function $f(n)$ which assumes *all prime values and only prime values,* i.e. which takes successively the values $p_1, p_2, \ldots$ when n takes the values 1, 2,.... This is the question which we discussed in § 1.5.

(ii) We may ask for a simple function of n which assumes *prime values only.* Fermat's conjecture, had it been right, would have supplied an answer to this question.† As it is, no satisfactory answer is known. But it is possible

† It had been suggested that Fermat's sequence should be replaced by

$$2+1, \quad 2^2+1, \quad 2^{2^2}+1, \quad 2^{2^{2^2}}+1, \ldots.$$

The first four numbers are prime, but F_{16}, the fifth member of this sequence, is now known to be composite. Another suggestion was that the sequence M_p, where p is confined to the Mersenne primes, would contain only primes. But $M_{13} = 8191$ and M_{8191} is composite.

to construct a polynomial (in several positive integral variables) whose positive values are all prime and include all the primes, though its negative values are composite. See § 2 of the Appendix.

(iii) We may moderate our demands and ask merely for a simple function of n which assumes *an infinity* of prime values. It follows from Euclid's theorem that $f(n) = n$ is such a function, and less trivial answers are given by Theorems 11–15. Apart from trivial solutions, Dirichlet's Theorem 15 is the only solution known. It has never been proved that n^2+1, or any other quadratic form in n, will represent an infinity of primes, and all such problems seem to be extremely difficult.

There are some simple negative theorems which contain a very partial reply to question (ii).

THEOREM 21. *No polynomial $f(n)$ with integral coefficients, not a constant, can be prime for all n, or for all sufficiently large n.*

We may assume that the leading coefficient in $f(n)$ is positive, so that $f(n) \to \infty$ when $n \to \infty$, and $f(n) > 1$ for $n > N$, say. If $x > N$ and

$$f(x) = a_0x^k + \cdots = y > 1,$$

then

$$f(ry + x) = a_0(ry + x)^k + \cdots$$

is divisible by y for every integral r; and $f(ry+x)$ tends to infinity with r. Hence there are infinitely many composite values of $f(n)$.

There are quadratic forms which assume prime values for considerable sequences of values of n. Thus $n^2 - n + 41$ is prime for $0 \leqslant n \leqslant 40$, and

$$n^2 - 79n + 1601 = (n - 40)^2 + (n - 40) + 41$$

for $0 \leqslant n \leqslant 79$.

A more general theorem, which we shall prove in § 6.4, is

THEOREM 22. *If*

$$f(n) = P(n, 2^n, 3^n, \ldots, k^n)$$

is a polynomial in its arguments, with integral coefficients, and $f(n) \to \infty$ when $n \to \infty$,[†] *then $f(n)$ is composite for an infinity of values of n.*

[†] Some care is required in the statement of the theorem, to avoid such an $f(n)$ as $2^n3^n - 6^n + 5$, which is plainly prime for all n.

2.8. Unsolved problems concerning primes. In § 1.4 we stated two conjectural theorems of which no proof is known, although empirical evidence makes their truth seem highly probable. There are many other conjectural theorems of the same kind.

There are infinitely many primes n^2+1. *More generally, if* a, b, c *are integers without a common divisor,* a *is positive,* $a+b$ *and* c *are not both even, and* b^2-4ac *is not a perfect square, then there are infinitely many primes* an^2+bn+c.

We have already referred to the form n^2+1 in § 2.7 (iii). If a, b, c have a common divisor, there can obviously be at most one prime of the form required. If $a+b$ and c are both even, then $N = an^2+bn+c$ is always even. If $b^2-4ac = k^2$, then

$$4aN = (2an+b)^2 - k^2.$$

Hence, if N is prime, either $2an+b+k$ or $2an+b-k$ divides $4a$, and this can be true for at most a finite number of values of n. The limitations stated in the conjecture are therefore essential.

There is always a prime between n^2 *and* $(n+1)^2$.

If $n > 4$ *is even, then* n *is the sum of two odd primes.*

This is 'Goldbach's theorem'.

If $n \geqslant 9$ *is odd, then* n *is the sum of three odd primes.*

Any n *from some point onwards is a square or the sum of a prime and a square.*

This is not true of all n; thus 34 and 58 are exceptions.

A more dubious conjecture, to which we referred in § 2.5, is

The number of Fermat primes F_n *is finite.*

2.9. Moduli of integers. We now give the proof of Theorems 3 and 2 which we postponed from § 1.3. Another proof will be given in § 2.11 and a third in § 12.4. Throughout this section integer means rational integer, positive or negative.

The proof depends upon the notion of a 'modulus' of numbers. A modulus is a system S of numbers such that *the sum and difference of any two members of* S *are themselves members of* S: i.e.

(2.9.1) $$m \in S \,.\, n \in S \to (m \pm n) \in S.$$

The numbers of a modulus need not necessarily be integers or even rational; they may be complex numbers, or quaternions: but here we are concerned only with moduli of integers.

The single number 0 forms a modulus (the *null modulus*).

It follows from the definition of S that

$$a \in S \rightarrow 0 = a - a \in S \,.\, 2a = a + a \in S.$$

Repeating the argument, we see that $na \in S$ for any integral n (positive or negative). More generally

(2.9.2) $$a \in S \,.\, b \in S \rightarrow xa + yb \in S$$

for any integral x, y. On the other hand, it is obvious that, if a and b are given, the aggregate of values of $xa+yb$ forms a modulus.

It is plain that any modulus S, except the null modulus, contains some positive numbers. Suppose that d is the smallest positive number of S. If n is any positive number of S, then $n-zd \in S$ for all z. If c is the remainder when n is divided by d and

$$n = zd + c,$$

then $c \in S$ and $0 \leqslant c < d$. Since d is the smallest positive number of S, we have $c = 0$ and $n = zd$. Hence

THEOREM 23. *Any modulus, other than the null modulus, is the aggregate of integral multiples of a positive number d.*

We define the *highest common divisor d* of two integers a and b, not both zero, as the largest positive integer which divides both a and b; and write

$$d = (a, b).$$

Thus $(0, a) = |a|$. We may define the highest common divisor

$$(a, b, c, \dots, k)$$

of any set of positive integers $a, b, c, \dots, k$ in the same way.

The aggregate of numbers of the form

$$xa + yb,$$

for integral x, y, is a modulus which, by Theorem 23, is the aggregate of multiples zc of a certain positive c. Since c divides every number of S, it divides a and b, and therefore

$$c \leqslant d.$$

On the other hand,

$$d|a \,.\, d|b \rightarrow d|xa + yb,$$

so that d divides every number of S, and in particular c. It follows that

$$c = d$$

and that S is the aggregate of multiples of d.

THEOREM 24. *The modulus $xa + yb$ is the aggregate of multiples of $d = (a, b)$.*

It is plain that we have proved incidentally

THEOREM 25. *The equation*

$$ax + by = n$$

is soluble in integers x, y if and only if $d \mid n$. In particular,

$$ax + by = d$$

is soluble.

THEOREM 26. *Any common divisor of a and b divides d.*

2.10. Proof of the fundamental theorem of arithmetic. We are now in a position to prove Euclid's theorem 3, and so Theorem 2.

Suppose that p is prime and $p|\, ab$. If $p \nmid a$ then $(a, p) = 1$, and therefore, by Theorem 24, there are an x and a y for which $xa + yp = 1$ or

$$xab + ypb = b.$$

But $p|ab$ and $p|pb$, and therefore $p|b$.

Practically the same argument proves

THEOREM 27:

$$(a, b) = d \,.\, c > 0 \rightarrow (ac, bc) = dc.$$

For there are an x and a y for which $xa + yb = d$ or

$$xac + ybc = dc.$$

Hence $(ac, bc) \mid dc$. On the other hand, $d|a \rightarrow dc \mid ac$ and $d\,|b \rightarrow dc \mid bc$; and therefore, by Theorem 26, $dc \mid (ac, bc)$. Hence $(ac, bc) = dc$.

2.11. Another proof of the fundamental theorem. We call numbers which can be factorized into primes in more than one way *abnormal*. Let n be the least abnormal number. The same prime P cannot appear in two different factorizations of n, for, if it did, n/P would be abnormal and $n/P < n$. We have then

$$n = p_1p_2p_3 \ldots = q_1q_2 \ldots,$$

where the p and q are primes, no p is a q and no q is a p.

We may take p_1 to be the least p; since n is composite, $p_1^2 \leqslant n$. Similarly, if q_1 is the least q, we have $q_1^2 \leqslant n$ and, since $p_1 \neq q_1$, it follows that $p_1q_1 < n$. Hence, if $N = n - p_1q_1$, we have $0 < N < n$ and N is not abnormal. Now $p_1 | n$ and so $p_1 | N$; similarly $q_1 | N$. Hence p_1 and q_1 both appear in the unique factorization of N and $p_1q_1 | N$. From this it follows that $p_1q_1 | n$ and hence that $q_1 | n/p_1$. But n/p_1 is less than n and so has the unique prime factorization $p_2p_3 \ldots$. Since q_1 is not a p, this is impossible. Hence there cannot be any abnormal numbers and this is the fundamental theorem.

NOTES

§ 2.2. Mr. Ingham tells us that the argument used here is due to Bohr and Littlewood: see Ingham, 2.

§ 2.3. For Theorems 11, 12, and 14, see Lucas, *Théorie des nombres*, i (1891), 353–4; and for Theorem 15 see Landau, *Handbuch*, 422–46, and *Vorlesungen*, i. 79–96.

An interesting extension of Theorem 15 has been obtained by Shiu (*J. London Math. Soc.* (2) 61 (2000), 359–73). This says that for a and b as in Theorem 15, the sequence of primes contains arbitrarily long strings of consecutive elements, all of which are of the form $an + b$. Taking $a = 1000$ and $b = 777$ for example, this means that one can find as many consecutive primes as desired, each of which ends in the digits 777.

§ 2.4. See Pólya and Szegő, No. 94.

§ 2.5. See Dickson, *History*, i, chs. i, xv, xvi, Rouse Ball *Mathematical recreations and essays*, Ch.2, and, for the earlier numerical results, Kraitchik, *Théorie des nombres*, i (Paris, 1922), 22, 218 and D. H. Lehmer, *Bulletin Amer. Math. Soc.* 38 (1932), 383–4. Miller and Wheeler (*Nature* 168 (1951), 838) give their large prime and Tuckerman (*Proc. Nat. Acad. Sci. U.S.A.* 68 (1971), 2319–20) gives the Mersenne prime M_p with $p = 19937$ and references to the other smaller ones found by electronic computing. The discovery of the prime M_p with $p = 21701$ was reported in the Times of 17th November, 1978. For factors of composite F_m see Hallyburton and Brillhart, *Math. Comp.* 29 (1975), 109–12 and, for a factor of F_8, see Brent, *American Math. Soc. Abstracts*, 1 (1980), 565.

By 2007, F_n was known to be composite and had been completely factored for the values $5 \leqslant n \leqslant 11$, while many factors had been discovered for larger n. It was known that F_n is composite for $4 \leqslant n \leqslant 32$. The smallest n for which no factor of F_n had been discovered was $n = 14$.

Similarly, by 2007, a total of 44 Mersenne primes had been discovered, the largest being $M_{32582657}$. The 39th Mersenne prime had been identified as $M_{13466917}$, but not all Mersenne numbers in between these two had been tested.

Ferrier's prime is $(2^{148}+1)/17$ and is the largest prime found without the use of electronic computing (and may well remain so).

The new large computers have made the subjects of factoring large numbers and of testing large numbers for primality very interesting and highly non-trivial. Guy (*Proc. 5th Manitoba Conf. Numerical Math.* 1975, 49–89) gives a full account of methods of factoring, some remarks about tests for primality and a substantial list of references on both topics. On tests for primality, see also, for example, Brillhart, Lehmer, and Selfridge, *Math. Comp.* 29 (1975), 620–47 and Selfridge and Wunderlich, *Proc. 4th Manitoba Conf. Numerical Math.* 1974, 109–20.

Our proof that $641 \mid F_5$ is due to Coxeter (*Introduction to geometry,* New York, Wiley, 1969), following Kraitchik and Bennett.

Ribenboim, *The new book of prime number records*, (Springer, New York, 1996) gives a full account of all the above work, and much besides.

§ 2.6. See Erdős, *Mathematica,* B 7 (1938), 1–2. Theorem 19 was proved by Euler in 1737.

§ 2.7. Theorem 21 is due to Goldbach (1752) and Theorem 22 to Morgan Ward, *Journal London Math. Soc.* 5 (1930), 106–7.

§ 2.8. See § 3 of the Appendix.

§§ 2.9–10. The argument follows the lines of Hecke, ch. i. The definition of a modulus is the natural one, but is redundant. It is sufficient to assume that

$$m \in S \,.\, n \in S \rightarrow m - n \in S.$$

For then

$$0 = n - n \in S, \quad -n = 0 - n \in S, \quad m + n = m - (-n) \in S.$$

§ 2.11. F. A. Lindemann, *Quart. J. of Math.* (Oxford), 4 (1933), 319–20, and Davenport, *Higher arithmetic*, 20. For somewhat similar proofs, see Zermelo, *Göttinger Nachrichten* (new series), i (1934), 43–4, and Hasse, *Journal für Math.* 159 (1928), 3–6.

III

FAREY SERIES AND A THEOREM OF MINKOWSKI

3.1. The definition and simplest properties of a Farey series. In this chapter we shall be concerned primarily with certain properties of the 'positive rationals' or 'vulgar fractions', such as $\frac{1}{2}$ or $\frac{7}{11}$. Such a fraction may be regarded as a relation between two positive integers, and the theorems which we prove embody properties of the positive integers.

The Farey series $\mathfrak{F}_n$ of order n is the ascending series of irreducible fractions between 0 and 1 whose denominators do not exceed n. Thus h/k belongs to $\mathfrak{F}_n$ if

$$0 \leqslant h \leqslant k \leqslant n, \quad (h,k) = 1; \tag{3.1.1}$$

the numbers 0 and 1 are included in the forms $\frac{0}{1}$ and $\frac{1}{1}$. For example, $\mathfrak{F}_5$ is

$$\frac{0}{1}, \frac{1}{5}, \frac{1}{4}, \frac{1}{3}, \frac{2}{5}, \frac{1}{2}, \frac{3}{5}, \frac{2}{3}, \frac{3}{4}, \frac{4}{5}, \frac{1}{1}.$$

The characteristic properties of Farey series are expressed by the following theorems.

THEOREM 28. *If h/k and h'/k' are two successive terms of $\mathfrak{F}_n$, then*

$$kh' - hk' = 1. \tag{3.1.2}$$

THEOREM 29. *If h/k, h''/k'', and h'/k' are three successive terms of $\mathfrak{F}_n$, then*

$$\frac{h''}{k''} = \frac{h+h'}{k+k'}. \tag{3.1.3}$$

We shall prove that the two theorems are equivalent in the next section, and then give three different proofs of both of them, in §§ 3.3, 3.4, and 3.7 respectively. We conclude this section by proving two still simpler properties of $\mathfrak{F}_n$.

THEOREM 30. *If h/k and h'/k' are two successive terms of $\mathfrak{F}_n$, then*

$$k + k' > n. \tag{3.1.4}$$

The 'mediant'

$$\frac{h+h'}{k+k'}\dagger$$

† Or the reduced form of this fraction.

of h/k and h'/k' falls in the interval

$$\left(\frac{h}{k}, \frac{h'}{k'}\right).$$

Hence, unless (3.1.4) is true, there is another term of $\mathfrak{F}_n$ between h/k and h'/k'.

Theorem 31. *If $n > 1$, then no two successive terms of $\mathfrak{F}_n$ have the same denominator.*

If $k > 1$ and h'/k succeeds h/k in $\mathfrak{F}_n$, then $h + 1 \leqslant h' < k$. But then

$$\frac{h}{k} < \frac{h}{k-1} < \frac{h+1}{k} \leqslant \frac{h'}{k};$$

and $h/(k-1)$[†] comes between h/k and h'/k in $\mathfrak{F}_n$, a contradiction.

3.2. The equivalence of the two characteristic properties. We now prove that each of Theorems 28 and 29 implies the other.

(1) *Theorem* 28 *implies Theorem* 29. If we assume Theorem 28, and solve the equations

$$kh'' - hk'' = 1, \quad k''h' - h''k' = 1 \tag{3.2.1}$$

for h'' and k'', we obtain

$$h''(kh' - hk') = h + h', \quad k''(kh' - hk') = k + k',$$

and so (3.1.3).

(2) *Theorem* 29 *implies Theorem* 28. We assume that Theorem 29 is true generally and that Theorem 28 is true for $\mathfrak{F}_{n-1}$, and deduce that Theorem 28 is true for $\mathfrak{F}_n$. It is plainly sufficient to prove that the equations (3.2.1) are satisfied when h''/k'' belongs to $\mathfrak{F}_n$ but not to $\mathfrak{F}_{n-1}$, so that $k'' = n$. In this case, after Theorem 31, both k and k' are less than k'', and h/k and h'/k' are consecutive terms in $\mathfrak{F}_{n-1}$.

Since (3.1.3) is true *ex hypothesi,* and h''/k'' is irreducible, we have

$$h + h' = \lambda h'', \; k + k' = \lambda k'',$$

where λ is an integer. Since k and k' are both less than k'', λ must be 1.

† Or the reduced form of this fraction.

Hence

$$h'' = h + h', \quad k'' = k + k',$$
$$kh'' - hk'' = kh' - hk' = 1;$$

and similarly

$$k''h' - h''k' = 1.$$

3.3. First proof of Theorems 28 and 29. Our first proof is a natural development of the ideas used in § 3.2.

The theorems are true for $n = 1$; we assume them true for $\mathfrak{F}_{n-1}$ and prove them true for $\mathfrak{F}_n$.

Suppose that h/k and h'/k' are consecutive in $\mathfrak{F}_{n-1}$ but separated by h''/k'' in $\mathfrak{F}_n$.† Let

(3.3.1) $$kh'' - hk'' = r > 0, \quad k''h' - h''k' = s > 0.$$

Solving these equations for h'' and k'', and remembering that

$$kh' - hk' = 1,$$

we obtain

(3.3.2) $$h'' = sh + rh', \quad k'' = sk + rk'.$$

Here $(r, s) = 1$, since $(h'', k'') = 1$.

Consider now the set S of all fractions

(3.3.3) $$\frac{H}{K} = \frac{\mu h + \lambda h'}{\mu k + \lambda k'}$$

in which λ and μ are positive integers and $(\lambda, \mu) = 1$. Thus h''/k'' belongs to S. Every fraction of S lies between h/k and h'/k', and is in its lowest terms, since any common divisor of H and K would divide

$$k(\mu h + \lambda h') - h(\mu k + \lambda k') = \lambda$$

† After Theorem 31, h''/k'' is the only term of $\mathfrak{F}_n$ between h/k and h'/k'; but we do not assume this in the proof.

and

$$h'(\mu k + \lambda k') - k'(\mu h + \lambda h') = \mu.$$

Hence every fraction of S appears sooner or later in some $\mathfrak{F}_q$; and plainly the first to make its appearance is that for which K is least, i.e. that for which $\lambda = 1$ and $\mu = 1$. This fraction must be h''/k'', and so

$$h'' = h + h', \quad k'' = k + k'. \tag{3.3.4}$$

If we substitute these values for h'', k'' in (3.3.1), we see that $r = s = 1$. This proves Theorem 28 for $\mathfrak{F}_n$. The equations (3.3.4) are not generally true for three successive fractions of $\mathfrak{F}_n$, but are (as we have shown) true when the central fraction has made its first appearance in $\mathfrak{F}_n$.

3.4. Second proof of the theorems. This proof is not inductive, and gives a rule for the construction of the term which succeeds h/k in $\mathfrak{F}_n$. Since $(h, k) = 1$, the equation

$$kx - hy = 1 \tag{3.4.1}$$

is soluble in integers (Theorem 25). If x_0, y_0 is a solution then

$$x_0 + rh, \quad y_0 + rk$$

is also a solution for any positive or negative integral r. We can choose r so that $n - k < y_0 + rk \leqslant n$. There is therefore a solution (x, y) of (3.4.1) such that

$$(x, y) = 1, \quad 0 \leqslant n - k < y \leqslant n. \tag{3.4.2}$$

Since x/y is in its lowest terms, and $y \leqslant n, x/y$ is a fraction of $\mathfrak{F}_n$. Also

$$\frac{x}{y} = \frac{h}{k} + \frac{1}{ky} > \frac{h}{k},$$

so that x/y comes later in $\mathfrak{F}_n$ than h/k. If it is not h'/k', it comes later than h'/k', and

$$\frac{x}{y} - \frac{h'}{k'} = \frac{k'x - h'y}{k'y} \geqslant \frac{1}{k'y};$$

while

$$\frac{h'}{k'} - \frac{h}{k} = \frac{kh' - hk'}{kk'} \geqslant \frac{1}{kk'}.$$

Hence

$$\frac{1}{ky} = \frac{kx - hy}{ky} = \frac{x}{y} - \frac{h}{k} \geqslant \frac{1}{k'y} + \frac{1}{kk'} = \frac{k+y}{kk'y}$$

$$> \frac{n}{kk'y} \geqslant \frac{1}{ky},$$

by (3.4.2). This is a contradiction, and therefore x/y must be h'/k', and $kh' - hk' = 1$.

Thus, to find the successor of $\frac{4}{9}$ in $\mathfrak{F}_{13}$, we begin by finding some solution (x_0, y_0) of $9x - 4y = 1$, e.g. $x_0 = 1$, $y_0 = 2$. We then choose r so that $2 + 9r$ lies between $13 - 9 = 4$ and 13. This gives $r = 1$, $x = 1 + 4r = 5$, $y = 2 + 9r = 11$, and the fraction required is $\frac{5}{11}$.

3.5. The integral lattice. Our third and last proof depends on simple but important geometrical ideas.

Suppose that we are given an origin O in the plane and two points P, Q not collinear with O. We complete the parallelogram *OPQR,* produce its sides indefinitely, and draw the two systems of equidistant parallels of which *OP, QR* and *OQ, PR* are consecutive pairs, thus dividing the plane into an infinity of equal parallelograms. Such a figure is called a *lattice* (*Gitter*).

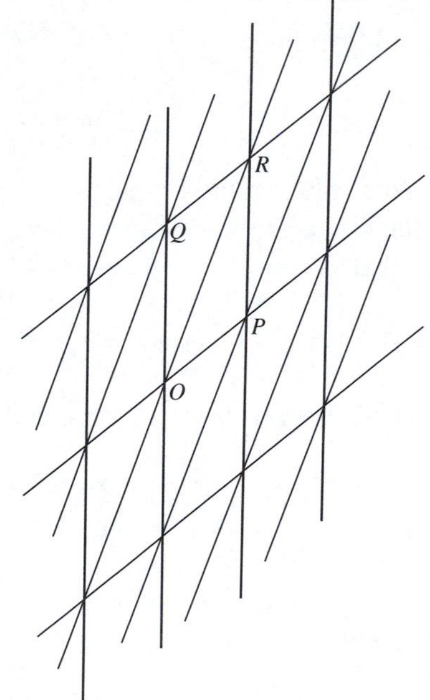

Fig. 1.

A lattice is a figure of lines. It defines a figure of points, viz. the system of points of intersection of the lines, or lattice points. Such a system we call a *point-lattice.*

Two different lattices may determine the same point-lattice; thus in Fig. 1 the lattices based on *OP, OQ* and on *OP, OR* determine the same system of points. Two lattices which determine the same point-lattice are said to be *equivalent*.

It is plain that any lattice point of a lattice might be regarded as the origin O, and that the properties of the lattice are independent of the choice of origin and symmetrical about any origin.

One type of lattice is particularly important here. This is the lattice which is formed (when the rectangular coordinate axes are given) by parallels to the axes at unit distances, dividing the plane into unit squares. We call this the *fundamental lattice* L, and the point-lattice which it determines, viz. the system of points (x, y) with integral coordinates, the *fundamental point-lattice* Λ.

Any point-lattice may be regarded as a system of numbers or vectors, the complex coordinates $x+iy$ of the lattice points or the vectors to these points from the origin. Such a system is plainly a modulus in the sense of § 2.9. If P and Q are the points (x_1,y_1) and (x_2,y_2), then the coordinates of any point S of the lattice based upon OP and OQ are

$$x = mx_1 + nx_2, \quad y = my_1 + ny_2,$$

where m and n are integers; or if z_1 and z_2 are the complex coordinates of P and Q, then the complex coordinate of S is

$$z = mz_1 + nz_2.$$

3.6. Some simple properties of the fundamental lattice. (1) We now consider the transformation defined by

$$x' = ax + by, \quad y' = cx + dy, \tag{3.6.1}$$

where a, b, c, d are given, positive or negative, integers with $ad - bc \neq 0$. It is plain that any point (x, y) of Λ is transformed into another point (x', y') of Λ.

Solving (3.6.1) for x and y, we obtain

$$x = \frac{dx' - by'}{ad - bc}, \quad y = -\frac{cx' - ay'}{ad - bc}. \tag{3.6.2}$$

If

$$\Delta = ad - bc = \pm 1, \tag{3.6.3}$$

then any integral values of x' and y' give integral values of x and y, and every lattice point (x', y') corresponds to a lattice point (x, y). In this case Λ is transformed into itself.

Conversely, if Λ is transformed into itself, every integral (x', y') must give an integral (x, y). Taking in particular (x', y') to be (1, 0) and (0, 1), we see that

$$\Delta|d, \quad \Delta|b, \quad \Delta|c, \quad \Delta|a,$$

and so

$$\Delta^2|ad - bc, \quad \Delta^2|\Delta.$$

Hence $\Delta = \pm 1$.

We have thus proved

THEOREM 32. *A necessary and sufficient condition that the transformation (3.6.1) should transform Λ into itself is that $\Delta = \pm 1$.*

We call such a transformation *unimodular.*

(2) Suppose now $P = (a, c)$ and $Q = (b, d)$ are points of Λ not collinear with O. The area of the parallelogram defined by OP and OQ is

$$\delta = \pm(ad - bc) = |ad - bc|,$$

the sign being chosen to make δ positive. The points (x', y') of the lattice Λ' based on OP and OQ are given by

$$x' = xa + yb, \quad y' = xc + yd,$$

where x and y are arbitrary integers. After Theorem 32, a necessary and sufficient condition that Λ' should be identical with Λ is that $\delta = 1$.

THEOREM 33. *A necessary and sufficient condition that the lattice L' based upon OP and OQ should be equivalent to L is that the area of the parallelogram defined by OP and OQ should be unity.*

(3) We call a point P of Λ *visible* (i.e. visible from the origin) if there is no point of Λ on OP between O and P. In order that (x, y) should be visible, it is necessary and sufficient that x/y should be in its lowest terms, or $(x, y) = 1$.

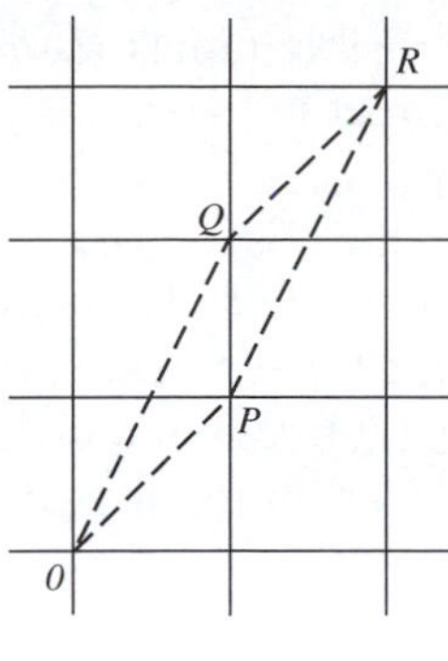

FIG. 2a.

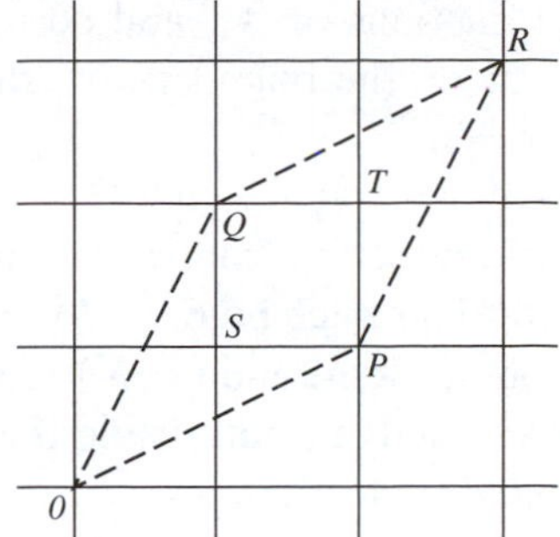

FIG. 2b.

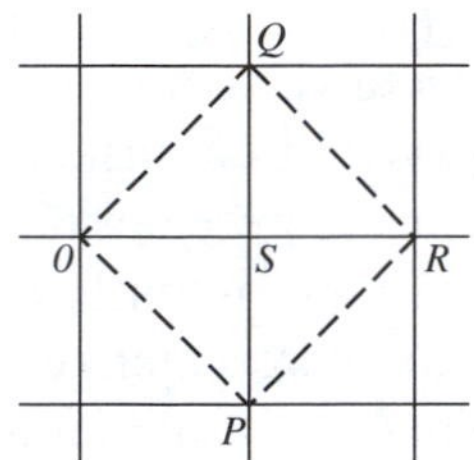

FIG. 2c.

THEOREM 34. *Suppose that P and Q are visible points of* Λ, *and that* δ *is the area of the parallelogram J defined by OP and OQ. Then*

(i) *if* $\delta = 1$, *there is no point of* Λ *inside J;*

(ii) *if* $\delta > 1$, *there is at least one point of* Λ *inside J, and, unless that point is the intersection of the diagonals of J, at least two, one in each of the triangles into which J is divided by PQ.*

There is no point of Λ inside J if and only if the lattice L' based on OP and OQ is equivalent to L, i.e. if and only if $\delta = 1$. If $\delta > 1$, there is at least one such point S. If R is the fourth vertex of the parallelogram J, and RT is parallel and equal to OS, but with the opposite sense, then (since the properties of a lattice are symmetrical, and independent of the particular lattice point chosen as origin) T is also a point of Λ, and there are at least two points of Λ inside J unless T coincides with S. This is the special case mentioned under (ii).

The different cases are illustrated in Figs. 2a, 2b, 2c.

3.7. Third proof of Theorems 28 and 29. The fractions h/k with

$$0 \leqslant h \leqslant k \leqslant n, \quad (h,k) = 1$$

are the fractions of $\mathfrak{F}_n$, and correspond to the visible points (k, h) of Λ inside, or on the boundary of, the triangle defined by the lines $y = 0$, $y = x, x = n$.

If we draw a ray through O and rotate it round the origin in the counter-clockwise direction from an initial position along the axis of x, it will pass in turn through each point (k, h) representative of a Farey fraction. If P and P' are points (k, h) and (k', h') representing consecutive fractions, there is no representative point inside the triangle OPP' or on the join PP', and therefore, by Theorem 34,

$$kh' - hk' = 1.$$

3.8. The Farey dissection of the continuum. It is often convenient to represent the real numbers on a circle instead of, as usual, on a straight line, the object of the circular representation being to eliminate integral parts. We take a circle C of unit circumference, and an arbitrary point O of the circumference as the representative of 0, and represent x by the point P_x whose distance from O, measured round the circumference in the counter-clockwise direction, is x. Plainly all integers are represented by the same point O, and numbers which differ by an integer have the same representative point.

It is sometimes useful to divide up the circumference of C in the following manner. We take the Farey series $\mathfrak{F}_n$, and form all the mediants

$$\mu = \frac{h + h'}{k + k'}$$

of successive pairs h/k, h'/k'. The first and last mediants are

$$\frac{0 + 1}{1 + n} = \frac{1}{n + 1}, \qquad \frac{n - 1 + 1}{n + 1} = \frac{n}{n + 1}.$$

The mediants naturally do not belong themselves to $\mathfrak{F}_n$.

We now represent each mediant μ by the point P_μ. The circle is thus divided up into arcs which we call *Farey arcs,* each bounded by two points P_μ and containing one *Farey point,* the representative of a term of $\mathfrak{F}_n$. Thus

$$\left(\frac{n}{n + 1}, \frac{1}{n + 1}\right)$$

is a Farey arc containing the one Farey point O. The aggregate of Farey arcs we call the *Farey dissection* of the circle.

In what follows we suppose that $n > 1$. If $P_{h/k}$ is a Farey point, and h_1/k_1, h_2/k_2 are the terms of $\mathfrak{F}_n$ which precede and follow h/k, then the Farey arc round $P_{h/k}$ is composed of two parts, whose lengths are

$$\frac{h}{k} - \frac{h+h_1}{k+k_1} = \frac{1}{k(k+k_1)}, \qquad \frac{h+h_2}{k+k_2} - \frac{h}{k} = \frac{1}{k(k+k_2)}$$

respectively. Now $k + k_1 < 2n$, since k and k_1 are unequal (Theorem 31) and neither exceeds n; and $k + k_1 > n$, by Theorem 30. We thus obtain

THEOREM 35. *In the Farey dissection of order n, where $n > 1$, each part of the arc which contains the representative of h/k has a length between*

$$\frac{1}{k(2n-1)} \quad \textit{and} \quad \frac{1}{k(n+1)}.$$

The dissection, in fact, has a certain 'uniformity' which explains its importance.

We use the Farey dissection here to prove a simple theorem concerning the approximation of arbitrary real numbers by rationals, a topic to which we shall return in Ch. XI.

THEOREM 36. *If ξ is any real number, and n a positive integer, then there is an irreducible fraction h/k such that*

$$0 < k \leqslant n, \qquad \left|\xi - \frac{h}{k}\right| \leqslant \frac{1}{k(n+1)}. \tag{3.8.1}$$

We may suppose that $0 < \xi < 1$. Then ξ falls in an interval bounded by two successive fractions of $\mathfrak{F}_n$, say h/k and h'/k', and therefore in one of the intervals

$$\left(\frac{h}{k}, \frac{h+h'}{k+k'}\right), \qquad \left(\frac{h+h'}{k+k'}, \frac{h'}{k'}\right).$$

Hence, after Theorem 35, either h/k or h'/k' satisfies the conditions: h/k if ξ falls in the first interval, h'/k' if it falls in the second.

3.9. A theorem of Minkowski. If P and Q are points of Λ, P' and Q' the points symmetrical to P and Q about the origin, and we add to the parallelogram J of Theorem 34 the three parallelograms based on OQ, OP', on OP', OQ', and on OQ', OP, we obtain a parallelogram K whose centre is the origin and whose area 4δ is four times that of J. If δ has the value 1 (its least possible value) there are points of Λ on the boundary of K, but none, except O, inside. If $\delta > 1$, *then there are points of* Λ, *other than* O, *inside* K. This is a very special case of a famous theorem of Minkowski, which asserts that the same property is possessed, not only by any parallelogram symmetrical about the origin (whether generated by points of Λ or not), but by any 'convex region' symmetrical about the origin.

An *open region* R is a set of points with the properties (1) if P belongs to R, then all points of the plane sufficiently near to P belong to R, (2) any two points of R can be joined by a continuous curve lying entirely in R. We may also express (1) by saying that any point of R is an *interior* point of R. Thus the inside of a circle or a parallelogram is an open region. The *boundary* C of R is the set of points which are limit points of R but do not themselves belong to R. Thus the boundary of a circle is its circumference. A *closed region* R^* is an open region R together with its boundary. We consider only bounded regions.

There are two natural definitions of a *convex* region, which may be shown to be equivalent. First, we may say that R (or R^*) is convex if every point of any chord of R, i.e. of any line joining two points of R, belongs to R. Secondly, we may say that R (or R^*) is convex if it is possible, through every point P of C, to draw at least one line l such that the whole of R lies on one side of l. Thus a circle and a parallelogram are convex; for the circle, l is the tangent at P, while for the parallelogram every line l is a side except at the vertices, where there are an infinity of lines with the property required.

It is easy to prove the equivalence of the two definitions. Suppose first that R is convex according to the second definition, that P and Q belong to R, and that a point S of PQ does not. Then there is a point T of C (which may be S itself) on PS, and a line l through T which leaves R entirely on one side; and, since all points sufficiently near to P or Q belong to R, this is a contradiction.

Secondly, suppose that R is convex according to the first definition and that P is a point of C; and consider the set L of lines joining P to points of R. If Y_1 and Y_2 are points of R, and Y is a point of Y_1Y_2, then Y is a point of R and PY a line of L. Hence there is an angle APB such that every line from P within APB, and no line outside APB, belongs to L. If $APB > \pi$, then there are points D, E of R such that DE passes through P, in which case P belongs to R and not to C, a contradiction. Hence $APB \leqslant \pi$. If $APB = \pi$, then AB is a line l; if $APB < \pi$, then any line through P, outside the angle, is a line l.

It is plain that convexity is invariant for translations and for magnifications about a point O.

A convex region R has an *area* (definable, for example, as the upper bound of the areas of networks of small squares whose vertices lie in R).

THEOREM 37. (MINKOWSKI'S THEOREM). *Any convex region R symmetrical about O, and of area greater than* 4, *includes points of* Λ *other than O.*

3.10. Proof of Minkowski's theorem. We begin by proving, a simple theorem whose truth is 'intuitive'.

THEOREM 38. *Suppose that* R_O *is an open region including O, that* R_P *is the congruent and similarly situated region about any point P of* Λ, *and that no two of the regions* R_P *overlap. Then the area of* R_O *does not exceed* 1.

The theorem becomes 'obvious' when we consider that, if R_O were the square bounded by the lines $x = \pm\frac{1}{2}, y = \pm\frac{1}{2}$, then the area of R_O would be 1 and the regions R_P, with their boundaries, would cover the plane. We may give an exact proof as follows.

Suppose that Δ is the area of R_O, and A the maximum distance of a point of C_O[†] from O; and that we consider the $(2n+1)^2$ regions R_P corresponding to points of Λ whose coordinates are not greater numerically than n. All these regions lie in the square whose sides are parallel to the axes and at a distance $n + A$ from O. Hence (since the regions do not overlap)

$$(2n+1)^2\Delta \leqslant (2n+2A)^2, \quad \Delta \leqslant \left(1 + \frac{A - \frac{1}{2}}{n + \frac{1}{2}}\right)^2,$$

and the result follows when we make n tend to infinity.

It is to be noticed that there is no reference to symmetry or to convexity in Theorem 38.

It is now easy to prove Minkowski's theorem. Minkowski himself gave two proofs, based on the two definitions of convexity.

(1) Take the first definition, and suppose that R_O is the result of contracting R about O to half its linear dimensions. Then the area of R_O is greater than 1, so that two of the regions R_P of Theorem 38 overlap, and there is a lattice-point P such that R_O and R_P overlap. Let Q (Fig. 3a) be a point common to R_O and R_P. If OQ' is equal and parallel to PQ, and Q'' is the image of Q' in O, then Q', and therefore Q'', lies in R_O; and therefore, by

[†] We use C systematically for the boundary of the corresponding R.

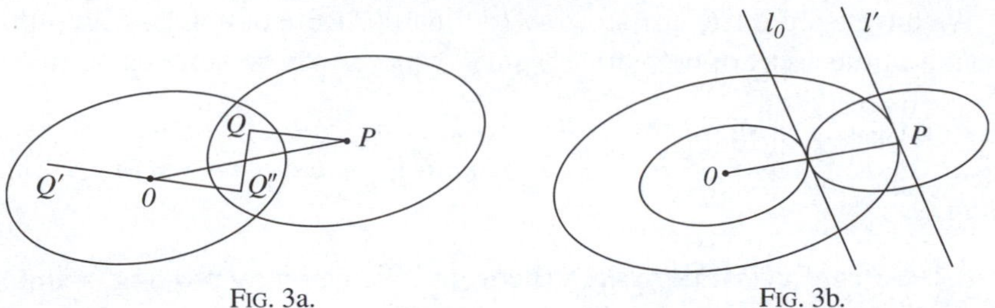

FIG. 3a. FIG. 3b.

the definition of convexity, the middle point of QQ' lies in R_O. But this point is the middle point of OP; and therefore P lies in R.

(2) Take the second definition, and suppose that there is no lattice point but O in R. Expand R^* about O until, as R'^*, it first includes a lattice point P. Then P is a point of C', and there is a line l, say l', through P (Fig. 3b). If R_O is R' contracted about O to half its linear dimensions, and l_O is the parallel to l through the middle point of OP, then l_O is a line l for R_O. It is plainly also a line l for R_P, and leaves R_O and R_P on opposite sides, so that R_O and R_P do not overlap. *A fortiori* R_O does not overlap any other R_P, and, since the area of R_O is greater than 1, this contradicts Theorem 38.

There are a number of interesting alternative proofs, of which perhaps the simplest is one due to Mordell.

If R is convex and symmetrical about O, and P_1 and P_2 are points of R with coordinates (x_1, y_1) and (x_2, y_2), then $(-x_2, -y_2)$, and therefore the point M whose coordinates are $\frac{1}{2}(x_1 - x_2)$ and $\frac{1}{2}(y_1 - y_2)$, is also a point of R.

The lines $x = 2p/t$, $y = 2q/t$, where t is a fixed positive integer and p and q arbitrary integers, divide up the plane into squares, of area $4/t^2$, whose corners are $(2p/t, 2q/t)$. If $N(t)$ is the number of corners in R, and A the area of R, then plainly $4t^{-2}N(t) \to A$ when $t \to \infty$; and if $A > 4$ then $N(t) > t^2$ for large t. But the pairs (p, q) give at most t^2 different pairs of remainders when p and q are divided by t; and therefore there are two points P_1 and P_2 of R, with coordinates $2p_1/t$, $2q_1/t$ and $2p_2/t$, $2q_2/t$, such that $p_1 - p_2$ and $q_1 - q_2$ are both divisible by t. Hence the point M, which belongs to R, is a point of Λ.

3.11. Developments of Theorem 37. There are some further developments of Theorem 37 which will be wanted in Ch. XXIV and which it is natural to prove here. We begin with a general remark which applies to all the theorems of §§ 3.6 and 3.9–10.

We have been interested primarily in the 'fundamental' lattice L (or Λ), but we can see in various ways how its properties may be restated as general properties of lattices. We use L or Λ now for any lattice of lines or points. If it is based upon the points O, P, Q, as in § 3.5, then we call the parallelogram $OPRQ$ the *fundamental parallelogram* of L or Λ.

(i) We may set up a system of oblique Cartesian coordinates with OP, OQ as axes, and agree that P and Q are the points (1, 0) and (0, 1). The area of the fundamental parallelogram is then

$$\delta = OP \cdot OQ \cdot \sin\omega,$$

where ω is the angle between OP and OQ. The arguments of § 3.6, interpreted in this system of coordinates, then prove

THEOREM 39. *A necessary and sufficient condition that the transformation (3.6.1) shall transform Λ into itself is that $\Delta = \pm 1$.*

THEOREM 40. *If P and Q are any two points of Λ, then a necessary and sufficient condition that the lattice L' based upon OP and OQ should be equivalent to L is that the area of the parallelogram defined by OP, OQ should be equal to that of the fundamental parallelogram of Λ.*

(ii) The transformation

$$x' = \alpha x + \beta y, \quad y' = \gamma x + \delta y$$

(where now α, β, γ, δ are any real numbers)† transforms the fundamental lattice of § 3.5 into the lattice based upon the origin and the points (α, γ), (β, δ). It transforms lines into lines and triangles into triangles. If the triangle $P_1P_2P_3$, where P_i is the point (x_i, y_i), is transformed into $Q_1Q_2Q_3$, then the areas of the triangles are

$$\pm\frac{1}{2}\begin{vmatrix} x_1 & y_1 & 1 \\ x_2 & y_2 & 1 \\ x_3 & y_3 & 1 \end{vmatrix}$$

and

$$\pm\frac{1}{2}\begin{vmatrix} \alpha x_1 + \beta y_1 & \gamma x_1 + \delta y_1 & 1 \\ \alpha x_2 + \beta y_2 & \gamma x_2 + \delta y_2 & 1 \\ \alpha x_3 + \beta y_3 & \gamma x_3 + \delta y_3 & 1 \end{vmatrix} = \pm\frac{1}{2}(\alpha\delta - \beta\gamma)\begin{vmatrix} x_1 & y_1 & 1 \\ x_2 & y_2 & 1 \\ x_3 & y_3 & 1 \end{vmatrix}.$$

† The δ of this paragraph has no connexion with the δ of (i), which reappears below.

Thus areas of triangles are multiplied by the constant factor $|\alpha\delta - \beta\gamma|$; and the same is true of areas in general, since these are sums, or limits of sums, of areas of triangles.

We can therefore generalize any property of the fundamental lattice by an appropriate linear transformation. The generalization of Theorem 38 is

THEOREM 41. *Suppose that* Λ *is any lattice with origin O, and that* R_O *satisfies* (*with respect to* Λ) *the conditions stated in Theorem* 38. *Then the area of* R_O *does not exceed that of the fundamental parallelogram of* Λ.

It is convenient also to give a proof *ab initio* which we state at length, since we use similar ideas in our proof of the next theorem. The proof, on the lines of (i) above, is practically the same as that in § 3.10.

The lines

$$x = \pm n, \quad y = \pm n$$

define a parallelogram Π of area $4n^2\delta$, with $(2n+1)^2$ points P of Λ inside it or on its boundary. We consider the $(2n+1)^2$ regions R_P corresponding to these points. If A is the greatest value of $|x|$ or $|y|$ on C_O, then all these regions lie inside the parallelogram Π', of area $4(n+A)^2\delta$, bounded by the lines

$$x = \pm(n+A), \quad y = \pm(n+A);$$

and

$$(2n+1)^2\Delta \leqslant 4(n+A)^2\delta.$$

Hence, making $n \to \infty$, we obtain.

$$\Delta \leqslant \delta.$$

We need one more theorem which concerns the limiting case $\Delta = \delta$. We suppose that R_O is a parallelogram; what we prove on this hypothesis will be sufficient for our purposes in Ch. XXIV.

We say that two points (x, y) and (x', y') are *equivalent with respect to* L if they have similar positions in two parallelograms of L (so that they would coincide if one parallelogram were moved into coincidence with the other by parallel displacement). If L is based upon OP and OQ, and P and

Q are (x_1, y_1) and (x_2, y_2), then the conditions that the points (x, y) and (x', y') should be equivalent are that

$$x' - x = rx_1 + sx_2, \quad y' - y = ry_1 + sy_2,$$

where r and s are integers.

THEOREM 42. *If R_O is a parallelogram whose area is equal to that of the fundamental parallelogram of L, and there are no two equivalent points inside R_O, then there is a point, inside R_O or on its boundary, equivalent to any given point of the plane.*

We denote the closed region corresponding to R_P by R_P^*.

The hypothesis that R_O includes no pair of equivalent points is equivalent to the hypothesis that no two R_P overlap. The conclusion that there is a point of R_O^* equivalent to any point of the plane is equivalent to the conclusion that the R_P^* cover the plane. Hence what we have to prove is that, *if $\Delta = \delta$ and the R_P do not overlap, then the R_P^* cover the plane.*

Suppose the contrary. Then there is a point Q outside all R_P^*. This point Q lies inside or on the boundary of some parallelogram of L, and there is a region D, in this parallelogram, and of positive area η outside all R_P; and a corresponding region in every parallelogram of L. Hence the area of all R_P, inside the parallelogram Π' of area $4(n + A)^2\delta$, does not exceed

$$4(\delta - \eta)(n + A + 1)^2.$$

It follows that

$$(2n + 1)^2\delta \leqslant 4(\delta - \eta)(n + A + 1)^2;$$

and therefore, making $n \to \infty$,

$$\delta \leqslant \delta - \eta,$$

a contradiction which proves the theorem.

Finally, we may remark that all these theorems may be extended to space of any number of dimensions. Thus if Λ is the fundamental point-lattice in three-dimensional space, i.e. the set of points (x, y, z) with integral coordinates, R is a convex region symmetrical about the origin, and of volume greater than 8, then there are points of Λ, other than O, in R. In n dimensions 8 must be replaced by 2^n. We shall say something about this generalization, which does not require new ideas, in Ch. XXIV.

NOTES

§ 3.1. The history of 'Farey series' is very curious. Theorems 28 and 29 seem to have been stated and proved first by Haros in 1802; see Dickson, *History*, i. 156. Farey did not publish anything on the subject until 1816, when he stated Theorem 29 in a note in the *Philosophical Magazine.* He gave no proof, and it is unlikely that he had found one, since he seems to have been at the best an indifferent mathematician.

Cauchy, however, saw Farey's statement, and supplied the proof (*Exercices de mathématiques,* i. 114–16). Mathematicians generally have followed Cauchy's example in attributing the results to Farey, and the series will no doubt continue to bear his name.

See Rademacher, *Lectures in elementary number theory* (New York, Blaisdell, 1964), for a fuller account of Farey series and Huxley, *Acta Arith.* 18 (1971), 281–7 and Hall, *J. London Math. Soc.* (2) 2 (1970), 139–48 for more details.

§ 3.3. Hurwitz, *Math. Annalen.* 44 (1894), 417–36. Professor H. G. Diamond drew my attention to the incompleteness of our proof in earlier editions.

§ 3.4. Landau, *Vorlesungen,* i. 98–100.

§§ 3.5–7. Here we follow the lines of a lecture by Professor Pólya.

§ 3.8. For Theorem 36 see Landau, *Vorlesungen*, i. 100.

§ 3.9. The reader need not pay much attention to the definitions of 'region', 'boundary', etc., given in this section if he does not wish to; he will not lose by thinking in terms of elementary regions such as parallelograms, polygons, or ellipses. Convex regions are simple regions involving no 'topological' difficulties. That a convex region has an area was first proved by Minkowski (*Geometrie der Zahlen,* Kap. 2).

§ 3.10. Minkowski's first proof will be found in *Geometrie der Zahlen*, 73–76, and his second in *Diophantische Approximationen*, 28–30. Mordell's proof was given in *Compositio Math.* 1 (1934), 248–53. Another interesting proof is that by Hajós, *Acta Univ. Hungaricae* (Szeged), 6 (1934), 224–5: this was set out in full in the first edition of this book.

IV

IRRATIONAL NUMBERS

4.1. Some generalities. The theory of 'irrational number', as explained in text books of analysis, falls outside the range of arithmetic. The theory of numbers is occupied, first with integers, then with rationals, as relations between integers, and then with irrationals, real or complex, of special forms, such as

$$r + s\sqrt{2}, \quad r + s\sqrt{(-5)},$$

where r and s are rational. It is not properly concerned with irrationals as a whole or with general criteria for irrationality (though this is a limitation which we shall not always respect).

There are, however, many problems of irrationality which may be regarded as part of arithmetic. Theorems concerning rationals may be restated as theorems about integers; thus the theorem

$$\text{`}r^3 + s^3 = 3 \text{ is insoluble in rationals'}$$

may be restated in the form

$$\text{`}a^3d^3 + b^3c^3 = 3b^3d^3 \text{ is insoluble in integers'}:$$

and the same is true of many theorems in which 'irrationality' intervenes. Thus

$$(P) \qquad \text{`}\sqrt{2} \text{ is irrational'}$$

means

$$(Q) \qquad \text{`}a^2 = 2b^2 \text{ is insoluble in integers'},$$

and then appears as a properly arithmetical theorem. We may ask 'is $\sqrt{2}$ irrational?' without trespassing beyond the proper bounds of arithmetic, and need not ask 'what is the meaning of $\sqrt{2}$?' We do not require any interpretation of the isolated symbol $\sqrt{2}$, since the meaning of (P) is defined as a whole and as being the same as that of (Q).†

In this chapter we shall be occupied with the problem

'is x rational or irrational?',

x being a number which, like $\sqrt{2}$, e, or π, makes its appearance naturally in analysis.

† In short $\sqrt{2}$ may be treated here as an 'incomplete symbol' in the sense of *Principia Mathematica*.

4.2. Numbers known to be irrational. The problem which we are considering is generally difficult, and there are few different types of numbers x for which the solution has been found. In this chapter we shall confine our attention to a few of the simplest cases, but it may be convenient to begin by a rough general statement of what is known. The statement must be rough because any more precise statement requires ideas which we have not yet defined.

There are, broadly, among numbers which occur naturally in analysis, two types of numbers whose irrationality has been established.

(*a*) *Algebraic irrationals.* The irrationality of $\sqrt{2}$ was proved by Pythagoras or his pupils, and later Greek mathematicians extended the conclusion to $\sqrt{3}$ and other square roots. It is now easy to prove that

$$\sqrt[m]{N}$$

is generally irrational for integral m and N. Still more generally, numbers defined by algebraic equations with integral coefficients, unless 'obviously' rational, can be shown to be irrational by the use of a theorem of Gauss. We prove this theorem (Theorem 45) in § 4.3.

(*b*) *The numbers e and π and numbers derived from them.* It is easy to prove e irrational (see § 4.7); and the proof, simple as it is, involves the ideas which are most fundamental in later extensions of the theorem. π is irrational, but of this there is no really simple proof. All powers of e or π, and polynomials in e or π with rational coefficients, are irrational. Numbers such as

$$e^{\sqrt{2}}, \quad e^{\sqrt{5}}, \quad \sqrt{7}e^{3\sqrt{2}}, \quad \log 2$$

are irrational. We shall return to this subject in Ch. XI (§§ 11.13–14).

It was not until 1929 that theorems were discovered which go beyond those of §§ 11.13–14 in any very important way. It has been shown recently that further classes of numbers, in which

$$e^{\pi}, \quad 2^{\sqrt{2}}, \quad e^{\pi\sqrt{2}}, \quad e^{\pi}+\pi$$

are included, are irrational. The irrationality of such numbers as

$$2^{e}, \quad \pi^{e}, \quad \pi^{\sqrt{2}}, \quad e+\pi$$

or 'Euler's constant'† γ is still unproved.

† $\gamma = \lim_{n\to\infty}\left(1+\frac{1}{2}+\ldots+\frac{1}{n}-\log n\right)$.

4.3. The theorem of Pythagoras and its generalizations. We shall begin by proving

THEOREM 43 (PYTHAGORAS' THEOREM). *$\sqrt{2}$ is irrational.*

We shall give two proofs of this theorem. The theorem and its simplest generalizations, though trivial now, deserve intensive study. The old Greek theory of proportion was based on the hypothesis that magnitudes of the same kind were necessarily commensurable, and it was the discovery of Pythagoras which, by exposing the inadequacy of this theory, opened the way for the more profound theory of Eudoxus which is set out in Euclid v.

(i) *First proof.* If $\sqrt{2}$ is rational, then the equation

$$a^2 = 2b^2 \tag{4.3.1}$$

is soluble in integers a, b with $(a, b) = 1$. Hence $b|a^2$ and therefore $p|a^2$ for any prime factor p of b. It follows that $p|a$. Since $(a, b) = 1$, this is impossible. Hence $b = 1$ and this also is clearly false.

(ii) *Second proof.* The traditional proof ascribed to Pythagoras runs as follows. From (4.3.1), we see that a^2 is even and therefore that a is even, i.e. $a = 2c$. Hence $b^2 = 2c^2$ and b is also even, contrary to the hypothesis that $(a, b) = 1$.

The two proofs are very similar but there is an important difference. In (ii) we consider divisibility by 2, a given number. Clearly, if $2|a^2$, then $2|a$, since the square of an odd number is certainly odd. In (i), on the other hand, we consider divisibility by the unknown prime p and, in fact, we assume Theorem 3. Thus (ii) is the logically simpler proof, while, as we shall see in a moment, (i) lends itself more readily to generalization.

We now prove the more general

THEOREM 44. *$\sqrt[m]{N}$ is irrational, unless N is the m-th power of an integer n.*

(iii) Suppose that

$$a^m = Nb^m, \tag{4.3.2}$$

where $(a, b) = 1$. Then $b|a^m$, and $p|a^m$ for every prime factor p of b. Hence $p|a$, and from this it follows as before that $b = 1$. It will be observed that this proof is almost the same as the first proof of Theorem 43.

(iv) To prove Theorem 44 for $m = 2$ without using Theorem 3, we suppose that

$$\sqrt{N} = a + \frac{b}{c},$$

where a, b, c are integers, $0 < b < c$ and b/c is the fraction with least numerator for which this is true. Hence

$$c^2N = (ca + b)^2 = a^2c^2 + 2abc + b^2$$

and so $c|b^2$, i.e. $b^2 = cd$. Hence

$$\sqrt{N} = a + \frac{b}{c} = a + \frac{d}{b}$$

and $0 < d < b$, a contradiction. It follows that $\sqrt{N}$ is integral or irrational.

A still more general theorem is

THEOREM 45. *If x is a root of an equation*

$$x^m + c_1x^{m-1} + \cdots + c_m = 0,$$

with integral coefficients of which the first is unity, then x is either integral or irrational.

In the particular case in which the equation is

$$x^m - N = 0,$$

Theorem 45 reduces to Theorem 44.

We may plainly suppose that $c_m \neq 0$. We argue as under (iii) above. If $x = a/b$, where $(a, b) = 1$, then

$$a^m + c_1a^{m-1}b + \ldots + c_mb^m = 0.$$

Hence $b|a^m$, and from this it follows as before that $b = 1$.

It is possible to prove Theorem 44 for general m and Theorem 45 also without using Theorem 3, but the argument is somewhat longer.

4.4. The use of the fundamental theorem in the proofs of Theorems 43–45. It is important, in view of the historical discussion in the next section, to observe what use is made, in the proofs of § 4.3, of the fundamental theorem of arithmetic or of the 'equivalent' Theorem 3.

The critical inference, in the proof (iii) of Theorem 44, is

$$\text{`}p|a^m \to p|a\text{'}.$$

Here we use Theorem 3. The same remark applies to the *first* proof of Theorem 43, the only simplification being that $m = 2$. In these proofs Theorem 3 plays an essential part.

The situation is different in the *second* proof of Theorem 43, since here we are considering divisibility by the special number 2. We need '$2|a^2 \to 2|a$', and this can be proved by 'enumeration of cases' and without an appeal to Theorem 3. Since

$$(2s+1)^2 = 4s^2 + 4s + 1,$$

the square of an odd number is odd, as we remarked, and the conclusion follows.

We can use a similar enumeration of cases to prove Theorem 44 for any special m and N. Suppose, for example, that $m = 2$, $N = 5$. We need '$5|a^2 \to 5|a$'. Now any number a which is not a multiple of 5 is of one of the forms $5m + 1$, $5m + 2$, $5m + 3$, $5m + 4$, and the squares of these numbers leave remainders 1, 4, 4, 1 after division by 5.

If $m = 2$, $N = 6$, we argue with 2, the smallest prime factor of 6, and the proof is almost identical with the second proof of Theorem 43. With $m = 2$ and

$$N = 2, 3, 5, 6, 7, 8, 10, 11, 12, 13, 14, 15, 17, 18,$$

we argue with the divisors

$$d = 2, 3, 5, 2, 7, 4, 2, 11, 3, 13, 2, 3, 17, 2,$$

the smallest prime factors of N which occur in odd multiplicity or, in the case of 8, an appropriate power of this prime factor. It is instructive to work through some of these cases; it is only when N is prime that the proof runs exactly according to the original pattern, and then it becomes tedious for the larger values of N.

We can deal similarly with cases such as $m = 3$, $N = 2$, 3, or 5; but we confine ourselves to those which are relevant in §§ 4.5–6.

4.5. A historical digression. It is unknown when, or by whom, the 'theorem of Pythagoras' was discovered. 'The discovery', says Heath,[†] 'can hardly have been made by Pythagoras himself, but it was certainly made in his school.' Pythagoras lived about 570–490 B.C. Democritus, born about 470, wrote 'on irrational lines and solids', and 'it is difficult to resist the conclusion that the irrationality of $\sqrt{2}$ was discovered before Democritus' time'.

It would seem that no extension of the theorem was made for over fifty years. There is a famous passage in Plato's *Theaetetus* in which it is stated that Theodorus (Plato's teacher) proved the irrationality of

$$\sqrt{3}, \sqrt{5}, \ldots,$$

'taking all the separate cases up to the root of 17 square feet, at which point, for some reason, he stopped'. We have no accurate information about this or other discoveries of Theodorus, but Plato lived 429–348, and it seems reasonable to date this discovery about 410–400.

The question how Theodorus proved his theorems has exercised the ingenuity of every historian. It would be natural to conjecture that he used some modification of the 'traditional' method of Pythagoras, such as those which we discussed in the last section. In that case, since he cannot have known the fundamental theorem,[‡] and it is unlikely that he knew even Euclid's Theorem 3, he may have argued much as we argued at the end of § 4.4. The objections to this (made by historians such as Zeuthen and Heath) are (i) that it is so obvious an adaptation of the proof for $\sqrt{2}$ that it would not be regarded as new and (ii) that it would be clear, long before $\sqrt{17}$ was reached, that it was generally applicable. Against this, however, it is fair to remark that Theodorus would have to consider each different d anew and that the work would become notably laborious at $\sqrt{11}$, $\sqrt{13}$, and $\sqrt{17}$ (and behind $\sqrt{17}$ lurk $\sqrt{19}$ and $\sqrt{23}$).

There are, however, two other hypotheses as to Theodorus' method of proof. These methods become notably more complicated, one at $\sqrt{17}$ and the other at $\sqrt{19}$. Which of these is to be preferred depends on the exact meaning of the Greek word $\mu\epsilon\chi\rho\iota$, translated as 'up to' by Heath; does it mean 'up to but not including' or 'up to and including' (the American usage of 'through')? Classical scholars tell me that the former is the more

[†] Sir Thomas Heath, *A manual of Greek mathematics,* 54–55. In what follows passages in inverted commas, unless attributed to other writers, are quotations from this book or from the same writer's *A history of Greek mathematics.*

[‡] See Ch. XII, § 12.5, for some further discussion of this point.

probable and, if so, the following method, proposed by McCabe, is a very likely one. It has the merit of depending essentially on the distinction between odd and even, a matter of great importance in Greek mathematics.

Considering $\sqrt{N}$ for successive values of N, Theodorus could ignore $N = 4n$, since he would already have dealt with $\sqrt{n}$. The other even values of N take the form $2(2n+1)$ and the proof for $\sqrt{2}$ extends to this at once. We have therefore only to consider odd N. For such N, if $\sqrt{N} = a/b$ and $(a, b) = 1$, we have $Nb^2 = a^2$ and a and b must both be odd. We write $a = 2A+1$ and $b = 2B+1$ and so obtain

$$N(2A + 1)^2 = (2B + 1)^2.$$

The number N must be of one of the forms

$$4n + 3, \quad 8n + 5, \quad 8n + 1.$$

If $N = 4n + 3$, we multiply out, divide by 2 and obtain

$$8nA(A + 1) + 6A(A + 1) + 2n + 1 = 2B(B + 1),$$

an impossibility, since one side is odd and the other even. If $N = 8n + 5$, we again multiply out, divide by 4 and have

$$8nA(A + 1) + 5A(A + 1) + 2n + 1 = B(B + 1),$$

again impossible, since $A(A + 1)$ and $B(B + 1)$ are each even.

There remain the numbers of the form $8n + 1$, which are $1, 9, 17, \ldots$. Of these, 1 and 9 are trivial and a difficulty first arises at $N = 17$. Arguing as before, we reach the equation

$$17(B^2 + B) + 4 = A^2 + A,$$

both sides being even. We have then to consider a variety of possibilities and the whole problem becomes much more complicated. (The reader may care to try them.) Hence, if this were Theodorus' method, he would very naturally stop just short of $\sqrt{17}$.

Zeuthen suggests an interesting method involving ratios which after a few transformations begin to cycle endlessly, thus leading to a proof by contradiction. This works well up to and including 17, while 18 is of course trivial, but 19 requires 8 ratios before an endless chain begins. We give his proof for $\sqrt{5}$ in § 4.6. But, even if μεχρι, means 'up to and including' in this passage, Plato might more reasonably have said 'up to and including 18'. On balance, McCabe's conjecture seems the most plausible.

4.6. Geometrical proof of the irrationality of $\sqrt{5}$. The proofs suggested by Zeuthen vary from number to number, and the variations depend at bottom on the form of the periodic continued fraction† which represents $\sqrt{N}$. We take as typical the simplest case ($N = 5$).

We argue in terms of

$$x = \frac{1}{2}(\sqrt{5} - 1).$$

Then

$$x^2 = 1 - x.$$

Geometrically, if $AB = 1, AC = x$, then

$$AC^2 = AB \, . \, CB$$

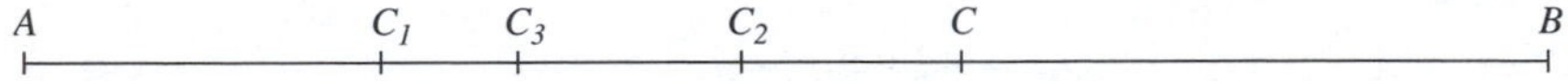

A C1 C3 C2 C B

FIG. 4.

and AB is divided 'in golden section' by C. These relations are fundamental in the construction of the regular pentagon inscribed in a circle (Euclid iv. 11).

If we divide 1 by x, taking the largest possible integral quotient, viz. 1,‡ the remainder is $1 - x = x^2$. If we divide x by x^2, the quotient is again 1 and the remainder is $x - x^2 = x^3$. We next divide x^2 by x^3, and continue the process indefinitely; at each stage the ratios of the number divided, the divisor, and the remainder are the same. Geometrically, if we take CC_1 equal and opposite to CB, CA is divided at C_1 in the same ratio as AB at C, i.e. in golden section; if we take C_1C_2 equal and opposite to C_1A, then C_1C is divided in golden section at C_2; and so on.‖ Since we are dealing at each stage with a segment divided in the same ratio, the process can never end.

It is easy to see that this contradicts the hypothesis of the rationality of x. If x is rational, then AB and AC are integral multiples of the same length δ, and the same is true of

$$C_1C = CB = AB - AC, \quad C_1C_2 = AC_1 = AC - C_1C, \ldots,$$

i.e. of all the segments in the figure. Hence we can construct an infinite sequence of descending integral multiples of δ, and this is plainly impossible.

† See Ch. X, § 10.12.

‡ Since $\frac{1}{2} < x < 1$.

‖ C_2C_3 equal and opposite to C_2C, C_3C_4 equal and opposite to $C_3C_1, \ldots$. The new segments defined are measured alternately to the left and the right.

4.7. Some more irrational numbers. We know, after Theorem 44, that $\sqrt{7}$, $\sqrt[3]{2}$, $\sqrt[4]{11}, \ldots$ are irrational. After Theorem 45, $x = \sqrt{2} + \sqrt{3}$ is irrational, since it is not an integer and satisfies

$$x^4 - 10x^2 + 1 = 0.$$

We can construct irrationals freely by means of decimals or continued fractions, as we shall see in Chs. IX and X; but it is not easy, without theorems such as we shall prove in §§ 11.13–14, to add to our list many of the numbers which occur naturally in analysis.

THEOREM 46. $\log_{10} 2$ *is irrational.*

This is trivial, since

$$\log_{10} 2 = \frac{a}{b}$$

involves $2^b = 10^a$, which is impossible. More generally $\log_n m$ is irrational if m and n are integers, one of which has a prime factor which the other lacks.

THEOREM 47. e *is irrational.*

Let us suppose e rational, so that $e = a/b$ where a and b are integers. If $k \geqslant b$ and

$$\alpha = k!\left(e - 1 - \frac{1}{1!} - \frac{1}{2!} - \ldots - \frac{1}{k!}\right),$$

then $b|k!$ and α is an integer. But

$$\begin{aligned} 0 < \alpha &= \frac{1}{k+1} + \frac{1}{(k+1)(k+2)} + \ldots \\ &< \frac{1}{k+1} + \frac{1}{(k+1)^2} + \ldots = \frac{1}{k} \end{aligned}$$

and this is a contradiction.

In this proof, we assumed the theorem false and deduced that α was (i) integral, (ii) positive, and (iii) less than one, an obvious contradiction. We prove two further theorems by more sophisticated applications of the same idea.

For any positive integer n, we write

$$f = f(x) = \frac{x^n(1-x)^n}{n!} = \frac{1}{n!}\sum_{m=n}^{2n} c_m x^m,$$

where the c_m are integers. For $0 < x < 1$, we have

$$0 < f(x) < \frac{1}{n!}. \tag{4.7.1}$$

Again $f(0) = 0$ and $f^{(m)}(0) = 0$ if $m < n$ or $m > 2n$. But, if $n \leqslant m \leqslant 2n$,

$$f^{(m)}(0) = \frac{m!}{n!}c_m,$$

an integer. Hence $f(x)$ and all its derivatives take integral values at $x = 0$. Since $f(1-x) = f(x)$, the same is true at $x = 1$.

THEOREM 48. *e^y is irrational for every rational $y \neq 0$.*

If $y = h/k$ and e^y is rational, so is $e^{ky} = e^h$. Again, if e^{-h} is rational, so is e^h. Hence it is enough to prove that, if h is a positive integer, e^h cannot be rational. Suppose this false, so that $e^h = a/b$ where a, b are positive integers. We write

$$F(x) = h^{2n}f(x) - h^{2n-1}f'(x) + \ldots - hf^{(2n-1)}(x) + f^{(2n)}(x),$$

so that $F(0)$ and $F(1)$ are integers. We have

$$\frac{d}{dx}\{e^{hx}F(x)\} = e^{hx}\{hF(x) + F'(x)\} = h^{2n+1}e^{hx}f(x).$$

Hence

$$b\int_0^1 h^{2n+1}e^{hx}f(x)dx = b[e^{hx}F(x)]_0^1 = aF(1) - bF(0),$$

an integer. But, by (4.7.1),

$$0 < b\int_0^1 h^{2n+1}e^{hx}f(x)dx < \frac{bh^{2n}e^h}{n!} < 1$$

for large enough n, a contradiction.

THEOREM 49. *π and π^2 are irrational.*

Suppose π^2 rational, so that $\pi^2 = a/b$, where a, b are positive integers. We write

$$G(x) = b^n \left\{ \pi^{2n} f(x) - \pi^{2n-2} f''(x) + \pi^{2n-4} f^{(4)}(x) - \cdots + (-1)^n f^{(2n)}(x) \right\},$$

so that $G(0)$ and $G(1)$ are integers. We have

$$\begin{aligned} &\frac{d}{dx}\{G'(x)\sin \pi x - \pi G(x)\cos \pi x\} \\ &\quad = \{G''(x) + \pi^2 G(x)\}\sin \pi x = b^n \pi^{2n+2} f(x) \sin \pi x \\ &\quad = \pi^2 a^n \sin \pi x f(x). \end{aligned}$$

Hence

$$\begin{aligned} \pi \int_0^1 a^n \sin \pi x\, f(x)dx &= \left[\frac{G'(x)\sin \pi x}{\pi} - G(x)\cos \pi x \right]_0^1 \\ &= G(0) + G(1), \end{aligned}$$

an integer. But, by (4.7.1),

$$0 < \pi \int_0^1 a^n \sin \pi x\, f(x)dx < \frac{\pi a^n}{n!} < 1$$

for large enough n, a contradiction.

NOTES

§ 4.2. The irrationality of e and π was proved by Lambert in 1761; and that of e^{π} by Gelfond in 1929. See the notes on Ch. XI.

§§ 4.3–6. A reader interested in Greek mathematics is referred to Heath's books mentioned on p. 42, to van der Waerden, *Science awakening* (Gronnigen, Nordhoff, 1954) and to Knorr, *Evolution of the Euclidean elements* (Boston, Reidel, 1975). See McCabe, *Math. Mag.* 49 (1976), 201–3 for his conjecture as to Theodorus' method of proof.

We do not give specific references, nor attempt to assign Greek theorems to their real discoverers. Thus we use 'Pythagoras' for 'some mathematician of the Pythagorean school'.

§ 4.3. Sir Alexander Oppenheim found the proof (iv) of Theorem 44 (improved by Prof. R. Rado) and the corresponding proof of Theorem 45 referred to at the end of § 4.3. Theorem 45 is proved, in a more general form, by Gauss, *D.A.*, § 42.

§ 4.7. Our proof of Theorem 48 is based on that of Hermite (*Œuvres*, 3, 154) and our proof of Theorem 49 on that of Niven (*Bulletin Amer. Math. Soc.* 53 (1947), 509).

By Theorem 49

$$\zeta(2) = \sum_{n=1}^{\infty} \frac{1}{n^2} = \frac{\pi^2}{6}$$

is irrational, and by Theorem 205, $\zeta(4) = \frac{\pi^4}{90}$ is also irrational, as are the values of $\zeta(m)$ for all even positive integers m. However when m is odd much less is known. Apéry (1978) showed that $\zeta(3)$ is irrational; for a short proof see Beukers (*Bull. London Math. Soc.* 11 (1979), 268–72). It is still unknown if $\zeta(5)$ is irrational. However Ball and Rivoal (*Inventiones Math.* 146 (2001), 193–207) proved that the sequence $\zeta(3), \zeta(5), \zeta(7), \zeta(9), \ldots$ contains infinitely many irrational numbers.

V
CONGRUENCES AND RESIDUES

5.1. Highest common divisor and least common multiple. We have already defined the highest common divisor (a, b) of two numbers a and b. There is a simple formula for this number.

We denote by $\min(x, y)$ and $\max(x, y)$ the lesser and the greater of x and y. Thus $\min(1, 2) = 1$, $\max(1, 1) = 1$.

THEOREM 50. *If*

$$a = \prod_p p^\alpha \quad (\alpha \geqslant 0),^\dagger$$

and

$$b = \prod_p p^\beta \quad (\beta \geqslant 0),$$

then

$$(a, b) = \prod_p p^{\min(\alpha,\beta)}.$$

This theorem is an immediate consequence of Theorem 2 and the definition of (a, b).

The *least common multiple* of two numbers a and b is the least positive number which is divisible by both a and b. We denote it by $\{a, b\}$, so that

$$a|\{a, b\}, \quad b|\{a, b\},$$

and $\{a, b\}$ is the least number which has this property.

† The symbol

$$\prod_p f(p)$$

denotes a product extended over all prime values of p. The symbol

$$\prod_{p|m} f(p)$$

denotes a product extended over all primes which divide m. In the first formula of Theorem 50, α is zero unless $p|a$ (so that the product is really a finite product). We might equally well write

$$a = \prod_{p|a} p^\alpha.$$

In this case every α would be positive.

THEOREM 51. *In the notation of Theorem* 50,

$$\{a, b\} = \prod_p p^{\max(\alpha,\beta)}.$$

From Theorems 50 and 51 we deduce

THEOREM 52:

$$\{a, b\} = \frac{ab}{(a, b)}.$$

If $(a, b) = 1$, a and b are said to be prime to one another or *coprime*. The numbers $a, b, c, \ldots, k$ are said to be coprime if every two of them are coprime. To say this is to say much more than to say that

$$(a, b, c, \ldots, k) = 1,$$

which means merely that there is no number but 1 which divides all of $a, b, c, \ldots, k$.

We shall sometimes say that 'a and b have no common factor' when we mean that they have no common factor greater than 1, i.e. that they are coprime.

5.2. Congruences and classes of residues. If m is a divisor of $x - a$, we say that x is congruent to a to modulus m, and write

$$x \equiv a \pmod{m}.$$

The definition does not introduce any new idea, since '$x \equiv a \pmod{m}$' and '$m|x-a$' have the same meaning, but each notation has its advantages. We have already used the word 'modulus' in a different sense in § 2.9, but the ambiguity will not cause any confusion.†

By $x \not\equiv a \pmod{m}$ we mean that x is not congruent to a.

If $x \equiv a \pmod{m}$, then a is called a *residue* of x to modulus m. If $0 \leqslant a \leqslant m - 1$, then a is the *least residue*‡ of x to modulus m. Thus two numbers a and b congruent (mod m) have the same residues (mod m). A *class of residues* (mod m) is the class of all the numbers congruent to a given

† The dual use has a purpose because the notion of a 'congruence with respect to a modulus of numbers' occurs at a later stage in the theory, though we shall not use it in this book.

‡ Strictly, least non-negative residue.

residue (mod m), and every member of the class is called a *representative* of the class. It is clear that there are in all m classes, represented by

$$0, 1, 2, \ldots, m-1.$$

These m numbers, or any other set of m numbers of which one belongs to each of the m classes, form a *complete system of incongruent residues to modulus m,* or, more shortly, a *complete system* (mod m).

Congruences are of great practical importance in everyday life. For example, 'today is Saturday' is a congruence property (mod 7) of the number of days which have passed since some fixed date. This property is usually much more important than the actual number of days which have passed since, say, the creation. Lecture lists or railway guides are tables of congruences; in the lecture list the relevant moduli are 365, 7, and 24.

To find the day of the week on which a particular event falls is to solve a problem in 'arithmetic (mod 7)'. In such an arithmetic congruent numbers are *equivalent,* so that the arithmetic is a strictly finite science, and all problems in it can be solved by trial. Suppose, for example, that a lecture is given on every alternate day (including Sundays), and that the first lecture occurs on a Monday. When will a lecture first fall on a Tuesday? If this lecture is the $(x+1)$th then

$$2x \equiv 1 \pmod{7};$$

and we find by trial that the least positive solution is

$$x = 4.$$

Thus the fifth lecture will fall on a Tuesday and this will be the first that will do so.

Similarly, we find by trial that the congruence

$$x^2 \equiv 1 \pmod{8}$$

has just four solutions, namely

$$x \equiv 1, 3, 5, 7 \pmod{8}.$$

It is sometimes convenient to use the notation of congruences even when the variables which occur in them are not integers. Thus we may write

$$x \equiv y \pmod{z}$$

whenever $x - y$ is an integral multiple of z, so that, for example,

$$\tfrac{3}{2} \equiv \tfrac{1}{2} \pmod 1, \qquad -\pi \equiv \pi \pmod{2\pi}.$$

5.3. Elementary properties of congruences. It is obvious that congruences to a given modulus m have the following properties:

(i) $a \equiv b \rightarrow b \equiv a$,

(ii) $a \equiv b \,.\, b \equiv c \rightarrow a \equiv c$,

(iii) $a \equiv a' \,.\, b \equiv b' \rightarrow a + b \equiv a' + b'$.

Also, if $a \equiv a'$, $b \equiv b'$, ... we have

(iv) $ka + lb + \ldots \equiv ka' + lb' + \ldots$,

(v) $a^2 \equiv a'^2, a^3 \equiv a'^3$,

and so on; and finally, if $\phi(a, b, \ldots)$ is any polynomial with integral coefficients, we have

(vi) $\phi(a, b, \ldots) \equiv \phi(a', b', \ldots)$.

THEOREM 53. *If* $a \equiv b \pmod m$ *and* $a \equiv b \pmod n$, *then*

$$a \equiv b \pmod{\{m, n\}}.$$

In particular, if $(m, n) = 1$, *then*

$$a \equiv b \pmod{mn},$$

This follows from Theorem 50. If p^c is the highest power of p which divides $\{m, n\}$, then $p^c|m$ or $p^c|n$ and so $p^c|(a - b)$. This is true for every prime factor of $\{m, n\}$, and so

$$a \equiv b \pmod{\{m, n\}}.$$

The theorem generalizes in the obvious manner to any number of congruences.

5.4. Linear congruences. The properties (i)–(vi) are like those of equations in ordinary algebra, but we soon meet with a difference. It is not true that

$$ka \equiv ka' \rightarrow a \equiv a';$$

for example

$$2\,.\,2 \equiv 2\,.\,4 \pmod 4,$$

but

$$2 \not\equiv 4 \pmod 4.$$

We consider next what is true in this direction.

THEOREM 54. *If* $(k, m) = d$, *then*

$$ka \equiv ka' \pmod m \rightarrow a \equiv a' \left(\bmod \frac{m}{d}\right),$$

and conversely.

Since $(k, m) = d$, we have

$$k = k_1 d, \quad m = m_1 d, \quad (k_1, m_1) = 1.$$

Then

$$\frac{ka - ka'}{m} = \frac{k_1(a - a')}{m_1},$$

and, since $(k_1, m_1) = 1$,

$$m|ka - ka' \equiv m_1|a - a'.^{\dagger}$$

This proves the theorem. A particular case is

THEOREM 55. *If* $(k, m) = 1$, *then*

$$ka \equiv ka' \pmod m \rightarrow a \equiv a' \pmod m$$

and conversely.

THEOREM 56. *If* $a_1, a_2, \ldots, a_m$ *is a complete system of incongruent residues* (*mod* m) *and* $(k, m) = 1$, *then* $ka_1, ka_2, \ldots, ka_m$ *is also such a system.*

For $ka_i - ka_j \equiv 0 \pmod m$ implies $a_i - a_j \equiv 0 \pmod m$, by Theorem 55, and this is impossible unless $i = j$. More generally, if

† '$\equiv$' is the symbol of logical equivalence: if P and Q are propositions, then $P \equiv Q$ if $P \rightarrow Q$ and $Q \rightarrow P$.

$(k, m) = 1$, then

$$ka_r + l \ (r = 1, 2, 3, \ldots, m)$$

is a complete system of incongruent residues (mod m).

THEOREM 57. *If $(k, m) = d$, then the congruence*

$$kx \equiv l \ (\text{mod } m) \tag{5.4.1}$$

is soluble if and only if $d|l$. It has then just d solutions. In particular, if $(k, m) = 1$, the congruence has always just one solution.

The congruence is equivalent to

$$kx - my = l,$$

so that the result is partly contained in Theorem 25. It is naturally to be understood, when we say that the congruence has 'just d' solutions, that congruent solutions are regarded as the same.

If $d = 1$, then Theorem 57 is a corollary of Theorem 56. If $d > 1$, the congruence (5.4.1) is clearly insoluble unless $d|l$. If $d|l$, then

$$m = dm', \quad k = dk', \quad l = dl',$$

and the congruence is equivalent to

$$k'x \equiv l' (\text{mod } m'). \tag{5.4.2}$$

Since $(k', m') = 1$, (5.4.2) has just one solution. If this solution is

$$x \equiv t \ (\text{mod } m'),$$

then

$$x = t + ym',$$

and the complete set of solutions of (5.4.1) is found by giving y all values which lead to values of $t + ym'$ incongruent to modulus m. Since

$$t + ym' \equiv t + zm' (\text{mod } m) \equiv m|m'(y - z) \equiv d|(y - z),$$

there are just d solutions, represented by

$$t, \quad t + tm', \quad t + 2m', \ldots, \quad t + (d - l)m'.$$

This proves the theorem.

5.5. Euler's function $\phi(m)$. We denote by $\phi(m)$ the number of positive integers not greater than and prime to m, that is to say the number of integers n such that

$$0 < n \leqslant m, \quad (n, m) = 1.^\dagger$$

If a is prime to m, then so is any number x congruent to a (mod m). There are $\phi(m)$ classes of residues prime to m, and any set of $\phi(m)$ residues, one from each class, is called a *complete set of residues prime to m*. One such complete set is the set of $\phi(m)$ numbers less than and prime to m.

THEOREM 58. *If $a_1, a_2, \ldots, a_{\phi(m)}$ is a complete set of residues prime to m, and $(k, m) = 1$, then*

$$ka_1,\ ka_2, \ldots,\ ka_{\phi(m)}$$

is also such a set.

For the numbers of the second set are plainly all prime to m, and, as in the proof of Theorem 56, no two of them are congruent.

THEOREM 59. *Suppose that $(m, m') = 1$, and that a runs through a complete set of residues* (mod m)*, and a' through a complete set of residues* (mod m')*. Then $a'm + am'$ runs through a complete set of residues* (mod mm').

There are mm' numbers $a'm + am'$. If

$$a'_1 m + a_1 m' \equiv a'_2 m + a_2 m' \pmod{mm'},$$

then

$$a_1 m' \equiv a_2 m' \pmod{m},$$

and so

$$a_1 \equiv a_2 \pmod{m};$$

and similarly

$$a'_1 \equiv a'_2 \pmod{m'}.$$

Hence the mm' numbers are all incongruent and form a complete set of residues (mod mm').

† n can be equal to m only when $n = 1$. Thus $\phi(1) = 1$.

A function $f(m)$ is said to be *multiplicative* if $(m, m') = 1$ implies

$$f(mm') = f(m)f(m').$$

THEOREM 60. *$\phi(n)$ is multiplicative.*

If $(m, m') = 1$, then, by Theorem 59, $a'm + am'$ runs through a complete set (mod mm') when a and a' both run through complete sets (mod m) and (mod m') respectively. Also

$$\begin{aligned}(a'm + am', mm') = 1 &\equiv (a'm + am', m) = 1 \,.\, (a'm + am', m') = 1\\ &\equiv (am', m) = 1 \,.\, (a'm, m') = 1\\ &\equiv (a, m) = 1 \,.\, (a', m') = 1.\end{aligned}$$

Hence the $\phi(mm')$ numbers less than and prime to mm' are the least positive residues of the $\phi(m)\phi(m')$ values of $a'm + am'$ for which a is prime to m and a' to m'; and therefore

$$\phi(mm') = \phi(m)\phi(m').$$

Incidentally we have proved

THEOREM 61. *If $(m, m') = 1$, a runs through a complete set of residues prime to m, and a' through a complete set of residues prime to m', then $am' + a'm$ runs through a complete set of residues prime to mm'.*

We can now find the value of $\phi(m)$ for any value of m. By Theorem 60, it is sufficient to calculate $\phi(m)$ when m is a power of a prime. Now there are $p^c - 1$ positive numbers less than p^c, of which $p^{c-1} - 1$ are multiples of p and the remainder prime to p. Hence

$$\phi(p^c) = p^c - 1 - (p^{c-1} - 1) = p^c\left(1 - \frac{1}{p}\right);$$

and the general value of $\phi(m)$ follows from Theorem 60.

THEOREM 62. *If $m = \Pi p^c$, then*

$$\phi(m) = m \prod_{p|m}\left(1 - \frac{1}{p}\right).$$

We shall also require

THEOREM 63:

$$\sum_{d|m} \phi(d) = m.$$

If $m = \Pi p^c$, then the divisors of m are the numbers $d = \Pi p^{c'}$, where $0 \leqslant c' \leqslant c$ for each p; and

$$\begin{aligned}\Phi(m) = \sum_{d|m} \phi(d) &= \sum_{p,c'} \prod \phi(p^{c'}) \\ &= \prod_p \left\{1 + \phi(p) + \phi(p^2) + \cdots + \phi(p^c)\right\},\end{aligned}$$

by the multiplicative property of $\phi(m)$. But

$$\begin{aligned}1 + \phi(p) + \cdots + \phi(p^c) &= 1 + (p-1) + p(p-1) + \cdots \\ &\quad + p^{c-1}(p-1) = p^c,\end{aligned}$$

so that

$$\Phi(m) = \prod_p p^c = m.$$

5.6. Applications of Theorems 59 and 61 to trigonometrical sums. There are certain trigonometrical sums which are important in the theory of numbers and which are either 'multiplicative' in the sense of § 5.5 or possess very similar properties.

We write†

$$e(\tau) = e^{2\pi i \tau}:$$

we shall be concerned only with rational values of τ. It is clear that

$$e\left(\frac{m}{n}\right) = e\left(\frac{m'}{n}\right)$$

when $m \equiv m' \pmod n$. It is this property which gives trigonometrical sums their arithmetical importance.

† Throughout this section e^{ζ} is the exponential function $e^{\zeta} = 1 + \zeta + \cdots$ of the complex variable ζ. We assume a knowledge of the elementary properties of the exponential function.

(1) *Multiplicative property of Gauss's sum.* Gauss's sum, which is particularly important in the theory of quadratic residues, is

$$S(m,n) = \sum_{h=0}^{n-1} e^{2\pi i h^2 m/n} = \sum_{h=0}^{n-1} e\left(\frac{h^2 m}{n}\right).$$

Since

$$e\left\{\frac{(h+rn)^2 m}{n}\right\} = e\left(\frac{h^2 m}{n}\right)$$

for any r, we have

$$e\left(\frac{h_1^2 m}{n}\right) = e\left(\frac{h_2^2 m}{n}\right)$$

whenever $h_1 \equiv h_2 \pmod{n}$. We may therefore write

$$S(m,n) = \sum_{h(n)} e\left(\frac{h^2 m}{n}\right),$$

the notation implying that h runs through any complete system of residues mod n. When there is no risk of ambiguity, we shall write h instead of $h(n)$.

THEOREM 64. *If* $(n, n') = 1$, *then*

$$S(m, nn') = S(mn', n)S(mn, n').$$

Let h, h' run through complete systems of residues to modulus n, n' respectively. Then, by Theorem 59,

$$H = hn' + h'n$$

runs through a complete set of residues to modulus nn'. Also

$$mH^2 = m(hn' + h'n)^2 \equiv mh^2n'^2 + mh'^2n^2 \pmod{nn'}.$$

Hence

$$S(mn', n)S(mn, n') = \left\{\sum_h e\left(\frac{h^2mn'}{n}\right)\right\}\left\{\sum_{h'} e\left(\frac{h'^2mn}{n'}\right)\right\}$$
$$= \sum_{h,h'} e\left(\frac{h^2mn'}{n} + \frac{h'^2mn}{n'}\right)$$
$$= \sum_{h,h'} e\left(\frac{m(h^2n'^2 + h'^2n^2)}{nn'}\right)$$
$$= \sum_H e\left(\frac{mH^2}{nn'}\right) = S(m, nn').$$

(2) *Multiplicative property of Ramanujan's sum.* Ramanujan's sum is

$$c_q(m) = \sum_{h^*(q)} e\left(\frac{hm}{q}\right),$$

the notation here implying that h runs only through residues prime to q. We shall sometimes write h instead of $h^*(q)$ when there is no risk of ambiguity.

We may write $c_q(m)$ in another form which introduces a notion of more general importance. We call ρ a *primitive q-th root* of unity if $\rho^q = 1$ but ρ^r is not 1 for any positive value of r less than q.

Suppose that $\rho^q = 1$ and that r is the least positive integer for which $p^r = 1$. Then $q = kr + s$, where $0 \leqslant s < r$. Also

$$\rho^s = \rho^{q-kr} = 1,$$

so that $s = 0$ and $r|q$. Hence

THEOREM 65. *Any q-th root of unity is a primitive r-th root, for some divisor r of q.*

THEOREM 66. *The q-th roots of unity are the numbers*

$$e\left(\frac{h}{q}\right) \quad (h = 0, 1, \ldots, q-1),$$

and a necessary and sufficient condition that the root should be primitive is that h should be prime to q.

We may now write Ramanujan's sum in the form

$$c_q(m) = \Sigma\rho^m,$$

where ρ runs through the primitive qth roots of unity.

THEOREM 67. *If* $(q, q') = 1$, *then*

$$c_{qq'}(m) = c_q(m)c_{q'}(m).$$

For

$$c_q(m)c_{q'}(m) = \sum_{h,h'} e\left\{m\left(\frac{h}{q} + \frac{h'}{q'}\right)\right\}$$

$$= \sum_{h,h'} e\left\{\frac{m(hq' + h'q)}{qq'}\right\} = c_{qq'}(m),$$

by Theorem 61.

(3) *Multiplicative property of Kloosterman's sum.* Kloosterman's sum (which is rather more recondite) is

$$S(u, v, n) = \sum_h e\left(\frac{uh + v\bar{h}}{n}\right),$$

where h runs through a complete set of residues prime to n, and $\bar{h}$ is defined by

$$h\bar{h} \equiv 1 (\text{mod } n).$$

Theorem 57 shows us that, given any h, there is a unique $\bar{h}$ (mod n) which satisfies this condition. We shall make no use of Kloosterman's sum, but the proof of its multiplicative property gives an excellent illustration of the ideas of the preceding sections.

THEOREM 68. *If* $(n, n') = 1$, *then*

$$S(u, v, n)S(u, v', n') = S(u, V, nn'),$$

where

$$V = vn'^2 + v'n^2.$$

If

$$h\bar{h} \equiv 1 (\text{mod } n), \quad h'\bar{h}' \equiv 1 (\text{mod } n')$$

then

$$S(u, v, n)S(u, v', n') = \sum_{h,h'} e\left(\frac{uh + v\bar{h}}{n} + \frac{uh' + v'\bar{h}'}{n'}\right)$$

$$= \sum_{h,h'} e\left\{u\left(\frac{hn + h'n}{nn'}\right) + \frac{v\bar{h}n' + v'\bar{h}'n}{nn'}\right\}$$

(5.6.1)
$$= \sum_{h,h'} e\left(\frac{uH + K}{nn'}\right),$$

where

$$H = hn' + h'n, \quad K = v\bar{h}n' + v'\bar{h}'n.$$

By Theorem 61, H runs through a complete system of residues prime to nn'. Hence, if we can show that

(5.6.2)
$$K \equiv V\bar{H} (\text{mod } nn'),$$

where $\bar{H}$ is defined by

$$H\bar{H} \equiv 1 (\text{mod } nn'),$$

then (5.6.1) will reduce to

$$S(u, v, n)S(u, v', n') = \sum_{H} e\left(\frac{uH + V\bar{H}}{nn'}\right) = S(u, V, nn').$$

Now

$$(hn' + h'n)\bar{H} = H\bar{H} \equiv 1 \ (\text{mod } nn').$$

Hence

$$hn'\bar{H} \equiv 1 (\text{mod } n), \quad n'\bar{H} \equiv \bar{h}hn'\bar{H} \equiv \bar{h} \ (\text{mod } n),$$

and so

$$n'^2\bar{H} \equiv n'\bar{h} \pmod{nn'}. \tag{5.6.3}$$

Similarly we see that

$$n^2\bar{H} \equiv n'\bar{h}' \pmod{nn'}; \tag{5.6.4}$$

and from (5.6.3) and (5.6.4) we deduce

$$V\bar{H} = (vn'^2 + v'n^2)\bar{H} \equiv vn'\bar{h}' + v'n\bar{h}' \equiv K \pmod{nn'}.$$

This is (5.6.2), and the theorem follows.

5.7. A general principle. We return for a moment to the argument which we used in proving Theorem 65. It will avoid a good deal of repetition later if we restate the theorem and the proof in a more general form. We use $P(a)$ to denote any proposition asserting a property of a non-negative integer a.

THEOREM 69. *If*

(i)*$P(a)$ and $P(b)$ imply $P(a + b)$ and $P(a - b)$, for every a and b (provided, in the second case, that $b \leqslant a$),*

(ii) *r is the least positive integer for which $P(r)$ is true, then*

(*a*) *$P(kr)$ is true for every non-negative integer k,*

(*b*) *any q for which $P(q)$ is true is a multiple of r.*

In the first place, (*a*) is obvious.

To prove (*b*) we observe that $0 < r \leqslant q$, by the definition of r. Hence we can write

$$q = kr + s, \quad s = q - kr,$$

where $k \geqslant 1$ and $0 \leqslant s < r$. But $P(r) \to P(kr)$, by (*a*), and

$$P(q) \,.\, P(kr) \to P(s),$$

by (i). Hence, again by the definition of r, s must be 0, and $q = kr$.

We can also deduce Theorem 69 from Theorem 23. In Theorem 65, $P(a)$ is $\rho^{\alpha} = 1$.

5.8. Construction of the regular polygon of 17 sides. We conclude this chapter by a short excursus on one of the famous problems of elementary geometry, that of the construction of a regular polygon of n sides, or of an angle $\alpha = 2\pi/n$.

Suppose that $(n_1, n_2) = 1$ and that the problem is soluble for $n = n_1$ and for $n = n_2$. There are integers r_1 and r_2 such that

$$r_1 n_1 + r_2 n_2 = 1$$

or

$$r_1 \alpha_2 + r_2 \alpha_1 = r_1 \frac{2\pi}{n_2} + r_2 \frac{2\pi}{n_1} = \frac{2\pi}{n_1 n_2}.$$

Hence, if the problem is soluble for $n = n_1$ and $n = n_2$, it is soluble for $n = n_1 n_2$. It follows that we need only consider cases in which n is a power of a prime. In what follows we suppose $n = p$ prime.

We can construct α if we can construct $\cos\alpha$ (or $\sin\alpha$); and the numbers

$$\cos k\alpha + i \sin k\alpha \quad (k = 1, 2, \ldots, n-1)$$

are the roots of

$$\frac{x^n - 1}{x - 1} = x^{n-1} + x^{n-2} + \cdots + 1 = 0. \tag{5.8.1}$$

Hence we can construct α if we can construct the roots of (5.8.1).

'Euclidean' constructions, by ruler and compass, are equivalent analytically to the solution of a series of linear or quadratic equations.† Hence our construction is possible if we can reduce the solution of (5.8.1) to that of such a series of equations.

The problem was solved by Gauss, who proved (as we stated in § 2.4) that the reduction is possible if and only if n is a 'Fermat prime'‡

$$n = p = 2^{2^h} + 1 = F_h.$$

The first five values of h, viz. 0, 1, 2, 3, 4, give

$$n = 3,\ 5,\ 17,\ 257,\ 65537,$$

all of which are prime, and in these cases the problem is soluble.

The constructions for $n = 3$ and $n = 5$ are familiar. We give here the construction for $n = 17$. We shall not attempt any systematic exposition

† See § 11.5. ‡ See § 2.5.

of Gauss's theory; but this particular construction gives a fair example of the working of his method, and should make it plain to the reader that (as is plausible from the beginning) success is to be expected when $n = p$ and $p - 1$ does not contain any prime but 2. This requires that p is a prime of the form $2^m + 1$, and the only such primes are the Fermat primes.[†]

Suppose then that $n = 17$. The corresponding equation is

$$\frac{x^{17} - 1}{x - 1} = x^{16} + x^{15} + \cdots + 1 = 0. \tag{5.8.2}$$

We write

$$\alpha = \frac{2\pi}{17}, \quad \epsilon_k = e\left(\frac{k}{17}\right) = \cos k\alpha + i \sin k\alpha,$$

so that the roots of (5.8.2) are

$$x = \epsilon_1, \epsilon_2, \ldots, \epsilon_{16}. \tag{5.8.3}$$

From these roots we form certain sums, known as *periods,* which are the roots of quadratic equations.

The numbers

$$3^m \ (0 \leqslant m \leqslant 15)$$

are congruent (mod 17), in some order, to the numbers $k = 1, 2, \ldots, 16$,[‡] as is shown by the table

$$m = 0, 1, 2, \ 3, \ 4, \ 5, \ 6, \ \ 7, \ 8, \ \ 9, 10, 11, 12, 13, 14, 15, \tag{5.8.4}$$

$$k = 1, 3, 9, 10, 13, \ 5, 15, 11, 16, 14, \ 8, \ 7, \ 4, 12, \ 2, \ 6. \tag{5.8.5}$$

We define x_1 and x_2 by

$$x_1 = \sum_{m \text{ even}} \epsilon_k = \epsilon_1 + \epsilon_9 + \epsilon_{13} + \epsilon_{15} + \epsilon_{16} + \epsilon_8 + \epsilon_4 + \epsilon_2,$$

$$x_2 = \sum_{m \text{ odd}} \epsilon_k = \epsilon_3 + \epsilon_{10} + \epsilon_5 + \epsilon_{11} + \epsilon_{14} + \epsilon_7 + \epsilon_{12} + \epsilon_6;$$

† See § 2.5, Theorem 17.

‡ In fact 3 is a 'primitive root of 17' in the sense which will be explained in § 6.8.

and y_1, y_2, y_3, y_4 by

$$y_1 = \sum_{m\equiv 0(\text{mod}4)} \epsilon_k = \epsilon_1 + \epsilon_{13} + \epsilon_{16} + \epsilon_4,$$

$$y_2 = \sum_{m\equiv 2(\text{mod}4)} \epsilon_k = \epsilon_9 + \epsilon_{15} + \epsilon_8 + \epsilon_2,$$

$$y_3 = \sum_{m\equiv 1(\text{mod}4)} \epsilon_k = \epsilon_3 + \epsilon_5 + \epsilon_{14} + \epsilon_{12},$$

$$y_4 = \sum_{m\equiv 3(\text{mod}4)} \epsilon_k = \epsilon_{10} + \epsilon_{11} + \epsilon_7 + \epsilon_6,$$

Since

$$\epsilon_k + \epsilon_{17-k} = 2\cos k\alpha$$

we have

$$x_1 = 2(\cos\alpha + \cos 8\alpha + \cos 4\alpha + \cos 2\alpha),$$
$$x_2 = 2(\cos 3\alpha + \cos 7\alpha + \cos 5\alpha + \cos 6\alpha),$$
$$y_1 = 2(\cos\alpha + \cos 4\alpha), \quad y_2 = 2(\cos 8\alpha + \cos 2\alpha),$$
$$y_3 = 2(\cos 3\alpha + \cos 5\alpha), \quad y_4 = 2(\cos 7\alpha + \cos 6\alpha).$$

We prove first that x_1 and x_2 are the roots of a quadratic equation with rational coefficients. Since the roots of (5.8.2) are the numbers (5.8.3), we have

$$x_1 + x_2 = 2\sum_{k=1}^{8} \cos k\alpha = 2\sum_{k=1}^{16} \epsilon_k = -1.$$

Again,

$$x_1x_2 = 4(\cos\alpha + \cos 8\alpha + \cos 4\alpha + \cos 2\alpha)$$
$$\times (\cos 3\alpha + \cos 7\alpha + \cos 5\alpha + \cos 6\alpha).$$

If we multiply out the right-hand side and use the identity

(5.8.6) $$2\cos m\alpha \cos n\alpha = \cos(m+n)\alpha + \cos(m-n)\alpha,$$

we obtain

$$x_1x_2 = 4(x_1 + x_2) = -4.$$

Hence x_1 and x_2 are the roots of

$$x^2 + x - 4 = 0. \tag{5.8.7}$$

Also

$$\cos\alpha + \cos 2\alpha > 2\cos\tfrac{1}{4}\pi = \sqrt{2} > -\cos 8\alpha, \quad \cos 4\alpha > 0.$$

Hence $x_1 > 0$ and therefore

$$x_1 > x_2. \tag{5.8.8}$$

We prove next that y_1, y_2 and y_3, y_4 are the roots of quadratic equations whose coefficients are rational in x_1 and x_2. We have

$$y_1 + y_2 = x_1,$$

and, using (5.8.4) again,

$$\begin{aligned} y_1y_2 &= 4(\cos\alpha + \cos 4\alpha)(\cos 8\alpha + \cos 2\alpha) \\ &= 2\sum_{k=1}^{8}\cos k\alpha = -1. \end{aligned}$$

Hence y_1, y_2 are the roots of

$$y^2 - x_1y - 1 = 0; \tag{5.8.9}$$

and it is plain that

$$y_1 > y_2. \tag{5.8.10}$$

Similarly

$$y_3 + y_4 = x_2, \quad y_3y_4 = -1,$$

and so y_3, y_4 are the roots of

$$y^2 - x_2y - 1 = 0, \tag{5.8.11}$$

and

$$y_3 > y_4. \tag{5.8.12}$$

Finally

$$2\cos\alpha + 2\cos 4\alpha = y_1,$$
$$4\cos\alpha\cos 4\alpha = 2(\cos 5\alpha + \cos 3\alpha) = y_3.$$

Also $\cos\alpha > \cos 4\alpha$. Hence $z_1 = 2\cos\alpha$ and $z_2 = 2\cos 4\alpha$ are the roots of the quadratic

(5.8.13) $$z^2 - y_1 z + y_3 = 0$$

and

(5.8.14) $$z_1 > z_2.$$

We can now determine $z_1 = 2\cos\alpha$ by solving the four quadratics (5.8.5), (5.8.7), (5.8.9), and (5.8.11), and remembering the associated inequalities. We obtain

$$\begin{aligned}2\cos\alpha = {} & \tfrac{1}{8}\{-1 + \surd 17 + \surd(34 - 2\surd 17)\} \\ & + \tfrac{1}{8}\surd\{68 + 12\surd 17 - 16\surd(34 + 2\surd 17) \\ & - 2(1 - \surd 17)\surd(34 - 2\surd 17)\},\end{aligned}$$

an expression involving only rationals and square roots. This number may now be constructed by the use of the ruler and compass only, and so α may be constructed.

There is a simpler geometrical construction. Let C be the least positive acute angle such that $\tan 4C = 4$, so that C, $2C$, and $4C$ are all acute. Then (5.8.5) may be written

$$x^2 + 4x\cot 4C - 4 = 0.$$

The roots of this equation are $2\tan 2C$, $-2\cot 2C$. Since $x_1 > x_2$, this gives $x_1 = 2\tan 2C$ and $x_2 = -2\cot 2C$. Substituting in (5.8.7) and (5.8.9) and solving, we obtain

$$y_1 = \tan\left(C + \tfrac{1}{4}\pi\right), \quad y_3 = \tan C,$$
$$y_2 = \tan\left(C - \tfrac{1}{4}\pi\right), \quad y_4 = -\cot C.$$

Hence

(5.8.15)
$$\begin{cases} 2\cos 3\alpha + 2\cos 5\alpha = y_3 = \tan C, \\ 2\cos 3\alpha \,.\, 2\cos 5\alpha = 2\cos 2\alpha + 2\cos 8\alpha = y_2 = \tan(C - \tfrac{1}{4}\pi). \end{cases}$$

Now let OA, OB (Fig. 5) be two perpendicular radii of a circle. Make OI one-fourth of OB and the angle OIE (with E in OA) one-fourth of the angle OIA. Find on AO produced a point F such that $EIF = \frac{1}{4}\pi$. Let the circle on AF as diameter cut OB in K, and let the circle whose centre is E and radius EK cut OA in N_3 and N_5 (N_3 on OA, N_5 on AO produced). Draw N_3P_3, N_5P_5 perpendicular to OA to cut the circumference of the original circle in P_3 and P_5.

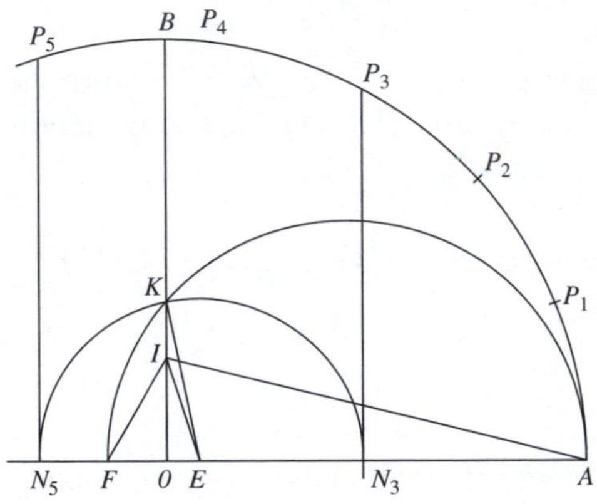

Fig. 5.

Then $OIA = 4C$ and $OIE = C$. Also

$$2\cos AOP_3 + 2\cos AOP_5 = 2\frac{ON_3 - ON_5}{OA} = \frac{4OE}{OA} = \frac{OE}{OI} = \tan C,$$

$$2\cos AOP_3 \,.\, 2\cos AOP_5 = -4\frac{ON_3 \cdot ON_5}{OA^2} = -4\frac{OK^2}{OA^2}$$

$$= -4\frac{OF}{OA} = -\frac{OF}{OI} = \tan(C - \tfrac{1}{4}\pi).$$

Comparing these equations with (5.8.13), we see that $AOP_3 = 3\alpha$ and $AOP_5 = 5\alpha$. It follows that A, P_3, P_5 are the first, fourth, and sixth vertices of a regular polygon of 17 sides inscribed in the circle; and it is obvious how the polygon may be completed.

NOTES

§ 5.1. The contents of this chapter are all 'classical' (except the properties of Ramanujan's and Kloosterman's sums proved in § 5.6), and will be found in text-books. The theory of congruences was first developed scientifically by Gauss, *D.A.*, though the main results must have been familiar to earlier mathematicians such as Fermat and Euler. We give occasional references, especially when some famous function or theorem is habitually associated with the name of a particular mathematician, but make no attempt to be systematic.

§ 5.5. Euler, *Novi Comm. Acad. Petrop.* 8 (1760–1), 74–104 [*Opera* (1), ii. 531–44].

It might seem more natural to say that $f(m)$ is multiplicative if

$$f(mm') = f(m)f(m')$$

for *all* m, m'. This definition would be too restrictive, and the less exacting definition of the text is much more useful.

§ 5.6. The sums of this section occur in Gauss, 'Summatio quarumdam serierum singularium' (1808), *Werke*, ii. 11–45; Ramanujan, *Trans. Camb. Phil. Soc.* 22 (1918), 259–76 (*Collected Papers*, 179–99); Kloosterman, *Acta Math.* 49 (1926), 407–64. 'Ramanujan's sum' may be found in earlier writings; see, for example, Jensen, *Beretning d. tredje Skand. Matematikercongres* (1913), 145, and Landau, *Handbuch*, 572: but Ramanujan was the first mathematician to see its full importance and use it systematically. It is particularly important in the theory of the representation of numbers by sums of squares. For the evaluation of Gauss's sums, their applications and their history, see Davenport, *Multiplicative number theory*, (Markham, Chicago, 1967) and for information and references about Kloostermann's sums, see Weil, *Proc. Nat. Acad. Sci. U.S.A.* 34 (1948), 204–7.

§ 5.8. The general theory was developed by Gauss, *D.A.*, §§ 335–66. The first explicit geometrical construction of the 17-agon was made by Erchinger (see Gauss, *Werke,* ii. 186–7). That in the text is due to Richmond, *Quarterly Journal of Math.* 26 (1893), 206–7, and *Math. Annalen*, 67 (1909), 459–61. Our figure is copied from Richmond's.

Gauss (*D.A.*, § 341) proved that the equation (5.8.1) is irreducible, i.e. that its left-hand side cannot be resolved into factors of lower degree with rational coefficients, when n is prime. Kronecker and Eisenstein proved, more generally, that the equation satisfied by the $\phi(n)$ primitive nth roots of unity is irreducible; see, for example, Mathews, *Theory of numbers* (Cambridge, Deighton Bell, 1892), 186–8. Grandjot has shown that the theorem can be deduced very simply from Dirichlet's Theorem 15: see Landau, *Vorlesungen*, iii. 219.

VI

FERMAT'S THEOREM AND ITS CONSEQUENCES

6.1. Fermat's theorem. In this chapter we apply the general ideas of Ch. V to the proof of a series of classical theorems, due mainly to Fermat, Euler, Legendre, and Gauss.

THEOREM 70. *If p is prime, then*

$$a^p \equiv a \pmod{p}. \tag{6.1.1}$$

THEOREM 71 (FERMAT'S THEOREM). *If p is prime, and $p \nmid a$, then*

$$a^{p-1} \equiv 1 \pmod{p}. \tag{6.1.2}$$

The congruences (6.1.1) and (6.1.2) are equivalent when $p \nmid a$; and (6.1.1) is trivial when $p|a$, since then $a^p \equiv 0 \equiv a$. Hence Theorems 70 and 71 are equivalent.

Theorem 71 is a particular case of the more general

THEOREM 72 (THE FERMAT–EULER THEOREM). *If $(a, m) = 1$, then*

$$a^{\phi(m)} \equiv 1 \pmod{m}.$$

If x runs through a complete system of residues prime to m, then, by Theorem 58, ax also runs through such a system. Hence, taking the product of each set, we have

$$\prod(ax) \equiv \prod x \pmod{m}$$

or

$$a^{\phi(m)} \prod x \equiv \prod x \pmod{m}.$$

Since every number x is prime to m, their product is prime to m; and hence, by Theorem 55,

$$a^{\phi(m)} \equiv 1 \pmod{m}.$$

The result is plainly false if $(a, m) > 1$.

6.2. Some properties of binomial coefficients. Euler was the first to publish a proof of Fermat's theorem. The proof, which is easily extended so as to prove Theorem 72, depends on the simplest arithmetical properties of the binomial coefficients.

THEOREM 73. *If m and n are positive integers, then the binomial coefficients*

$$\binom{m}{n} = \frac{m(m-1)\ldots(m-n+1)}{n!},$$

$$\binom{-m}{n} = (-1)^n \frac{m(m+1)\ldots(m+n-1)}{n!}$$

are integers.

It is the first part of the theorem which we need here, but, since

$$\binom{-m}{n} = (-1)^n \binom{m+n-1}{n},$$

the two parts are equivalent. Either part may be stated in a more striking form, viz.

THEOREM 74. *The product of any n successive positive integers is divisible by $n!$.*

The theorems are obvious from the genesis of the binomial coefficients as the coefficients of powers of x in $(1+x)(1+x)\ldots$ or in

$$(1-x)^{-1}(1-x)^{-1}\ldots = (1+x+x^2+\cdots)(1+x+x^2+\cdots)\ldots.$$

We may prove them by induction as follows. We choose Theorem 74, which asserts that

$$(m)_n = m(m+1)\ldots(m+n-1)$$

is divisible by $n!$. This is plainly true for $n=1$ and all m, and also for $m=1$ and all n. We assume that it is true (a) for $n=N-1$ and all m and (b) for $n=N$ and $m=M$. Then

$$(M+1)_N - M_N = N(M+1)_{N-1},$$

and $(M+1)_{N-1}$ is divisible by $(N-1)!$. Hence $(M+1)_N$ is divisible by $N!$, and the theorem is true for $n=N$ and $m=M+1$. It follows that the theorem is true for $n = N$ and all m. Since it is also true for $n = N + 1$ and $m = 1$, we can repeat the argument; and the theorem is true generally.

THEOREM 75. *If p is prime, then*

$$\binom{p}{1}, \binom{p}{2}, \ldots, \binom{p}{p-1}$$

are divisible by p.

If $1 \leqslant n \leqslant p - 1$, then

$$n! \mid p(p-1)\ldots(p-n+1),$$

by Theorem 74. But $n!$ is prime to p, and therefore

$$n! \mid (p-1)(p-2)\ldots(p-n+1).$$

Hence

$$\binom{p}{n} = p\frac{(p-1)(p-2)\ldots(p-n+1)}{n!}$$

is divisible by p.

THEOREM 76. *If p is prime, then all the coefficients in* $(1 - x)^{-p}$ *are divisible by p, except those of* $1, x^p, x^{2p}, \ldots,$ *which are congruent to* 1 (mod p).

By Theorem 73, the coefficients in

$$(1-x)^{-p} = 1 + \sum_{n=1}^{\infty} \binom{p+n-1}{n} x^n$$

are all integers. Since

$$(1-x^p)^{-1} = 1 + x^p + x^{2p} + \ldots,$$

we have to prove that every coefficient in the expansion of

$$(1-x^p)^{-1} - (1-x)^{-p} = (1-x)^{-p}(1-x^p)^{-1}\{(1-x)^p - 1 + x^p\}$$

is divisible by p. Since the coefficients in the expansions of $(1-x)^{-p}$ and $(1-x^p)^{-1}$ are integers it is enough to prove that every coefficient in the polynomial $(1-x)^p - 1 + x^p$ is divisible by p. For $p = 2$ this is trivial and, for $p \geqslant 3$, it follows from Theorem 75 since

$$(1-x)^p - 1 + x^p = \sum_{r=1}^{p-1} (-1)^r \binom{p}{r} x^r.$$

We shall require this theorem in Ch. XIX.

THEOREM 77. *If p is prime, then*

$$(x + y + \cdots + w)^p \equiv x^p + y^p + \cdots + w^p \pmod{p}.$$

For

$$(x + y)^p \equiv x^p + y^p \pmod{p},$$

by Theorem 75, and the general result follows by repetition of the argument.

Another useful corollary of Theorem 75 is

THEOREM 78. *If $\alpha > 0$ and*

$$m \equiv 1 \pmod{p^\alpha},$$

then

$$m^p \equiv 1 \pmod{p^{\alpha+1}}.$$

For $m = 1 + kp^\alpha$, where k is an integer, and $\alpha p \geqslant \alpha + 1$. Hence

$$m^p = (1 + kp^\alpha)^p = 1 + lp^{\alpha+1},$$

where l is an integer.

6.3. A second proof of Theorem 72. We can now give Euler's proof of Theorem 72. Suppose that $m = \Pi p^\alpha$. Then it is enough, after Theorem 53, to prove that

$$a^{\phi(m)} \equiv 1 \pmod{p^\alpha}.$$

But

$$\phi(m) = \prod \phi(p^\alpha) = \prod p^{\alpha-1}(p-1),$$

and so it is sufficient to prove that

$$a^{p^{\alpha-1}(p-1)} \equiv 1 \pmod{p^{\alpha}}$$

when $p \nmid a$.

By Theorem 77,

$$(x+y+\ldots)^p \equiv x^p + y^p + \ldots (\bmod p).$$

Taking $x = y = z = \ldots = 1$, and supposing that there are a numbers, we obtain

$$a^p \equiv a \pmod{p},$$

or

$$a^{p-1} \equiv 1 \pmod{p}.$$

Hence, by Theorem 78,

$$a^{p(p-1)} \equiv 1 \pmod{p^2}, \quad a^{p^2(p-1)} \equiv 1 \pmod{p^3}, \quad \ldots,$$
$$a^{p^{\alpha-1}(p-1)} \equiv 1 \pmod{p^{\alpha}}.$$

6.4. Proof of Theorem 22. Before proceeding to the more important applications of Fermat's theorem, we use it to prove Theorem 22 of Ch. II.

We can write $f(n)$ in the form

$$f(n) = \sum_{r=1}^{m} Q_r(n)\, a_r^n = \sum_{r=1}^{m} \left(\sum_{S=0}^{q_r} c_{r,s} n^s \right) a_r^n,$$

where the a and c are integers and

$$1 \leqslant a_1 < a_2 < \ldots < a_m.$$

The terms of $f(n)$ are thus arranged in increasing order of magnitude for large n, and $f(n)$ is dominated by its last term

$$c_{m,q_m} n^{q_m} a_m^n$$

for large n (so that the last c is positive).

If $f(n)$ is prime for all large n, then there is an n for which

$$f(n) = p > a_m$$

and p is prime. Then

$$\{n+kp(p-1)\}^s \equiv n^s \pmod{p},$$

for all integral k and s. Also, by Fermat's theorem,

$$a_r^{p-1} \equiv 1 \pmod{p}$$

and so

$$a_r^{n+kp(p-1)} \equiv a_r^n \pmod{p}$$

for all positive integral k. Hence

$$\{n + kp\,(p-1)\}^s\, a_r^{n+kp(p-1)} \equiv n^s a_r^n \pmod{p}$$

and therefore

$$f\{n+kp(p-1)\} \equiv f(n) \equiv 0 \pmod{p}$$

for all positive integral k; a contradiction.

6.5. Quadratic residues. Let us suppose that p is an odd prime, that $p \nmid a$, and that x is one of the numbers

$$1, 2, 3, \ldots, p-1.$$

Then, by Theorem 58, just one of the numbers

$$1\,.\,x, 2\,.\,x, \ldots, (p-1)x$$

is congruent to $a \pmod{p}$. There is therefore a unique x' such that

$$xx' \equiv a \pmod{p}, \quad 0 < x' < p.$$

We call x' the *associate* of x. There are then two possibilities: either there is at least one x associated with itself, so that $x' = x$, or there is no such x.

(1) Suppose that the first alternative is the true one and that x_1 is associated with itself. In this case the congruence

$$x^2 \equiv a \pmod{p}$$

has the solution $x = x_1$; and we say that a is a *quadratic residue* of p, or (when there is no danger of a misunderstanding) simply a *residue* of p, and write $a \mathrm{R} p$. Plainly

$$x = p - x_1 \equiv -x_1 \pmod{p}$$

is another solution of the congruence. Also, if $x' = x$ for any other value x_2 of x, we have

$$x_1^2 \equiv a, \quad x_2^2 \equiv a, \quad (x_1 - x_2)(x_1 + x_2) = x_1^2 - x_2^2 \equiv 0 \pmod{p}.$$

Hence either $x_2 \equiv x_1$ or

$$x_2 \equiv -x_1 \equiv p - x_1;$$

and there are just two solutions of the congruence, namely x_1 and $p - x_1$.

In this case the numbers

$$1, 2, \ldots, p-1$$

may be grouped as x_1, $p - x_1$, and $\frac{1}{2}(p-3)$ pairs of unequal associated numbers. Now

$$x_1(p - x_1) \equiv -x_1^2 \equiv -a \pmod{p},$$

while

$$xx' \equiv a \pmod{p}$$

for any associated pair x, x'. Hence

$$(p-1)! = \prod x \equiv -a.a^{\frac{1}{2}(p-3)} \equiv -a^{\frac{1}{2}(p-1)} \pmod{p}.$$

(2) If the second alternative is true and no x is associated with itself, we say that a is a *quadratic non-residue* of p, or simply a *non-residue* of p, and write $a \mathrm{N} p$. In this case the congruence

$$x^2 \equiv a \pmod{p}$$

has no solution, and the numbers

$$1, 2, \ldots, p-1$$

may be arranged in $\frac{1}{2}(p-1)$ associated unequal pairs. Hence

$$(p-1)! = \prod x \equiv a^{\frac{1}{2}(p-1)} \pmod{p}.$$

We define 'Legendre's symbol' $\left(\frac{a}{p}\right)$, where p is an odd prime and a is any number not divisible by p, by

$$\left(\frac{a}{p}\right) = +1, \quad \text{if} \quad a\,\mathrm{R}\,p,$$
$$\left(\frac{a}{p}\right) = -1, \quad \text{if} \quad a\,\mathrm{N}\,p.$$

It is plain that

$$\left(\frac{a}{p}\right) = \left(\frac{b}{p}\right)$$

if $a \equiv b \pmod{p}$. We have then proved

Theorem 79. *If p is an odd prime and a is not a multiple of p, then*

$$(p-1)! \equiv -\left(\frac{a}{p}\right) a^{\frac{1}{2}(p-1)} \pmod{p}.$$

We have supposed p odd. It is plain that $0 = 0^2, 1 = 1^2$, and so all numbers, are quadratic residues of 2. We do not define Legendre's symbol when $p = 2$, and we ignore this case in what follows. Some of our theorems are true (but trivial) when $p = 2$.

6.6. Special cases of Theorem 79: Wilson's theorem. The two simplest cases are those in which $a = 1$ and $a = -1$.

(1) First let $a = 1$. Then

$$x^2 \equiv 1 \pmod{p}$$

has the solutions $x = \pm 1$; hence 1 is a quadratic residue of p and

$$\left(\frac{1}{p}\right) = 1.$$

If we put $a = 1$ in Theorem 79, it becomes

Theorem 80 (Wilson's theorem):

$$(p-1)! \equiv -1 \pmod{p}.$$

Thus $11 \mid 3628801$.

The congruence

$$(p-1)!+1 \equiv 0 \pmod{p^2}$$

is true for

$$p = 5, \quad p = 13, \quad p = 563,$$

but for no other value of p less than 200000. Apparently no general theorem concerning the congruence is known.

If m is composite, then

$$m \mid (m-1)!+1$$

is false, for there is a number d such that

$$d \mid m, \quad 1 < d < m,$$

and d does not divide $(m-1)!+1$. Hence we derive

THEOREM 81. *If $m > 1$, then a necessary and sufficient condition that m should be prime is that*

$$m \mid (m-1)!+1.$$

The theorem is of course quite useless as a practical test for the primality of a given number m.

(2) Next suppose $a = -1$. Then Theorems 79 and 80 show that

$$\left(\frac{-1}{p}\right) \equiv -(-1)^{\frac{1}{2}(p-1)}(p-1)! \equiv (-1)^{\frac{1}{2}(p-1)}.$$

THEOREM 82. *The number -1 is a quadratic residue of primes of the form $4k+1$ and a non-residue of primes of the form $4k+3$, i.e.*

$$\left(\frac{-1}{p}\right) = (-1)^{\frac{1}{2}(p-1)}.$$

More generally, combination of Theorems 79 and 80 gives

THEOREM 83:

$$\left(\frac{a}{p}\right) \equiv a^{\frac{1}{2}(p-1)} \pmod{p}.$$

6.7. Elementary properties of quadratic residues and non-residues. The numbers

$$1^2, 2^2, 3^2, \ldots, \left\{\tfrac{1}{2}(p-1)\right\}^2 \tag{6.7.1}$$

are all incongruent; for $r^2 \equiv s^2$ implies $r \equiv s$ or $r \equiv -s \pmod p$, and the second alternative is impossible here. Also

$$r^2 \equiv (p-r)^2 \pmod p.$$

It follows that there are $\frac{1}{2}(p-1)$ residues and $\frac{1}{2}(p-1)$ non-residues of p.

Theorem 84. *There are $\frac{1}{2}(p-1)$ residues and $\frac{1}{2}(p-1)$ non-residues of an odd prime p.*

We next prove

Theorem 85. *The product of two residues, or of two non-residues, is a residue, while the product of a residue and a non-residue is a non-residue.*

(1) Let us write α, α', $\alpha_1, \ldots$ for residues and β, β', $\beta_1, \ldots$ for non-residues. Then every $\alpha\alpha'$ is an α, since

$$x^2 \equiv \alpha \cdot y^2 \equiv \alpha' \to (xy)^2 \equiv \alpha\alpha' \pmod p.$$

(2) If α_1 is a fixed residue, then

$$1.\alpha_1, 2.\alpha_1, 3.\alpha_1, \ldots, (p-1)\alpha_1$$

is a complete system $(\operatorname{mod} p)$. Since every $\alpha\alpha_1$ is a residue, every $\beta\alpha_1$ must be a non-residue.

(3) Similarly, if β_1 is a fixed non-residue, every $\beta\beta_1$ is a residue. For

$$1.\beta_1, 2.\beta_1, \ldots, (p-1)\beta_1$$

is a complete system $(\operatorname{mod} p)$, and every $\alpha\beta_1$ is a non-residue, so that every $\beta\beta_1$ is a residue.

Theorem 85 is also a corollary of Theorem 83.

We add two theorems which we shall use in Ch. XX. The first is little but a restatement of part of Theorem 82.

Theorem 86. *If p is a prime $4k+1$, then there is an x such that*

$$1 + x^2 = mp,$$

where $0 < m < p$.

For, by Theorem 82, -1 is a residue of p, and so congruent to one of the numbers (6.7.1), say x^2; and

$$0 < 1 + x^2 < 1 + (\tfrac{1}{2}p)^2 < p^2.$$

THEOREM 87. *If p is an odd prime, then there are numbers x and y such that*

$$1 + x^2 + y^2 = mp,$$

where $0 < m < p$.

The $\frac{1}{2}(p+1)$ numbers

$$x^2 \ \left(0 \leqslant x \leqslant \tfrac{1}{2}(p-1)\right) \tag{6.7.2}$$

are incongruent, and so are the $\frac{1}{2}(p+1)$ numbers

$$-1 - y^2 \ \left(0 \leqslant y \leqslant \tfrac{1}{2}(p-1)\right). \tag{6.7.3}$$

But there are $p + 1$ numbers in the two sets together, and only p residues $(\bmod p)$; and therefore some number (6.7.2) must be congruent to some number (6.7.3). Hence there are an x and a y, each numerically less than $\frac{1}{2}p$, such that

$$x^2 \equiv -1 - y^2, \quad 1 + x^2 + y^2 = mp.$$

Also

$$0 < 1 + x^2 + y^2 < 1 + 2(\tfrac{1}{2}p)^2 < p^2,$$

so that $0 < m < p$.

Theorem 86 shows that we may take $y = 0$ when $p = 4k + 1$.

6.8. The order of *a* (mod *m*). We know, by Theorem 72, that

$$a^{\phi(m)} \equiv 1 \pmod{m}$$

if $(a, m) = 1$. We denote by d the smallest positive value of x for which

$$a^x \equiv 1 \pmod{m}, \tag{6.8.1}$$

so that $d \leqslant \phi(m)$.

We call the congruence (6.8.1) the proposition $P(x)$. Then it is obvious that $P(x)$ and $P(y)$ imply $P(x + y)$. Also, if $y \leqslant x$ and

$$a^{x-y} \equiv b \pmod{m},$$

then

$$a^x \equiv ba^y \pmod{m},$$

so that $P(x)$ and $P(y)$ imply $P(x-y)$. Hence $P(x)$ satisfies the conditions of Theorem 69, and

$$d|\phi(m).$$

We call d the *order*[†] of a (mod m), and say that *a belongs to d* (mod m). Thus

$$2 \equiv 2, \quad 2^2 \equiv 4, \quad 2^3 \equiv 1 \pmod{7},$$

and so 2 belongs to 3 (mod 7). If $d = \phi(m)$, we say that a is a *primitive root* of m. Thus 2 is a primitive root of 5, since

$$2 \equiv 2, \quad 2^2 \equiv 4, \quad 2^3 \equiv 3, \quad 2^4 \equiv 1 \pmod{5};$$

and 3 is a primitive root of 17. The notion of a primitive root of m bears some analogy to the algebraical notion, explained in § 5.6, of a primitive root of unity. We shall prove in § 7.5 that there are primitive roots of every odd prime p.

We can sum up what we have proved in the form

THEOREM 88. *Any number a prime to m belongs* (mod m) *to a divisor of* $\phi(m)$: *if d is the order of a* (mod m), *then* $d\,|\phi(m)$. *If m is a prime p, then* $d\,|(p-1)$. *The congruence* $a^x \equiv 1 \pmod{m}$ *is true or false according as x is or is not a multiple of d.*

6.9. The converse of Fermat's theorem. The direct converse of Fermat's theorem is false; it is not true that, if $m \nmid a$ and

$$a^{m-1} \equiv 1 \pmod{m}, \tag{6.9.1}$$

then m is necessarily a prime. It is not even true that, if (6.9.1) is true for all a prime to m, then m is prime. Suppose, for example, that $m = 561 = 3.\,11.\,17$. If $3 \nmid a$, $11 \nmid a$, $17 \nmid a$, we have

$$a^2 \equiv 1 \pmod{3}, \quad a^{10} \equiv 1 \pmod{11}, \quad a^{16} \equiv 1 \pmod{17}$$

by Theorem 71. But $2\,|\,560$, $10\,|\,560$, $16\,|\,560$ and so $a^{560} \equiv 1$ to each of the moduli 3, 11, 17 and so to the modulus $3.11.17 = 561$.

If (6.9.1) is true for a particular a and a composite m, we say that m is a *pseudo-prime* with respect to a. If m is a pseudo-prime with respect

[†] Often called the *index*; but this word has a quite different meaning in the theory of groups.

to every a such that $(a, m) = 1$, we call m a *Carmichael number*. It is not known whether there is an infinity of Carmichael numbers,† nor even whether there is an infinity of composite m such that $2^m \equiv 2$ and $3^m \equiv 3 \pmod{m}$. But we can prove.

THEOREM 89. *There is an infinity of pseudo-primes with respect to every* $a > 1$.

Let p be any odd prime which does not divide $a(a^2 - 1)$. We take

$$m = \frac{a^{2p} - 1}{a^2 - 1} = \left(\frac{a^p - 1}{a - 1}\right)\left(\frac{a^p + 1}{a + 1}\right), \tag{6.9.2}$$

so that m is clearly composite. Now

$$(a^2 - 1)(m - 1) = a^{2p} - a^2 = a(a^{p-1} - 1)(a^p + a).$$

Since a and a^p are both odd or both even, $2|(a^p + a)$. Again $a^{p-1} - 1$ is divisible by p (after Theorem 71) and by a^2-1, since $p-1$ is even. Since $p \nmid (a^2 - 1)$, this means that $p(a^2 - 1)|(a^{p-1} - 1)$. Hence

$$2p(a^2 - 1)|(a^2 - 1)(m - 1),$$

so that $2p|(m-1)$ and $m = 1+2pu$ for some integral u. Now, to modulus m,

$$a^{2p} = 1 + m(a^2 - 1) \equiv 1, \quad a^{m-1} = a^{2pu} \equiv 1,$$

and this is (6.9.1). Since we have a different value of m for every odd p which does not divide $a(a^2 - 1)$, the theorem is proved.

A correct converse of Theorem 71 is

THEOREM 90. *If* $a^{m-1} \equiv 1 \pmod{m}$ *and* $a^x \not\equiv 1 \pmod{m}$ *for any divisor* x *of* $m - 1$ *less than* $m - 1$, *then* m *is prime.*

Clearly $(a, m) = 1$. If d is the order of $a \pmod{m}$, then $d|(m - 1)$ and $d|\phi(m)$ by Theorem 88. Since $a^d \equiv 1$, we must have $d = m - 1$ and so $(m - 1)|\phi(m)$. But

$$\phi(m) = m\prod_{p|m}\left(1 - \frac{1}{p}\right) < m - 1$$

if m is composite, and therefore m must be prime.

† This has now been settled, see the end of chapter notes.

6.10. Divisibility of $2^{p-1}-1$ by p^2. By Fermat's theorem

$$2^{p-1}-1 \equiv 0 \pmod{p}$$

if $p > 2$. Is it ever true that

$$2^{p-1}-1 \equiv 0 \pmod{p^2}?$$

This question is of importance in the theory of 'Fermat's last theorem' (see Ch. XIII). The phenomenon does occur, but very rarely.

THEOREM 91. *There is a prime p for which*

$$2^{p-1}-1 \equiv 0 \pmod{p^2}.$$

In fact this is true when $p = 1093$, as can be shown by straightforward calculation. We give a shorter proof, in which all congruences are to modulus $p^2 = 1194649$.

In the first place,

$$3^7 = 2187 = 2p+1, \quad 3^{14} = (2p+1)^2 \equiv 4p+1. \tag{6.10.1}$$

Next

$$2^{14} = 16384 = 15p - 11, \quad 2^{28} \equiv -330p + 121,$$
$$3^2.2^{28} \equiv -2970p + 1089 = -2969p - 4 \equiv -1876p - 4,$$

and so

$$3^2.2^{26} \equiv -469p - 1.$$

Hence, by the binomial theorem,

$$3^{14}.2^{182} \equiv -(469p+1)^7 \equiv -3283p - 1 \equiv -4p - 1 \equiv -3^{14}$$

by (6.10.1). It follows that

$$2^{182} \equiv -1, \quad 2^{1092} \equiv 1 \pmod{1093^2}.$$

The same result is true for $p = 3511$ but for no other $p < 3 \times 10^7$.

6.11. Gauss's lemma and the quadratic character of 2. If p is an odd prime, there is just one residue[†] of n (mod p) between $-\frac{1}{2}p$ and $\frac{1}{2}p$. We call this residue the *minimal* residue of n (mod p); it is positive or negative according as the least non-negative residue of n lies between 0 and $\frac{1}{2}p$ or between $\frac{1}{2}p$ and p.

We now suppose that m is an integer, positive or negative, not divisible by p, and consider the minimal residues of the $\frac{1}{2}(p-1)$ numbers

$$m,\ 2m,\ 3m, \dots,\ \tfrac{1}{2}(p-1)m. \tag{6.11.1}$$

We can write these residues in the form.

$$r_1, r_2, \dots, r_\lambda, \quad -r'_1, -r'_2, \dots, -r'_\mu,$$

where

$$\lambda + \mu = \tfrac{1}{2}(p-1), \quad 0 < r_i < \tfrac{1}{2}p, \quad 0 < r'_i < \tfrac{1}{2}p.$$

Since the numbers (6.11.1) are incongruent, no two r can be equal, and no two r'. If an r and an r' are equal, say $r_i = r'_j$, let am, bm be the two of the numbers (6.11.1) such that

$$am \equiv r_i, \quad bm \equiv -r'_j \qquad (\text{mod } p).$$

Then

$$am + bm \equiv 0 \pmod{p},$$

and so

$$a + b \equiv 0 \pmod{p},$$

which is impossible because $0 < a < \frac{1}{2}p$, $0 < b < \frac{1}{2}p$.

It follows that the numbers r_i, r'_j are a rearrangement of the numbers

$$1,\ 2, \dots,\ \tfrac{1}{2}(p-1);$$

and therefore that

$$m.2m \dots \tfrac{1}{2}(p-1)m \equiv (-1)^\mu 1.2 \dots \tfrac{1}{2}(p-1) \pmod{p},$$

and so

$$m^{\frac{1}{2}(p-1)} \equiv (-1)^\mu \pmod{p}.$$

[†] Here, of course, 'residue' has its usual meaning and is not an abbreviation of 'quadratic residue'.

But

$$\left(\frac{m}{p}\right) \equiv m^{\frac{1}{2}(p-1)} \pmod{p},$$

by Theorem 83. Hence we obtain

THEOREM 92 (GAUSS'S LEMMA). $\left(\frac{m}{p}\right) = (-1)^{\mu}$, *where* μ *is the number of members of the set*

$$m,\ 2m,\ 3m,\ \ldots,\ \tfrac{1}{2}(p-1)m,$$

whose least positive residues (mod p) *are greater than* $\frac{1}{2}p$.

Let us take in particular $m = 2$, so that the numbers (6.11.1) are

$$2, 4, \ldots, p-1.$$

In this case λ is the number of positive even integers less than $\frac{1}{2}p$.

We introduce here a notation which we shall use frequently later. We write $[x]$ for the 'integral part of x', the largest integer which does not exceed x. Thus

$$x = [x] + f,$$

where $0 \leqslant f < 1$. For example,

$$\left[\tfrac{5}{2}\right] = 2, \quad \left[\tfrac{1}{2}\right] = 0, \quad \left[-\tfrac{3}{2}\right] = -2.$$

With this notation

$$\lambda = \left[\tfrac{1}{4}p\right]$$

But

$$\lambda + \mu = \tfrac{1}{2}(p-1),$$

and so

$$\mu = \tfrac{1}{2}(p-1) - \left[\tfrac{1}{4}p\right].$$

If $p \equiv 1 \pmod 4$, then

$$\mu = \tfrac{1}{2}(p-1) - \tfrac{1}{4}(p-1) = \tfrac{1}{4}(p-1) = \left[\tfrac{1}{4}(p+1)\right],$$

and if $p \equiv 3 \pmod 4$, then

$$\mu = \tfrac{1}{2}(p-1) - \tfrac{1}{4}(p-3) = \tfrac{1}{4}(p+1) = \left[\tfrac{1}{4}(p+1)\right].$$

Hence

$$\left(\frac{2}{p}\right) \equiv 2^{\frac{1}{2}(p-1)} \equiv (-1)^{\left[\frac{1}{4}(p+1)\right]} \pmod{p},$$

that is to say $\left(\frac{2}{p}\right) = 1$, if $p = 8n + 1$ or $8n - 1$,

$$\left(\frac{2}{p}\right) = 1, \text{ if } p = 8n + 3 \text{ or } 8n - 3.$$

If $p = 8n \pm 1$, then $\frac{1}{8}(p^2 - 1)$ is even, while if $p = 8n \pm 3$, it is odd. Hence

$$(-1)^{\left[\frac{1}{4}(p+1)\right]} = (-1)^{\left[\frac{1}{8}(p^2+1)\right]}.$$

Summing up, we have the following theorems.

THEOREM 93:

$$\left(\frac{2}{p}\right) = (-1)^{\left[\frac{1}{4}(p+1)\right]}.$$

THEOREM 94:

$$\left(\frac{2}{p}\right) = (-1)^{\left[\frac{1}{8}(p^2-1)\right]}.$$

THEOREM 95. *2 is a quadratic residue of primes of the form* $8n \pm 1$ *and a quadratic non-residue of primes of the form* $8n \pm 3$.

Gauss's lemma may be used to determine the primes of which any given integer m is a quadratic residue. For example, let us take $m = -3$, and suppose that $p > 3$. The numbers (6.11.1) are

$$-3a \quad (1 \leqslant a < \tfrac{1}{2}p),$$

and μ is the number of these numbers whose least positive residues lie between $\frac{1}{2}p$ and p. Now

$$-3a \equiv p - 3a \pmod{p},$$

and $p - 3a$ lies between $\frac{1}{2}p$ and p if $1 \leqslant a < \frac{1}{6}p$. If $\frac{1}{6}p < a < \frac{1}{3}p$, then $p - 3a$ lies between 0 and $\frac{1}{2}p$. If $\frac{1}{3}p < a\frac{1}{2}p$ then

$$-3a \equiv 2p - 3a \pmod{p},$$

and $2p - 3a$ lies between $\frac{1}{2}p$ and p. Hence the values of a which satisfy the condition are

$$1, 2, \ldots, \left[\tfrac{1}{6}p\right], \left[\tfrac{1}{3}p\right] + 1, \left[\tfrac{1}{3}p\right] + 2, \ldots, \left[\tfrac{1}{2}p\right],$$

and

$$\mu = \left[\frac{1}{6}p\right] + \left[\frac{1}{2}p\right] - \left[\frac{1}{3}p\right].$$

If $p = 6n + 1$ then $\mu = n + 3n - 2n$ is even, and if $p = 6n + 5$ then

$$\mu = n + (3n + 2) - (2n + 1)$$

is odd.

THEOREM 96. *-3 is a quadratic residue of primes of the form $6n + 1$ and a quadratic non-residue of primes of the form $6n + 5$.*

A further example, which we leave for the moment† to the reader, is

THEOREM 97. *7 is a quadratic residue of primes of the form $10n \pm 1$ and a quadratic non-residue of primes of the form $10n \pm 3$.*

6.12. The law of reciprocity. The most famous theorem in this field is Gauss's 'law of reciprocity'.

THEOREM 98. *If p and q are odd primes, then*

$$\left(\frac{p}{q}\right)\left(\frac{q}{p}\right) = (-1)^{p'q'},$$

where

$$p' = \tfrac{1}{2}(p - 1), \quad q' = \tfrac{1}{2}(q - 1).$$

Since $p'q'$ is even if either p or q is of the form $4n + 1$, and odd if both are of the form $4n + 3$, we can also state the theorem as

THEOREM 99. *If p and q are odd primes, then*

$$\left(\frac{p}{q}\right) = \left(\frac{q}{p}\right),$$

unless both p and q are of the form $4n + 3$, in which case

$$\left(\frac{p}{q}\right) = -\left(\frac{q}{p}\right).$$

We require a lemma.

† See § 6.13 for a proof depending on Gauss's law of reciprocity.

THEOREM 100. † *If*

$$S(q, p) = \sum_{s=1}^{p'} \left[\frac{sq}{p}\right],$$

then

$$S(q, p) + S(p, q) = p'q'.$$

The proof may be stated in a geometrical form. In the figure (Fig. 6) AC and BC are $x = p, y = q$, and KM and LM are $x = p', y = q'$.

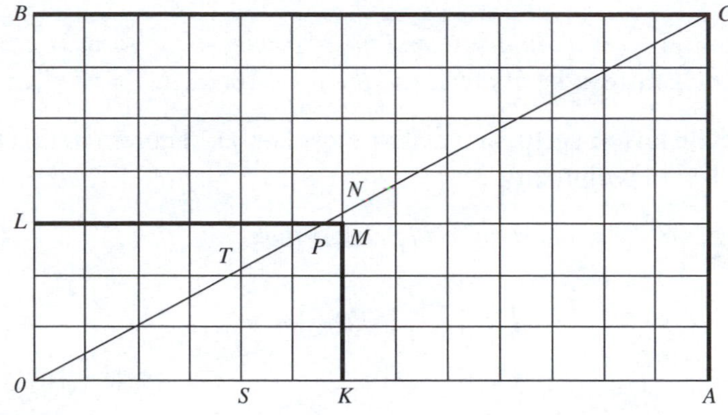

FIG. 6.

If (as in the figure) $p > q$, then $q'/p' < q/p$, and M falls below the diagonal OC. Since

$$q' < \frac{qp'}{p} < q' + 1,$$

there is no integer between $KM = q'$ and $KN = qp'/p$.

We count up, in two different ways, the number of lattice points in the rectangle $OKML$, counting the points on KM and LM but not those on the axes. In the first place, this number is plainly $p'q'$. But there are no lattice points on OC (since p and q are prime), and none in the triangle PMN except perhaps on PM. Hence the number of lattice points in $OKML$ is the sum of those in the triangles OKN and OLP (counting those on KN and LP but not those on the axes).

† The notation has no connection with that of § 5.6.

The number on ST, the line $x = s$, is $[sq/p]$, since sq/p is the ordinate of T. Hence the number in OKN is

$$\sum_{s=1}^{p'} \left[\frac{sq}{p}\right] = S(q, p).$$

Similarly, the number in OLP is $S(p, q)$, and the conclusion follows.

6.13. Proof of the law of reciprocity. We can write

$$kq = p\left[\frac{kp}{p}\right] + u_k, \tag{6.13.1}$$

where

$$1 \leqslant k \leqslant p', \quad 1 \leqslant u_k \leqslant p-1.$$

Here u_k is the least positive residue of $kq \pmod{p}$. If $u_k = v_k \leqslant p'$, then u_k is one of the minimal residues r_i of § 6.11, while if $u_k = w_k > p'$, then $u_k - p$ is one of the minimal residues $-r'_j$. Thus

$$r_i = v_k, \quad r'_j = p - w_k$$

for every i, j, and some k.

The r_i and r'_j are (as we saw in § 6.11) the numbers $1, 2, \ldots, p'$ in some order. Hence, if

$$R = \sum r_i = \sum v_k, \quad R' = \sum r'_j = \sum (p - w_k) = \mu p - \sum w_k$$

(where μ is, as in § 6.11, the number of the r'_j), we have

$$R + R' = \sum_{v=1}^{p'} v = \frac{1}{2}\frac{p-1}{2}\frac{p+1}{2} = \frac{p^2-1}{8},$$

and so

$$\mu p + \sum v_k - \sum w_k = \frac{1}{8}(p^2 - 1). \tag{6.13.2}$$

On the other hand, summing (6.13.1) from $k = 1$ to $k = p'$, we have

(6.13.3)

$$\tfrac{1}{8}q(p^2 - 1) = pS(q,p) + \sum u_k = pS(q,p) + \sum v_k + \sum w_k.$$

From (6.13.2) and (6.13.3) we deduce

$$\tfrac{1}{8}(p^2-1)(q-1) = pS(q,p) + 2\sum w_k - \mu p. \tag{6.13.4}$$

Now $q-1$ is even, and $p^2-1 \equiv 0 \pmod 8$;† so that the left-hand side of (6.13.4) is even, and also the second term on the right. Hence (since p is odd)

$$S(q,p) \equiv \mu \pmod 2,$$

and therefore, by Theorem 92,

$$\left(\frac{q}{p}\right) = (-1)^{\mu} = (-1)^{S(q,p)}.$$

Finally,

$$\left(\frac{q}{p}\right)\left(\frac{p}{q}\right) = (-1)^{S(q,p)+S(p,q)} = (-1)^{p'q'},$$

by Theorem 100.

We now use the law of reciprocity to prove Theorem 97. If

$$p = 10n + k,$$

where k is 1, 3, 7, or 9, then (since 5 is of the form $4n+1$)

$$\left(\frac{5}{p}\right) = \left(\frac{p}{5}\right) = \left(\frac{10n+k}{5}\right) = \left(\frac{k}{5}\right).$$

The residues of 5 are 1 and 4. Hence 5 is a residue of primes $5n+1$ and $5n+4$, i.e. of primes $10n+1$ and $10n+9$, and a non-residue of the other odd primes.

6.14. Tests for primality. We now prove two theorems which provide tests for the primality of numbers of certain special forms. Both are closely related to Fermat's Theorem.

THEOREM 101. *If $p > 2$, $h < p$, $n = hp + 1$ or $hp^2 + 1$ and*

$$2^h \not\equiv 1, \quad 2^{n-1} \equiv 1 \pmod n, \tag{6.14.1}$$

then n is prime.

We write $n = hp^b + 1$, where $b = 1$ or 2, and suppose d to be the order of 2 (mod n). After Theorem 88, it follows from (6.14.1) that $d \nmid h$ and

† If $p = 2n+1$ then $p^2-1 = 4n(n+1) \equiv 0 \pmod 8$.

$d|(n-1)$, i.e. $d|hp^b$. Hence $p|d$. But, by Theorem 88 again, $d|\phi(n)$ and so $p|\phi(n)$. If

$$n = p_1^{a_1} \dots p_k^{a_k},$$

we have

$$\phi(n) = p_1^{a_1-1} \dots p_k^{a_k-1}(p_1 - 1)\dots(p_k - 1)$$

and so, since $p \nmid n, p$ divides at least one of $p_1 - 1, p_2 - 1, \dots, p_k - 1$. Hence n has a prime factor $P \equiv 1 \pmod{p}$.

Let $n = Pm$. Since $n \equiv 1 \equiv P \pmod{p}$, we have $m \equiv 1 \pmod{p}$. If $m > 1$, then

$$n = (up+1)(vp+1), \quad 1 \leqslant u \leqslant v \tag{6.14.2}$$

and

$$hp^{b-1} = uvp + u + v.$$

If $b = 1$, this is $h = uvp + u + v$ and so

$$p \leqslant uvp < h < p,$$

a contradiction. If $b = 2$,

$$hp = uvp + u + v, \quad p|(u+v), \quad u + v \geqslant p$$

and so

$$2v \geqslant u + v \geqslant p, \quad v > \frac{1}{2}p$$

and

$$uv < h < p, \quad uv \leqslant p - 2, \quad u \leqslant \frac{p-2}{v} < \frac{2(p-2)}{p} < 2.$$

Hence $u = 1$ and so

$$v \geqslant p - 1, \quad uv \geqslant p - 1,$$

a contradiction. Hence (6.14.2) is impossible and $m = 1$ and $n = P$.

THEOREM 102. *Let $m \geqslant 2$, $h < 2^m$ and $n = h2^m + 1$ be a quadratic non-residue* (mod p) *for some odd prime p. Then the necessary and sufficient condition for n to be a prime is that*

$$p^{\frac{1}{2}(n-1)} \equiv -1 \pmod{n}. \tag{6.14.3}$$

First let us suppose n prime. Since $n \equiv 1 \pmod 4$, we have

$$\left(\frac{p}{n}\right) = \left(\frac{n}{p}\right) = -1$$

by Theorem 99. Then (6.14.3) follows at once by Theorem 83. Hence the condition is necessary.

Now let us suppose (6.14.3) true. Let P be any prime factor of n and let d be the order of $p \pmod P$. We have

$$p^{\frac{1}{2}(n-1)} \equiv -1, \quad p^{n-1} \equiv 1, \quad p^{P-1} \equiv 1 \pmod P$$

and so, by Theorem 88,

$$d \nmid \tfrac{1}{2}(n-1), \quad d|(n-1), \quad d|(P-1),$$

that is

$$d \nmid 2^{m-1}h, \quad d|2^m h, \quad d|(P-1),$$

so that $2^m|d$ and $2^m|(P-1)$. Hence $P = 2^m x + 1$.

Since $n \equiv 1 \equiv P \pmod{2^m}$, we have $n/P \equiv 1 \pmod{2^m}$ and so

$$n = (2^m x + 1)(2^m y + 1), \quad x \geqslant 1,\ y \geqslant 0.$$

Hence

$$2^m xy < 2^m xy + x + y = h < 2^m, \quad y = 0,$$

and $n = P$. The condition is therefore sufficient.

If we put $h = 1$, $m = 2^k$, we have $n = F_k$ in the notation of § 2.4. Since $1^2 \equiv 2^2 \equiv 1 \pmod 3$ and $F_k \equiv 2 \pmod 3$, F_k is a non-residue (mod 3). Hence a necessary and sufficient condition that F_k be prime is that $F_k|(3^{\frac{1}{2}(F_k-1)} + 1)$.

6.15. Factors of Mersenne numbers; a theorem of Euler. We return for the moment to the problem of Mersenne's numbers, mentioned in § 2.5. There is one simple criterion, due to Euler, for the factorability of $M_p = 2^p - 1$.

THEOREM 103. *If $k > 1$ and $p = 4k + 3$ is prime, then a necessary and sufficient condition that $2p + 1$ should be prime is that*

(6.15.1) $$2^p \equiv 1 \pmod{2p + 1}.$$

Thus, if $2p + 1$ is prime, $(2p + 1) \mid M_p$ and M_p is composite.

First let us suppose that $2p + 1 = P$ is prime. By Theorem 95, since $P \equiv 7 \pmod 8$, 2 is a quadratic residue (mod P) and

$$2^p = 2^{\frac{1}{2}(P-1)} \equiv 1 \pmod P$$

by Theorem 83. The condition (6.15.1) is therefore necessary and $P|M_p$. But $k > 1$ and so $p > 3$ and $M_p = 2^p - 1 > 2p + 1 = P$. Hence M_p is composite.

Next, suppose that (6.15.1) is true. In Theorem 101, put $h = 2$, $n = 2p + 1$. Clearly $h < p$ and $2^h = 4 \not\equiv 1 \pmod n$ and, by (6.15.1),

$$2^{n-1} = 2^{2p} \equiv 1 \pmod n.$$

Hence n is prime and the condition (6.15.1) is sufficient.

Theorem 103 contains the simplest criterion known for the character of Mersenne numbers. The first eight cases in which this test gives a factor of M_p are those for which

$$p = 11,\ 23,\ 83,\ 131,\ 179,\ 191,\ 239,\ 251.$$

NOTES

§ 6.1. Fermat stated his theorem in 1640 (*Œuvres*, ii. 209). Euler's first proof dates from 1736, and his generalization from 1760. See Dickson, *History*, i, ch. iii, for full information.

§ 6.5. Legendre introduced 'Legendre's symbol' in his *Essai sur la théorie des nombres,* first published in 1798. See, for example, § 135 of the second edition (1808).

§ 6.6. Wilson's theorem was first published by Waring, *Meditationes algebraicae* (1770), 288. There is evidence that it was known long before to Leibniz. Goldberg (*Journ. London Math. Soc.* 28 (1953), 252–6) gives the residue of $(p-1)!+1$ to modulus p^2 for $p < 10000$. See E. H. Pearson [*Math. Computation* 17 (1963), 194–5] for the statement about the congruence (mod p^2). By 2007, the computation had been extended to 5×10^8 without finding further examples.

§ 6.7. We can use Theorem 85 to find an upper bound for q, the least positive quadratic non-residue (mod p). Let $m = [p/q] + 1$, so that $p < mq < p + q$. Since $0 < mq - p < q$, we see that $mq - p$ must be a quadratic residue and so must mq. Hence m is a quadratic non-residue and so $q < m$. Hence $q^2 < p + q$ and $q < \surd(p + \frac{1}{4} + \frac{1}{2})$ Burgess (*Mathematatika* 4 (1957), 106–12) proved that $q = O(p^a)$ as $p \to \infty$ for any fixed $a > \frac{1}{4}e^{-1/2}$.

§ 6.9. Theorem 89 is due to Cipolla, *Annali di Mat.* (3), 9 (1903), 139–60. Amongst others the following are Carmichael numbers, viz. 3.11.17, 5.13.17, 5.17.29, 5.29.73, 7.13.19. Apart from these, the pseudo-primes with respect to 2 which are less than 2000 are

$$341 = 11.31,\ 645 = 3.5.43,\ 1387 = 19.73,\ 1905 = 3.5.127.$$

See Dickson, *History*, i. 91–95, Lehmer, *Amer. Math. Monthly*, 43 (1936), 347–54, and Leveque, *Reviews*, 1, 47–53 for further references.

It has been shown by Alford, Granville, and Pomerance, (*Ann. of Math.* (2) 139 (1994), 703–22) that there are in fact infinitely many Carmichael numbers. Indeed the numbers they construct are coprime to 6, yielding composite integers m for which $2^m \equiv 2$ and $3^m \equiv 3 \pmod{m}$. It had been shown in 1899 by Korselt (*L'inermédiaire des math.* 6 (1899), 142–3) that n is a Carmichael number if and only if n is square-free and $p-1 \mid n-1$ for every prime $p|n$.

Theorem 90 is due to Lucas, *Amer. Journal of Math.* 1 (1878), 302. It has been modified in various ways by D. H. Lehmer and others in order to obtain practicable tests for the prime or composite character of a given large m. See Lehmer, *loc. cit.*, and *Bulletin Amer. Math. Soc.* 33 (1927), 327–40, and 34 (1928), 54–56, and Duparc, *Simon Stevin* 29 (1952), 21–24.

§ 6.10. The proof is that of Landau, *Vorlesungen,* iii. 275, improved by R. F. Whitehead. Theorem 91 for $p = 3511$ is due to Beeger. See also Pearson (*loc. cit.* above) and Fröberg (*Computers in Math. Research,* (North Holland, 1968), 84–88) for the numerical statement at the end. It is now (2007) known that there are no further primes below 1.25×10^{15} with the property described.

§§ 6.11–13. Theorem 95 was first proved by Euler. Theorem 98 was stated by Euler and Legendre, but the first satisfactory proofs were by Gauss. See Bachmann, *Niedere Zahlentheorie,* i, ch. 6, for the history of the subject, and many other proofs.

§ 6.14. Miller and Wheeler took the known prime $2^{127}-1$ as p in Theorem 101 and found $n = 190p^2 + 1$ to satisfy the test. See our note to § 2.5. Theorem 101 is also true when $n = hp^3 + 1$, provided that $h < \sqrt{p}$ and that h is not a cube. See Wright, *Math. Gazette,* 37 (1953), 104–6.

Robinson extended Theorem 102 (*Amer. Math. Monthly*, 64 (1957), 703–10) and he and Selfridge used the case $p = 3$ of the theorem to find a large number of primes of the form $h.\, 2^m + 1$ (*Math. tables and other aids to computation*, 11 (1957), 21–22). Amongst these primes are several factors of Fermat numbers. See also the note to § 15.5.

Lucas [*Théorie des nombres,* i (1891), p. xii] stated the test for the primality of F_k. Hurwitz [*Math. Werke*, ii. 747] gave a proof. F_7 and F_{10} were proved composite by this test, though actual factors were subsequently found.

The most important development in this area is undoubtedly the result of Agrawal, Kayal, and Saxena (*Ann. of Math.* (2) 160 (2004), 781–93), which gives a primality test, based ultimately on Fermat's Theorem, which takes time of order $(\log n)^c$ to test the number n. Here c is a numerical constant, which one can take to be 6 according to work of Lenstra and Pomerance.

§ 6.15. Theorem 103; Euler, *Comm. Acad. Petrop.* 6 (1732–3), 103 [*Opera* (1), ii. 3].

VII

GENERAL PROPERTIES OF CONGRUENCES

7.1. Roots of congruences. An integer x which satisfies the congruence

$$f(x) = c_0x^n + c_1x^{n-1} + \ldots + c_n \equiv 0 \pmod{m}$$

is said to be a *root* of the congruence or a *root of* $f(x) \pmod{m}$. If a is such a root, then so is any number congruent to $a \pmod{m}$. Congruent roots are considered equivalent; when we say that the congruence has l roots, we mean that it has l incongruent roots.

An algebraic equation of degree n has (with appropriate conventions) just n roots, and a polynomial of degree n is the product of n linear factors. It is natural to inquire whether there are analogous theorems for congruences, and the consideration of a few examples shows at once that they cannot be so simple. Thus

$$x^{p-1} - 1 \equiv 0 \pmod{p} \tag{7.1.1}$$

has $p - 1$ roots, viz.

$$1, 2, \ldots, p - 1,$$

by Theorem 71;

$$x^4 - 1 \equiv 0 \pmod{16} \tag{7.1.2}$$

has 8 roots, viz. 1, 3, 5, 7, 9, 11, 13, 15; and

$$x^4 - 2 \equiv 0 \pmod{16} \tag{7.1.3}$$

has no root. The possibilities are plainly much more complex than they are for an algebraic equation.

7.2. Integral polynomials and identical congruences. If $c_0, c_1, \ldots, c_n$ are integers then

$$c_0x^n + c_1x^{n-1} + \cdots + c_n$$

is called an *integral polynomial*. If

$$f(x) = \sum_{r=0}^{n} c_r x^{n-r}, \quad g(x) = \sum_{r=0}^{n} c'_r x^{n-r},$$

and $c_r \equiv c'_r$ (mod m) for every r, then we say that $f(x)$ and $g(x)$ are *congruent* to modulus m, and write

$$f(x) \equiv g(x) \pmod{m}.$$

Plainly

$$f(x) \equiv g(x) \to f(x)h(x) \equiv g(x)h(x)$$

if $h(x)$ is any integral polynomial.

In what follows we shall use the symbol '$\equiv$' in two different senses, the sense of § 5.2, in which it expresses a relation between numbers, and the sense just defined, in which it expresses a relation between polynomials. There should be no confusion because, except in the phrase 'the congruence $f(x) \equiv 0$', the variable x will occur only when the symbol is used in the second sense. When we assert that $f(x) \equiv g(x)$, or $f(x) \equiv 0$, we are using it in this sense, and there is no reference to any numerical value of x. But when we make an assertion about 'the roots of the congruence $f(x) \equiv 0$', or discuss 'the solution of the congruence', it is naturally the first sense which we have in mind.

In the next section we introduce a similar double use of the symbol '$|$'.

THEOREM 104. (i) *If p is prime and*

$$f(x)g(x) \equiv 0 \pmod{p},$$

then either $f(x) \equiv 0$ or $g(x) \equiv 0$ (mod p).

(ii) *More generally, if*

$$f(x)g(x) \equiv 0 \pmod{p^a}$$

and

$$f(x) \not\equiv 0 \pmod{p},$$

then

$$g(x) \equiv 0 \pmod{p^a}.$$

(i) We form $f_1(x)$ from $f(x)$ by rejecting all terms of $f(x)$ whose coefficients are divisible by p, and $g_1(x)$ similarly. If $f(x) \not\equiv 0$ and $g(x) \not\equiv 0$, then the first coefficients in $f_1(x)$ and $g_1(x)$ are not divisible by p, and therefore the first coefficient in $f_1(x)g_1(x)$ is not divisible by p. Hence

$$f(x)g(x) \equiv f_1(x)g_1(x) \not\equiv 0 \pmod{p}.$$

(ii) We may reject multiples of p from $f(x)$, and multiples of p^a from $g(x)$, and the result follows in the same way. This part of the theorem will be required in Ch. VIII.

If $f(x) \equiv g(x)$, then $f(a) \equiv g(a)$ for all values of a. The converse is not true; thus

$$a^p \equiv a \pmod{p}$$

for all a, by Theorem 70, but

$$x^p \equiv x \pmod{p}$$

is false.

7.3. Divisibility of polynomials (mod *m*). We say that $f(x)$ is *divisible by* $g(x)$ *to modulus* m if there is an integral polynomial $h(x)$ such that

$$f(x) \equiv g(x)h(x) \pmod{m}.$$

We then write

$$g(x) | f(x) \pmod{m}.$$

THEOREM 105. *A necessary and sufficient condition that*

$$(x-a) | f(x) \pmod{m}$$

is that

$$f(a) \equiv 0 \pmod{m}.$$

If

$$(x-a) | f(x) \pmod{m},$$

then

$$f(x) \equiv (x-a)h(x) \pmod{m}$$

for some integral polynomial $h(x)$, and so

$$f(a) \equiv 0 \pmod{m}.$$

The condition is therefore necessary.

It is also sufficient. If

$$f(a) \equiv 0 \pmod{m},$$

then

$$f(x) \equiv f(x) - f(a) \pmod{m}.$$

But

$$f(x) = \sum c_r x^{n-r}$$

and

$$f(x) - f(a) = (x-a)h(x),$$

where

$$h(x) = \frac{f(x) - f(a)}{x-a} = \sum c_r(x^{n-r-1} + x^{n-r-2}a + \cdots + a^{n-r-1})$$

is an integral polynomial. The degree of $h(x)$ is one less than that of $f(x)$.

7.4. Roots of congruences to a prime modulus. In what follows we suppose that the modulus m is prime; it is only in this case that there is a simple general theory. We write p for m.

THEOREM 106. *If p is prime and*

$$f(x) \equiv g(x)h(x) \pmod{p},$$

then any root of $f(x) \pmod{p}$ is a root either of $g(x)$ or of $h(x)$.

If a is any root of $f(x) \pmod{p}$, then

$$f(a) \equiv 0 \pmod{p},$$

or

$$g(a)h(a) \equiv 0 \pmod{p}.$$

Hence $g(a) \equiv 0 \pmod{p}$ or $h(a) \equiv 0 \pmod{p}$, and so a is a root of $g(x)$ or of $h(x) \pmod{p}$.

The condition that the modulus is prime is essential. Thus

$$x^2 \equiv x^2 - 4 \equiv (x-2)(x+2) \pmod{4},$$

and 4 is a root of $x^2 \equiv 0 \pmod{4}$ but not of $x - 2 \equiv 0 \pmod{4}$ or of $x + 2 \equiv 0 \pmod{4}$.

Theorem 107. *If $f(x)$ is of degree n, and has more than n roots* $\pmod{p}$, *then*

$$f(x) \equiv 0 \pmod{p}.$$

The theorem is significant only when $n < p$. It is true for $n = 1$, by Theorem 57; and we may therefore prove it by induction.

We assume then that the theorem is true for a polynomial of degree less than n. If $f(x)$ is of degree n, and $f(a) \equiv 0 \pmod{p}$, then

$$f(x) \equiv (x-a)g(x) \pmod{p},$$

by Theorem 105; and $g(x)$ is at most of degree $n-1$. By Theorem 106, any root of $f(x)$ is either a or a root of $g(x)$. If $f(x)$ has more than n roots, then $g(x)$ must have more than $n-1$ roots, and so

$$g(x) \equiv 0 \pmod{p},$$

from which it follows that

$$f(x) \equiv 0 \pmod{p}.$$

The condition that the modulus is prime is again essential. Thus

$$x^4 - 1 \equiv 0 \pmod{16}$$

has 8 roots.

The argument proves also

Theorem 108. *If $f(x)$ has its full number of roots*

$$a_1, a_2, \ldots, a_n \pmod{p},$$

then

$$f(x) \equiv c_0(x-a_1)(x-a_2)\ldots(x-a_n) \pmod{p}.$$

7.5. Some applications of the general theorems. (1) Fermat's theorem shows that the binomial congruence

$$x^d \equiv 1 \pmod{p} \tag{7.5.1}$$

has its full number of roots when $d = p - 1$. We can now prove that this is true when d is any divisor of $p - 1$.

THEOREM 109. *If p is prime and $d \mid p - 1$, then the congruence* (7.5.1) *has d roots.*

We have

$$x^{p-1} - 1 = (x^d - 1)g(x),$$

where

$$g(x) = x^{p-1-d} + x^{p-1-2d} + \cdots + x^d + 1.$$

Now $x^{p-1} - 1 \equiv 0$ has $p - 1$ roots, and $g(x) \equiv 0$ has at most $p - 1 - d$. It follows, by Theorem 106, that $x^d - 1 \equiv 0$ has at least d roots, and therefore exactly d.

Of the d roots of (7.5.1), some will belong to d in the sense of § 6.8, but others (for example 1) to smaller divisors of $p - 1$. The number belonging to d is given by the next theorem.

THEOREM 110. *Of the d roots of* (7.5.1), *$\phi(d)$ belong to d. In particular, there are $\phi(p - 1)$ primitive roots of p.*

If $\psi(d)$ is the number of roots belonging to d, then

$$\sum_{d|p-1} \psi(d) = p - 1,$$

since each of $1, 2, \ldots, p - 1$ belongs to some d; and also

$$\sum_{d|p-1} \phi(d) = p - 1,$$

by Theorem 63. If we can show that $\psi(d) \leqslant \phi(d)$, it will follow that $\psi(d) = \phi(d)$, for each d.

If $\psi(d) > 0$, then one at any rate of $1, 2, \ldots, p-1$, say f, belongs to d. We consider the d numbers

$$f_h = f^h \quad (0 \leqslant h \leqslant d-1).$$

Each of these numbers is a root of (7.5.1), since $f^d \equiv 1$ implies $f^{hd} \equiv 1$. They are incongruent (mod p), since $f^h \equiv f^{h'}$, where $h' < h < d$, would imply $f^k \equiv 1$, where $0 < k = h - h' < d$, and then f would not belong to d; and therefore, by Theorem 109, they are *all* the roots of (7.5.1). Finally, if f_h belongs to d, then $(h, d) = 1$; for $k|h, k|d$, and $k > 1$ would imply

$$(f^h)^{d/k} = (f^d)^{h/k} \equiv 1,$$

in which case f_h would belong to a smaller index than d. Thus h must be one of the $\phi(d)$ numbers less than and prime to d, and therefore $\psi(d) \leqslant \phi(d)$.

We have plainly proved incidentally

THEOREM 111. *If p is an odd prime, then there are numbers g such that $1, g, g^2, \ldots, g^{p-2}$ are incongruent mod p.*

(2) The polynomial

$$f(x) = x^{p-1} - 1$$

is of degree $p-1$ and, by Fermat's theorem, has the $p-1$ roots $1, 2, 3, \ldots, p-1$ (mod p). Applying Theorem 108, we obtain

THEOREM 112. *If p is prime, then*

$$x^{p-1} - 1 \equiv (x-1)(x-2)\ldots(x-p+1) \pmod{p}. \tag{7.5.2}$$

If we compare the constant terms, we obtain a new proof of Wilson's theorem. If we compare the coefficients of $x^{p-2}, x^{p-3}, \ldots, x$, we obtain

THEOREM 113. *If p is an odd prime, $1 \leqslant l < p-1$, and A_l is the sum of the products of l different members of the set $1, 2, \ldots, p-1$, then $A_l \equiv 0$ (mod p).*

We can use Theorem 112 to prove Theorem 76. We suppose p odd. Suppose that

$$n = rp - s \quad (r \geqslant 1, 0 \leqslant s < p).$$

Then

$$\binom{p+n-1}{n} = \frac{(rp-s+p-1)!}{(rp-s)!(p-1)!}$$
$$= \frac{(rp-s+1)(rp-s+2)...(rp-s+p-1)}{(p-1)!}$$

is an integer i, and

$$(rp-s+1)(rp-s+2)\dots(rp-s+p-1) = (p-1)!i \equiv -i \pmod{p},$$

by Wilson's theorem (Theorem 80). But the left-hand side is congruent to

$$(s-1)(s-2)\dots(s-p+1) \equiv s^{p-1} - 1 \pmod{p},$$

by Theorem 112, and is therefore congruent to -1 when $s = 0$ and to 0 otherwise.

7.6. Lagrange's proof of Fermat's and Wilson's theorems. We based our proof of Theorem 112 on Fermat's theorem and on Theorem 108. Lagrange, the discoverer of the theorem, proved it directly, and his argument contains another proof of Fermat's theorem.

We suppose p odd. Then

$$(7.6.1)\quad (x-1)(x-2)\dots(x-p+1) = x^{p-1} - A_1x^{p-2} + \dots + A_{p-1},$$

where $A_1, \dots$ are defined as in Theorem 113. If we multiply both sides by x and change x into $x-1$, we have

$$(x-1)^p - A_1(x-1)^{p-1} + \dots + A_{p-1}(x-1) = (x-1)(x-2)\dots(x-p)$$
$$= (x-p)(x^{p-1} - A_1x^{p-2} + \dots + A_{p-1}).$$

Equating coefficients, we obtain

$$\binom{p}{1} + A_1 = p + A_1,\quad \binom{p}{2} + \binom{p-1}{1}A_1 + A_2 = pA_1 + A_2,$$
$$\binom{p}{3} + \binom{p-1}{2}A_1 + \binom{p-2}{1}A_2 + A_3 = pA_2 + A_3,$$

and so on. The first equation is an identity; the others yield in succession

$$A_1 = \binom{p}{2}, \; 2A_2 = \binom{p}{3} + \binom{p-1}{2} A_1,$$

$$3A_2 = \binom{p}{4} + \binom{p-1}{3} A_1 + \binom{p-2}{2} A_2,$$

$$\cdot \quad \cdot \quad \cdot \quad \cdot \quad \cdot \quad \cdot \quad \cdot \quad \cdot \quad \cdot \quad \cdot \quad \cdot$$

$$(p-1)A_{p-1} = 1 + A_1 + A_2 + \ldots + A_{p-2}.$$

Hence we deduce successively

$$p|A_1, \quad p|A_2, \quad \ldots, \quad p|A_{p-2}, \tag{7.6.2}$$

and finally

$$(p-1)A_{p-1} \equiv 1 \pmod{p}$$

or

$$A_{p-1} \equiv -1 \pmod{p}. \tag{7.6.3}$$

Since $A_{p-1} = (p-1)!$, (7.6.3) is Wilson's theorem; and (7.6.2) and (7.6.3) together give Theorem 112. Finally, since

$$(x-1)(x-2)\ldots(x-p+1) \equiv 0 \pmod{p}$$

for any x which is not a multiple of p, Fermat's theorem follows as a corollary.

7.7. The residue of $\{\frac{1}{2}(p-1)\}!$. Suppose that p is an odd prime and

$$\varpi = \tfrac{1}{2}(p-1).$$

From

$$\begin{aligned}(p-1)! &= 1.2\ldots\tfrac{1}{2}(p-1)\left\{p-\tfrac{1}{2}(p-1)\right\}\left\{p-\tfrac{1}{2}(p-3)\right\}\ldots(p-1)\\ &\equiv (-1)^{\varpi}(\varpi!)^2 \pmod{p}\end{aligned}$$

it follows, by Wilson's theorem, that

$$(\varpi!)^2 \equiv (-1)^{\varpi-1} \pmod{p}.$$

We must now distinguish the two cases $p = 4n+1$ and $p = 4n+3$. If $p = 4n + 1$, then

$$(\varpi!)^2 \equiv -1 \pmod{p},$$

so that (as we proved otherwise in § 6.6) -1 is a quadratic residue of p. In this case $\varpi!$ is congruent to one or other of the roots of $x^2 \equiv -1 \pmod{p}$.

If $p = 4n + 3$, then

$$(\varpi!)^2 \equiv 1 \pmod{p}, \tag{7.7.1}$$

$$\varpi! \equiv \pm 1 \pmod{p}. \tag{7.7.2}$$

Since -1 is a non-residue of p, the sign in (7.7.2) is positive or negative according as $\varpi!$ is a residue or non-residue of p. But $\varpi!$ is the product of the positive integers less than $\frac{1}{2}p$, and therefore, by Theorem 85, the sign in (7.7.2) is positive or negative according as the number of non-residues of p less than $\frac{1}{2}p$ is even or odd.

Theorem 114. *If p is a prime $4n + 3$, then*

$$\left\{\tfrac{1}{2}(p-1)\right\}! \equiv (-1)^{\nu} \pmod{p},$$

where ν is the number of quadratic non-residues less than $\frac{1}{2}p$

7.8. A theorem of Wolstenholme. It follows from Theorem 113 that the numerator of the fraction

$$1 + \frac{1}{2} + \frac{1}{3} + \cdots + \frac{1}{p-1}$$

is divisible by p; in fact the numerator is the A_{p-2} of that theorem. We can, however, go farther.

Theorem 115. *If p is a prime greater than 3, then the numerator of the fraction*

$$1 + \frac{1}{2} + \frac{1}{3} + \cdots + \frac{1}{p-1} \tag{7.8.1}$$

is divisible by p^2.

The result is false when $p = 3$. It is irrelevant whether the fraction is or is not reduced to its lowest terms, since in any case the denominator cannot be divisible by p.

The theorem may be stated in a different form. If i is prime to m, the congruence

$$ix \equiv 1 \pmod{m}$$

has just one root, which we call the *associate* of $i \pmod{m}$.[†] We may denote this associate by $\bar{\imath}$, but it is often convenient, when it is plain that we are concerned with an integer, to use the notation

$$\frac{1}{i}$$

(or $1/i$). More generally we may, in similar circumstances, use

$$\frac{b}{a}$$

(or b/a) for the solution of $ax \equiv b$.

We may then (as we shall see in a moment) state Wolstenholme's theorem in the form

THEOREM 116. *If $p > 3$, and $1/i$ is the associate of $i \pmod{p^2}$, then*

$$1 + \frac{1}{2} + \frac{1}{3} + \cdots + \frac{1}{p-1} \equiv 0 \pmod{p^2}.$$

We may elucidate the notation by proving first that

$$1 + \frac{1}{2} + \frac{1}{3} + \cdots + \frac{1}{p-1} \equiv 0 \pmod{p}.\text{‡} \tag{7.8.2}$$

For this, we have only to observe that, if $0 < i < p$, then

$$i.\frac{1}{i} \equiv 1, \quad (p-i)\frac{1}{p-i} \equiv 1 \pmod{p}.$$

Hence

$$i\left(\frac{1}{i} + \frac{1}{p-i}\right) \equiv i.\frac{1}{i} - (p-i)\frac{1}{p-i} \equiv 0 \pmod{p},$$

$$\frac{1}{i} + \frac{1}{p-i} \equiv 0 \pmod{p},$$

and the result follows by summation.

† As in § 6.5, the a of § 6.5 being now 1.

‡ Here, naturally, $1/i$ is the associate of $i \pmod{p}$. This is determinate $\pmod{p}$, but indeterminate $\pmod{p^2}$ to the extent of an arbitrary multiple of p.

We show next that the two forms of Wolstenholme's theorem (Theorems 115 and 116) are equivalent. If $0 < x < p$ and $\bar{x}$ is the associate of x (mod p^2), then

$$\bar{x}(p-1)! = x\bar{x}\frac{(p-1)!}{x} \equiv \frac{(p-1)}{x} \pmod{p^2}.$$

Hence

$$(p-1)!(\bar{1}+\bar{2}+\cdots+\overline{p-1}) \equiv (p-1)!\left(1+\frac{1}{2}+\cdots+\frac{1}{p-1}\right) \pmod{p^2},$$

the fractions on the right having their common interpretation; and the equivalence follows.

To prove the theorem itself we put $x = p$ in the identity (7.6.1). This gives

$$(p-1)! = p^{p-1} - A_1 p^{p-2} + \ldots - A_{p-2}p + A_{p-1}.$$

But $A_{p-1} = (p-1)!$, and therefore

$$p^{p-2} - A_1 p^{p-3} + \ldots + A_{p-3}p - A_{p-2} = 0.$$

Since $p > 3$ and

$$p|A_1,\ p|A_2,\ \ldots,\ p|A_{p-3},$$

by Theorem 113, it follows that $p^2|A_{p-2}$, i.e.

$$p^2|(p-1)!\left(1+\frac{1}{2}+\ldots+\frac{1}{p-1}\right).$$

This is equivalent to Wolstenholme's theorem.

The numerator of

$$C_p = 1 + \frac{1}{2^2} + \ldots + \frac{1}{(p-1)^2}$$

is $A_{p-2}^2 - 2A_{p-1}A_{p-3}$, and is therefore divisible by p. Hence

THEOREM 117. *If* $p > 3$, *then* $C_p \equiv 0 \pmod{p}$.

7.9. The theorem of von Staudt. We conclude this chapter by proving a famous theorem of von Staudt concerning Bernoulli's numbers.

Bernoulli's numbers are usually defined as the coefficients in the expansion[†]

$$\frac{x}{e^x - 1} = 1 - \tfrac{1}{2}x + \frac{B_1}{2!}x^2 - \frac{B_2}{4!}x^4 + \frac{B_3}{6!}x^6 - \dots .$$

We shall find it convenient to write

$$\frac{x}{e^x - 1} = \beta_0 + \frac{\beta_1}{1!}x + \frac{\beta_2}{2!}x^2 + \frac{\beta_3}{3!}x^3 + \dots ,$$

so that $\beta_0 = 1$, $\beta_1 = -\frac{1}{2}$ and

$$\beta_{2k} = (-1)^{k-1}B_k, \quad \beta_{2k+1} = 0 \quad (k \geqslant 1).$$

The importance of the numbers comes primarily from their occurrence in the 'Euler–Maclaurin sum-formula' for $\sum m^k$. In fact

$$(7.9.1) \qquad 1^k + 2^k + \dots + (n-1)^k = \sum_{r=0}^{k} \frac{1}{k+1-r}\binom{k}{r} n^{k+1-r}\beta_r$$

for $k \geqslant 1$. For the left-hand side is the coefficient of x^{k+1} in

$$\begin{aligned} &k!x(1 + e^x + e^{2x} + \dots + e^{(n-1)x}) \\ &= k!x\frac{1 - e^{nx}}{1 - e^x} = k!\frac{x}{e^x - 1}(e^{nx} - 1) \\ &= k!\left(1 + \frac{\beta_1}{1!}x + \frac{\beta_2}{2!}x^2 + \cdots\right)\left(nx\frac{n^2x^2}{2!} + \dots\right); \end{aligned}$$

and (7.9.1) follows by picking out the coefficient in this product.

Von Staudt's theorem determines the fractional part of B_k.

THEOREM 118. *If $k \geqslant 1$, then*

$$(7.9.2) \qquad (-1)^k B_k \equiv \sum \frac{1}{p} \pmod{1},$$

the summation being extended over the primes p such that $(p-1)|2k$.

[†] This expansion is convergent whenever $|x| < 2\pi$.

For example, if $k = 1$, then $(p-1)|2$, which is true if $p = 2$ or $p = 3$. Hence $-B_1 \equiv \frac{1}{2} + \frac{1}{3} = \frac{5}{6}$; and in fact $B_1 = \frac{1}{6}$ When we restate (7.9.2) in terms of the β, it becomes

$$\beta_k + \sum_{(p-1)|k} \frac{1}{p} = i, \tag{7.9.3}$$

where

$$k = 1, 2, 4, 6, \ldots \tag{7.9.4}$$

and i is an integer. If we define $\epsilon_k(p)$ by

$$\epsilon_k(p) = 1 \quad ((p-1) \mid k), \quad \epsilon_k(p) = 0 \ \big((p-1) \nmid k\big),$$

then (7.9.3) takes the form

$$\beta_k + \sum \frac{\epsilon_k(p)}{p} = i, \tag{7.9.5}$$

where p now runs through all primes.

In particular von Staudt's theorem shows that there is no squared factor in the denominator of any Bernoullian number.

7.10. Proof of von Staudt's theorem. The proof of Theorem 118 depends upon the following lemma.

THEOREM 119:

$$\sum_{1}^{p-1} m^k \equiv -\epsilon_k(p) \pmod{p}.$$

If $(p-1)|k$, then $m^k \equiv 1$, by Fermat's theorem, and

$$\sum m^k \equiv p - 1 \equiv -1 \equiv -\epsilon_k(p) \pmod{p}.$$

If $(p-1) \nmid k$, and g is a primitive root of p, then

$$g^k \not\equiv 1 \pmod{p}, \tag{7.10.1}$$

by Theorem 88. The sets $g, 2g, \ldots, (p-1)g$ and $1, 2, \ldots, p-1$ are equivalent (mod p), and therefore

$$\sum (mg)^k \equiv \sum m^k \pmod{p},$$
$$(g^k - 1)\sum m^k \equiv 0 \pmod{p},$$

and

$$\sum m^k \equiv 0 = -\epsilon_k(p) \pmod{p},$$

by (7.10.1). Thus $\sum m^k \equiv -\epsilon_k(p)$ in any case.

We now prove Theorem 118 by induction, assuming that it is true for any number l of the sequence (7.9.4) less than k, and deducing that it is true for k. In what follows k and l belong to (7.9.4), r runs from 0 to k, $\beta_0 = 1$, and $\beta_3 = \beta_5 = \ldots = 0$. We have already verified the theorem when $k = 2$, and we may suppose $k > 2$.

It follows from (7.9.1) and Theorem 119 that, if ϖ is any prime,

$$\epsilon_k(\varpi) + \sum_{r=0}^{k} \frac{1}{k+1-r}\binom{k}{r}\varpi^{k+1-r}\beta_r \equiv 0 \pmod{\varpi}$$

or

(7.10.2)
$$\beta_k + \frac{\epsilon_k(\varpi)}{\varpi} + \sum_{r=0}^{k-2} \frac{1}{k+1-r}\binom{k}{r}\varpi^{k-1-r}(\varpi\beta_r) \equiv 0 \pmod{1};$$

there is no term in β_{k-1}, since $\beta_{k-1} = 0$. We consider whether the denominator of

$$u_{k,r} = \frac{1}{k+1-r}\binom{k}{r}\varpi^{k-1-r}(\varpi\beta_r)$$

can be divisible by ϖ.

If r is not an l, β_r is 1 or 0. If r is an l, then, by the inductive hypothesis, the denominator of β_r has no squared factor,† and that of $\varpi\beta_r$ is not divisible by

† It will be observed that we do not need the full force of the inductive hypothesis.

ϖ. The factor $\binom{k}{r}$ is integral. Hence the denominator of $u_{k,r}$ is divisible by ϖ only if that of

$$\frac{\varpi^{k-1-r}}{k+1-r} = \frac{\varpi^{s-1}}{s+1}$$

is divisible by ϖ. In this case

$$s+1 \geqslant \varpi^s.$$

But $s = k - r \geqslant 2$, and therefore

$$s+1 < 2^s \leqslant \varpi^s.$$

a contradiction. It follows that the denominator of $u_{k,r}$ is not divisible by ϖ.

Hence

$$\beta_k + \frac{\epsilon_k(\varpi)}{\varpi} = \frac{a_k}{b_k},$$

where $\varpi \nmid b_k$; and

$$\frac{\epsilon_k(p)}{p} \ (p \neq \varpi)$$

is obviously of the same form. It follows that

$$\beta_k + \sum \frac{\epsilon_k(p)}{p} = \frac{A_k}{B_k}, \tag{7.10.3}$$

where B_k is not divisible by ϖ. Since ϖ is an arbitrary prime, B_k must be 1. Hence the right-hand side of (7.10.3) is an integer; and this proves the theorem.

Suppose in particular that k is a prime of the form $3n+1$. Then $(p-1)|2k$ only if p is one of $2, 3, k+1, 2k+1$. But $k+1$ is even, and $2k+1 = 6n+3$ is divisible by 3, so that 2 and 3 are the only permissible values of p. Hence

THEOREM 120: *If k is a prime of the form $3n+1$, then*

$$B_k \equiv \tfrac{1}{6} \pmod{1}.$$

The argument can be developed to prove that if k is given, there are an infinity of l for which B_l has the same fractional part as B_k; but for this we need Dirichlet's Theorem 15 (or the special case of the theorem in which $b = 1$).

NOTES

§§ 7.2–4. For the most part we follow Hecke, § 3.

§ 7.6. Lagrange, *Nouveaux mémoires de l'Académie royale de Berlin*, 2 (1773), 125 (*Œuvres*, iii. 425). This was the first published proof of Wilson's theorem.

§ 7.7. Dirichlet, *Journal für Math.* 3 (1828), 407–8 (*Werke*, i. 107–8).

§ 7.8. Wolstenholme, *Quarterly Journal of Math.* 5 (1862), 35–39. There are many generalizations of Theorem 115, some of which are also generalizations of Theorem 113. See § 8.7.

The theorem has generally been described as 'Wolstenholme's theorem', and we follow the usual practice. But N. Rama Rao [*Bull. Calcutta Math. Soc.* 29 (1938), 167–70] has pointed out that it, and a good many of its extensions, had been anticipated by Waring, *Meditationes algebraicae*, ed. 2 (1782), 383.

§§ 7.9–10. von Staudt, *Journal für Math.* 21 (1840), 372–4. The theorem was discovered independently by Clausen, *Astronomische Nachrichten*, 17 (1840), 352. We follow a proof by R. Rado, *Journal London Math. Soc.* 9 (1934), 85–8.

Many authors use the notation

$$\frac{x}{e^x - 1} = \sum_{n=0}^{\infty} B_n \frac{x^n}{n!},$$

so that their B_n is our β_n.

Theorem 120, and the more general theorem referred to in connexion with it, are due to Rado (ibid. 88–90). Indeed Erdős and Wagstaff (*Illinois J. Math.* 24 (1980), 104–12) have shown, for given k, that one has $B_m \equiv B_k \pmod 1$ for a positive proportion of values of m.

VIII

CONGRUENCES TO COMPOSITE MODULI

8.1. Linear congruences. We have supposed since § 7.4 (apart from a momentary digression in § 7.8) that the modulus m is prime. In this chapter we prove a few theorems concerning congruences to general moduli. The theory is much less simple when the modulus is composite, and we shall not attempt any systematic discussion.

We considered the general linear congruence

$$ax \equiv b \pmod{m} \tag{8.1.1}$$

in § 5.4, and it will be convenient to recall our results. The congruence is insoluble unless

$$d = (a, m) \mid b. \tag{8.1.2}$$

If this condition is satisfied, then (8.1.1) has just d solutions, viz.

$$\xi, \xi + \frac{m}{d}, \xi + 2\frac{m}{d}, \ldots, \xi + (d-1)\frac{m}{d},$$

where ξ is the unique solution of

$$\frac{a}{d}x \equiv \frac{b}{d}\left(\operatorname{mod} \frac{m}{d}\right).$$

We consider next a system

$$a_1x \equiv b_1 \pmod{m_1},\quad a_2x \equiv b_2 \pmod{m_2},\quad \ldots, a_kx \equiv b_k \pmod{m_k}. \tag{8.1.3}$$

of linear congruences to coprime moduli $m_1, m_2, \ldots, m_k$. The system will be insoluble unless $(a_i, m_i)|b_i$ for every i. If this condition is satisfied, we can solve each congruence separately, and the problem is reduced to that of the solution of the system

$$x \equiv c_1 \pmod{m_1},\quad x \equiv c_2 \pmod{m_2}, \ldots, x \equiv c_k \pmod{m_k}. \tag{8.1.4}$$

The m_i here are not the same as in (8.1.3); in fact the m_i of (8.1.4) is $m_i/(a_i, m_i)$ in the notation of (8.1.3).

We write

$$m = m_1 m_2 \dots m_k = m_1 M_1 = m_2 M_2 = \dots = m_k M_k.$$

Since $(m_i, M_i) = 1$, there is an n_i (unique to modulus m_i) such that

$$n_i M_i \equiv 1 \pmod{m_i}.$$

If

$$x = n_1 M_1 c_1 + n_2 M_2 c_2 + \cdots + n_k M_k c_k, \tag{8.1.5}$$

then $x \equiv n_i M_i c_i \equiv c_i \pmod{m_i}$ for every i, so that x satisfies (8.1.4). If y satisfies (8.1.4), then

$$y \equiv c_i \equiv x \pmod{m_i}$$

for every i, and therefore (since the m_i are coprime), $y \equiv x \pmod{m}$. Hence the solution x is unique $\pmod{m}$.

THEOREM 121. *If $m_1, m_2, \dots, m_k$ are coprime, then the system* (8.1.4) *has a unique solution* $\pmod{m}$ *given by* (8.1.5).

The problem is more complicated when the moduli are not coprime. We content ourselves with an illustration.

Six professors begin courses of lectures on Monday, Tuesday, Wednesday, Thursday, Friday, and Saturday, and announce their intentions of lecturing at intervals of two, three, four, one, six, and five days respectively. The regulations of the university forbid Sunday lectures (*so that a Sunday lecture must be omitted*). *When first will all six professors find themselves compelled to omit a lecture?*

If the day in question is the xth (counting from and including the first Monday), then

$$\begin{aligned} x &= 1 + 2k_1 = 2 + 3k_2 = 3 + 4k_3 = 4 + k_4 \\ &= 5 + 6k_5 = 6 + 5k_6 = 7k_7, \end{aligned}$$

where the k are integers; i.e.

$$\begin{gathered} (1)\ x \equiv 1 \pmod{2},\ (2)\ x \equiv 2 \pmod{3},\ (3)\ x \equiv 3 \pmod{4}, \\ (4)\ x \equiv 4 \pmod{1},\ (5)\ x \equiv 5 \pmod{6},\ (6)\ x \equiv 6 \pmod{5}, \\ (7)\ x \equiv 0 \pmod{7}. \end{gathered}$$

Of these congruences, (4) is no restriction, and (1) and (2) are included in (3) and (5). Of the two latter, (3) shows that x is congruent to 3, 7, or 11 (mod 12), and (5) that x is congruent to 5 or 11, so that (3) and (5) together are equivalent to $x \equiv 11 \pmod{12}$. Hence the problem is that of solving

$$x \equiv 11 \pmod{12}, \quad x \equiv 6 \pmod{5}, \quad x \equiv 0 \pmod{7}$$

or

$$x \equiv -1 \pmod{12}, \quad x \equiv 1 \pmod{5}, \quad x \equiv 0 \pmod{7}.$$

This is a case of the problem solved by Theorem 121. Here

$$m_1 = 12, \quad m_2 = 5, \quad m_3 = 7, \quad m = 420,$$
$$M_1 = 35, \quad M_2 = 84, \quad M_3 = 60.$$

The n are given by

$$35n_1 \equiv 1 \pmod{12}, \quad 84n_2 \equiv 1 \pmod{5}, \quad 60n_3 \equiv 1 \pmod{7},$$

or

$$-n_1 \equiv 1 \pmod{12}, \quad -n_2 \equiv 1 \pmod{5}, \quad 4n_3 \equiv 1 \pmod{7};$$

and we can take $n_1 = -1$, $n_2 = -1$, $n_3 = 2$. Hence

$$x \equiv (-1)(-1)35 + (-1)1.84 + 2.0.60 = -49 \equiv 371 \pmod{420}.$$

The first x satisfying the condition is 371.

8.2. Congruences of higher degree. We can now reduce the solution of the general congruence[†]

$$(8.2.1) \qquad f(x) \equiv 0 \pmod{m},$$

where $f(x)$ is any integral polynomial, to that of a number of congruences whose moduli are powers of primes.

Suppose that

$$m = m_1 m_2 \dots m_k,$$

no two m_i having a common factor. Every solution of (8.2.1) satisfies

$$(8.2.2) \qquad f(x) \equiv 0 \pmod{m_i} \quad (i = 1, 2, \dots, k).$$

[†] See § 7.2.

If $c_1, c_2, \ldots, c_k$ is a set of solutions of (8.2.2), and x is the solution of

$$x \equiv c_i \pmod{m_i} \quad (i = 1, 2, \ldots, k), \tag{8.2.3}$$

given by Theorem 121, then

$$f(x) \equiv f(c_i) \equiv 0 \pmod{m_i}$$

and therefore $f(x) \equiv 0 \pmod{m}$. Thus every set of solutions of (8.2.2) gives a solution of (8.2.1), and conversely. In particular

THEOREM 122. *The number of roots of* (8.2.1) *is the product of the numbers of roots of the separate congruences* (8.2.2).

If $m = p_1^{a_1} p_2^{a_2} \ldots p_k^{a_k}$, we may take $m_i = p_i^{a_i}$.

8.3. Congruences to a prime-power modulus. We have now to consider the congruence

$$f(x) \equiv 0 \pmod{p^a} \tag{8.3.1}$$

where p is prime and $a > 1$.

Suppose first that x is a root of (8.3.1) for which

$$0 \leqslant x < p^a. \tag{8.3.2}$$

Then x satisfies

$$f(x) \equiv 0 \pmod{p^{a-1}}, \tag{8.3.3}$$

and is of the form

$$\xi + sp^{a-1} \quad (0 \leqslant s < p), \tag{8.3.4}$$

where ξ is a root of (8.3.3) for which

$$0 \leqslant \xi < p^{a-1}. \tag{8.3.5}$$

Next, if ξ is a root of (8.3.3) satisfying (8.3.5), then

$$\begin{aligned} f(\xi + sp^{a-1}) &= f(\xi) + sp^{a-1}f'(\xi) + \tfrac{1}{2}s^2p^{2a-2}f''(\xi) + \cdots \\ &\equiv f(\xi) + sp^{a-1}f'(\xi) \pmod{p^a}, \end{aligned}$$

since $2a-2 \geqslant a, 3a-3 \geqslant a, \ldots$, and the coefficients in

$$\frac{f^{(k)}(\xi)}{k!}$$

are integers. We have now to distinguish two cases.

(1) Suppose that

$$f'(\xi) \not\equiv 0 \pmod{p}. \tag{8.3.6}$$

Then $\xi + sp^{a-1}$ is a root of (8.3.1) if and only if

$$f(\xi) + sp^{a-1}f'(\xi) \equiv 0 \pmod{p^a}$$

or

$$sf'(\xi) \equiv -\frac{f(\xi)}{p^{a-1}} \pmod{p},$$

and there is just one $s \pmod{p}$ satisfying this condition. Hence the number of roots of (8.3.3) is the same as the number of roots of (8.3.1).

(2) Suppose that

$$f'(\xi) \equiv 0 \pmod{p}. \tag{8.3.7}$$

Then

$$f(\xi + sp^{a-1}) \equiv f(\xi) \pmod{p^a}.$$

If $f(\xi) \not\equiv 0 \pmod{p^a}$, then (8.3.1) is insoluble. If $f(\xi) = 0 \pmod{p^a}$, then (8.3.4) is a solution of (8.3.1) for every s, and there are p solutions of (8.3.1) corresponding to every solution of (8.3.3).

THEOREM 123. *The number of solutions of* (8.3.1) *corresponding to a solution* ξ *of* (8.3.3) *is*

(*a*) *none, if* $f'(\xi) \equiv 0 \pmod{p}$ *and* ξ *is not a solution of* (8.3.1);

(*b*) *one, if* $f'(\xi) \not\equiv 0 \pmod{p}$;

(*c*) p, *if* $f'(\xi) \equiv 0 \pmod{p}$ *and* ξ *is a solution of* (8.3.1).

The solutions of (8.3.1) *corresponding to* ξ *may be derived from* ξ, *in case* (*b*) *by the solution of a linear congruence, in case* (*c*) *by adding any multiple of* p^{a-1} *to* ξ.

8.4. Examples. (1) The congruence

$$f(x) = x^{p-1} - 1 \equiv 0 \pmod{p}$$

has the $p-1$ roots $1, 2, \ldots, p-1$; and if ξ is any one of these, then

$$f'(\xi) = (p-1)\xi^{p-2} \not\equiv 0 \pmod{p}.$$

Hence $f(x) \equiv 0 \pmod{p^2}$ has just $p-1$ roots. Repeating the argument, we obtain

THEOREM 124. *The congruence*

$$x^{p-1} - 1 \equiv 0 \pmod{p^a}$$

has just $p-1$ *roots for every* a.

(2) We consider next the congruence

$$f(x) = x^{\frac{1}{2}p(p-1)} - 1 \equiv 0 \pmod{p^2}, \tag{8.4.1}$$

where p is an odd prime. Here

$$f'(\xi) = \tfrac{1}{2}p(p-1)\xi^{\frac{1}{2}p(p-1)-1} \equiv 0 \pmod{p}$$

for every ξ. Hence there are p roots of (8.4.1) corresponding to every root of $f(x) \equiv 0 \pmod{p}$.

Now, by Theorem 83,

$$x^{\frac{1}{2}(p-1)} \equiv \pm 1 \pmod{p}$$

according as x is a quadratic residue or non-residue of p, and

$$x^{\frac{1}{2}p(p-1)} \equiv \pm 1 \pmod{p}$$

in the same cases. Hence there are $\frac{1}{2}(p-1)$ roots of $f(x) \equiv 0 \pmod{p}$, and $\frac{1}{2}p(p-1)$ of (8.4.1).

We define the quadratic residues and non-residues of p^2 as we defined those of p in § 6.5. We consider only numbers prime to p. We say that x is a residue of p^2 if (i) $(x, p) = 1$ and (ii) there is a y for which

$$y^2 \equiv x \pmod{p^2},$$

and a non-residue if (i) $(x, p) = 1$ and (ii) there is no such y.

If x is a quadratic residue of p^2, then, by Theorem 72,

$$x^{\frac{1}{2}p(p-1)} \equiv y^{p(p-1)} \equiv 1 \pmod{p^2},$$

so that x is one of the $\frac{1}{2}p(p-1)$ roots of (8.4.1). On the other hand, if y_1 and y_2 are two of the $p(p-1)$ numbers less than and prime to p^2, and $y_1^2 \equiv y_2^2$, then either $y_2 = p^2 - y_1$ or $y_1 - y_2$ and $y_1 + y_2$ are both divisible by p, which is impossible because y_1 and y_2 are not divisible by p. Hence the numbers y^2 give just $\frac{1}{2}p(p-1)$ incongruent residues (mod p^2), and there are $\frac{1}{2}p(p-1)$ quadratic residues of p^2, namely the roots of (8.4.1).

THEOREM 125. *There are* $\frac{1}{2}p(p-1)$ *quadratic residues of* p^2, *and these residues are the roots of* (8.4.1).

(3) We consider finally the congruence

$$f(x) = x^2 - c \equiv 0 \pmod{p^a}, \tag{8.4.2}$$

where $p \nmid c$. If p is odd, then

$$f'(\xi) = 2\xi \not\equiv 0 \pmod{p}$$

for any ξ not divisible by p. Hence the number of roots of (8.4.2) is the same as that of the similar congruences to moduli $p^{a-1}, p^{a-2}, \ldots, p$; that is to say, two or none, according as c is or is not a quadratic residue of p. We could use this argument as a substitute for the last paragraph of (2).

The situation is a little more complex when $p = 2$, since then $f'(\xi) \equiv 0 \pmod{p}$ for every ξ. We leave it to the reader to show that there are two roots or none when $a = 2$ and four or none when $a \geqslant 3$.

8.5. Bauer's identical congruence. We denote by t one of the $\phi(m)$ numbers less than and prime to m, by $t(m)$ the set of such numbers, and by

$$f_m(x) = \prod_{t(m)} (x - t) \tag{8.5.1}$$

a product extended over all the t of $t(m)$. Lagrange's Theorem 112 states that

$$f_m(x) = x^{\phi(m)} - 1 \pmod{m} \tag{8.5.2}$$

when m is prime. Since

$$x^{\phi(m)} - 1 \equiv 0 \pmod{m}$$

has always the $\phi(m)$ roots t, we might expect (8.5.2) to be true for all m; but this is false. Thus, when $m = 9$, t has the 6 values $\pm 1, \pm 2, \pm 4 \pmod 9$, and

$$f_m(x) \equiv (x^2 - 1^2)(x^2 - 2^2)(x^2 - 4^2) \equiv x^6 - 3x^4 + 3x^2 - 1 \pmod 9.$$

The correct generalization was found comparatively recently by Bauer, and is contained in the two theorems which follow.

Theorem 126. *If p is an odd prime divisor of m, and p^a is the highest power of p which divides m, then*

$$(8.5.3) \qquad f_m(x) = \prod_{t(m)} (x - t) \equiv (x^{p-1} - 1)^{\phi(m)/(p-1)} \pmod{p^a}.$$

In particular

$$(8.5.4) \qquad f_{p^a}(x) = \prod_{t(p^a)} (x - t) \equiv (x^{p-1} - 1)^{p^{a-1}} \pmod{p^a}.$$

Theorem 127. *If m is even, $m > 2$, and 2^a is the highest power of 2 which divides m, then*

$$(8.5.5) \qquad f_m(x) \equiv (x^2 - 1)^{\frac{1}{2}\phi(m)} \pmod{2^a}.$$

In particular

$$(8.5.6) \qquad f_{2^a}(x) \equiv (x^2 - 1)^{2^{a-2}} \pmod{2^a}.$$

when $a > 1$.

In the trivial case $m = 2, f_2(x) = x - 1$. This falls under (8.5.3) and not under (8.5.5).

We suppose first that $p > 2$, and begin by proving (8.5.4). This is true when $a = 1$. If $a > 1$, the numbers in $t(p^a)$ are the numbers

$$t + vp^{a-1} \ (0 \leqslant v < p),$$

where t is a number included in $t(p^{a-1})$. Hence

$$f_{p^a}(x) = \prod_{v=0}^{p-1} f_{p^{a-1}}(x - vp^{a-1}).$$

But

$$f_{p^{a-1}}(x - vp^{a-1}) \equiv f_{p^{a-1}}(x) - vp^{a-1}f'_{p^{a-1}}(x) \pmod{p^a};$$

and

$$\begin{aligned} f_{p^a}(x) &\equiv \{f_{p^{a-1}}(x)\}^p - \sum v.p^{a-1}\{f_{p^{a-1}}(x)\}^{p-1}f'_{p^{a-1}}(x) \\ &\equiv \{f_{p^{a-1}}(x)\}^p \pmod{p^a}, \end{aligned}$$

since $\sum v = \frac{1}{2}p(p-1) \equiv 0 \pmod{p}$.

This proves (8.5.4) by induction.

Suppose now that $m = p^a M$ and that $p \nmid M$. Let t run through the $\phi(p^a)$ numbers of $t(p^a)$ and T through the $\phi(M)$ numbers of $t(M)$. By Theorem 61, the resulting set of $\phi(m)$ numbers

$$tM + Tp^a,$$

reduced mod m, is just the set $t(m)$. Hence

$$f_m(x) = \prod_{t(m)} (x - t) \equiv \prod_{T \in t(M)} \prod_{t \in t(p^a)} (x - tM - Tp^a) \pmod{m}.$$

For any fixed T, since $(p^a, M) = 1$,

$$\begin{aligned} \prod_{t \in t(p^a)} (x - tM - Tp^a) &\equiv \prod_{t \in t(p^a)} (x - tM) \\ &\equiv \prod_{t \in t(p^a)} (x - t) \equiv f_{p^a}(x) \pmod{p^a}. \end{aligned}$$

Hence, since there are $\phi(M)$ members of $t(M)$,

$$f_m(x) \equiv (x^{p-1} - 1)^{p^{a-1}\phi(M)} \pmod{p^a}$$

by (8.5.4). But (8.5.3) follows at once, since

$$p^{a-1}\phi(M) = \frac{\phi(p^a)}{p-1}\phi(M) = \frac{\phi(m)}{p-1}.$$

8.6. Bauer's congruence: the case $p = 2$. We have now to consider the case $p = 2$. We begin by proving (8.5.6).

If $a = 2$,

$$f_4(x) = (x-1)(x-3) \equiv x^2 - 1 \pmod{4},$$

which is (8.5.6). When $a > 2$, we proceed by induction. If

$$f_{2^{a-1}}(x) \equiv (x^2-1)^{2^{a-3}} \pmod{2^{a-1}},$$

then

$$f'_{2^{a-1}}(x) \equiv 0 \pmod{2}.$$

Hence

$$\begin{aligned} f_{2^a}(x) &= f_{2^{a-1}}(x) f_{2^{a-1}}(x - 2^{a-1}) \\ &\equiv \{f_{2^{a-1}}(x)\}^2 - 2^{a-1} f_{2^{a-1}}(x) f'_{2^{a-1}}(x) \\ &\equiv \{f_{2^{a-1}}(x)\}^2 \equiv (x^2-1)^{2^{a-2}} \pmod{2^a}. \end{aligned}$$

Passing to the proof of (8.5.5), we have now to distinguish two cases.

(1) If $m = 2M$ and $M > 1$, where M is odd, then

$$f_m(x) \equiv (x-1)^{\phi(m)} \equiv (x^2-1)^{\frac{1}{2}\phi(m)} \pmod{2},$$

because $(x-1)^2 \equiv x^2 - 1 \pmod{2}$.

(2) If $m = 2^a M$, where M is odd and $a > 1$, we argue as in § 8.5, but use (8.5.6) instead of (8.5.4). The set of $\phi(m) = 2^{a-1}\,\phi(M)$ numbers

$$tM + T2^a,$$

reduced mod m, is just the set $t(m)$. Hence

$$\begin{aligned} f_m(x) &= \prod_{t(m)} (x - t) \equiv \prod_{T \in t(M)} \prod_{t \in t(2^a)} (x - tM - 2^a T) \pmod{m} \\ &\equiv \{f_{2^a}(x)\}^{\phi(M)} \pmod{2}, \end{aligned}$$

just as in § 8.5. (8.5.5) follows at once from this and (8.5.6).

8.7. A theorem of Leudesdorf. We can use Bauer's theorem to obtain a comprehensive generalization of Wolstenholme's Theorem 115.

THEOREM 128. *If*

$$S_m = \sum_{t(m)} \frac{1}{t},$$

then

$$S_m \equiv 0 \pmod{m^2} \tag{8.7.1}$$

if $2 \nmid m, 3 \nmid m$;

$$S_m \equiv 0 \pmod{\tfrac{1}{3}m^2} \tag{8.7.2}$$

if $2 \nmid m, 3|m$;

$$S_m \equiv 0 \pmod{\tfrac{1}{2}m^2} \tag{8.7.3}$$

if $2|m, 3 \nmid m$, *and m is not a power of* 2;

$$S_m \equiv 0 \pmod{\tfrac{1}{6}m^2} \tag{8.7.4}$$

if $2|m, 3|\ m$; *and*

$$S_m \equiv 0 \pmod{\tfrac{1}{4}m^2} \tag{8.7.5}$$

if $m = 2^a$.

We use Σ, Π for sums or products over the range $t(m)$, and Σ', Π' for sums or products over the part of the range in which t is less than $\frac{1}{2}m$; and we suppose that $m = p^a q^b r^c \ldots$.

If $p > 2$ then, by Theorem 126,

$$(x^{p-1} - 1)^{\phi(m)/(p-1)} \equiv \prod (x - t) \tag{8.7.6}$$

$$= \prod{}' \{(x-t)(x-m+t)\} \equiv \prod{}' \{x^2 + t(m-t)\} \pmod{p^a}.$$

We compare the coefficients of x^2 on the two sides of (8.7.6). If $p > 3$, the coefficient on the left is 0, and

(8.7.7)

$$0 \equiv \prod{}' \{t(m-t)\} \sum{}' \frac{1}{t(m-t)} = \tfrac{1}{2} \prod t \sum \frac{1}{t(m-t)} \pmod{p^a}.$$

Hence

$$S_m \prod t = \prod t \sum \frac{1}{t} = \tfrac{1}{2} \prod t \sum \left(\frac{1}{t} + \frac{1}{m-t} \right)$$
$$= \tfrac{1}{2} m \prod t \sum \frac{1}{t(m-t)} \equiv 0 \pmod{p^{2a}},$$

or

(8.7.8) $$S_m \equiv 0 \pmod{p^{2a}}.$$

If $2 \nmid m, 3 \nmid m$, and we apply (8.7.8) to every prime factor of m, we obtain (8.7.1).

If $p = 3$, then (8.7.7) must be replaced by

$$(-1)^{\frac{1}{2}\phi(m)-1} \tfrac{1}{2}\phi(m) \equiv \tfrac{1}{2} \prod t \sum \frac{1}{t(m-t)} \pmod{3^a};$$

so that

$$S_m \prod t \equiv (-1)^{\frac{1}{2}\phi(m)-1} \tfrac{1}{2} m\phi(m) \pmod{3^{2a}}.$$

Since $\phi(m)$ is even, and divisible by 3^{a-1}, this gives

$$S_m \equiv 0 \pmod{3^{2a-1}}.$$

Hence we obtain (8.7.2).

If $p = 2$, then, by Theorem 127,

$$(x^2 - 1)^{\frac{1}{2}\phi(m)} \equiv \prod{}' \{x^2 + t(m-t)\} \pmod{2^a}$$

and so

$$(-1)^{\frac{1}{2}\phi(m)-1} \tfrac{1}{2}\phi(m) \equiv \tfrac{1}{2} \prod t \sum \frac{1}{t(m-t)},$$
$$S_m \prod t = \tfrac{1}{2} m \prod t \sum \frac{1}{t(m-t)} \equiv (-1)^{\frac{1}{2}\phi(m)-1} \tfrac{1}{2} m\phi(m) \pmod{2^{2a}}.$$

If $m = 2^a M$, where M is odd and greater than 1, then

$$\tfrac{1}{2}\phi(m) = 2^{a-2}\phi(M)$$

is divisible by 2^{a-1}, and

$$S_m \equiv 0 \pmod{2^{2a-1}}.$$

This, with the preceding results, gives (8.7.3) and (8.7.4).

Finally, if $m = 2^a$, $\frac{1}{2}\phi(m) = 2^{a-2}$, and

$$S_m \equiv 0 \pmod{2^{2a-2}}.$$

This is (8.7.5).

8.8. Further consequences of Bauer's theorem. (1) Suppose that

$$m > 2, \quad m = \prod p^a, \quad u_2 = \tfrac{1}{2}\phi(m), \quad u_p = \frac{\phi(m)}{p-1} \ (p > 2).$$

Then $\phi(m)$ is even and, when we equate the constant terms in (8.5.3) and (8.5.5), we obtain

$$\prod_{t(m)} t \equiv (-1)^{u_p} \pmod{p^a}.$$

It is easily verified that the numbers u_2 and u_p are all even, except when m is of one of the special forms 4, p^a, or $2p^a$; so that $\Pi t \equiv 1 \pmod m$ except in these cases. If $m = 4$, then $\Pi t = 1.3 \equiv -1 \pmod 4$. If m is p^a or $2p^a$, then u_p is odd, so that $\Pi t \equiv -1 \pmod{p^a}$ and therefore (since Πt is odd) $\Pi t = -1 \pmod m$.

THEOREM 129.

$$\prod_{t(m)} t \equiv \pm 1 \pmod m,$$

where the negative sign is to be chosen when m is 4, p^a, *or* $2p^a$, *where p is an odd prime, and the positive sign in all other cases.*

The case $m = p$ is Wilson's theorem.

(2) If $p > 2$ and

$$f(x) = \prod_{t(p^a)} (x - t) = x^{\phi(p^a)} - A_1 x^{\phi(p^a)-1} + \cdots,$$

then $f(x) = f(p^a - x)$. Hence

$$2A_1 x^{\phi(p^a)-1} + 2A_3 x^{\phi(p^a)-3} + \cdots = f(-x) - f(x) = f(p^a + x) - f(x)$$
$$\equiv p^a f'(x) \pmod{p^{2a}}.$$

But

$$p^a f'(x) \equiv p^{2a-1}(p-1)x^{p-2}(x^{p-1}-1)^{p^{a-1}-1} \pmod{p^{2a}}$$

by Theorem 126. It follows that $A_{2\nu+1}$ is a multiple of p^{2a} except when

$$\phi(p^a) - 2\nu - 1 \equiv p - 2 \pmod{p-1},$$

i.e. when

$$2\nu \equiv 0 \pmod{p-1}.$$

THEOREM 130. *If $A_{2\nu+1}$ is the sum of the homogeneous products, $2\nu+1$ at a time, of the numbers of $t(p^a)$, and 2ν is not a multiple of $p-1$, then*

$$A_{2\nu+1} \equiv 0 \pmod{p^{2a}}.$$

Wolstenholme's theorem is the case

$$a = 1, \quad 2\nu + 1 = p - 2, \quad p > 3.$$

(3) There are also interesting theorems concerning the sums

$$S_{2\nu+1} = \sum \frac{1}{t^{2\nu+1}}.$$

We confine ourselves for simplicity to the case $a = 1, m = p$,† and suppose $p > 2$. Then $f(x) = f(p-x)$ and

$$\begin{aligned} f(-x) &= f(p+x) \equiv f(x) + pf'(x), \\ f'(-x) &= -f'(p+x) \equiv -f'(x) - pf''(x), \\ f(x)f'(-x) + f'(x)f(-x) &\equiv p\{f'^2(x) - f(x)f''(x)\} \end{aligned}$$

to modulus p^2. Since $f(x) \equiv x^{p-1} - 1 \pmod{p}$,

$$f'^2(x) - f(x)f''(x) \equiv 2x^{p-3} - x^{2p-4} \pmod{p}$$

and so

$$f(x)f'(-x) + f'(x)f(-x) \equiv p(2x^{p-3} - x^{2p-4}) \pmod{p^2}. \tag{8.8.1}$$

† In this case Theorem 112 is sufficient for our purpose, and we do not require the general form of Bauer's theorem.

Now

$$\frac{f'(x)}{f(x)} = \sum \frac{1}{x-t} = -S_1 - xS_2 - x^2S_3 - \cdots .\dagger$$

$$\frac{f(x)f'(-x)+f(-x)f'(x)}{f(x)f(-x)} = -2S_1 - 2x^2S_3 - \cdots . \tag{8.8.2}$$

Also

$$f(x) = \prod (x-t) = \prod (t-x) = \varpi\left(1 + \frac{a_1x}{\varpi} + \frac{a_2x^2}{\varpi^2} + \cdots\right),$$

$$\frac{1}{f(x)} = \frac{1}{\varpi}\left(1 + \frac{b_1x}{\varpi} + \frac{b_2x^2}{\varpi^2} + \cdots\right),$$

$$\frac{1}{f(x)f(-x)} = \frac{1}{\varpi^2}\left(1 + \frac{c_1x^2}{\varpi^2} + \frac{c_2x^4}{\varpi^4} + \cdots\right), \tag{8.8.3}$$

where $\varpi = (p-1)!$ and the a, b, and c are integers. It follows from (8.8.1), (8.8.2), and (8.8.3) that

$$-2S_1 - 2x^2S_3 - \cdots = \frac{p(2x^{p-3} - x^{2p-4}) + p^2g(x)}{\varpi^2} \times \left(1 + \frac{c_1x^2}{\varpi^2} + \frac{c_2x^4}{\varpi^4} + \cdots\right),$$

where $g(x)$ is an integral polynomial. Hence, if $2v < p-3$, the numerator of S_{2v+1} is divisible by p^2.

THEOREM 131. *If p is prime, $2v < p-3$, and*

$$S_{2v+1} = 1 + \frac{1}{2^{2v+1}} + \cdots + \frac{1}{(p-1)^{2v+1}},$$

then the numerator of S_{2v+1} is divisible by p^2.

The case $v = 0$ is Wolstenholme's theorem. When $v = 1$, p must be greater than 5. The numerator of

$$1 + \frac{1}{2^3} + \frac{1}{3^3} + \frac{1}{4^3}$$

is divisible by 5 but not by 5^2.

There are many more elaborate theorems of the same character.

† The series which follow are ordinary power series in the variable x.

8.9. The residues of 2^{p-1} and $(p-1)!$ to modulus p^2. Fermat's and Wilson's theorems show that 2^{p-1} and $(p-1)!$ have the residues 1 and $-1 \pmod{p}$. Little is known about their residues $\pmod{p^2}$, but they can be transformed in interesting ways.

THEOREM 132. *If p is an odd prime, then*

$$\frac{2^{p-1}-1}{p} \equiv 1 + \frac{1}{3} + \frac{1}{5} + \cdots + \frac{1}{p-2} \pmod{p}. \tag{8.9.1}$$

In other words, the residue of $2^{p-1} \pmod{p^2}$ is

$$1 + p\left(\frac{1}{3} + \frac{1}{5} + \cdots + \frac{1}{p-2}\right),$$

where the fractions indicate associates $\pmod{p}$.

We have

$$2^p = (1+1)^p = 1 + \binom{p}{1} + \cdots + \binom{p}{p} = 2 + \sum_{1}^{p-1} \binom{p}{l}.$$

Every term on the right, except the first, is divisible by p,[†] and

$$\binom{p}{l} = px_l,$$

where

$$l!x_l = (p-1)(p-2)\ldots(p-l+1) \equiv (-1)^{l-1}(l-1)! \pmod{p},$$

or $lx_l \equiv (-1)^{l-1} \pmod{p}$. Hence

$$x_l \equiv (-1)^{l-1}\frac{1}{l} \pmod{p},$$

$$\binom{p}{l} = px_l \equiv (-1)^{l-1}p\frac{1}{l} \pmod{p^2},$$

$$\frac{2^{p-2}}{p} = \sum_{1}^{p-1} x_l \equiv 1 - \frac{1}{2} + \frac{1}{3} - \cdots - \frac{1}{p-1} \pmod{p}. \tag{8.9.2}$$

[†] By Theorem 75.

But

$$1-\frac{1}{2}+\frac{1}{3}-\cdots-\frac{1}{p-1}=2\left(1+\frac{1}{3}+\frac{1}{5}+\cdots+\frac{1}{p-2}\right)$$
$$-\left(1+\frac{1}{2}+\frac{1}{3}+\cdots+\frac{1}{p-1}\right)$$
$$\equiv 2\left(1+\frac{1}{3}+\cdots+\frac{1}{p-2}\right) \pmod{p},$$

by Theorem 116,[†] so that (8.9.2) is equivalent to (8.9.1).

Alternatively, after Theorem 116, the residue in (8.9.1) is

$$-\frac{1}{2}-\frac{1}{4}-\cdots-\frac{1}{p-1} \pmod{p}.$$

THEOREM 133. *If p is an odd prime, then*

$$(p-1)! \equiv (-1)^{\frac{1}{2}(p-1)}2^{2p-2}\left(\frac{p-1}{2}!\right)^2 \pmod{p^2}.$$

Let $p = 2n+1$. Then

$$\frac{(2n)!}{2^n n!} = 1.3\ldots(2n-1) = (p-2)(p-4)\ldots(p-2n),$$
$$(-1)^n\frac{(2n)!}{2^n n!} \equiv 2^n n! - 2^n n! p\left(\frac{1}{2}+\frac{1}{4}+\cdots+\frac{1}{2n}\right) \pmod{p^2}$$
$$\equiv 2^n n! + 2^n n!(2^{2n}-1) \pmod{p^2},$$

by Theorems 116 and 132; and

$$(2n)! \equiv (-1)^n 2^{4n}(n!)^2 \pmod{p^2}.$$

† We need only (7.8.2).

NOTES

§ 8.1. Theorem 121 (Gauss, *D.A.*, § 36) was known to the Chinese mathematician Sun-Tsu in the first century A.D. See Bachmann, *Niedere Zahlentheorie,* i. 83.

§ 8.5. Bauer, *Nouvelles annales* (4), 2 (1902), 256–64. Rear-Admiral C. R. Darlington suggested the method by which I deduce (8.5.3) from (8.5.4). This is much simpler than that used in earlier editions, which was given by Hardy and Wright, *Journal London Math. Soc.* 9 (1934), 38–41 and 240.

Dr. Wylie points out to us that (8.5.5) is equivalent to (8.5.3), with 2 for p, except when m is a power of 2, since it may easily be verified that

$$(x^2 - 1)^{\frac{1}{2}\phi(m)} \equiv (x-1)^{\phi(m)} \pmod{2^a}$$

when $m = 2^a M, M$ is odd, and $M > 1$.

§ 8.7. Leudesdorf, *Proc. London Math. Soc.* (1) 20 (1889), 199–212. See also S. Chowla, *Journal London Math. Soc.* 9 (1934), 246; N. Rama Rao, ibid. 12 (1937), 247–50; and E. Jacobstal, *Forhand. K. Norske Vidensk. Selskab,* 22 (1949), nos. 12, 13, 41.

§ 8.8. Theorem 129 (Gauss, *D.A.*, § 78) is sometimes called the 'generalized Wilson's theorem'.

Many theorems of the type of Theorems 130 and 131 will be found in Leudesdorf's paper quoted above, and in papers by Glaisher in vols. 31 and 32 of the *Quarterly Journal of Mathematics.*

§ 8.9. Theorem 132 is due to Eisenstein (1850). Full references to later proofs and generalizations will be found in Dickson, *History*, i, ch. iv. See also the note to § 6.6.

IX

THE REPRESENTATION OF NUMBERS BY DECIMALS

9.1. The decimal associated with a given number. There is a process for expressing any positive number ξ as a 'decimal' which is familiar in elementary arithmetic.

We write

$$\xi = [\xi] + x = X + x, \tag{9.1.1}$$

where X is an integer and $0 \leqslant x < 1$,[†] and consider X and x separately.

If $X > 0$ and

$$10^s \leqslant X < 10^{s+1},$$

and A_1 and X_1 are the quotient and remainder when X is divided by 10^s, then

$$X = A_1.10^s + X_1,$$

where

$$0 < A_1 = [10^{-s}X] < 10, \quad 0 \leqslant X_1 < 10^s.$$

Similarly

$$\begin{aligned} X_1 &= A_2.10^{s-1} + X_2 \quad (0 \leqslant A_2 < 10,\ 0 \leqslant X_2 < 10^{s-1}),\\ X_2 &= A_3.10^{s-2} + X_3 \quad (0 \leqslant A_3 < 10,\ 0 \leqslant X_3 < 10^{s-2}),\\ &\cdots\cdots\cdots\cdots\cdots\cdots\\ X_{s-1} &= A_s.10 + X_s \quad (0 \leqslant A_s < 10,\ 0 \leqslant X_s < 10),\\ X_s &= A_{s+1} \quad (0 \leqslant A_{s+1} < 10). \end{aligned}$$

Thus X may be expressed uniquely in the form

$$X = A_1.10^s + A_2.10^{s-1} + \cdots + A_s.10 + A_{s+1}, \tag{9.1.2}$$

where every A is one of $0, 1, 2, \ldots, 9$, and A_1 is not 0. We abbreviate this expression to

$$X = A_1 A_2 \ldots A_s A_{s+1}, \tag{9.1.3}$$

the ordinary representation of X in decimal notation.

† Thus $[\xi]$ has the same meaning as in § 6.11.

Passing to x, we write

$$x = f_1 \quad (0 \leqslant f_1 < 1).$$

We suppose that $a_1 = [10f_1]$, so that

$$\frac{a_1}{10} \leqslant f_1 < \frac{a_1 + 1}{10};$$

a_1 is one of $0, 1, \ldots, 9$, and

$$a_1 = [10f_1], \quad 10f_1 = a_1 + f_2 \quad (0 \leqslant f_2 < 1).$$

Similarly, we define $a_2, a_3, \ldots$ by

$$a_2 = [10f_2], \quad 10f_2 = a_2 + f_3 \quad (0 \leqslant f_3 < 1),$$
$$a_3 = [10f_3], \quad 10f_3 = a_3 + f_4 \quad (0 \leqslant f_4 < 1),$$
$$\cdots\cdots\cdots\cdots\cdots\cdots$$

Every a_n is one of $0, 1, 2, \ldots, 9$. Thus

(9.1.4) $$x = x_n + g_{n+1},$$

where

(9.1.5) $$x_n = \frac{a_1}{10} + \frac{a_2}{10^2} + \cdots + \frac{a_n}{10^n},$$

(9.1.6) $$0 \leqslant g_{n+1} = \frac{f_{n+1}}{10^n} < \frac{1}{10^n}.$$

We thus define a decimal

$$\cdot a_1 a_2 a_3 \ldots a_n \ldots$$

associated with x. We call $a_1, a_2, \ldots$ the first, second, ... *digits* of the decimal.

Since $a_n < 10$, the series

(9.1.7) $$\sum_1^{\infty} \frac{a_n}{10^n}$$

is convergent; and since $g_{n+1} \to 0$, its sum is x. We may therefore write

(9.1.8) $$x = \cdot a_1 a_2 a_3 \ldots,$$

the right-hand side being an abbreviation for the series (9.1.7).

If $f_{n+1} = 0$ for some n, i.e. if $10^n x$ is an integer, then

$$a_{n+1} = a_{n+2} = \ldots = 0.$$

In this case we say that the decimal *terminates*. Thus

$$\frac{17}{400} = \cdot 0425000\ldots,$$

and we write simply

$$\frac{17}{400} = \cdot 0425.$$

It is plain that the decimal for x will terminate if and only if x is a rational fraction whose denominator is of the form $2^\alpha 5^\beta$.

Since

$$\frac{a_{n+1}}{10^{n+1}} + \frac{a_{n+2}}{10^{n+2}} + \cdots = g_{n+1} < \frac{1}{10^n}$$

and

$$\frac{9}{10^{n+1}} + \frac{9}{10^{n+2}} + \cdots = \frac{9}{10^{n+1}\left(1 - \frac{1}{10}\right)} = \frac{1}{10^n},$$

it is impossible that every a_n from a certain point on should be 9. With this reservation, every possible sequence (a_n) will arise from some x. We define x as the sum of the series (9.1.7), and x_n and g_{n+1} as in (9.1.4) and (9.1.5). Then $g_{n+1} < 10^{-n}$ for every n, and x yields the sequence required.

Finally, if

$$\sum_1^\infty \frac{a_n}{10^n} = \sum_1^\infty \frac{b_n}{10^n}, \tag{9.1.9}$$

and the b_n satisfy the conditions already imposed on the a_n, then $a_n = b_n$ for every n. For if not, let a_N and b_N be the first pair which differ, so that $|a_N - b_N| \geqslant 1$. Then

$$\left|\sum_1^\infty \frac{a_n}{10^n} - \sum_1^\infty \frac{b_n}{10^n}\right| \geqslant \frac{1}{10^N} - \sum_{N+1}^\infty \frac{|a-b|}{10^n} \geqslant \frac{1}{10^N} - \sum_{N+1}^\infty \frac{9}{10^n} = 0.$$

This contradicts (9.1.9) unless there is equality. If there is equality, then all of $a_{N+1} - b_{N+1}, a_{N+2} - b_{N+2}, \ldots$ must have the same sign and the absolute value 9. But then either $a_n = 9$ and $b_n = 0$ for $n > N$, or else $a_n = 0$ and $b_n = 9$, and we have seen that each of these alternatives is impossible. Hence $a_n = b_n$ for all n. In other words, different decimals correspond to different numbers.

We now combine (9.1.1), (9.1.3), and (9.1.8) in the form

$$\xi = X + x = A_1A_2 \ldots A_{s+1} \cdot a_1a_2a_3 \ldots; \tag{9.1.10}$$

and we can sum up our conclusions as follows.

THEOREM 134. *Any positive number ξ may be expressed as a decimal*

$$A_1A_2 \ldots A_{s+1} \cdot a_1a_2a_3 \ldots,$$

where

$$0 \leqslant A_1 < 10,\ 0 \leqslant A_2 < 10, \ldots, 0 \leqslant a_n < 10,$$

not all A and a are 0, *and an infinity of the a_n are less than* 9. *If $\xi \geqslant 1$, then $A_1 > 0$. There is a* (1, 1) *correspondence between the numbers and the decimals, and*

$$\xi = A_1.10^s + \ldots + A_{s+1} + \frac{a_1}{10} + \frac{a_2}{10^2} + \cdots.$$

In what follows we shall usually suppose that $0 \leqslant \xi < 1$ so that $X = 0$, $\xi = x$. In this case all the A are 0. We shall sometimes save words by ignoring the distinction between the number x and the decimal which represents it, saying, for example, that the second digit of $\frac{17}{400}$ is 4.

9.2. Terminating and recurring decimals. A decimal which does not terminate may *recur*. Thus

$$\tfrac{1}{3} = \cdot 3333 \ldots, \quad \tfrac{1}{7} = \cdot 14285714285714 \ldots;$$

equations which we express more shortly as

$$\tfrac{1}{3} = \cdot\dot{3}, \quad \tfrac{1}{7} = \cdot\dot{1}4285\dot{7}.$$

These are *pure recurring* decimals in which the period reaches back to the beginning. On the other hand,

$$\tfrac{1}{6} = \cdot 1666 \ldots = \cdot 1\dot{6},$$

a *mixed recurring* decimal in which the period is preceded by one non-recurrent digit.

We now determine the conditions for termination or recurrence.

(1) If

$$x = \frac{p}{q} = \frac{p}{2^\alpha 5^\beta},$$

where $(p, q) = 1$, and

$$\mu = \max(\alpha, \beta), \tag{9.2.1}$$

then $10^n x$ is an integer for $n = \mu$ and for no smaller value of n, so that x terminates at a_μ. Conversely,

$$\frac{a_1}{10} + \frac{a_2}{10^2} + \cdots + \frac{a_\mu}{10^\mu} = \frac{P}{10^\mu} = \frac{p}{q},$$

where q has the prime factors 2 and 5 only.

(2) Suppose next that $x = p/q$, $(p, q) = 1$, and $(q, 10) = 1$, so that q is not divisible by 2 or 5. Our discussion of this case depends upon the theorems of Ch. VI.

By Theorem 88,

$$10^\nu \equiv 1 \pmod{q}$$

for some ν, the least such ν being a divisor of $\phi(q)$. We suppose that ν has this smallest possible value, i.e. that, in the language of § 6.8, 10 belongs to ν (mod q) or ν is the order of 10 (mod q). Then

$$10^\nu x = \frac{10^\nu p}{q} = \frac{(mq+1)p}{q} = mp + \frac{p}{q} = mp + x, \tag{9.2.2}$$

where m is an integer. But

$$10^\nu x = 10^\nu x_\nu + 10^\nu g_{\nu+1} = 10^\nu x_\nu + f_{\nu+1},$$

by (9.1.4). Since $0 < x < 1$, $f_{\nu+1} = x$, and the process by which the decimal was constructed repeats itself from $f_{\nu+1}$ onwards. Thus x is a pure recurring decimal with a period of at most ν figures.

On the other hand, a pure recurring decimal $\cdot\dot{a}_1 a_2 \ldots \dot{a}_\lambda$ is equal to

$$\left(\frac{a_1}{10} + \frac{a_2}{10^2} + \cdots + \frac{a_\lambda}{10^\lambda}\right)\left(1 + \frac{1}{10^\lambda} + \frac{1}{10^{2\lambda}} + \cdots\right)$$
$$= \frac{10^{\lambda-1}a_1 + 10^{\lambda-2}a_2 + \cdots + a_\lambda}{10^\lambda - 1} = \frac{p}{q},$$

when reduced to its lowest terms. Here $q | 10^\lambda - 1$, and so $\lambda \geqslant \nu$. It follows that if $(q, 10) = 1$, and the order of 10 (mod q) is ν, then x is a pure recurring decimal with a period of just ν digits; and conversely.

(3) Finally, suppose that

$$x = \frac{p}{q} = \frac{p}{2^\alpha 5^\beta Q}, \tag{9.2.3}$$

where $(p, q) = 1$ and $(Q, 10) = 1$; that μ is defined as in (9.2.1); and that ν is the order of 10 (mod Q). Then

$$10^\mu x = \frac{p'}{Q} = X + \frac{P}{Q},$$

where p', X, P are integers and

$$0 \leqslant X < 10^\mu, \quad 0 < P < Q, \quad (P, Q) = 1.$$

If $X > 0$ then $10^s \leqslant X < 10^{s+1}$, for some $s < \mu$, and $X = A_1 A_2 \ldots A_{s+1}$; and the decimal for P/Q is pure recurring and has a period of ν digits.

Hence

$$10^\mu x = A_1 A_2 \ldots A_{s+1} \cdot \dot{a}_1 a_2 \ldots \dot{a}_\nu$$

and

$$x = \cdot b_1 b_2 \ldots b_\mu \dot{a}_1 a_2 \ldots \dot{a}_\nu, \tag{9.2.4}$$

the last $s + 1$ of the b being $A_1, A_2, \ldots, A_{s+1}$ and the rest, if any, 0.

Conversely, it is plain that any decimal (9.2.4) represents a fraction (9.2.3). We have thus proved

THEOREM 135. *The decimal for a rational number p/q between* 0 *and* 1 *is terminating or recurring, and any terminating or recurring decimal is equal to a rational number. If $(p, q) = 1, q = 2^\alpha 5^\beta$, and* $\max(\alpha, \beta) = \mu$, *then the decimal terminates after μ digits. If $(p, q) = 1, q = 2^\alpha 5^\beta Q$, where*

$Q > 1, (Q, 10) = 1$, and ν is the order of 10 (mod *Q*), *then the decimal contains μ non-recurring and ν recurring digits.*

9.3. Representation of numbers in other scales. There is no reason except familiarity for our special choice of the number 10; we may replace 10 by 2 or by any greater number r. Thus

$$\frac{1}{8} = \frac{0}{2} + \frac{0}{2^2} + \frac{1}{2^3} = \cdot 001,$$

$$\frac{2}{3} = \frac{1}{2} + \frac{0}{2^2} + \frac{1}{2^3} + \frac{0}{2^4} + \ldots = \cdot\dot{1}\dot{0},$$

$$\frac{2}{3} = \frac{4}{7} + \frac{4}{7^2} + \frac{4}{7^3} + \ldots = \cdot\dot{4},$$

the first two decimals being 'binary' decimals or 'decimals in the scale of 2', the third a 'decimal in the scale of 7'.[†] Generally, we speak of 'decimals in the scale of r'.

The arguments of the preceding sections may be repeated with certain changes, which are obvious if r is a prime or a product of different primes (like 2 or 10), but require a little more consideration if r has square divisors (like 12 or 8). We confine ourselves for simplicity to the first case, when our arguments require only trivial alterations. In § 9.1, 10 must be replaced by r and 9 by $r - 1$. In § 9.2, the part of 2 and 5 is played by the prime divisors of r.

THEOREM 136. *Suppose that r is a prime or a product of different primes. Then any positive number ξ may be represented uniquely as a decimal in the scale of r. An infinity of the digits of the decimal are less than $r - 1$; with this reservation, the correspondence between the numbers and the decimals is* (1, 1).

Suppose further that

$$0 < x < 1, \quad x = \frac{p}{q}, \quad (p, q) = 1.$$

If

$$q = s^{\alpha} t^{\beta} \ldots u^{\gamma},$$

[†] We ignore the verbal contradiction involved in the use of 'decimal'; there is no other convenient word.

where $s, t, \ldots, u$ are the prime factors of r, and

$$\mu = \max(\alpha, \beta, \ldots, \gamma),$$

then the decimal for x terminates at the μth digit. If q is prime to r, and ν is the order of r (mod q), then the decimal is pure recurring and has a period of ν digits. If

$$q = s^{\alpha} t^{\beta} \ldots u^{\gamma} Q \quad (Q > 1),$$

Q is prime to r, and ν is the order of r (mod Q), then the decimal is mixed recurring, and has μ non-recurring and ν recurring digits.†

9.4. Irrationals defined by decimals. It follows from Theorem 136 that a decimal (in any scale‡) which neither terminates nor recurs must represent an irrational number. Thus

$$x = \cdot 0100100010\ldots$$

(the number of 0's increasing by 1 at each stage) is irrational. We consider some less obvious examples.

Theorem 137:

$$\cdot 011010100010\ldots,$$

where the digit a_n is 1 if n is prime and 0 otherwise, is irrational.

Theorem 4 shows that the decimal does not terminate. If it recurs, there is a function $An + B$ which is prime for all n from some point onwards; and Theorem 21 shows that this also is impossible.

This theorem is true in any scale. We state our next theorem for the scale of 10, leaving the modifications required for other scales to the reader.

Theorem 138.

$$\cdot 2357111317192329\ldots,$$

† Generally, when $r \equiv s^A t^B \ldots, u^C$, we must define μ as

$$\max\left(\frac{\alpha}{A}, \frac{\beta}{B}, \ldots, \frac{\gamma}{C}\right)$$

if this number is an integer, and otherwise as the first greater integer.

‡ Strictly, any 'quadratfrei' scale (scale whose base is a prime or a product of different primes). This is the only case actually covered by the theorems, but there is no difficulty in the extension.

where the sequence of digits is formed by the primes in ascending order, is irrational.

The proof of Theorem 138 is a little more difficult. We give two alternative proofs.

(1) Let us assume that *any arithmetical progression of the form*

$$k.10^{s+1}+1 \quad (k=1,2,3,\dots)$$

contains primes. Then there are primes whose expressions in the decimal system contain an arbitrary number s of 0's, followed by a 1. Since the decimal contains such sequences, it does not terminate or recur.

(2) Let us assume that *there is a prime between* N *and* $10N$ *for every* $N \geqslant 1$. Then, given s, there are primes with just s digits. If the decimal recurs, it is of the form

$$\dots|a_1a_2\dots a_k|a_1a_2\dots a_k|\dots, \tag{9.4.1}$$

the bars indicating the period, and the first being placed where the first period begins. We can choose $l>1$ so that all primes with $s=kl$ digits stand later in the decimal than the first bar. If p is the first such prime, then it must be of one of the forms

$$p=a_1a_2\dots a_k|a_1a_2\dots a_k|\dots|a_1a_2\dots a_k$$

or

$$p=a_{m+1}\dots a_k|a_1a_2\dots a_k|\dots|a_1a_2\dots a_k|a_1a_2\dots a_m$$

and is divisible by $a_1a_2\dots a_k$ or by $a_{m+1}\dots a_ka_1a_2\dots a_m$; a contradiction.

In our first proof we assumed a special case of Dirichlet's Theorem 15. This special case is easier to prove than the general theorem, but we shall not prove it in this book, so that (1) will remain incomplete. In (2) we assumed a result which follows at once from Theorem 418 (which we shall prove in Chapter XXII). The latter theorem asserts that, for every $N \geqslant 1$, there is at least one prime satisfying $N<p\leqslant 2N$. It follows, *a fortiori*, that $N<p<10N$.

9.5. Tests for divisibility. In this and the next few sections we shall be concerned for the most part with trivial but amusing puzzles.

There are not very many useful tests for the divisibility of an integer by particular integers such as $2,3,5,\dots$. A number is divisible by 2 if its last digit is even. More generally, it is divisible by 2^ν if and only if the number represented by its last ν digits is divisible by 2^ν. The reason, of course, is that $2^\nu | 10^\nu$; and there are similar rules for 5 and 5^ν.

Next

$$10^\nu \equiv 1 (\text{mod } 9)$$

for every ν, and therefore

$$A_1.10^s + A_2.10^{s-1} + \cdots + A_s.10 + A_{s+1}$$
$$\equiv A_1 + A_2 + \cdots + A_{s+1} \ (\text{mod } 9).$$

A fortiori this is true mod 3. Hence we obtain the well-known rule 'a number is divisible by 9 (or by 3) if and only if the sum of its digits is divisible by 9 (or by 3)'.

There is a rather similar rule for 11. Since $10 \equiv -1$ (mod 11), we have

$$10^{2r} \equiv 1,\ 10^{2r+1} \equiv -1 (\text{mod} 11),$$

so that

$$A_1.10^s + A_2.10^{s-1} + \cdots + A_s.10 + A_{s+1}$$
$$\equiv A_{s+1} - A_s + A_{s-1} - \cdots (\text{mod} 11).$$

A number is divisible by 11 if and only if the difference between the sums of its digits of odd and even ranks is divisible by 11.

We know of only one other rule of any practical use. This is a test for divisibility by any one of 7, 11, or 13, and depends on the fact that $7.11.13 = 1001$. Its working is best illustrated by an example: if 29310478561 is divisible by 7, 11 or 13, so is

$$561 - 478 + 310 - 29 = 364 = 4.7.13.$$

Hence the original number is divisible by 7 and by 13 but not by 11.

9.6. Decimals with the maximum period. We observe when learning elementary arithmetic that

$$\tfrac{1}{7} = \cdot\dot{1}4285\dot{7}, \quad \tfrac{2}{7} = \cdot\dot{2}8571\dot{4}, \quad \ldots, \quad \tfrac{6}{7} = \cdot\dot{8}5714\dot{2},$$

the digits in each of the periods differing only by a cyclic permutation.

Consider, more generally, the decimal for the reciprocal of a prime q. The number of digits in the period is the order of 10 (mod q), and is a divisor of $\phi(q) = q - 1$. If this order is $q - 1$, i.e. if 10 is a primitive root of q, then the period has $q - 1$ digits, the maximum number possible.

We convert $1/q$ into a decimal by dividing successive powers of 10 by q; thus

$$\frac{10^n}{q} = 10^n x_n + f_{n+1},$$

in the notation of § 9.1. The later stages of the process depend only upon the value of f_{n+1}, and the process recurs so soon as f_{n+1} repeats a value. If, as here, the period contains $q - 1$ digits, then the remainders

$$f_2, f_3, \ldots, f_q$$

must all be different, and must be a permutation of the fractions

$$\frac{1}{q}, \frac{2}{q}, \ldots, \frac{q-1}{q}.$$

The last remainder f_q is $1/q$.

The corresponding remainders when we convert p/q into a decimal are

$$pf_2, pf_3, \ldots, pf_q,$$

reduced (mod 1). These are, by Theorem 58, the same numbers in a different order, and the sequence of digits, after the occurrence of a particular remainder s/q, is the same as it was after the occurrence of s/q before. Hence the two decimals differ only by a cyclic permutation of the period.

What happens with 7 will happen with any q of which 10 is a primitive root. Very little is known about these q, but the q below 50 which satisfy the condition are

$$7, 17, 19, 23, 29, 47.$$

THEOREM 139. *If q is a prime, and 10 is a primitive root of q, then the decimals for*

$$\frac{p}{q} \, (p = 1, 2, \ldots, q-1)$$

have periods of length $q - 1$ and differing only by cyclic permutation.

9.7. Bachet's problem of the weights. What is the least number of weights which will weigh any integral number of pounds up to 40 (*a*) when weights may be put into one pan only and (*b*) when weights may be put into either pan?

The second problem is the more interesting. We can dispose of the first by proving

THEOREM 140. *Weights* $1, 2, 4, \dots, 2^{n-1}$ *will weigh any integral weight up to* $2^n - 1$; *and no other set of so few as n weights is equally effective* (*i.e. will weigh so long an unbroken sequence of weights from* 1).

Any positive integer up to $2^n - 1$ inclusive can be expressed uniquely as a binary decimal of n figures, i.e. as a sum

$$\sum_{0}^{n-1} a_s 2^s,$$

where every a_s is 0 or 1. Hence our weights will do what is wanted, and 'without waste' (no two arrangements of them producing the same result). Since there is no waste, no other selection of weights can weigh a longer sequence.

Finally, one weight must be 1 (to weigh 1); one must be 2 (to weigh 2); one must be 4 (to weigh 4); and so on. Hence $1, 2, 4, \dots, 2^{n-1}$ is the only system of weights which will do what is wanted.

It is to be observed that Bachet's number 40, not being of the form $2^n - 1$, is not chosen appropriately for this problem. The weights 1, 2, 4, 8, 16, 32 will weigh up to 63, and no combination of 5 weights will weigh beyond 31. But the solution for 40 is not unique; the weights 1, 2, 4, 8, 9, 16 will also weigh any weight up to 40.

Passing to the second problem, we prove

THEOREM 141. *Weights* 1, 3, $3^2, \dots, 3^{n-1}$ *will weigh any weight up to* $\frac{1}{2}(3^n - 1)$, *when weights may be placed in either pan; and no other set of so few as n weights is equally effective.*

(1) Any positive integer up to $3^n - 1$ inclusive can be expressed uniquely by n digits in the ternary scale, i.e. as a sum

$$\sum_{0}^{n-1} a_s 3^s,$$

where every a_s is 0, 1, or 2. Subtracting

$$1 + 3 + 3^2 + \cdots + 3^{n-1} = \tfrac{1}{2}(3^n - 1),$$

we see that every positive or negative integer between $-\frac{1}{2}(3^n - 1)$ and $\frac{1}{2}(3^n - 1)$ inclusive can be expressed uniquely in the form

$$\sum_0^{n-1} b_s 3^s,$$

where every b_s is -1, 0, or 1. Hence our weights, placed in either pan, will weigh any weight between these limits.† Since there is no waste, no other combination of n weights can weigh a longer sequence.

(2) The proof that no other combination will weigh so long a sequence is a little more troublesome. It is plain, since there must be no waste, that the weights must all differ. We suppose that they are

$$w_1 < w_2 < \cdots < w_n.$$

The two largest weighable weights are plainly

$$W = w_1 + w_2 + \cdots + w_n, \quad W_1 = w_2 + \cdots + w_n.$$

Since $W_1 = W - 1$, w_1 must be 1.

The next weighable weight is

$$-w_1 + w_2 + w_3 + \cdots + w_n = W - 2,$$

and the next must be

$$w_1 + w_3 + w_4 + \cdots + w_n.$$

Hence $w_1 + w_3 + \cdots + w_n = W - 3$ and $w_2 = 3$.

† Counting the weight to be weighed positive if it is placed in one pan and negative if it is placed in the other.

Suppose now that we have proved that

$$w_1 = 1, w_2 = 3, \ldots, w_s = 3^{s-1}.$$

If we can prove that $w_{s+1} = 3^s$, the conclusion will follow by induction.

The largest weighable weight W is

$$W = \sum_{1}^{s} w_t + \sum_{s+1}^{n} w_t.$$

Leaving the weights $w_{s+1}, \ldots, w_n$ undisturbed, and removing some of the other weights, or transferring them to the other pan, we can weigh every weight down to

$$-\sum_{1}^{s} w_t + \sum_{s+1}^{n} w_t = W - (3^s - 1),$$

but none below. The next weight less than this is $W - 3^s$, and this must be

$$w_1 + w_2 + \cdots + w_s + w_{s+2} + w_{s+3} + \cdots + w_n.$$

Hence

$$w_{s+1} = 2(w_1 + w_2 + \cdots + w_s) + 1 = 3^s,$$

the conclusion required.

Bachet's problem corresponds to the case $n = 4$.

9.8. The game of Nim. The game of Nim is played as follows. Any number of matches are arranged in heaps, the number of heaps, and the number of matches in each heap, being arbitrary. There are two players, *A* and *B*. The first player *A* takes any number of matches from a heap; he may take one only, or any number up to the whole of the heap, but he must touch one heap only. *B* then makes a move conditioned similarly, and the players continue to take alternately. The player who takes the last match wins the game.

The game has a precise mathematical theory, and one or other player can always force a win.

We define a *winning* position as a position such that if one player *P* (*A* or *B*) can secure it by his move, leaving his opponent *Q* (*B* or *A*) to move next, then, whatever *Q* may do, *P* can play so as to win the game. Any other position we call a *losing* position.

For example, the position

$$\cdot\,\cdot \mid \cdot\,\cdot\,,$$

or (2, 2), is a winning position. If A leaves this position to B, B must take one match from a heap, or two. If B takes two, A takes the remaining two. If B takes one, A takes one from the other heap; and in either case A wins. Similarly, as the reader will easily verify,

$$\cdot \mid \cdot\,\cdot \mid \cdot\,\cdot\,\cdot\,,$$

or (1, 2, 3), is a winning position.

We next define a *correct* position. We express the number of matches in each heap in the binary scale, and form a figure F by writing them down one under the other. Thus (2, 2), (1, 2, 3), and (2, 3, 6, 7) give the figures

$$\begin{array}{cccl} 10 & 01 & 010 & ; \\ 10 & 10 & 011 & \\ \text{---} & 11 & 110 & \\ 20 & \text{---} & 111 & \\ & 22 & \text{---} & \\ & & 242 & \end{array}$$

it is convenient to write 01, 010, ... for 1, 10, ... so as to equalize the number of figures in each row. We then add up the columns, as indicated in the figures. If the sum of each column is even (as in the cases shown) then the position is 'correct'. An *incorrect* position is one which is not correct: thus (1, 3, 4) is incorrect.

THEOREM 142. *A position in Nim is a winning position if and only if it is correct.*

(1) Consider first the special case in which no heap contains more than one match. It is plain that the position is winning if the number of matches left is even, and losing if it is odd; and that the same conditions define correct and incorrect positions.

(2) Suppose that P has to take from a correct position. He must replace one number defining a row of F by a smaller number. If we replace any number, expressed in the binary scale, by a smaller number, we change the parity of at least one of its digits. Hence *when P takes from a correct position, he necessarily transforms it into an incorrect position.*

(3) If a position is incorrect, then the sum of at least one column of F is odd. Suppose, to fix our ideas, that the sums of the columns are

$$\textit{even, even, odd, even, odd, even.}$$

Then there is at least one 1 in the third column (the first with an odd sum). Suppose (again to fix our ideas) that one row in which this happens is

$$0\,1\,\overset{*}{1}\,1\,\overset{*}{0}\,1,$$

the asterisks indicating that the numbers below them are in columns whose sum is odd. We can replace this number by the smaller number

$$0\,1\,\overset{*}{0}\,1\,\overset{*}{1}\,0,$$

in which the digits with an asterisk, and those only, are altered. Plainly this change corresponds to a possible move, and makes the sum of every column even; and the argument is general. Hence P, *if presented with an incorrect position, can always convert it into a correct position.*

(4) If A leaves a correct position, B is compelled to convert it into an incorrect position, and A can then move so as to restore a correct position. This process will continue until every heap is exhausted or contains one match only. The theorem is thus reduced to the special case already proved.

The issue of the game is now clear. In general, the original position will be incorrect, and the first player wins if he plays properly. But he loses if the original position happens to be correct and the second player plays properly.†

† When playing against an opponent who does not know the theory of the game, there is no need to play strictly according to rule. The experienced player can play at random until he recognizes a winning position of a comparatively simple type. It is quite enough to know that

$$1, 2n, 2n+1, \qquad n, 7-n, 7, \qquad 2, 3, 4, 5$$

are winning positions; that

$$1, 2n+1, 2n+2$$

is a losing position; and that a combination of two winning positions is a winning position. The winning move is not always unique. The position

$$1, 3, 9, 27$$

is incorrect, and the only move which makes it correct is to take 16 from the 27. The position

$$3, 5, 7, 8, 11$$

is also incorrect, but may be made correct by taking 2 from the 3, the 7, or the 11.

There is a variation in which the player who takes the last match *loses*. The theory is the same so long as a heap remains containing more than one match; thus (2, 2) and (1, 2, 3) are still winning positions. We leave it to the reader to think out for himself the small variations in tactics at the end of the game.

9.9. Integers with missing digits. There is a familiar paradox[†] concerning integers from whose expression in the decimal scale some particular digit such as 9 is missing. It might seem at first as if this restriction should only exclude 'about one-tenth' of the integers, but this is far from the truth.

THEOREM 143. *Almost all numbers*[‡] *contain a* 9, *or any given sequence of digits such as* 937. *More generally, almost all numbers, when expressed in any scale, contain every possible digit, or possible sequence of digits.*

Suppose that the scale is r, and that v is a number whose decimal misses the digit b. The number of v for which $r^{l-1} \leqslant v < r^l$ is $(r-1)^l$ if $b = 0$ and $(r-2)(r-1)^{l-1}$ if $b \neq 0$, and in any case does not exceed $(r-1)^l$. Hence, if

$$r^{k-1} \leqslant n < r^k,$$

the number $N(n)$ of v up to n does not exceed

$$r - 1 + (r-1)^2 + \cdots + (r-1)^k \leqslant k(r-1)^k;$$

and

$$\frac{N(n)}{n} \leqslant k\frac{(r-1)^k}{r^{k-1}} \leqslant kr\left(\frac{r-1}{r}\right)^k,$$

which tends to 0 when $n \to \infty$.

The statements about sequences of digits need no additional proof, since, for example, the sequence 937 in the scale of 10 may be regarded as a single digit in the scale of 1000.

The 'paradox' is usually stated in a slightly stronger form, viz.

THEOREM 144. *The sum of the reciprocals of the numbers which miss a given digit is convergent.*

[†] Relevant in controversies about telephone directories.

[‡] In the sense of § 1.6.

The number of v between r^{k-1} and r^k is at most $(r-1)^k$. Hence

$$\sum \frac{1}{v} = \sum_{k=1}^{\infty} \sum_{r^{k-1} \leqslant v < r^k} \frac{1}{v}$$
$$< \sum_{k=1}^{\infty} \frac{(r-1)^k}{r^{k-1}} = (r-1) \sum_{k=1}^{\infty} \left(\frac{r-1}{r}\right)^{k-1} = r(r-1).$$

We shall discuss next some analogous, but more interesting, properties of infinite decimals. We require a few elementary notions concerning the measure of point-sets or sets of real numbers.

9.10. Sets of measure zero. A real number x defines a 'point' of the continuum. In what follows we use the words 'number' and 'point' indifferently, saying, for example, that 'P is the point x'.

An aggregate of real numbers is called a *set of points*. Thus the set T defined by

$$x = \frac{1}{n} \quad (n = 1, 2, 3, \ldots),$$

the set R of all rationals between 0 and 1 inclusive, and the set C of all real numbers between 0 and 1 inclusive, are sets of points.

An interval $(x - \delta, x + \delta)$, where δ is positive, is called a *neighbourhood* of x. If S is a set of points, and every neighbourhood of x includes an infinity of points of S, then x is called a *limit point* of S. The limit point may or may not belong to S, but there are points of S as near to it as we please. Thus T has one limit point, $x = 0$, which does not belong to T. Every x between 0 and 1 is a limit point of R.

The set S' of limit points of S is called the derived set or *derivative* of S. Thus C is the derivative of R. If S includes S', i.e. if every limit point of S belongs to S, then S is said to be *closed*. Thus C is closed. If S' includes S, i.e., if every point of S is a limit point of S, then S is said to be *dense in itself*. If S and S' are identical (so that S is both closed and dense in itself), then S is said to be *perfect*. Thus C is perfect. A less trivial example will be found in § 9.11.

A set S is said to be *dense in an interval* (a, b) if every point of (a, b) belongs to S'. Thus R is dense in (0, 1).

If S can be included in a set J of intervals, finite or infinite in number, whose total length is as small as we please, then S is said to be *of measure zero*. Thus T is of measure zero. We include the point $1/n$ in the interval

$$\frac{1}{n} - 2^{-n-1}\delta, \quad \frac{1}{n} + 2^{-n-1}\delta$$

of length $2^{-n}\delta$, and the sum of all these intervals (without allowance for possible overlapping) is

$$\delta \sum_{1}^{\infty} 2^{-n} = \delta,$$

which we may suppose as small as we please.

Generally, *any enumerable set is of measure zero*. A set is *enumerable* if its members can be correlated, as

(9.10.1) $$x_1, x_2, \ldots, x_n, \ldots,$$

with the integers $1, 2, \ldots, n, \ldots$. We include x_n in an interval of length $2^{-n}\delta$, and the conclusion follows as in the special case of T.

A subset of an enumerable set is finite or enumerable. The sum of an enumerable set of enumerable sets is enumerable.

The rationals may be arranged as

$$\frac{0}{1}, \frac{1}{1}, \frac{1}{2}, \frac{1}{3}, \frac{2}{3}, \frac{1}{4}, \frac{3}{4}, \frac{1}{5}, \frac{2}{5}, \frac{3}{5}, \ldots$$

and so in the form (9.10.1). Hence R is enumerable, and therefore of measure zero. A set of measure zero is sometimes called a *null* set; thus R is null. Null sets are negligible for many mathematical purposes, particularly in the theory of integration.

The sum S of an enumerable infinity of null sets S_n (i.e. the set formed by all the points which belong to some S_n) is null. For we may include S_n in a set of intervals of total length $2^{-n}\delta$, and so S in a set of intervals of total length not greater than $\delta \sum 2^{-n} = \delta$.

Finally, we say that *almost all* points of an interval I possess a property if the set of points which do not possess the property is null. This sense of the phrase should be compared with the sense defined in § 1.6 and used in § 9.9. It implies in either case that 'most' of the numbers under consideration (the positive integers in §§ 1.6 and 9.9, the real numbers here) possess the property, and that other numbers are 'exceptional'.†

† Our explanations here contain the minimum necessary for the understanding of §§ 9.11–13 and a few later passages in the book. In particular, we have not given any general definition of the measure of a set. There are fuller accounts of all these ideas in the standard treatises on analysis.

9.11. Decimals with missing digits. The decimal

$$\tfrac{1}{7} = \cdot\dot{1}4285\dot{7}$$

has four missing digits, viz. 0, 3, 6, 9. But it is easy to prove that decimals which miss digits are exceptional.

We define S as the set of points between 0 (inclusive) and 1 (exclusive) whose decimals, in the scale of r, miss the digit b. This set may be generated as follows.

We divide (0, 1) into r equal parts

$$\frac{s}{r} \leqslant x < \frac{s+1}{r} \qquad (s = 0, 1, \ldots, r-1);$$

the left-hand end point, but not the right-hand one, is included. The sth part contains just the numbers whose decimals begin with $s - 1$, and if we remove the $(b + 1)$th part, we reject the numbers whose first digit is b.

We next divide each of the $r - 1$ remaining intervals into r equal parts and remove the $(b + 1)$th part of each of them. We have then rejected all numbers whose first or second digit is b. Repeating the process indefinitely, we reject all numbers in which any digit is b; and S is the set which remains.

In the first stage of the construction we remove one interval of length $1/r$; in the second, $r - 1$ intervals of length $1/r^2$, i.e. of total length $(r - 1)/r^2$; in the third, $(r - 1)^2$ intervals of total length $(r - 1)^2/r^3$; and so on. What remains after k stages is a set J_k of intervals whose total length is

$$1 - \sum_{l=1}^{k} \frac{(r-1)^{l-1}}{r^l},$$

and this set includes S for every k. Since

$$1 - \sum_{l=1}^{k} \frac{(r-1)^{l-1}}{r^l} \to 1 - \left\{\frac{1}{r} \Big/ \left(1 - \frac{r-1}{r}\right)\right\} = 0$$

when $k \to \infty$, the total length of J_k is small when k is large; and S is therefore null.

THEOREM 145. *The set of points whose decimals, in any scale, miss any digit is null: almost all decimals contain all possible digits.*

The result may be extended to cover combinations of digits. If the sequence 937 never occurs in the ordinary decimal for x, then the digit '937' never occurs in the decimal in the scale of 1000. Hence

THEOREM 146. *Almost all decimals, in any scale, contain all possible sequences of any number of digits.*

Returning to Theorem 145, suppose that $r = 3$ and $b = 1$. The set S is formed by rejecting the middle third $\left(\frac{1}{3}, \frac{2}{3}\right)$ of $(0, 1)$, then the middle thirds $\left(\frac{1}{9}, \frac{2}{9}\right)$, $\left(\frac{7}{9}, \frac{8}{9}\right)$ of $\left(0, \frac{1}{3}\right)$, and $\left(\frac{2}{3}, 1\right)$ and so on. The set which remains is null.

It is immaterial for this conclusion whether we reject or retain the *end points* of rejected intervals, since their aggregate is enumerable and therefore null. In fact our definition rejects some, such as $1/3 = \cdot 1$, and includes others, such as $2/3 = \cdot 2$.

The set becomes more interesting if we retain all end points. In this case (if we wish to preserve the arithmetical definition) we must allow ternary decimals ending in 2 (and excluded in our account of decimals at the beginning of the chapter). All fractions $p/3^n$ have then two representations, such as

$$\tfrac{1}{3} = \cdot 1 = \cdot 0\dot{2}$$

(and it was for this reason that we made the restriction); and an end point of a rejected interval has always one without a 1.

The set S thus defined is called *Cantor's ternary set.*

Suppose that x is any point of $(0, 1)$, except 0 or 1. If x does not belong to S, it lies inside a rejected interval, and has neighbourhoods free from points of S, so that it does not belong to S'. If x does belong to S, then all its neighbourhoods contain other points of S; for otherwise there would be one containing x only, and two rejected intervals would abut. Hence x belongs to S'. Thus S and S' are identical, and x is perfect.

THEOREM 147. *Cantor's ternary set is a perfect set of measure zero.*

9.12. Normal numbers. The theorems proved in the last section express much less than the full truth. Actually it is true, for example, not only that almost all decimals contain a 9, but that, in almost all decimals, 9 occurs with the proper frequency, that is to say in about one-tenth of the possible places.

Suppose that x is expressed in the scale of r, and that the digit b occurs n_b times in the first n places. If

$$\frac{n_b}{n} \to \beta$$

when $n \to \infty$, then we say that b has *frequency* β. It is naturally not necessary that such a limit should exist; n_b/n may oscillate, and one might expect that usually it would. The theorems which follow prove that, contrary to our expectation, there is usually a definite frequency. The existence of the limit is in a sense the ordinary event.

We say that x is *simply normal* in the scale of r if

$$\frac{n_b}{n} \to \frac{1}{r} \tag{9.12.1}$$

for each of the r possible values of b. Thus

$$x = \cdot\dot{0}12345678\dot{9}$$

is simply normal in the scale of 10. The same x may be expressed in the scale of 10^{10}, when its expression is

$$x = \cdot\dot{b},$$

where $b = 123456789$. It is plain that in this scale x is not simply normal, $10^{10} - 1$ digits being missing.

This remark leads us to a more exacting definition. We say that x is *normal* in the scale of r if all of the numbers

$$x, rx, r^2x, \ldots\dagger$$

are simply normal in all of the scales

$$r, r^2, r^3, \ldots.$$

It follows at once that, when x is expressed in the scale of r, every combination

$$b_1 b_2 \ldots b_k$$

† Strictly, the fractional parts of these numbers (since we have been considering numbers between 0 and 1). A number greater than 1 is simply normal, or normal, if its fractional part is simply normal, or normal.

of digits occurs with the proper frequency; i.e. that, if n_b is the number of occurrences of this sequence in the first n digits of x, then

$$\frac{n_b}{n} \to \frac{1}{r^k} \tag{9.12.2}$$

when $n \to \infty$.

Our main theorem, which includes and goes beyond those of § 9.11, is

THEOREM 148. *Almost all numbers are normal in any scale.*

9.13. Proof that almost all numbers are normal. It is sufficient to prove that almost all numbers are *simply* normal in a given scale. For suppose that this has been proved, and that $S(x, r)$ is the set of numbers x which are not simply normal in the scale of r. Then $S(x, r)$, $S(x, r^2)$, $S(x, r^3), \ldots$ are null, and therefore their sum is null. Hence the set $T(x, r)$ of numbers which are not simply normal in all the scales $r, r^2, \ldots$ is null. The set $T(rx, r)$ of numbers such that rx is not simply normal in all these scales is also null; and so are $T(r^2x, r)$, $T(r^3x, r), \ldots$. Hence again the sum of these sets, i.e. the set $U(x, r)$ of numbers which are not normal in the scale of r, is null. Finally, the sum of $U(x, 2)$, $U(x, 3), \ldots$ is null; and this proves the theorem.

We have therefore only to prove that (9.12.1) is true for almost all numbers x. We may suppose that n tends to infinity through multiples of r, since (9.12.1) is true generally if it is true for n so restricted.

The numbers of r-ary decimals of n figures, with just m b's in assigned places, is $(r-1)^{n-m}$. Hence the number of such decimals which contain just m b's, in one place or another, is[†]

$$p(n, m) = \frac{n!}{m!(n-m)!}(r-1)^{n-m}.$$

We consider any decimal, and the incidence of b's among its first n digits, and call

$$\mu = m - \frac{n}{r} = m - n^*$$

[†] $p(n, m)$ is the term in $(r-1)^{n-m}$ in the binomial expansion of

$$\{1 + (r-1)\}^n.$$

the *n-excess* of b (the excess of the actual number of b's over the number to be expected). Since n is a multiple of r, n^* and μ are integers. Also

$$-\frac{1}{r} \leqslant \frac{\mu}{n} \leqslant 1 - \frac{1}{r}. \tag{9.13.1}$$

We have

$$\frac{p(n,\, m+1)}{p(n,\, m)} = \frac{n-m}{(r-1)(m+1)} = \frac{(r-1)n - r\mu}{(r-1)n + r(r-1)(\mu+1)}. \tag{9.13.2}$$

Hence

$$\frac{p(n,\, m+1)}{p(n,\, m)} > 1 \;\; (\mu = -1, -2, \ldots), \qquad \frac{p(n, m+1)}{p(n, m)} < 1 \;\; (\mu = 0, 1, 2, \ldots);$$

so that $p(n, m)$ is greatest when

$$\mu = 0, \quad m = n^*.$$

If $\mu \geqslant 0$, then, by (9.13.2)

$$\frac{p(n,\, m+1)}{p(n,\, m)} = \frac{(r-1)n - r\mu}{(r-1)n + r(r-1)(\mu+1)} < 1 - \frac{r}{r-1}\frac{\mu}{n} \leqslant \exp\left(-\frac{r}{r-1}\frac{\mu}{n}\right). \tag{9.13.3}$$

If $\mu < 0$ and $v = |\mu|$, then

$$\frac{p(n,\, m+1)}{p(n,\, m)} = \frac{(r-1)m}{n-m+1} = \frac{(r-1)n - r(r-1)v}{(r-1)n + r(v+1)} < 1 - \frac{rv}{n} < \exp\left(-\frac{rv}{n}\right) = \exp\left(-\frac{r|\mu|}{n}\right). \tag{9.13.4}$$

We now fix a positive δ, and consider the decimals for which

$$|\mu| \geqslant \delta n \tag{9.13.5}$$

for a given n. Since n is to be large, we may suppose that $|\mu| \geqslant 2$. If μ is positive then, by (9.13.3),

$$\frac{p(n, m)}{p(n, m-\mu)} = \frac{p(n, m)}{p(n, m-1)} \frac{p(n, m-1)}{p(n, m-2)} \cdots \frac{p(n, m-\mu+1)}{p(n, m-\mu)}$$
$$< \exp\left\{-\frac{r}{r-1}\frac{(\mu-1)+(\mu-2)+\cdots+1}{n}\right\}$$
$$= \exp\left\{-\frac{r(\mu-1)\mu}{2(r-1)n}\right\} < e^{-K\mu^2/n},$$

where K is a positive number which depends only on r. Since

$$p(n, m-\mu) = p(n, n^*) < r^n,\dagger$$

it follows that

(9.13.6) $$p(n, m) < r^n e^{-K\mu^2/n}.$$

Similarly it follows from (9.13.4) that (9.13.6) is true also for negative μ.

Let $S_n(\mu)$ be the set of numbers whose n-excess is μ. There are $p = p(n, m)$ numbers $\xi_1, \xi_2, \ldots, \xi_p$ represented by terminating decimals of n figures and excess μ, and the numbers of $S_n(\mu)$ are included in the intervals

$$\xi_s, \xi_s + r^{-n} \quad (s = 1, 2, \ldots, p).$$

Hence $S_n(\mu)$ is included in a set of intervals whose total length does not exceed

$$r^{-n}p(n, m) < e^{-K\mu^2/n}.$$

And if $T_n(\delta)$ is the set of numbers whose n-excess satisfies (9.13.5), then $T_n(\delta)$ can be included in a set of intervals whose length does not exceed

$$\sum_{|\mu|\geqslant\delta n} e^{-K\mu^2/n} = 2\sum_{\mu\geqslant\delta n} e^{-K\mu^2/n} \leqslant 2\sum_{\mu\geqslant\delta n} e^{-\frac{1}{2}K\mu^2/n} e^{-\frac{1}{2}K\mu/n}$$
$$\leqslant 2e^{-\frac{1}{2}K\delta^2 n}\sum_{\mu=0}^{\infty} e^{-\frac{1}{2}K\mu/n} = \frac{2e^{-\frac{1}{2}K\delta^2 n}}{1-e^{-\frac{1}{2}K/n}} < Lne^{-\frac{1}{2}K\delta^2 n},$$

where L, like K, depends only on r.

† Indeed $p(n, m) < r^n$ for all m.

We now fix N (a multiple N^*r of r), and consider the set $U_N(\delta)$ of numbers such that (9.13.5) is true for some

$$n = n^*r \geqslant N = N^*r.$$

Then $U_N(\delta)$ is the sum of the sets

$$T_N(\delta), T_{N+r}(\delta), T_{N+2r}(\delta), \ldots,$$

i.e. the sets $T_n(\delta)$ for which $n = kr$ and $k \geqslant N^*$. It can therefore be included in a set of intervals whose length does not exceed

$$L \sum_{k=N^*}^{\infty} kre^{-\frac{1}{2}K\delta^2 kr} = \eta(N^*);$$

and $\eta(N^*) \to 0$ when n^* and N^* tend to infinity.

If $U(\delta)$ is the set of numbers whose n-excess satisfies (9.13.5) for *an infinity* of n (all multiples of r), then $U(\delta)$ is included in $U_N(\delta)$ for every N, and can therefore be included in a set of intervals whose total length is as small as we please. That is to say, $U(\delta)$ is null.

Finally, if x is not simply normal, (9.12.1) is false (even when n is restricted to be a multiple of r), and

$$|\mu| \geqslant \zeta n$$

for some positive ζ and an infinity of multiples n of r. This ζ is greater than some one of the sequence δ, $\frac{1}{2}\delta$, $\frac{1}{4}\delta, \ldots$, and so x belongs to some one of the sets

$$U(\delta),\ U(\tfrac{1}{2}\delta),\ U\left(\tfrac{1}{4}\delta\right), \ldots,$$

all of which are null. Hence the set of all such x is null.

It might be supposed that, since almost all numbers are normal, it would be easy to construct examples of normal numbers. There are in fact simple constructions; thus the number

$$\cdot 123456789101112\ldots,$$

formed by writing down all the positive integers in order, in decimal notation, is normal. But the proof that this is so is more troublesome than might be expected.

NOTES

§ 9.4. For Theorem 138 see Pólya and Szegő, No. 257. The result is stated without proof in W. H. and G. C. Youngs' *The theory of sets of points*, 3.

§ 9.5. See Dickson, *History*, i, ch. xii. The test for 7, 11, and 13 is not mentioned explicitly. It is explained by Grunert, *Archiv der Math. und Phys.* 42 (1864), 478–82. Grunert gives slightly earlier references to Brilka and V. A. Lebesgue.

§§ 9.7–8. See Ahrens, ch. iii.

There is an interesting logical point involved in the definition of a 'losing' position in Nim. We define a losing position as one which is not a winning position, i.e. as a position such that P cannot force a win by leaving it to Q. It follows from our analysis of the game that a losing position in this sense is also a losing position in the sense that Q can force a win if P leaves such a position to Q. This is a case of a general theorem (due to Zermelo and von Neumann) true of any game in which there are only two possible results and only a finite choice of 'moves' at any stage. See D. König, *Acta Univ. Hungaricae* (Szeged), 3 (1927), 121–30.

§ 9.10. Our 'limit point' is the 'limiting point' of Hobson's *Theory of functions of a real variable* or the 'Häufungspunkt' of Hausdorff's *Mengenlehre*.

§§ 9.12–13. Niven and Zuckerman (*Pacific Journal of Math.* 1 (1951), 103–9) and Cassels (ibid. 2 (1952), 555–7) give proofs that, if (9.12.2) holds for every sequence of digits, then x is normal. This is the converse of our statement that (9.12.2) follows from the definition; the proof of this converse is not trivial.

For the substance of these sections see Borel, *Leçons sur la théorie des fonctions* (2nd ed., 1914), 182–216. Theorem 148 has been developed in various ways since it was originally proved by Borel in 1909. For an account and bibliography, see Kuipers and Niederreiter, 69–78.

Champernowne (*Journal London Math. Soc.* 8 (1933), 254–60) proved that $\cdot 123\ldots$ is normal. Copeland and Erdős (*Bulletin Amer. Math. Soc.* 52 (1946), 857–60) proved that, if $a_1, a_2, \ldots$ is any increasing sequence of integers such that $a_n < n^{1+\epsilon}$ for every $\epsilon > 0$ and $n > n_0(\epsilon)$, then the decimal

$$\cdot a_1 a_2 a_3 \ldots$$

(formed by writing out the digits of the a_n in any scale in order) is normal in that scale.

X

CONTINUED FRACTIONS

10.1. Finite continued fractions. We shall describe the function

$$a_0 + \cfrac{1}{a_1 + \cfrac{1}{a_2 + \cfrac{1}{a_3 + \cdots \genfrac{}{}{0pt}{}{}{+\frac{1}{a_N}}}}} \tag{10.1.1}$$

of the $N+1$ variables

$$a_0, a_1, \ldots, a_n, \ldots, a_N,$$

as a *finite continued fraction,* or, when there is no risk of ambiguity, simply as a *continued fraction.* Continued fractions are important in many branches of mathematics, and particularly in the theory of approximation to real numbers by rationals. There are more general types of continued fractions in which the 'numerators' are not all 1's, but we shall not require them here.

The formula (10.1.1) is cumbrous, and we shall usually write the continued fraction in one of the two forms

$$a_0 + \frac{1}{a_1+}\,\frac{1}{a_2+}\cdots\frac{1}{a_N}$$

or

$$[a_0, a_1, a_2, \ldots, a_N].$$

We call $a_0, a_1, \ldots, a_N$ the *partial quotients*, or simply the *quotients*, of the continued fraction.

We find by calculation that†

$$[a_0] = \frac{a_0}{1}, \quad [a_0, a_1] = \frac{a_1a_0+1}{a_1},$$

$$[a_0, a_1, a_2] = \frac{a_2a_1a_0 + a_2 + a_0}{a_2a_1+1};$$

† There is a clash between our notation here and that of § 6.11, which we shall use again later in the chapter (for example in § 10.5). In § 6.11, $[x]$ was defined as the integral part of x; while here $[a_0]$ means simply a_0. The ambiguity should not confuse the reader, since we use $[a_0]$ here merely as a special case of $[a_0, a_1, \ldots, a_n]$. The square bracket in this sense will seldom occur with a single letter inside it, and will not then be important.

and it is plain that

(10.1.2) $$[a_0, a_1] = a_0 + \frac{1}{a_1},$$

(10.1.3) $$[a_0, a_1, \ldots, a_{n-1}, a_n] = \left[a_0, a_1, \ldots, a_{n-2}, a_{n-1} + \frac{1}{a_n}\right],$$

(10.1.4)
$$[a_0, a_1, \ldots, a_n] = a_0 + \frac{1}{[a_0, a_1, \ldots, a_n]} = [a_0, [a_0, a_1, \ldots, a_n]],$$

for $1 \leqslant n \leqslant N$. We could define our continued fraction by (10.1.2) and either (10.1.3) or (10.1.4). More generally

(10.1.5) $$[a_0, a_1, \ldots, a_n] = [a_0, a_1, \ldots, a_{m-1}, [a_m, a_{m+1}, \ldots, a_n]]$$

for $1 \leqslant m < n \leqslant N$.

10.2. Convergents to a continued fraction. We call

$$[a_0, a_1, \ldots, a_n] \; (0 \leqslant n \leqslant N)$$

the nth *convergent* to $[a_0, a_1, \ldots, a_N]$. It is easy to calculate the convergents by means of the following theorem.

THEOREM 149. *If p_n and q_n are defined by*

(10.2.1)
$$p_0 = a_0, \quad p_1 = a_1 a_0 + 1, \quad p_n = a_n p_{n-1} + p_{n-2} \quad (2 \leqslant n \leqslant N),$$

(10.2.2)
$$q_0 = 1, \quad q_1 = a_1, \quad q_n = a_n q_{n-1} + q_{n-2} \quad (2 \leqslant n \leqslant N),$$

then

(10.2.3) $$[a_0, a_1, \ldots, a_n] = \frac{p_n}{q_n}.$$

We have already verified the theorem for $n = 0$ and $n = 1$. Let us suppose it to be true for $n \leqslant m$, where $m < N$. Then

$$[a_0, a_1, \ldots, a_{m-1}, a_m] = \frac{p_m}{q_m} = \frac{a_m p_{m-1} + p_{m-2}}{a_m q_{m-1} + q_{m-2}},$$

and $p_{m-1}, p_{m-2}, q_{m-1}, q_{m-2}$ depend only on

$$a_0, a_1, \ldots, a_{m-1}.$$

Hence, using (10.1.3), we obtain

$$\begin{aligned}
[a_0, a_1, \ldots, a_{m-1}, a_m, a_{m+1}] &= \left[a_0, a_1, \ldots, a_{m-1}, a_m + \frac{1}{a_{m+1}}\right] \\
&= \frac{\left(a_m + \frac{1}{a_{m+1}}\right) p_{m-1} + p_{m-2}}{\left(a_m + \frac{1}{a_{m+1}}\right) q_{m-1} + q_{m-2}} \\
&= \frac{a_{m+1}(a_m p_{m-1} + p_{m-2}) + p_{m-1}}{a_{m+1}(a_m q_{m-1} + q_{m-1}) + q_{m-1}} \\
&= \frac{a_{m+1} p_m + p_{m-1}}{a_{m+1} q_m + q_{m-1}} = \frac{p_{m+1}}{q_{m+1}};
\end{aligned}$$

and the theorem is proved by induction.

It follows from (10.2.1) and (10.2.2) that

$$\frac{p_n}{q_n} = \frac{a_n p_{n-1} + p_{n-2}}{a_n q_{n-1} + q_{n-2}}. \tag{10.2.4}$$

Also

$$\begin{aligned}
p_n q_{n-1} - p_{n-1} q_n &= (a_n p_{n-1} + p_{n-2}) q_{n-1} - p_{n-1}(a_n q_{n-1} + q_{n-2}) \\
&= -(p_{n-1} q_{n-2} - p_{n-2} q_{n-1}).
\end{aligned}$$

Repeating the argument with $n-1, n-2, \ldots, 2$ in place of n, we obtain

$$p_n q_{n-1} - p_{n-1} q_n = (-1)^{n-1}(p_1 q_0 - p_0 q_1) = (-1)^{n-1}.$$

Also

$$\begin{aligned}
p_n q_{n-2} - p_{n-2} q_n &= (a_n p_{n-1} + p_{n-2}) q_{n-2} - p_{n-2}(a_n q_{n-1} + q_{n-2}) \\
&= a_n(p_{n-1} q_{n-2} - p_{n-2} q_{n-1}) = (-1)^n a_n.
\end{aligned}$$

THEOREM 150. *The functions p_n and q_n satisfy*

$$p_n q_{n-1} - p_{n-1} q_n = (-1)^{n-1} \tag{10.2.5}$$

or

$$\frac{p_n}{q_n} - \frac{p_{n-1}}{q_{n-1}} = \frac{(-1)^{n-1}}{q_{n-1}q_n}. \tag{10.2.6}$$

THEOREM 151. *They also satisfy*

$$p_nq_{n-2} - p_{n-2}q_n = (-1)^n a_n \tag{10.2.7}$$

or

$$\frac{p_n}{q_n} - \frac{p_{n-2}}{q_{n-2}} = \frac{(-1)^n a_n}{q_{n-2}q_n}. \tag{10.2.8}$$

10.3. Continued fractions with positive quotients. We now assign numerical values to the quotients a_n, and so to the fraction (10.1.1) and to its convergents. We shall always suppose that

$$a_1 > 0, \dots, a_N > 0,^{\dagger} \tag{10.3.1}$$

and usually also that a_n is *integral,* in which case the continued fraction is said to be *simple.* But it is convenient first to prove three theorems (Theorems 152–4 below) which hold for all continued fractions in which the quotients satisfy (10.3.1). We write

$$x_n = \frac{p_n}{q_n}, \qquad x = x_N,$$

so that the value of the continued fraction is x_N or x.

It follows from (10.1.5) that

$$\begin{aligned}[a_0, a_1, \dots, a_N] &= [a_0, a_1, \dots, a_{n-1}, [a_n, a_{n+1}, \dots, a_N]] \\ &= \frac{[a_n, a_{n+1}, \dots, a_N]p_{n-1} + p_{n-2}}{[a_n, a_{n+1}, \dots, a_N]q_{n-1} + q_{n-2}}\end{aligned} \tag{10.3.2}$$

for $2 \leqslant n \leqslant N$.

THEOREM 152. *The even convergents x_{2n} increase strictly with n, while the odd convergents x_{2n+1} decrease strictly.*

THEOREM 153. *Every odd convergent is greater than any even convergent.*

† a_0 may be negative.

THEOREM 154. *The value of the continued fraction is greater than that of any of its even convergents and less than that of any of its odd convergents (except that it is equal to the last convergent, whether this be even or odd).*

In the first place every q_n is positive, so that, after (10.2.8) and (10.3.1), $x_n - x_{n-2}$ has the sign of $(-1)^n$. This proves Theorem 152.

Next, after (10.2.6), $x_n - x_{n-1}$ has the sign of $(-1)^{n-1}$, so that

$$x_{2m+1} > x_{2m}. \tag{10.3.3}$$

If Theorem 153 were false, we should have $x_{2m+1} \leqslant x_{2\mu}$ for some pair m, μ. If $\mu < m$, then, after Theorem 152, $x_{2m+1} < x_{2m}$, and if $\mu > m$, then $x_{2\mu+1} < x_{2\mu}$; and either inequality contradicts (10.3.3).

Finally, $x = x_N$ is the greatest of the even, or the least of the odd convergents, and Theorem 154 is true in either case.

10.4. Simple continued fractions. We now suppose that the a_n are integral and the fraction simple. The rest of the chapter will be concerned with the special properties of simple continued fractions, and other fractions will occur only incidentally. It is plain that p_n and q_n are integers, and q_n positive. If

$$[a_0, a_1, a_2, \ldots, a_N] = \frac{p_N}{q_N} = x,$$

we say that the number x (which is necessarily rational) is represented by the continued fraction. We shall see in a moment that, with one reservation, the representation is unique.

THEOREM 155. $q_n \geqslant q_{n-1}$ *for* $n \geqslant 1$, *with inequality when* $n > 1$.

THEOREM 156. $q_n \geqslant n$, *with inequality when* $n > 3$.

In the first place, $q_0 = 1$, $q_1 = a_1 \geqslant 1$. If $n \geqslant 2$, then

$$q_n = a_n q_{n-1} + q_{n-2} \geqslant q_{n-1} + 1,$$

so that $q_n > q_{n-1}$ and $q_n \geqslant n$. If $n > 3$, then

$$q_n \geqslant q_{n-1} + q_{n-2} > q_{n-1} + 1 \geqslant n,$$

and so $q_n > n$.

A more important property of the convergents is

THEOREM 157. *The convergents to a simple continued fraction are in their lowest terms.*

For, by Theorem 150,

$$d|p_n \,.\, d|q_n \to d|(-1)^{n-1} \to d|1.$$

10.5. The representation of an irreducible rational fraction by a simple continued fraction. Any simple continued fraction $[a_0, a_1, \ldots, a_N]$ represents a rational number

$$x = x_N.$$

In this and the next section we prove that, conversely, every positive rational x is representable by a simple continued fraction, and that, apart from one ambiguity, the representation is unique.

THEOREM 158. *If x is representable by a simple continued fraction with an odd (even) number of convergents, it is also representable by one with an even (odd) number.*

For, if $a_n \geqslant 2$,

$$[a_0, a_1, \ldots, a_n] = [a_0, a_1, \ldots, a_n - 1, 1],$$

while, if $a_n = 1$, $[a_0, a_1, \ldots, a_{n-1}, 1] = [a_0, a_1, \ldots, a_{n-2}, a_{n-1} + 1]$.

For example

$$[2, 2, 3] = [2, 2, 2, 1].$$

This choice of alternative representations is often useful.

We call

$$a'_n = [a_n, a_{n+1}, \ldots, a_N] \quad (0 \leqslant n \leqslant N)$$

the *n-th complete quotient* of the continued fraction

$$[a_0, a_1, \ldots, a_n, \ldots, a_N].$$

Thus

$$x = a'_0, \qquad x = \frac{a'_1 a_0 + 1}{a'_1}$$

and

$$x = \frac{a'_n p_{n-1} + p_{n-2}}{a'_n q_{n-1} + q_{n-2}} \quad (2 \leqslant n \leqslant N). \tag{10.5.1}$$

THEOREM 159. *$a_n = [a'_n]$, the integral part of a'_n,*[†] *except that*

$$a_{N-1} = [a_{N-1}] - 1$$

when $a_N = 1$.

If $N = 0$, then $a_0 = a'_0 = [a'_0]$. If $N > 0$, then

$$a'_n = a_n + \frac{1}{a'_{n+1}} \quad (0 \leqslant n \leqslant N-1).$$

Now

$$a'_{n+1} > 1 \quad (0 \leqslant n \leqslant N-1)$$

except that $a'_{n+1} = 1$ when $n = N-1$ and $a_N = 1$. Hence

$$a_n < a'_n < a_n + 1 \quad (0 \leqslant n \leqslant N-1) \tag{10.5.2}$$

and

$$a_n = [a'_n] \quad (0 \leqslant n \leqslant N-1)$$

except in the case specified. And in any case

$$a_N = a'_N = [a'_N].$$

THEOREM 160. *If two simple continued fractions*

$$[a_0, a_1, \ldots, a_N], \quad [b_0, b_1, \ldots, b_M]$$

have the same value x, and $a_N > 1, b_M > 1$, then $M = N$ and the fractions are identical.

When we say that two continued fractions are identical we mean that they are formed by the same sequence of partial quotients.

By Theorem 159, $a_0 = [x] = b_0$. Let us suppose that the first n partial quotients in the continued fractions are identical, and that a'_n, b'_n are the nth complete quotients. Then

$$x = [a_0, a_1, \ldots, a_{n-1}, a'_n] = [a_0, a_1, \ldots, a_{n-1}, b'_n].$$

If $n = 1$, then

$$a_0 + \frac{1}{a'_1} = a_0 + \frac{1}{b'_1},$$

[†] We revert here to our habitual use of the square bracket in accordance with the definition of § 6.11.

$a'_1 = b'_1$, and therefore, by Theorem 159, $a_1 = b_1$. If $n > 1$, then, by (10.5.1),

$$\frac{a'_n p_{n-1} + p_{n-2}}{a'_n q_{n-1} + q_{n-2}} = \frac{b'_n p_{n-1} + p_{n-2}}{b'_n q_{n-1} + q_{n-2}},$$
$$(a'_n - b'_n)(p_{n-1}q_{n-2} - p_{n-2}q_{n-1}) = 0.$$

But $p_{n-1}q_{n-2} - p_{n-2}q_{n-1} = (-1)^n$, by Theorem 150, and so $a'_n = b'_n$. It follows from Theorem 159 that $a_n = b_n$.

Suppose now, for example, that $N \leqslant M$. Then our argument shows that

$$a_n = b_n$$

for $n \leqslant N$. If $M > N$, then

$$\frac{p_N}{q_N} = [a_0, a_1, \ldots, a_N] = [a_0, a_1, \ldots, a_N, b_{N+1}, \ldots, b_M] = \frac{b'_{N+1}p_N + p_{N-1}}{b'_{N+1}q_N + q_{N-1}},$$

by (10.5.1); or

$$p_N q_{N-1} - p_{N-1} q_N = 0,$$

which is false. Hence $M = N$ and the fractions are identical.

10.6. The continued fraction algorithm and Euclid's algorithm. Let x be any real number, and let $a_0 = [x]$. Then

$$x = a_0 + \xi_0, \quad 0 \leqslant \xi_0 < 1.$$

If $\xi_0 \neq 0$, we can write

$$\frac{1}{\xi_0} = a'_1, \quad [a'_1] = a_1, \quad a'_1 = a_1 + \xi_1, \quad 0 \leqslant \xi_1 < 1.$$

If $\xi_1 \neq 0$, we can write

$$\frac{1}{\xi_1} = a'_2 = a_2 + \xi_2, \quad 0 \leqslant \xi_2 < 1,$$

and so on. Also $a'_n = 1/\xi_{n-1} > 1$, and so $a_n \geqslant 1$, for $n \geqslant 1$. Thus

$$x = [a_0, a'_1] = \left[a_0, a_1 + \frac{1}{a'_2}\right] = [a_0, a_1, a'_2] = [a_0, a_1, a_2, a'_3] = \ldots,$$

where $a_0, a_1, \ldots$ are integers and

$$a_1 > 0, \quad a_2 > 0, \ldots.$$

The system of equations

$$\begin{aligned} x &= a_0 + \xi_0 & (0 \leqslant \xi_0 < 1),\\ \frac{1}{\xi_0} &= a_1' = a_1 + \xi_1 & (0 \leqslant \xi_1 < 1),\\ \frac{1}{\xi_1} &= a_2' = a + \xi_2 & (0 \leqslant \xi_2 < 1),\\ &\cdot \quad \cdot \quad \cdot \quad \cdot \quad \cdot \quad \cdot \quad \cdot \end{aligned}$$

is known as the *continued fraction algorithm*. The algorithm continues so long as $\xi_n \neq 0$. If we eventually reach a value of n, say N, for which $\xi_N = 0$, the algorithm terminates and

$$x = [a_0, a_1, a_2, \ldots, a_N].$$

In this case x is represented by a simple continued fraction, and is rational. The numbers a_n' are the complete quotients of the continued fraction.

THEOREM 161. *Any rational number can be represented by a finite simple continued fraction.*

If x is an integer, then $\xi_0 = 0$ and $x = a_0$. If x is not integral, then

$$x = \frac{h}{k},$$

where h and k are integers and $k > 1$. Since

$$\frac{h}{k} = a_0 + \xi_0, \quad h = a_0 k + \xi_0 k,$$

a_0 is the quotient, and $k_1 = \xi_0 k$ the remainder, when h is divided by k.†

† The 'remainder', here and in what follows, is to be non-negative (here positive). If $a_0 \geqslant 0$, then x and h are positive and k_1 is the remainder in the ordinary sense of arithmetic. If $a_0 < 0$, then x and h are negative and the 'remainder' is

$$(x - [x])k.$$

Thus if $h = -7, k = 5$, the 'remainder' is

$$\left(-\frac{7}{5} - \left[-\frac{7}{5}\right]\right)5 = \left(-\frac{7}{5} + 2\right)5 = 3.$$

If $\xi_0 \neq 0$, then

$$a'_1 = \frac{1}{\xi_0} = \frac{k}{k_1}$$

and

$$\frac{k}{k_1} = a_1 + \xi_1, \quad k = a_1 k_1 + \xi_1 k_1;$$

thus a_1 is the quotient, and $k_2 = \xi_1 k_1$ the remainder, when k is divided by k_1. We thus obtain a series of equations

$$h = a_0 k + k_1, \quad k = a_1 k_1 + k_2, \quad k_1 = a_2 k_2 + k_3, \ldots$$

continuing so long as $\xi_n \neq 0$, or, what is the same thing, so long as $k_{n+1} \neq 0$.

The non-negative integers $k, k_1, k_2, \ldots$ form a strictly decreasing sequence, and so $k_{N+1} = 0$ for some N. It follows that $\xi_N = 0$ for some N, and that the continued fraction algorithm terminates. This proves Theorem 161.

The system of equations

$$\begin{aligned}
h &= a_0 k + k_1 & (0 < k_1 < k),\\
k &= a_1 k_1 + k_2 & (0 < k_2 < k_1),\\
&\cdots\cdots & \\
k_{N-2} &= a_{N-1} k_{N-1} + k_N & (0 < k_N < k_{N-1}),\\
k_{N-1} &= a_N k_N &
\end{aligned}$$

is known as *Euclid's algorithm*. The reader will recognize the process as that adopted in elementary arithmetic to determine the greatest common divisor k_N of h and k.

Since $\xi_N = 0$, $a'_N = a_N$; also

$$0 < \frac{1}{a_N} = \frac{1}{a'_N} = \xi_{N-1} < 1,$$

and so $a_N \geqslant 2$. Hence the algorithm determines a representation of the type which was shown to be unique in Theorem 160. We may always make the variation of Theorem 158.

Summing up our results we obtain

THEOREM 162. *A rational number can be expressed as a finite simple continued fraction in just two ways, one with an even and the other with an odd number of convergents. In one form the last partial quotient is* 1, *in the other it is greater than* 1.

10.7. The difference between the fraction and its convergents. Throughout this section we suppose that $N > 1$ and $n > 0$. By (10.5.1)

$$x = \frac{a'_{n+1}p_n + p_{n-1}}{a'_{n+1}q_n + q_{n-1}},$$

for $1 \leqslant n \leqslant N-1$, and so

$$x - \frac{p_n}{q_n} = -\frac{p_n q_{n-1} - p_{n-1}q_n}{q_n(a'_{n+1}q_n + q_{n-1})} = \frac{(-1)^n}{q_n(a'_{n+1}q_n + q_{n-1})}.$$

Also

$$x - \frac{p_0}{q_0} = x - a_0 = \frac{1}{a'_1}.$$

If we write

(10.7.1) $$q'_1 = a'_1, \quad q'_n = a'_n q_{n-1} + q_{n-2} \quad (1 < n \leqslant N)$$

(so that, in particular, $q'_N = q_N$), we obtain

THEOREM 163. *If* $1 \leqslant n \leqslant N-1$, *then*

$$x - \frac{p_n}{q_n} = \frac{(-1)^n}{q_n q'_{n+1}}.$$

This formula gives another proof of Theorem 154.

Next,

$$a_{n+1} < a'_{n+1} < a_{n+1} + 1$$

for $n \leqslant N-2$, by (10.5.2), except that

$$a'_{N-1} = a_{N-1} + 1$$

when $a_N = 1$. Hence, if we ignore this exceptional case for the moment, we have

(10.7.2) $$q_1 = a_1 < a'_1 + 1 \leqslant q_2$$

and

(10.7.3) $$q'_{n+1} = a'_{n+1}q_n + q_{n-1} > a_{n+1}q_n + q_{n-1} = q_{n+1},$$

(10.7.4) $$q'_{n+1} < a_{n+1}q_n + q_{n-1} + q_n = q_{n+1} + q_n \leqslant a_{n+2}q_{n+1} + q_n = q_{n+2},$$

for $1 \leqslant n \leqslant N-2$. It follows that

$$\frac{1}{q_{n+2}} < |p_n - q_n x| < \frac{1}{q_{n+1}} \ (n \leqslant N-2), \tag{10.7.5}$$

while

$$|p_{N-1} - q_{N-1}x| = \frac{1}{q_N}, \quad p_N - q_N x = 0. \tag{10.7.6}$$

In the exceptional case, (10.7.4) must be replaced by

$$q'_{N-1} = (a_{N-1}+1)q_{N-2} + q_{N-3} = q_{N-1} + q_{N-2} = q_N$$

and the first inequality in (10.7.5) by an equality. In any case (10.7.5) shows that $|p_n - q_n x|$ decreases steadily as n increases; *a fortiori*, since q_n increases steadily,

$$\left| x - \frac{p_n}{q_n} \right|$$

decreases steadily.

We may sum up the most important of our conclusions in

THEOREM 164. *If $N > 1$, $n > 0$, then the differences*

$$x - \frac{p_n}{q_n}, \quad q_n x - p_n$$

decrease steadily in absolute value as n increases. Also

$$q_n x - p_n = \frac{(-1)^n \delta_n}{q_{n+1}},$$

where

$$0 < \delta_n < 1 \ (1 \leqslant n \leqslant N-2), \quad \delta_{N-1} = 1,$$

and

$$\left| x - \frac{p_n}{q_n} \right| \leqslant \frac{1}{q_n q_{n+1}} < \frac{1}{q_n^2} \tag{10.7.7}$$

for $n \leqslant N-1$, with inequality in both places except when $n = N-1$.

10.8. Infinite simple continued fractions. We have considered so far only finite continued fractions; and these, when they are simple, represent rational numbers. The chief interest of continued fractions, however, lies in their application to the representation of irrationals, and for this *infinite* continued fractions are needed.

Suppose that $a_0, a_1, a_2, \ldots$ is a sequence of integers satisfying (10.3.1), so that

$$x_n = [a_0, a_1, \ldots, a_n]$$

is, for every n, a simple continued fraction representing a rational number x_n. If, as we shall prove in a moment, x_n tends to a limit x when $n \to \infty$, then it is natural to say that *the simple continued fraction*

$$[a_0, a_1, a_2, \ldots] \tag{10.8.1}$$

converges to the value x, and to write

$$x = [a_0, a_1, a_2, \ldots]. \tag{10.8.2}$$

THEOREM 165. *If* $a_0,\ a_1,\ a_2, \ldots$ *is a sequence of integers satisfying* (10.3.1), *then* $x_n = [a_0, a_1, \ldots, a_n]$ *tends to a limit* x *when* $n \to \infty$.

We may express this more shortly as

THEOREM 166. *All infinite simple continued fractions are convergent.*

We write

$$x_n = \frac{p_n}{q_n} = [a_0, a_1, \ldots, a_n],$$

as in § 10.3, and call these fractions the convergents to (10.8.1). We have to show that the convergents tend to a limit.

If $N \geqslant n$, the convergent x_n is also a convergent to $[a_0,\ a_1, \ldots, a_N]$. Hence, by Theorem 152, the even convergents form an increasing and the odd convergents a decreasing sequence.

Every even convergent is less than x_1, by Theorem 153, so that the increasing sequence of even convergents is bounded above; and every odd convergent is greater than x_0, so that the decreasing sequence of odd convergents is bounded below. Hence the even convergents tend to a limit ξ_1, and the odd convergents to a limit ξ_2, and $\xi_1 \leqslant \xi_2$.

Finally, by Theorems 150 and 156,

$$\left|\frac{p_{2n}}{q_{2n}} - \frac{p_{2n-1}}{q_{2n-1}}\right| = \frac{1}{q_{2n}q_{2n-1}} \leqslant \frac{1}{2n(2n-1)} \to 0,$$

so that $\xi_1 = \xi_2 = x$, say, and the fraction (10.8.1) converges to x.

Incidentally we see that

THEOREM 167. *An infinite simple continued fraction is less than any of its odd convergents and greater than any of its even convergents.*

Here, and often in what follows, we use 'the continued fraction' as an abbreviation for 'the value of the continued fraction'.

10.9. The representation of an irrational number by an infinite continued fraction. We call

$$a'_n = [a_n, a_{n+1}, \ldots]$$

the *n-th complete quotient* of the continued fraction

$$x = [a_0, a_1, \ldots].$$

Clearly

$$\begin{aligned} a'_n &= \lim_{N\to\infty} [a_n, a_{n+1}, \ldots, a_N] \\ &= a_n + \lim_{N\to\infty} \frac{1}{[a_{n+1}, \ldots, a_N]} = a_n + \frac{1}{a'_{n+1}}, \end{aligned}$$

and in particular

$$x = a'_0 = a_0 + \frac{1}{a'_1}.$$

Also

$$a'_n > a_n, \quad a'_{n+1} > a_{n+1} > 0, \quad 0 < \frac{1}{a'_{n+1}} < 1;$$

and so $a_n = [a'_n]$.

THEOREM 168. *If* $[a_0, a_1, a_2, \ldots] = x$, *then*

$$a_0 = [x], \quad a_n = [a'_n] \ (n \geqslant 0).$$

From this we deduce, as in §10.5,

THEOREM 169. *Two infinite simple continued fractions which have the same value are identical.*

We now return to the continued fraction algorithm of § 10.6. If x is irrational the process cannot terminate. Hence it defines an infinite sequence of integers

$$a_0, a_1, a_2, \dots,$$

and as before

$$x = [a_0, a_1'] = [a_0, a_1, a_2'] = \dots = [a_0, a_1, a_2, \dots, a_n, a_{n+1}'],$$

where

$$a_{n+1}' = a_{n+1} + \frac{1}{a_{n+2}'} > a_{n+1}.$$

Hence

$$x = \frac{a_{n+1}'p_n + p_{n-1}}{a_{n+1}'q_n + q_{n-1}},$$

by (10.5.1), and so

$$x - \frac{p_n}{q_n} = \frac{p_{n-1}q_n - p_nq_{n-1}}{q_n(a_{n+1}'q_n + q_{n-1})} = \frac{(-1)^n}{q_n(a_{n+1}'q_n + q_{n-1})},$$

$$\left|x - \frac{p_n}{q_n}\right| < \frac{1}{q_n(a_{n+1}q_n + q_{n-1})} = \frac{1}{q_nq_{n+1}} \leqslant \frac{1}{n(n+1)} \to 0$$

when $n \to \infty$. Thus

$$x = \lim_{n\to\infty} \frac{p_n}{q_n} = [a_0, a_1, \dots, a_n, \dots],$$

and the algorithm leads to the continued fraction whose value is x, and which is unique by Theorem 169.

THEOREM 170. *Every irrational number can be expressed in just one way as an infinite simple continued fraction.*

Incidentally we see that the value of an infinite simple continued fraction is necessarily irrational, since the algorithm would terminate if x were rational.

We define

$$q_n' = a_n'q_{n-1} + q_{n-2}$$

as in § 10.7. Repeating the argument of that section, we obtain

THEOREM 171. *The results of Theorems* 163 *and* 164 *hold also (except for the references to N) for infinite continued fractions. In particular*

$$\left|x - \frac{p_n}{q_n}\right| < \frac{1}{q_n q_{n+1}} < \frac{1}{q_n^2}. \tag{10.9.1}$$

10.10. A lemma. We shall need the theorem which follows in § 10.11.

THEOREM 172. *If*

$$x = \frac{P\zeta + R}{Q\zeta + S},$$

where $\zeta > 1$ *and P, Q, R, and S are integers such that*

$$Q > S > 0, \quad PS - QR = \pm 1,$$

then R/S and P/Q are two consecutive convergents to the simple continued fraction whose value is x. If R/S is the $(n-1)$*th convergent, and P/Q the n-th, then* ζ *is the* $(n+1)$*th complete quotient.*

We can develop P/Q in a simple continued fraction

$$\frac{P}{Q} = [a_0, a_1, \ldots, a_n] = \frac{p_n}{q_n}. \tag{10.10.1}$$

After Theorem 158, we may suppose n odd or even as we please. We shall choose n so that

$$PS - QR = \pm 1 = (-1)^{n-1}. \tag{10.10.2}$$

Now $(P, Q) = 1$ and $Q > 0$, and p_n and q_n satisfy the same conditions. Hence (10.10.1) and (10.10.2) imply $P = p_n$, $Q = q_n$, and

$$p_n S - q_n R = PS - QR = (-1)^{n-1} = p_n q_{n-1} - p_{n-1} q_n,$$

or

$$p_n(S - q_{n-1}) = q_n(R - p_{n-1}). \tag{10.10.3}$$

Since $(p_n, q_n) = 1$, (10.10.3) implies

$$q_n \mid (S - q_{n-1}). \tag{10.10.4}$$

But

$$q_n = Q > S > 0, \quad q_n \geqslant q_{n-1} > 0,$$

and so

$$|S - q_{n-1}| < q_n,$$

and this is inconsistent with (10.10.4) unless $S - q_{n-1} = 0$. Hence

$$S = q_{n-1}, \quad R = p_{n-1}$$

and

$$x = \frac{p_n \zeta + p_{n-1}}{q_n \zeta + q_{n-1}}$$

or

$$x = [a_0, a_1, \ldots, a_n, \zeta].$$

If we develop ζ as a simple continued fraction, we obtain

$$\zeta = [a_{n+1}, a_{n+2}, \ldots]$$

where $a_{n+1} = [\zeta] \geqslant 1$. Hence

$$x = [a_0, a_1, \ldots, a_n, a_{n+1}, a_{n+2}, \ldots],$$

a simple continued fraction. But p_{n-1}/q_{n-1} and p_n/q_n, that is R/S and P/Q, are consecutive convergents of this continued fraction, and ζ is its (n+1)th complete quotient.

10.11. Equivalent numbers. If ξ and η are two numbers such that

$$\xi = \frac{a\eta + b}{c\eta + d},$$

where a, b, c, d are integers such that $ad - bc = \pm 1$, then ξ is said to be *equivalent* to η. In particular, ξ is equivalent to itself.†

If ξ is equivalent to η, then

$$\eta = \frac{-d\xi + b}{c\xi - a}, \quad (-d)(-a) - bc = ad - bc = \pm 1,$$

and so η is equivalent to ξ. Thus the relation of equivalence is symmetrical.

THEOREM 173. *If ξ and η are equivalent, and η and ζ are equivalent, then ξ and ζ are equivalent.*

† $a = d = 1, b = c = 0$.

For

$$\xi = \frac{a\eta + b}{c\eta + d}, \quad ad - bc = \pm 1,$$

$$\eta = \frac{a'\zeta + b'}{c'\zeta + d'}, \quad a'd' - b'c' = \pm 1,$$

and

$$\xi = \frac{A\zeta + B}{C\zeta + D},$$

where

$$A = aa' + bc', \quad B = ab' + bd', \quad C = ca' + dc', \quad D = cb' + dd',$$
$$AD - BC = (ad - bc)(a'd' - b'c') = \pm 1.$$

We may also express Theorem 173 by saying that the relation of equivalence is transitive. The theorem enables us to arrange irrationals in classes of equivalent irrationals.

If h and k are coprime integers, then, by Theorem 25, there are integers h' and k' such that

$$hk' - h'k = 1;$$

and then

$$\frac{h}{k} = \frac{h'.0 + h}{k'.0 + k} = \frac{a.0 + b}{c.0 + d},$$

with $ad - bc = -1$. Hence any rational h/k is equivalent to 0, and therefore, by Theorem 173, to any other rational.

THEOREM 174. *Any two rational numbers are equivalent.*

In what follows we confine our attention to irrational numbers, represented by infinite continued fractions.

THEOREM 175. *Two irrational numbers ξ and η are equivalent if and only if*

(10.11.1)
$$\xi = [a_0, a_1, \ldots, a_m, c_0, c_1, c_2, \ldots], \quad \eta = [b_0, b_1, \ldots, b_n, c_0, c_1, c_2, \ldots],$$

the sequence of quotients in ξ after the m-th being the same as the sequence in η after the n-th.

Suppose first that ξ and η are given by (10.11.1) and write

$$\omega = [c_0, c_1, c_2, \ldots].$$

Then

$$\xi = [a_0, a_1, \ldots, a_m, \omega] = \frac{p_m\omega + p_{m-1}}{q_m\omega + q_{m-1}};$$

and $p_m q_{m-1} - p_{m-1} q_m = \pm 1$, so that ξ and ω are equivalent. Similarly, η and ω are equivalent, and so ξ and η are equivalent. The condition is therefore sufficient.

On the other hand, if ξ and η are two equivalent numbers, we have

$$\eta = \frac{a\xi + b}{c\xi + d}, \quad ab - bc = \pm 1.$$

We may suppose $c\xi + d > 0$, since otherwise we may replace the coefficients by their negatives. When we develop ξ by the continued fraction algorithm, we obtain

$$\begin{aligned} \xi &= [a_0, a_1, \ldots, a_k, a_{k+1}, \ldots] \\ &= [a_0, \ldots, a_{k-1}, a'_k] = \frac{p_{k-1}a'_k + p_{k-2}}{q_{k-1}a'_k + q_{k-2}}. \end{aligned}$$

Hence

$$\eta = \frac{Pa'_k + R}{Qa'_k + S},$$

where

$$\begin{aligned} P &= ap_{k-1} + bq_{k-1}, \quad & R &= ap_{k-2} + bq_{k-2}, \\ Q &= cp_{k-1} + dq_{k-1}, \quad & S &= cp_{k-2} + dq_{k-2}, \end{aligned}$$

so that P, Q, R, S are integers and

$$PS - QR = (ad - bc)(p_{k-1}q_{k-2} - p_{k-2}q_{k-1}) = \pm 1.$$

By Theorem 171,

$$p_{k-1} = \xi q_{k-1} + \frac{\delta}{q_{k-1}}, \quad p_{k-2} = \xi q_{k-2} + \frac{\delta'}{q_{k-2}},$$

where $|\delta| < 1, |\delta'| < 1$. Hence

$$Q = (c\xi + d)q_{k-1} + \frac{c\delta}{d_{k-1}}, \quad S = (c\xi + d)q_{k-2} + \frac{c\delta'}{q_{k-2}}.$$

Now $c\xi + d > 0, q_{k-1} > q_{k-2} > 0$, and q_{k-1} and q_{k-2} tend to infinity; so that

$$Q > S > 0$$

for sufficiently large k. For such k

$$\eta = \frac{P\zeta + R}{Q\zeta + S},$$

where

$$PS - QR = \pm 1, \quad Q > S > 0, \quad \zeta = a'_k > 1;$$

and so, by Theorem 172,

$$\eta = [b_0, b_1, \ldots, b_l, \zeta] = [b_0, b_1, \ldots, b_l, a_k, a_{k+1}, \ldots],$$

for some $b_0, b_1, \ldots, b_l$. This proves the necessity of the condition.

10.12. Periodic continued fractions. A *periodic continued fraction* is an infinite continued fraction in which

$$a_l = a_{l+k}$$

for a fixed positive k and all $l \geqslant L$. The set of partial quotients

$$a_L, a_{L+1}, \ldots, a_{L+k-1}$$

is called the *period*, and the continued fraction may be written

$$[a_0, a_1, \ldots, a_{L-1}, \dot{a}_L, a_{L+1}, \ldots, \dot{a}_{L+k-1}].$$

We shall be concerned only with *simple* periodic continued fractions.

THEOREM 176. *A periodic continued fraction is a quadratic surd, i.e. an irrational root of a quadratic equation with integral coefficients.*

If a'_L is the Lth complete quotient of the periodic continued fraction x, we have

$$a'_L = [a_L, a_{L+1}, \ldots, a_{L+k-1}, a_L, a_{L+1}, \ldots]$$
$$= [a_L, a_{L+1}, \ldots, a_{L+k-1}, a'_L],$$

$$a'_L = \frac{p' a'_L + p''}{q' a'_L + q''},$$

(10.12.1) $$q' a'^2_L + (q'' - p') a'_L - p'' = 0,$$

where the fractions p''/q'' and p'/q' are the last two convergents to $[a_L, a_{L+1}, \ldots, a_{L+k-1}]$.

But

$$x = \frac{p_{L-1} a'_L + p_{L-2}}{q_{L-1} a'_L + q_{L-2}}, \quad a'_L = \frac{p_{L-2} - q_{L-2} x}{q_{L-1} x - p_{L-1}}.$$

If we substitute for a'_L in (10.12.1), and clear of fractions, we obtain an equation

(10.12.2) $$ax^2 + bx + c = 0$$

with integral coefficients. Since x is irrational, $b^2 - 4ac \neq 0$.

The converse of the theorem is also true, but its proof is a little more difficult.

THEOREM 177. *The continued fraction which represents a quadratic surd is periodic.*

A quadratic surd satisfies a quadratic equation with integral coefficients, which we may write in the form (10.12.2). If

$$x = [a_0, a_1, \ldots, a_n, \ldots],$$

then

$$x = \frac{p_{n-1} a'_n + p_{n-2}}{q_{n-1} a'_n + q_{n-2}};$$

and if we substitute this in (10.12.2) we obtain

(10.12.3) $$A_n a'^2_n + B_n a'_n + C_n = 0,$$

where

$$A_n = ap_{n-1}^2 + bp_{n-1}q_{n-1} + cq_{n-1}^2,$$
$$B_n = 2ap_{n-1}p_{n-2} + b(p_{n-1}q_{n-2} + p_{n-2}q_{n-1}) + 2cq_{n-1}q_{n-2},$$
$$C_n = ap_{n-2}^2 + bp_{n-2}q_{n-2} + cq_{n-2}^2.$$

If

$$A_n = ap_{n-2}^2 + bq_{n-1}q_{n-1} + cq_{n-1}^2 = 0,$$

then (10.12.2) has the rational root p_{n-1}/q_{n-1}, and this is impossible because x is irrational. Hence $A_n \neq 0$ and

$$A_n y^2 + B_n y + C = 0$$

is an equation one of whose roots is a_n'. A little calculation shows that

$$\text{(10.12.4)} \qquad B_n^2 - 4A_nC_n = (b^2 - 4ac)(p_{n-1}q_{n-2} - p_{n-2}q_{n-1})^2 = b^2 - 4ac.$$

By Theorem 171,

$$p_{n-1} = xq_{n-1} + \frac{\delta_{n-1}}{q_{n-1}} \quad (|\delta_{n-1}| < 1).$$

Hence

$$A_n = a\left(xq_{n-1} + \frac{\delta_{n-1}}{q_{n-1}}\right)^2 + bq_{n-1}\left(xq_{n-1} + \frac{\delta_{n-1}}{q_{n-1}}\right) + cq_{n-1}^2$$
$$= (ax^2 + bx + c)q_{n-1}^2 + 2ax\delta_{n-1} + a\frac{\delta_{n-1}^2}{q_{n-1}^2} + b\delta_{n-1}$$
$$= 2ax\delta_{n-1} + a\frac{\delta_{n-1}^2}{q_{n-1}^2} + b\delta_{n-1},$$

and

$$|A_n| < 2|ax| + |a| + |b|.$$

Next, since $C_n = A_{n-1}$,

$$|C_n| < 2|ax| + |a| + |b|.$$

Finally, by (10.12.4),

$$B_n^2 \leqslant 4\,|A_nC_n| + \left|b^2 - 4ac\right|$$
$$< 4(2\,|ax| + |a| + |b|)^2 + \left|b^2 - 4ac\right|.$$

Hence the absolute values of A_n, B_n, and C_n are less than numbers independent of n.

It follows that there are only a finite number of different triplets (A_n, B_n, C_n); and we can find a triplet (A, B, C) which occurs at least three times, say as $(A_{n_1}, B_{n_1}, C_{n_1})$, $(A_{n_2}, B_{n_2}, C_{n_2})$, and $(A_{n_3}, B_{n_3}, C_{n_3})$. Hence $a'_{n_1}, a'_{n_2}, a'_{n_3}$, are all roots of

$$Ay^2 + By + C = 0,$$

and at least two of them must be equal. But if, for example, $a'_{n_1} = a'_{n_2}$, then

$$a_{n_2} = a_{n_1}, \quad a_{n_2+1} = a_{n_1+1} \ldots,$$

and the continued fraction is periodic.

10.13. Some special quadratic surds. It is easy to find the continued fraction for a special surd such as $\sqrt{2}$ or $\sqrt{3}$ by carrying out the algorithm of § 10.6 until it recurs. Thus

(10.13.1) $$\sqrt{2} = 1 + (\sqrt{2} - 1) = 1 + \frac{1}{\sqrt{2}+1} = 1 + \frac{1}{2 + (\sqrt{2} - 1)}$$
$$= 1 + \frac{1}{2+}\,\frac{1}{\sqrt{2}+1} = 1 + \frac{1}{2+}\,\frac{1}{2+\ldots} = [1, \dot{2}],$$

and, similarly,

(10.13.2) $$\sqrt{3} = 1 + \frac{1}{1+}\,\frac{1}{2+}\,\frac{1}{1+}\,\frac{1}{2+\ldots} = [1, \dot{1}, \dot{2}],$$

(10.13.3) $$\sqrt{5} = 2 + \frac{1}{4+}\,\frac{1}{4+\ldots} = [2, \dot{4}],$$

(10.13.4) $$\sqrt{7} = 2 + \frac{1}{1+}\,\frac{1}{1+}\,\frac{1}{1+}\,\frac{1}{4+\ldots} = [2, \dot{1}, 1, 1, \dot{4}].$$

But the most interesting special continued fractions are not usually 'pure' surds.

A particular simple type is

$$x = b + \frac{1}{a+}\frac{1}{b+}\frac{1}{a+}\frac{1}{b+\dots} = [\dot{b}, \dot{a}], \tag{10.13.5}$$

where $a|b$, so that $b = ac$, where c is an integer. In this case

$$x = b + \frac{1}{a+}\frac{1}{x} = \frac{(ab+1)x+b}{ax+1},$$

$$x^2 - bx - c = 0, \tag{10.13.6}$$

$$x = \tfrac{1}{2}\{b + \surd(b^2+4c)\}. \tag{10.13.7}$$

In particular

$$\alpha = 1 + \frac{1}{1+}\frac{1}{1+} = [\dot{1}] = \frac{\surd 5 + 1}{2}, \tag{10.13.8}$$

$$\beta = 2 + \frac{1}{2+}\frac{1}{2+} = [\dot{2}] = \surd 2 + 1, \tag{10.13.9}$$

$$\gamma = 2 + \frac{1}{1+}\frac{1}{2+\dots} = [\dot{2}, \dot{1}] = \surd 3 + 1. \tag{10.13.10}$$

It will be observed that β and γ are equivalent, in the sense of § 10.11, to $\surd 2$ and $\surd 3$ respectively, but that α is not equivalent to $\surd 5$.

It is easy to find a general formula for the convergents to (10.13.5).

THEOREM 178. *The* $(n+1)$*th convergent to* (10.13.5) *is given by*

$$p_n = c^{-\left[\frac{1}{2}(n+1)\right]}u_{n+2}, \quad q_n = c^{-\left[\frac{1}{2}(n+1)\right]}u_{n+1},\dagger \tag{10.13.11}$$

where

$$u_n = \frac{x^n - y^n}{x - y} \tag{10.13.12}$$

and x *and* y *are the roots of* (10.13.6).

† The power of c is c^{-m} when $n = 2m$ and c^{-m-1} when $n = 2m+1$.

In the first place

$$q_0 = 1 = u_1, \quad q_1 = a = \frac{b}{c} = \frac{x+y}{c} = \frac{u_2}{c},$$

$$p_0 = b = x + y = u_2,$$

$$p_1 = ab + 1 = \frac{b^2 + c}{c} = \frac{(x+y)^2 - xy}{c} = \frac{u_3}{c},$$

so that the formulae (10.13.11) are true for $n = 0$ and $n = 1$. We prove the general formulae by induction.

We have to prove that

$$p_n = c^{-\left[\frac{1}{2}(n+1)\right]} u_{n+2} = w_{n+2},$$

say. Now

$$x^{n+2} = bx^{n+1} + cx^n, \quad y^{n+2} = by^{n+1} + cy^n,$$

and so

(10.13.13) $$u_{n+2} = bu_{n+1} + cu_n.$$

But

$$u_{2m+2} = c^m w_{2m+2}, \quad u_{2m+1} = c^m w_{2m+1}.$$

Substituting into (10.13.13), and distinguishing the cases of even and odd n, we find that

$$w_{2m+2} = bw_{2m+1} + w_{2m}, \quad w_{2m+1} = aw_{2m} + w_{2m-1}.$$

Hence w_{n+2} satisfies the same recurrence formulae as p_n, and so $p_n = w_{n+2}$. Similarly we prove that $q_n = w_{n+1}$.

The argument is naturally a little simpler when $a = b$, $c = 1$. In this case p_n and q_n satisfy

$$u_{n+2} = bu_{n+1} + u_n$$

and are of the form

$$Ax^n + By^n,$$

where A and B are independent of n and may be determined from the values of the first two convergents. We thus find that

$$p_n = \frac{x^{n+2} - y^{n+2}}{x - y}, \quad q_n = \frac{x^{n+1} - y^{n+1}}{x - y},$$

in agreement with Theorem 178.

10.14. The series of Fibonacci and Lucas. In the special case $a = b = 1$ we have

(10.14.1) $$x = \frac{\sqrt{5}+1}{2}, \quad y = -\frac{1}{x} = -\frac{\sqrt{5}-1}{2},$$

$$p_n = u_{n+2} = \frac{x^{n+2} - y^{n+2}}{\sqrt{5}}, \quad q_n = u_{n+1} = \frac{x^{n+1} - y^{n+1}}{\sqrt{5}}.$$

The series (u_n) or

(10.14.2) $$1, 1, 2, 3, 5, 8, 13, 21, \ldots$$

in which the first two terms are u_1 and u_2, and each term after is the sum of the two preceding, is usually called Fibonacci's series. There are, of course, similar series with other initial terms, the most interesting being the series (v_n) or

(10.14.3) $$1, 3, 4, 7, 11, 18, 29, 47, \ldots$$

defined by

(10.14.4) $$v_n = x^n + y^n.$$

Such series have been studied in great detail by Lucas and later writers, in particular D. H. Lehmer, and have very interesting arithmetical properties. We shall come across the series (10.14.3) again in Ch. XV in connexion with the Mersenne numbers.

We note here some arithmetical properties of these series, and particularly of (10.14.2).

THEOREM 179. *The numbers u_n and v_n defined by* (10.14.2) *and* (10.14.3) *have the following properties:*

(i) $(u_n, u_{n+1}) = 1, \quad (v_n, v_{n+1}) = 1;$

(ii) *u_n and v_n are both odd or both even, and*

$$(u_n, v_n) = 1, \quad (u_n, v_{n+1}) = 2$$

in these two cases;

(iii) *$u_n | u_{rn}$ for every r;*

(iv) *if $(m, n) = d$ then*

$$(u_m, u_n) = u_d,$$

and, in particular, u_m *and* u_n *are coprime if* m *and* n *are coprime;*

(v) *if* $(m, n) = 1$, *then*

$$u_m u_n | u_{mn}.$$

It is convenient to regard (10.13.12) and (10.14.4) as defining u_n and v_n for all integral n. Then

$$u_0 = 0, \quad v_0 = 2$$

and

(10.14.5) $$u_{-n} = -(xy)^{-n} u_n = (-1)^{n-1} u_n, \quad v_{-n} = (-1)^n v_n.$$

We can verify at once that

(10.14.6) $$2u_{m+n} = u_m v_n + u_n v_m,$$

(10.14.7) $$v_n^2 - 5u_n^2 = (-1)^n 4,$$

(10.14.8) $$u_n^2 - u_{n-1}u_{n+1} = (-1)^{n-1},$$

(10.14.9) $$v_n^2 - v_{n-1}v_{n+1} = (-1)^n 5.$$

Proceeding to the proof of the theorem, we observe first that (i) follows from the recurrence formulae, or from (10.14.8), (10.14.9), and (10.14.7), and (ii) from (10.14.7).

Next, suppose (iii) true for $r = 1, 2, \ldots, R-1$. By (10.14.6),

$$2u_{Rn} = u_n v_{(R-1)n} + u_{(R-1)n} v_n.$$

If u_n is odd, then $u_n | 2u_{Rn}$ and so $u_n | u_{Rn}$. If u_n is even, then v_n is even by (ii), $u_{(R-1)n}$ by hypothesis, and $v_{(R-1)n}$ by (ii). Hence we may write

$$u_{Rn} = u_n \cdot \tfrac{1}{2} v_{(R-1)_n} + u_{(R-1)n} \cdot \tfrac{1}{2} v_n,$$

and again $u_n | u_{Rn}$.

This proves (iii) for all positive r. The formulae (10.14.5) then show that it is also true for negative r.

To prove (iv) we observe that, if $(m, n) = d$, there are integers r, s (positive or negative) for which

$$rm + sn = d,$$

and that

(10.14.10) $$2u_d = u_{rm} v_{sn} + u_{sn} v_{rm},$$

by (10.14.6). Hence, if $(u_m, u_n) = h$, we have

$$h|u_m.h|u_n \rightarrow h|u_{rm}.h|u_{sn} \rightarrow h|2u_d.$$

If h is odd, $h|u_d$. If h is even, then u_m and u_n are even, and so $u_{rm}, u_{sn}, v_{rm}, v_{sn}$ are all even, by (ii) and (iii). We may therefore write (10.14.10) as

$$u_d = u_{rm}\left(\tfrac{1}{2}v_{sn}\right) + u_{sn}\left(\tfrac{1}{2}v_{rm}\right),$$

and it follows as before that $h|u_d$. Thus $h|u_d$ in any case. Also $u_d|u_m, u_d|u_n$, by (iii), and so

$$u_d|(u_m, u_n) = h.$$

Hence

$$h = u_d,$$

which is (iv).

Finally, if $(m, n) = 1$, we have

$$u_m|u_{mn}, \quad u_n|u_{mn}$$

by (iii), and $(u_m, u_n) = 1$ by (iv). Hence

$$u_m u_n|u_{mn}.$$

In particular it follows from (iii) that u_m can be prime only when m is 4 (when $u_4 = 3$) or an odd prime p. But u_p is not necessarily prime: thus

$$u_{53} = 53316291173 = 953 \,.\, 55945741.$$

THEOREM 180. *Every prime p divides some Fibonacci number (and therefore an infinity of the numbers). In particular*

$$u_{p-1} \equiv 0 \pmod{p}$$

if $p = 5m \pm 1$, and

$$u_{p+1} \equiv 0 \pmod{p}$$

if $p = 5m \pm 2$.

Since $u_3 = 2$ and $u_5 = 5$, we may suppose that $p \neq 2, p \neq 5$. It follows from (10.13.12) and (10.14.1) that

$$2^{n-1}u_n = n + \binom{n}{3}5 + \binom{n}{5}5^2 + \dots, \tag{10.14.11}$$

where the last term is $5^{\frac{1}{2}(n-1)}$ if n is odd and $n.\,5^{\frac{1}{2}n-1}$ if n is even. If $n = p$ then

$$2^{p-1} \equiv 1, \quad 5^{\frac{1}{2}(p-1)} \equiv \left(\frac{5}{p}\right) \pmod{p},$$

by Theorems 71 and 83; and the binomial coefficients are all divisible by p, except the last which is 1. Hence

$$u_p \equiv \left(\frac{5}{p}\right) = \pm 1 \pmod{p}$$

and therefore, by (10.14.8),

$$u_{p-1}u_{p+1} \equiv 0 \pmod{p}.$$

Also $(p-1, p+1) = 2$, and so

$$(u_{p-1}, u_{p+1}) = u_2 = 1,$$

by Theorem 179 (iv). Hence one and only one of u_{p-1} and u_{p+1} is divisible by p.

To distinguish the two cases, take $n = p + 1$ in (10.14.11). Then

$$2^p u_{p+1} = (p+1) + \binom{p+1}{3} 5 + \ldots + (p+1)\,5^{\frac{1}{2}(p-1)}.$$

Here all but the first and last coefficients are divisible by p,† and so

$$2^p u_{p+1} \equiv 1 + \left(\frac{5}{p}\right) \pmod{p}.$$

Hence $u_{p+1} \equiv 0 \pmod{p}$ if $\left(\frac{5}{p}\right) = -1$, i.e. if $p \equiv \pm 2 \pmod{5}$,‡ and $u_{p-1} \equiv 0 \pmod{p}$ in the contrary case.

We shall give another proof of Theorem 180 in § 15.4.

† $\binom{p+1}{\nu}$, where $3 \leqslant \nu \leqslant p-1$, is an integer, by Theorem 73; the numerator contains p, and the denominator does not.

‡ By Theorem 97.

10.15. Approximation by convergents. We conclude this chapter by proving some theorems whose importance will become clearer in Ch. XI.

By Theorem 171,

$$\left|\frac{p_n}{q_n} - x\right| < \frac{1}{q_n^2},$$

so that p_n/q_n provides a good approximation to x. The theorem which follows shows that p_n/q_n is the fraction, among all fractions of no greater complexity, i.e. all fractions whose denominator does not exceed q_n, which provides the *best* approximation.

THEOREM 181. *If* $n > 1$,† $0 < q \leqslant q_n$, *and* $p/q \neq p_n/q_n$, *then*

$$\left|\frac{p_n}{q_n} - x\right| < \left|\frac{p}{q} - x\right|. \tag{10.15.1}$$

This is included in a stronger theorem, viz.

THEOREM 182. *If* $n > 1, 0 < q \leqslant q_n$, *and* $p/q \neq p_n/q_n$ *then*

$$|p_n - q_n x| < |p - qx|. \tag{10.15.2}$$

We may suppose that $(p, q) = 1$. Also, by Theorem 171,

$$|p_n - q_n x| < |p_{n-1} - q_{n-1}x|,$$

and it is sufficient to prove the theorem on the assumption that $q_{n-1} < q \leqslant q_n$, the complete theorem then following by induction.

Suppose first that $q = q_n$. Then

$$\left|\frac{p_n}{q_n} - \frac{p}{q_n}\right| \geqslant \frac{1}{q_n}$$

† We state Theorems 181 and 182 for $n > 1$ in order to avoid a trivial complication. The proof is valid for $n = 1$ unless $q_2 = q_{n+1} = 2$, which is possible only if $a_1 = a_2 = 1$. In this case

$$x = a_0 + \frac{1}{1+}\,\frac{1}{1+}\,\frac{1}{a_3 + \ldots}, \quad \frac{p_1}{q_1} = a_0 + 1,$$

and

$$a_0 + \tfrac{1}{2} < x < a_0 + 1$$

unless the fraction ends at the second 1. If this is not so then p_1/q_1 is nearer to x than any other integer. But in the exceptional case $x = a_0 + \frac{1}{2}$ there are two integers equidistant from x, and (10.15.1) may become an equality.

if $p \neq p_n$. But

$$\left|\frac{p_n}{q_n} - x\right| \leqslant \frac{1}{q_n q_{n+1}} < \frac{1}{2q_n},$$

by Theorems 171 and 156; and therefore

$$\left|\frac{p_n}{q_n} - x\right| < \left|\frac{p}{q_n} - x\right|,$$

which is (10.15.2).

Next suppose that $q_{n-1} < q < q_n$, so that p/q is not equal to either of p_{n-1}/q_{n-1} or p_n/q_n. If we write

$$\mu p_n + \nu p_{n-1} = p, \quad \mu q_n + \nu q_{n-1} = q,$$

then

$$\mu(p_n q_{n-1} - p_{n-1} q_n) = p q_{n-1} - q p_{n-1},$$

so that

$$\mu = \pm(p q_{n-1} - q p_{n-1});$$

and similarly

$$\nu = \pm(p q_n - q p_n).$$

Hence μ and ν are integers and neither is zero.

Since $q = \mu q_n + \nu q_{n-1} < q_n$, μ and ν must have opposite signs. By Theorem 171,

$$p_n - q_n x, \quad p_{n-1} - q_{n-1} x$$

have opposite signs. Hence

$$\mu(p_n - q_n x), \quad \nu(p_{n-1} - q_{n-1} x)$$

have the same sign. But

$$p - qx = \mu(p_n - q_n x) + \nu(p_{n-1} - q_{n-1} x),$$

and therefore

$$|p - qx| > |p_{n-1} - q_{n-1} x| > |p_n - q_n x|.$$

Our next theorem gives a refinement on the inequality (10.9.1) of Theorem 171.

THEOREM 183. *Of any two consecutive convergents to x, one at least satisfies the inequality*

$$\left|\frac{p}{q} - x\right| < \frac{1}{2q^2}. \tag{10.15.3}$$

Since the convergents are alternately less and greater than x, we have

$$\left|\frac{p_{n+1}}{q_{n+1}} - \frac{p_n}{q_n}\right| = \left|\frac{p_n}{q_n} - x\right| + \left|\frac{p_{n+1}}{q_{n+1}} - x\right|. \tag{10.15.4}$$

If (10.15.3) were untrue for both p_n/q_n and p_{n+1}/q_{n+1}, then (10.15.4) would imply

$$\frac{1}{q_n q_{n+1}} = \left|\frac{p_{n+1}q_n - p_n q_{n+1}}{q_n q_{n+1}}\right| = \left|\frac{p_{n+1}}{q_{n+1}} - \frac{p_n}{q_n}\right| \geqslant \frac{1}{2q_n^2} + \frac{1}{2q_{n+1}^2},$$

or

$$(q_{n+1} - q_n)^2 \leqslant 0,$$

which is false except in the special case

$$n = 0, \quad a_1 = 1, \quad q_1 = q_0 = 1.$$

In this case

$$0 < \frac{p_1}{q_1} - x = 1 - \frac{1}{1+}\,\frac{1}{a_2 + \dots} < 1 - \frac{a_2}{a_2 + 1} \leqslant \frac{1}{2},$$

so that the theorem is still true.

It follows that, when x is irrational, there are an infinity of convergents p_n/q_n which satisfy (10.15.3). Our last theorem in this chapter shows that this inequality is characteristic of convergents.

THEOREM 184. *If*

$$\left|\frac{p}{q} - x\right| < \frac{1}{2q^2}, \tag{10.15.5}$$

then p/q is a convergent.

If (10.15.5) is true, then

$$\frac{p}{q} - x = \frac{\epsilon\theta}{q^2},$$

where

$$\epsilon = \pm 1, \quad 0 < \theta < \tfrac{1}{2}.$$

We can express p/q as a finite continued fraction

$$[a_0, a_1, \ldots, a_n];$$

and since, by Theorem 158, we can make n odd or even at our discretion, we may suppose that

$$\epsilon = (-1)^{n-1}.$$

We write

$$x = \frac{\omega p_n + p_{n-1}}{\omega q_n + q_{n-1}},$$

where $p_n/q_n, p_{n-1}/q_{n-1}$ are the last and the last but one convergents to the continued fraction for p/q. Then

$$\frac{\epsilon\theta}{q_n^2} = \frac{p_n}{q_n} - x = \frac{p_n q_{n-1} - p_{n-1} q_n}{q_n(\omega q_n + q_{n-1})} = \frac{(-1)^{n-1}}{q_n(\omega q_n + q_{n-1})},$$

and so

$$\frac{q_n}{\omega q_n + q_{n-1}} = \theta.$$

Hence

$$\omega = \frac{1}{\theta} - \frac{q_{n-1}}{q_n} > 1$$

(since $0 < \theta < \frac{1}{2}$); and so, by Theorem 172, p_{n-1}/q_{n-1} and p_n/q_n are consecutive convergents to x. But $p_n/q_n = p/q$.

NOTES

§ 10.1. Many proofs in this and the next chapter are modelled on those given in Perron's *Kettenbrüche* and *Irrationalzahlen;* the former contains full references to the early history of the subject. There are accounts in English in Cassels, *Diophantine approximation*, Olds, *Continued fractions* and Wall, *Analytic theory of continued fractions* (New York, van Norstrand, 1948). Stark, *Number theory*, also gives additional references and material.

§ 10.12. Theorem 177 is Lagrange's most famous contribution to the theory. The proof given here (Perron, *Kettenbrüche*, 77) due to Charves.

§§ 10.13–14. There is a large literature concerned with Fibonacci's and similar series. See Bachmann, *Niedere Zahlentheorie*, ii, ch. ii; Dickson, *History*, i, ch. xvii; D. H. Lehmer, *Annals of Math.* (2), 31 (1930), 419–48.

XI

APPROXIMATION OF IRRATIONALS BY RATIONALS

11.1. Statement of the problem. The problem considered in this chapter is that of the approximation of a given number ξ, usually irrational, by a rational fraction

$$r = \frac{p}{q}.$$

We suppose throughout that $0 < \xi < 1$ and that p/q is irreducible.[†]

Since the rationals are dense in the continuum, there are rationals as near as we please to any ξ. Given ξ and any positive number ϵ, there is an $r = p/q$ such that

$$|r - \xi| = \left|\frac{p}{q} - \xi\right| \leqslant \epsilon;$$

any number can be approximated by a rational with any assigned degree of accuracy. We ask now how *simply* or, what is essentially the same thing, how *rapidly* can we approximate to ξ? Given ξ and ϵ, how complex must p/q be (i.e. how large q) to secure an approximation with the measure of accuracy ϵ? Given ξ and q, or some upper bound for q, how small can we make ϵ?

We have already done something to answer these questions. We proved, for example, in Ch. III (Theorem 36) that, given ξ and n,

$$\exists p, q \,.\, 0 < q \leqslant n \,.\, \left|\frac{p}{q} - \xi\right| \leqslant \frac{1}{q(n+1)},$$

and *a fortiori*

$$\left|\frac{p}{q} - \xi\right| < \frac{1}{q^2}; \tag{11.1.1}$$

and in Ch. X we proved a number of similar theorems by the use of continued fractions.[‡] The inequality (11.1.1), or stronger inequalities of the same type, will recur continually throughout this chapter.

When we consider (11.1.1) more closely, we find at once that we must distinguish two cases.

[†] Except in § 11.12.

[‡] See Theorems 171 and 183.

(1) ξ *is a rational a/b.* If $r \neq \xi$, then

$$|r - \xi| = \left|\frac{p}{q} - \frac{a}{b}\right| = \frac{|bp - aq|}{bq} \geqslant \frac{1}{bq}, \tag{11.1.2}$$

so that (11.1.1) involves $q < b$. There are therefore only a finite number of solutions of (11.1.1).

(2) ξ *is irrational.* Then there are an infinity of solutions of (11.1.1). For, if p_n/q_n is any one of the convergents to the continued fraction to ξ, then, by Theorem 171,

$$\left|\frac{p_n}{q_n} - \xi\right| < \frac{1}{q_n^2},$$

and p_n/q_n is a solution.

Theorem 185. *If ξ is irrational, then there is an infinity of fractions p/q which satisfy* (11.1.1).

In § 11.3 we shall give an alternative proof, independent of the theory of continued fractions.

11.2. Generalities concerning the problem. We can regard our problem from two different points of view. We suppose ξ irrational.

(1) We may think first of ϵ. Given ξ, for what functions

$$\Phi = \Phi\left(\xi, \frac{1}{\epsilon}\right)$$

is it true that

$$\exists p, q \,.\, q \leqslant \Phi \,.\, \left|\frac{p}{q} - \xi\right| \leqslant \epsilon, \tag{11.2.1}$$

for the given ξ and every positive ϵ? Or for what functions

$$\Phi = \Phi\left(\frac{1}{\epsilon}\right),$$

independent of ξ, is (11.2.1) true for every ξ and every positive ϵ? It is plain that any Φ with these properties must tend to infinity when ϵ tends to zero, but the more slowly it does so the better.

There are certainly *some* functions Φ which have the properties required. Thus we may take

$$\Phi = \left[\frac{1}{2\epsilon}\right] + 1,$$

and $q = \Phi$. There is then a p for which

$$\left|\frac{p}{q} - \xi\right| \leqslant \frac{1}{2q} < \epsilon,$$

and so this Φ satisfies our requirements. The problem remains of finding, if possible, more advantageous forms of Φ.

(2) We may think first of q. Given ξ, for what functions

$$\phi = \phi(\xi, q),$$

tending to infinity with q, is it true that

$$\exists p \, . \left|\frac{p}{q} - \xi\right| \leqslant \frac{1}{\phi}? \tag{11.2.2}$$

Or for what functions $\phi = \phi(q)$ independent of ξ, is (11.2.2) true for every ξ? Here, naturally, the *larger* ϕ the better. If we put the question in its second and stronger form, it is substantially the same as the second form of question (1). If ϕ is the function inverse to Φ, it is substantially the same thing to assert that (11.2.1) is true (with Φ independent of ξ) or that (11.2.2) is true for all ξ and q.

These questions, however, are not the questions most interesting to us now. We are not so much interested in approximations to ξ with an *arbitrary* denominator q, as in approximations with *an appropriately selected* q. For example, there is no great interest in approximations to π with denominator 11; what is interesting is that two particular denominators, 7 and 113, give the very striking approximations $\frac{22}{7}$ and $\frac{335}{113}$. We should ask, not how closely we can approximate to ξ with an arbitrary q, but how closely we can approximate for *an infinity of values* of q.

We shall therefore be occupied, throughout the rest of this chapter, with the following problem: *for what $\phi = \phi(\xi, q)$, or $\phi = \phi(q)$, is it true, for a given ξ, or for all ξ, or for all ξ of some interesting class, that*

$$\left|\frac{p}{q} - \xi\right| \leqslant \frac{1}{\phi} \tag{11.2.3}$$

for an infinity of q and appropriate p? We know already, after Theorem 171, that we can take $\phi = q^2$ for all irrational ξ.

11.3. An argument of Dirichlet. In this section we prove Theorem 185 by a method independent of the theory of continued fractions. The method gives nothing new, but is of great importance because it can be extended to multi-dimensional problems.†

We have already defined $[x]$, the greatest integer in x. We define (x) by

$$(x) = x - [x];$$

and $\bar{x}$ as the difference between x and the nearest integer, with the convention that $\bar{x} = \frac{1}{2}$ when x is $n + \frac{1}{2}$. Thus

$$\left[\frac{5}{3}\right] = 1, \quad \left(\frac{5}{3}\right) = \frac{2}{3}, \quad \overline{\frac{5}{3}} = -\frac{1}{3}.$$

Suppose ξ and ϵ given. Then the Q+1 numbers

$$0, (\xi), (2\xi), \ldots, (Q\xi)$$

define Q+1 points distributed among the Q intervals or 'boxes'

$$\frac{s}{Q} \leqslant x < \frac{s+1}{Q} \ (s = 0, 1, \ldots, Q-1).$$

There must be one box which contains at least two points, and therefore two numbers q_1 and q_2, not greater than Q, such that $(q_1\xi)$ and $(q_2\xi)$ differ by less than $1/Q$. If q_2 is the greater, and $q = q_2 - q_1$, then $0 < q \leqslant Q$ and $|\overline{q\xi}| < 1/Q$. There is therefore a p such that

$$|q\xi - p| < \frac{1}{Q}.$$

Hence, taking

$$Q = \left[\frac{1}{\epsilon}\right] + 1,$$

we obtain

$$\exists p, q \,.\, q \leqslant \left[\frac{1}{\epsilon}\right] + 1 \,.\, \left|\frac{p}{q} - \xi\right| < \frac{\epsilon}{q}$$

† See § 11.12.

(which is nearly the same as the result of Theorem 36) and

$$(11.3.1) \qquad \left|\frac{p}{q} - \xi\right| < \frac{1}{qQ} \leqslant \frac{1}{q^2},$$

which is (11.1.1).

If ξ is rational, then there is only a finite number of solutions.[†] We have to prove that there is an infinity when ξ is irrational. Suppose that

$$\frac{p_1}{q_1}, \frac{p_2}{q_2}, \ldots, \frac{p_k}{q_k}$$

exhaust the solutions. Since ξ is irrational, there is a Q such that

$$\left|\frac{p_s}{q_s} - \xi\right| > \frac{1}{Q} \quad (s = 1, 2, \ldots, k).$$

But then the p/q of (11.3.1) satisfies

$$\left|\frac{p_s}{q_s} - \xi\right| < \frac{1}{qQ} \leqslant \frac{1}{Q},$$

and is not one of p_s/q_s; a contradiction. Hence the number of solutions of (11.1.1) is infinite.

Dirichlet's argument proves that $q\xi$ is nearly an integer, so that $(q\xi)$ is nearly 0 or 1, but does not distinguish between these cases. The argument of § 11.1 gives rather more: for

$$\frac{p_n}{q_n} - \xi = \frac{(-1)^{n-1}}{q_n q'_{n+1}}$$

is positive or negative according as n is odd or even, and $q_n\xi$ is alternately a little less and a little greater than p_n.

11.4. Orders of approximation. We shall say that ξ *is approximable by rationals to order n* if there is a $K(\xi)$, depending only on ξ, for which

$$(11.4.1) \qquad \left|\frac{p}{q} - \xi\right| < \frac{K(\xi)}{q^n}$$

has an infinity of solutions.

We can dismiss the trivial case in which ξ is rational. If we look back at (11.1.2), and observe that the equation $bp - aq = 1$ has an infinity of

[†] The proof of this in § 11.1 was independent of continued fractions.

solutions, we obtain

THEOREM 186. *A rational is approximable to order* 1, *and to no higher order.*

We may therefore suppose ξ irrational. After Theorem 171, we have

THEOREM 187. *Any irrational is approximable to order* 2.

We can go farther when ξ is a quadratic surd (i.e. the root of a quadratic equation with integral coefficients). We shall sometimes describe such a ξ as a quadratic irrational, or simply as 'quadratic'.

THEOREM 188. *A quadratic irrational is approximable to order* 2 *and to no higher order.*

The continued fraction for a quadratic ξ is periodic, by Theorem 177. In particular its quotients are bounded, so that

$$0 < a_n < M,$$

where M depends only on ξ. Hence, by (10.5.2),

$$q'_{n+1} = a'_{n+1}q_n + q_{n-1} < (a_{n+1} + 1)q_{n-1} < (M + 2)q_n$$

and *a fortiori* $q_{n+1} < (M+2)q_n$. Similarly $q_n < (M+2)q_{n-1}$.

Suppose now that $q_{n-1} < q \leqslant q_n$. Then $q_n < (M+2)q$ and, by Theorem 181,

$$\left|\frac{p}{q} - \xi\right| \geqslant \left|\frac{p_n}{q_n} - \xi\right| = \frac{1}{q_n q'_{n+1}} > \frac{1}{(M+2)q_n^2} > \frac{1}{(M+2)^3 q_{n-1}^2} > \frac{K}{q^2},$$

where $K = (M+2)^{-3}$; and this proves the theorem.

The negative half of Theorem 188 is a special case of a theorem (Theorem 191) which we shall prove in § 11.7 without the use of continued fractions. This requires some preliminary explanations and some new definitions.

11.5. Algebraic and transcendental numbers. An *algebraic number* is a number x which satisfies an *algebraic equation,* i.e. an equation

$$a_0x^n + a_1x^{n-1} + \cdots + a_n = 0, \tag{11.5.1}$$

where $a_0, a_1, \ldots$ are integers, not all zero.

A number which is not algebraic is called *transcendental.*

If $x = a/b$, then $bx - a = 0$, so that any rational x is algebraic. Any quadratic surd is algebraic; thus $i = \sqrt{(-1)}$ is algebraic. But in this chapter we are concerned with *real* algebraic numbers.

An algebraic number satisfies any number of algebraic equations of different degrees; thus $x = \sqrt{2}$ satisfies $x^2-2 = 0$, $x^4-4 = 0$,.... If x satisfies an algebraic equation of degree n, but none of lower degree, then we say that x is *of degree n.* Thus a rational is of degree 1.

A number is *Euclidean* if it measures a length which can be constructed, starting from a given unit length, by a Euclidean construction, i.e. a finite construction with ruler and compasses only. Thus $\sqrt{2}$ is Euclidean. It is plain that we can construct any finite combination of real quadratic surds, such as

$$\sqrt{(11+2\sqrt{7})} - \sqrt{(11-2\sqrt{7})} \tag{11.5.2}$$

by Euclidean methods. We may describe such a number as of real quadratic type.

Conversely, any Euclidean construction depends upon a series of points defined as intersections of lines and circles. The coordinates of each point in turn are defined by two equations of the types

$$lx + my + n = 0$$

or

$$x^2 + y^2 + 2gx + 2fy + c = 0,$$

where l, m, n, g, f, c are measures of lengths already constructed; and two such equations define x and y as real quadratic combinations of $l, m,\ldots$. Hence every Euclidean number is of real quadratic type.

The number (11.5.2) is defined by

$$x = y - z, \quad y^2 = 11 + 2t, \quad z^2 = 11 - 2t, \quad t^2 = 7$$

and we obtain

$$x^4 - 44x^2 + 112 = 0$$

on eliminating y, z, and t. Thus x is algebraic. It is not difficult to prove that any Euclidean number is algebraic, but the proof demands a little knowledge of the general theory of algebraic numbers.†

† In fact any number defined by an equation $\alpha_0 x^n + \alpha_1 x^{n-1} + \cdots + \alpha_n = 0$, where $\alpha_0, \alpha_1, \ldots, \alpha_n$ are algebraic, is algebraic. For the proof see Hecke 66, or Hardy, *Pure mathematics* (ed. 9, 1944), 39.

11.6. The existence of transcendental numbers. It is not immediately obvious that there are any transcendental numbers, though actually, as we shall see in a moment, almost all real numbers are transcendental.

We may distinguish three different problems. The first is that of proving the existence of transcendental numbers (without necessarily producing a specimen). The second is that of giving an example of a transcendental number by a construction specially designed for the purpose. The third, which is much more difficult, is that of proving that some number given independently, some one of the 'natural' numbers of analysis, such as e or π, is transcendental.

We may define the *rank* of the equation (11.5.1) as

$$N = n + |a_0| + |a_1| + \cdots + |a_n|.$$

The minimum value of N is 2. It is plain that there are only a finite number of equations

$$E_{N,1},\ E_{N,2},\ \ldots,\ E_{N,k_N}$$

of rank N. We can arrange the equations in the sequence

$$E_{2,1},\ E_{2,2},\ \ldots,\ E_{2,k_2}, E_{3,1},\ E_{3,2},\ \ldots,\ E_{3,k_3}, E_{4,1}, \ldots$$

and so correlate them with the numbers 1, 2, 3,.... Hence the aggregate of equations is enumerable. But every algebraic number corresponds to at least one of these equations, and the number of algebraic numbers corresponding to any equation is finite. Hence

THEOREM 189. *The aggregate of algebraic numbers is enumerable.*

In particular, the aggregate of real algebraic numbers has measure zero.

THEOREM 190. *Almost all real numbers are transcendental.*

Cantor, who had not the more modern concept of measure, arranged his proof of the existence of transcendental numbers differently. After Theorem 189, it is enough to prove that *the continuum* $0 \leqslant x < 1$ *is not enumerable.* We represent x by its decimal

$$x = \cdot a_1 a_2 a_3 \ldots$$

(9 being excluded, as in § 9.1). Suppose that the continuum is enumerable, as $x_1, x_2, x_3, \ldots$, and let

$$\begin{aligned} x_1 &= \cdot a_{11} a_{12} a_{13} \ldots \\ x_2 &= \cdot a_{21} a_{22} a_{23} \ldots \\ x_3 &= \cdot a_{31} a_{32} a_{33} \ldots \\ &\quad \cdot\ \cdot\ \cdot\ \cdot\ \cdot \end{aligned}$$

If now we define a_n by

$$a_n = a_{nn} + 1 \quad (\text{if } a_{nn} \text{ is neither 8 nor 9}),$$
$$a_n = 0 \quad (\text{if } a_{nn} \text{ is 8 or 9}),$$

then $a_n \neq a_{nn}$ for any n; and x cannot be any of $x_1, x_2, \ldots$, since its decimal differs from that of any x_n in the nth digit. This is a contradiction.

11.7. Liouville's theorem and the construction of transcendental numbers. Liouville proved a theorem which enables us to produce as many examples of transcendental numbers as we please. It is the generalization to algebraic numbers of any degree of the negative half of Theorem 188.

THEOREM 191. *A real algebraic number of degree n is not approximable to any order greater than n.*

An algebraic number ξ satisfies an equation

$$f(\xi) = a_0\xi^n + a_1\xi^{n-1} + \cdots + a_n = 0$$

with integral coefficients. There is a number $M(\xi)$ such that

(11.7.1) $$|f'(x)| < M \qquad (\xi - 1 < x < \xi + 1).$$

Suppose now that $p/q \neq \xi$ is an approximation to ξ. We may assume the approximation close enough to ensure that p/q lies in $(\xi-1, \xi+1)$, and is nearer to ξ than any other root of $f(x) = 0$, so that $f(p/q) \neq 0$. Then

(11.7.2) $$\left|f\left(\frac{p}{q}\right)\right| = \frac{|a_0p^n + a_1p^{n-1}q + \cdots|}{q^n} \geqslant \frac{1}{q^n},$$

since the numerator is a positive integer; and

(11.7.3) $$f\left(\frac{p}{q}\right) = f\left(\frac{p}{q}\right) - f(\xi) = \left(\frac{p}{q} - \xi\right)f'(x),$$

where x lies between p/q and ξ. It follows from (11.7.2) and (11.7.3) that

$$\left|\frac{p}{q} - \xi\right| = \frac{|f(p/q)|}{|f'(x)|} > \frac{1}{Mq^n} = \frac{K}{q^n},$$

so that ξ is not approximable to any order higher than n.

The cases $n = 1$ and $n = 2$ are covered by Theorems 186 and 188. These theorems, of course, included a positive as well as a negative statement.

(*a*) Suppose, for example, that

$$\xi = \cdot 110001000\ldots = 10^{-1!} + 10^{-2!} + 10^{-3!} + \ldots,$$

that $n > N$, and that ξ_n is the sum of the first n terms of the series. Then

$$\xi_n = \frac{p}{10^{n!}} = \frac{p}{q},$$

say. Also

$$0<\xi - \frac{p}{q} = \xi - \xi_n = 10^{-(n+1)!} + 10^{-(n+2)!} + \cdots < 2.10^{-(n+1)!}<2q^{-N}.$$

Hence ξ is not an algebraic number of degree less than N. Since N is arbitrary, ξ is transcendental.

(*b*) Suppose that

$$\xi = \frac{1}{10+}\,\frac{1}{10^{2!}+}\,\frac{1}{10^{3!}+\cdots},$$

that $n > N$, and that

$$\frac{p}{q} = \frac{p_n}{q_n},$$

the nth convergent to ξ. Then

$$\left|\frac{p}{q} - \xi\right| = \frac{1}{q_n q'_{n+1}} < \frac{1}{a_{n+1}q_n^2} < \frac{1}{a_{n+1}}.$$

Now $a_{n+1} = 10^{(n+1)!}$ and

$$q_1 < a_1 + 1, \qquad \frac{q_{n+1}}{q_n} = a_{n+1} + \frac{q_{n-1}}{q_n} < a_{n+1} + 1 \quad (n \geqslant 1);$$

so that

$$\begin{aligned}
q_n &< (a_1+1)(a_2+1)\cdots(a_n+1) \\
&< \left(1+\frac{1}{10}\right)\left(1+\frac{1}{10^2}\right)\cdots\left(1+\frac{1}{10^n}\right)a_1a_2\cdots a_n \\
&< 2a_1\,a_2\cdots a_n = 2.10^{1!+\cdots+n!} < 10^{2(n!)} = a_n^2, \\
\left|\frac{p}{q} - \xi\right| &< \frac{1}{a_{n+1}} = \frac{1}{a_n^{n+1}} < \frac{1}{a_n^n} < \frac{1}{q_n^{\frac{1}{2}n}} < \frac{1}{q_n^{\frac{1}{2}N}}.
\end{aligned}$$

We conclude, as before, that ξ is transcendental.

THEOREM 192. *The numbers*

$$\xi = 10^{-1!} + 10^{-2!} + 10^{-3!} + \cdots$$

and

$$\xi = \frac{1}{10^{1!}+}\,\frac{1}{10^{2!}+}\,\frac{1}{10^{3!}+\cdots}$$

are transcendental.

It is plain that we could replace 10 by other integers, and vary the construction in many other ways. The general principle of the construction is simply that *a number defined by a sufficiently rapid sequence of rational approximations is necessarily transcendental.* It is the *simplest* irrationals, such as $\sqrt{2}$ or $\frac{1}{2}(\sqrt{5}-1)$, which are the least rapidly approximable.

It is much more difficult to prove that a number given 'naturally' is transcendental. We shall prove e and π transcendental in §§ 11.13–14. Few classes of transcendental numbers are known even now. These classes include, for example, the numbers

$$e,\ \pi,\ \sin 1, J_0(1),\ \log 2,\ \frac{\log 3}{\log 2},\ e^{\pi},\ 2^{\sqrt{2}}$$

but not 2^e, 2^{π}, π^e, or Euler's constant γ. It has never been proved even that any of these last numbers are irrational.

11.8. The measure of the closest approximations to an arbitrary irrational. We know that every irrational has an infinity of approximations satisfying (11.1.1), and indeed, after Theorem 183 of Ch. X, of rather better approximations. We know also that an algebraic number, which is an irrational of a comparatively simple type, cannot be 'too rapidly' approximable, while the transcendental numbers of Theorem 192 have approximations of abnormal rapidity.

The best approximations to ξ are given, after Theorem 181, by the convergents p_n/q_n of the continued fraction for ξ; and

$$\left|\frac{p_n}{q_n} - \xi\right| = \frac{1}{q_n q'_{n+1}} < \frac{1}{a_{n+1}q_n^2},$$

so that we get a particularly good approximation when a_{n+1} is large. It is plain that, to put the matter roughly, ξ will or will not be rapidly approximable according as its continued fraction does or does not contain a sequence of rapidly increasing quotients. The second ξ of Theorem 192, whose quotients increase with great rapidity, is a particularly instructive example.

One may say, again very roughly, that the structure of the continued fraction for ξ affords a measure of the 'simplicity' or 'complexity' of ξ. Thus the second ξ of Theorem 192 is a 'complicated' number. On the other hand, if a_n behaves regularly, and does not become too large, then ξ may reasonably be regarded as a 'simple' number; and in this case the rational approximations to ξ cannot be too good. From the point of view of rational approximation, *the simplest numbers are the worst.*

The 'simplest' of all irrationals, from this point of view, is the number

$$\xi = \frac{1}{2}\left(\sqrt{5}-1\right) = \frac{1}{1+}\,\frac{1}{1+}\,\frac{1}{1+\cdots}, \tag{11.8.1}$$

in which every a_n has the smallest possible value. The convergents to this fraction are

$$\frac{0}{1}, \frac{1}{1}, \frac{1}{2}, \frac{2}{3}, \frac{3}{5}, \frac{5}{8}, \ldots$$

so that $q_{n-1} = p_n$ and $$\frac{q_{n-1}}{q_n} = \frac{p_n}{q_n} \to \xi.$$

Hence

$$\left|\frac{p_n}{q_n} - \xi\right| = \frac{1}{q_n q'_{n+1}} = \frac{1}{q_n\{(1+\xi)q_n + q_{n-1}\}}$$

$$= \frac{1}{q_n^2}\left(1 + \xi + \frac{q_{n-1}}{q_n}\right)^{-1} \sim \frac{1}{q_n^2}\,\frac{1}{1+2\xi} = \frac{1}{q_n^2\sqrt{5}},$$

when $n \to \infty$.

These considerations suggest the truth of the following theorem.

THEOREM 193. *Any irrational ξ has an infinity of approximations which satisfy*

$$\left|\frac{p}{q} - \xi\right| < \frac{1}{q^2\sqrt{5}}. \tag{11.8.2}$$

The proof of this theorem requires some further analysis of the approximations given by the convergents to the continued fraction. This we give in the next section, but we prove first a complement to the theorem which shows that it is in a certain sense a 'best possible' theorem.

THEOREM 194. *In Theorem 193, the number $\sqrt{5}$ is the best possible number: the theorem would become false if any larger number were substituted for $\sqrt{5}$.*

It is enough to show that, if $A > \sqrt{5}$, and ξ is the particular number (11.8.1), then the inequality

$$\left|\frac{p}{q} - \xi\right| < \frac{1}{Aq^2}$$

has only a finite number of solutions.

Suppose the contrary. Then there are infinitely many q and p such that

$$\xi = \frac{p}{q} + \frac{\delta}{q^2}, \quad |\delta| < \frac{1}{A} < \frac{1}{\sqrt{5}}.$$

Hence
$$\frac{\delta}{q} = q\xi - p, \quad \frac{\delta}{q} - \frac{1}{2}q\sqrt{5} = -\frac{1}{2}q - p,$$

$$\frac{\delta^2}{q^2} - \delta\sqrt{5} = \left(\frac{1}{2}q + p\right)^2 - \frac{5}{4}q^2 = p^2 + pq - q^2.$$

The left-hand side is numerically less than 1 when q is large, while the right-hand side is integral. Hence $p^2 + pq - q^2 = 0$ or $(2p + q)^2 = 5q^2$, which is plainly impossible.

11.9. Another theorem concerning the convergents to a continued fraction. Our main object in this section is to prove

THEOREM 195. *Of any three consecutive convergents to ξ, one at least satisfies* (11.8.2).

This theorem should be compared with Theorem 183 of Ch. X.

We write

(11.9.1) $$\frac{q_{n-1}}{q_n} = b_{n+1}.$$

Then

$$\left|\frac{p_n}{q_n} - \xi\right| = \frac{1}{q_n q'_{n+1}} = \frac{1}{q_n^2}\frac{1}{a'_{n+1} + b_{n+1}};$$

and it is enough to prove that

(11.9.2) $$a'_i + b_i \leqslant \sqrt{5}$$

cannot be true for the three values $n-1, n, n+1$ of i.

Suppose that (11.9.2) is true for $i = n-1$ and $i = n$. We have

$$a'_{n-1} = a_{n-1} + \frac{1}{a'_n}$$

and

(11.9.3) $$\frac{1}{b_n} = \frac{q_{n-1}}{q_{n-2}} = a_{n-1} + b_{n-1}.$$

Hence

$$\frac{1}{a_n'} + \frac{1}{b_n} = a'_{n-1} + b_{n-1} \leqslant \sqrt{5},$$

and

$$1 - a_n' \frac{1}{a_n'} \leqslant (\sqrt{5} - b_n)\left(\sqrt{5} - \frac{1}{b_n}\right)$$

or

$$b_n + \frac{1}{b_n} \leqslant \sqrt{5}.$$

Equality is excluded, since b_n is rational, and $b_n < 1$. Hence

$$b_n^2 - b_n\sqrt{5} + 1 < 0, \quad \left(\frac{1}{2}\sqrt{5} - b_n\right)^2 < \frac{1}{4},$$

(11.9.4) $$b_n > \frac{1}{2}(\sqrt{5} - 1).$$

If (11.9.2) were true also for $i = n + 1$, we could prove similarly that

(11.9.5) $$b_{n+1} > \frac{1}{2}\left(\sqrt{5} - 1\right);$$

and (11.9.3),† (11.9.4), and (11.9.5) would give

$$a_n = \frac{1}{b_{n+1}} - b_n < \frac{1}{2}\left(\sqrt{5} + 1\right) - \frac{1}{2}\left(\sqrt{5} - 1\right) = 1,$$

a contradiction. This proves Theorem 195, and Theorem 193 is a corollary.

† With $n + 1$ for n.

11.10. Continued fractions with bounded quotients. The number $\sqrt{5}$ has a special status, in Theorems 193 and 195, which depends upon the particular properties of the number (11.8.1). For this ξ, every a_n is 1; for a ξ equivalent to this one, in the sense of § 10.11, every a_n from a certain point is 1; but, for any other ξ, a_n is at least 2 for infinitely many n. It is natural to suppose that, if we excluded ξ equivalent to (11.8.1), the $\sqrt{5}$ of Theorem 193 could be replaced by some larger number; and this is actually true. *Any irrational ξ not equivalent to* (11.8.1) *has an infinity of rational approximations for which*

$$\left|\frac{p}{q} - \xi\right| < \frac{1}{2q^2\sqrt{2}}.$$

There are other numbers besides $\sqrt{5}$ and $2\sqrt{2}$ which play a special part in problems of this character, but we cannot discuss these problems further here.

If a_n is not bounded, i.e. if

$$\overline{\lim_{n\to\infty}}\, a_n = \infty, \tag{11.10.1}$$

then q'_{n+1}/q_n assumes arbitrarily large values, and

$$\left|\frac{p}{q} - \xi\right| < \frac{\epsilon}{q^2} \tag{11.10.2}$$

for every positive ϵ and an infinity of p and q. Our next theorem shows that this is the general case, since (11.10.1) is true for 'almost all' ξ in the sense of § 9.10.

THEOREM 196. *a_n is unbounded for almost all ξ; the set of ξ for which a_n is bounded is null.*

We may confine our attention to ξ of (0,1), so that $a_0 = 0$, and to irrational ξ, since the set of rationals is null. It is enough to show that the set F_k of irrational ξ for which

$$a_n \leqslant k \tag{11.10.3}$$

is null; for the set for which a_n is bounded is the sum of $F_1, F_2, F_3, \ldots$.

We denote by

$$E_{a_1, a_2, \ldots, a_n}$$

the set of irrational ξ for which the first n quotients have given values a_1, $a_2, \ldots, a_n$. The set E_{a_1} lies in the interval

$$\frac{1}{a_1+1}, \frac{1}{a_1},$$

which we call I_{a_1}. The set E_{a_1,a_2} lies in

$$\frac{1}{a_1+}\frac{1}{a_2}, \quad \frac{1}{a_1+}\frac{1}{a_2+1},$$

which we call I_{a_1,a_2}. Generally, $E_{a_1,a_2,\ldots,a_n}$ lies in the interval $I_{a_1,a_2,\ldots,a_n}$ whose end points are

$$[a_1, a_2, \ldots, a_{n-1}, a_n+1], \quad [a_1, a_2, \ldots, a_{n-1}, a_n]$$

(the first being the left-hand end point when n is odd). The intervals corresponding to different sets $a_1, a_2, \ldots, a_n$ are mutually exclusive (except that they may have end points in common), the choice of $a_{\nu+1}$ dividing up $I_{a_1,a_2,\ldots,a_\nu}$ into exclusive intervals. Thus $I_{a_1,a_2,\ldots,a_n}$ is the sum of

$$I_{a_1,a_2,\ldots,a_n,1}, \quad I_{a_1,a_2,\ldots,a_n,2}, \ldots.$$

The end points of $I_{a_1,a_2,\ldots,a_n}$, can also be expressed as

$$\frac{(a_n+1)p_{n-1}+p_{n-2}}{(a_n+1)q_{n-1}+q_{n-2}}, \quad \frac{a_np_{n-1}+p_{n-2}}{a_nq_{n-1}+q_{n-2}};$$

and its length (for which we use the same symbol as for the interval) is

$$\frac{1}{\{(a_n+1)q_{n-1}+q_{n-2}\}(a_nq_{n-1}+q_{n-2})} = \frac{1}{(q_n+q_{n-1})q_n}.$$

Thus

$$I_{a_1} = \frac{1}{(a_1+1)a_1}.$$

We denote by

$$E_{a_1,a_2,\ldots,a_n;k}$$

the sub-set of $E_{a_1,a_2,\ldots,a_n}$ for which $a_{n+1} \leqslant k$. The set is the sum of

$$E_{a_1,a_2,\ldots,a_n,a_{n+1}} \quad (a_{n+1} = 1, 2, \ldots, k).$$

The last set lies in the interval $I_{a_1, a_2, \dots, a_n, a_{n+1}}$, whose end points are

$$[a_1, a_2, \dots, a_n, a_{n+1} + 1], \qquad [a_1, a_2, \dots, a_n, a_{n+1}];$$

and so $E_{a_1, a_2, \dots, a_n; k}$ lies in the interval $I_{a_1, a_2, \dots, a_n; k}$ whose end points are

$$[a_1, a_2, \dots, a_n, k + 1], \qquad [a_1, a_2, \dots, a_n, 1],$$

or

$$\frac{(k+1)p_n + p_{n-1}}{(k+1)q_n + q_{n-1}}, \qquad \frac{p_n + p_{n-1}}{q_n + q_{n-1}}.$$

The length of $I_{a_1, a_2, \dots, a_n; k}$ is

$$\frac{k}{\{(k+1)q_n + q_{n-1}\}(q_n + q_{n-1})};$$

and

$$\frac{I_{a_1, a_2, \dots, a_n; k}}{I_{a_1, a_2, \dots, a_n}} = \frac{kq_n}{(k+1)q_n + q_{n-1}} < \frac{k}{k+1}, \tag{11.10.4}$$

for all $a_1, a_2, \dots, a_n$.

Finally, we denote by

$$I_k^{(n)} = \sum_{a_1 \leqslant k, \dots, a_n \leqslant k} I_{a_1, a_2, \dots, a_n}$$

the sum of the $I_{a_1, \dots, a_n}$ for which $a_1 \leqslant k, \dots, a_n \leqslant k$; and by $F_k^{(n)}$ the set of irrational ξ for which $a_1 \leqslant k, \dots, a_n \leqslant k$. Plainly $F_k^{(n)}$ is included in $I_k^{(n)}$

First, $I_k^{(1)}$ is the sum of I_{a_1} for $a_1 = 1, 2, \dots, k$, and

$$I_k^{(1)} = \sum_{a_1=1}^{k} \frac{1}{a_1(a_1+1)} = 1 - \frac{1}{k+1} = \frac{k}{k+1}.$$

Generally, $I_k^{(n+1)}$ is the sum of the parts of the $I_{a_1, a_2, \dots, a_n}$, included in $I_k^{(n)}$, for which $a_{n+1} \leqslant k$, i.e. is

$$\sum_{a_1 \leqslant k, \dots, a_n \leqslant k} I_{a_1, a_2, \dots, a_n; k}.$$

Hence, by (11.10.4),

$$I_k^{(n+1)} < \frac{k}{k+1} \sum_{a_1 \leqslant k, \ldots, a_n \leqslant k} I_{a_1, a_2, \ldots, a_n} = \frac{k}{k+1} I_k^{(n)};$$

and so

$$I_k^{(n+1)} < \left(\frac{k}{k+1}\right)^{n+1}.$$

It follows that $F_k^{(n)}$ can be included in a set of intervals of length less than

$$\left(\frac{k}{k+1}\right)^n,$$

which tends to zero when $n \to \infty$. Since F_k is part of $F_k^{(n)}$ for every n, the theorem follows.

It is possible to prove a good deal more by the same kind of argument. Thus Borel and F. Bernstein proved

THEOREM 197*. *If $\phi(n)$ is an increasing function of n for which*

(11.10.5) $$\sum \frac{1}{\phi(n)}$$

is divergent, then the set of ξ for which

(11.10.6) $$a_n \leqslant \phi(n),$$

for all sufficiently large n, is null. On the other hand, if

(11.10.7) $$\sum \frac{1}{\phi(n)}$$

is convergent, then (11.10.6) *is true for almost all ξ and sufficiently large n.*

Theorem 196 is the special case of this theorem in which $\phi(n)$ is a constant. The proof of the general theorem is naturally a little more complex, but does not involve any essentially new idea.

11.11. Further theorems concerning approximation. Let us suppose, to fix our ideas, that a_n tends steadily, fairly regularly, and not too rapidly, to infinity. Then

$$\left|\frac{p_n}{q_n} - x\right| = \frac{1}{q_n q'_{n+1}} \sim \frac{1}{a_{n+1} q_n^2} = \frac{1}{q_n \chi(q_n)},$$

where

$$\chi(q_n) = a_{n+1} q_n.$$

There is a certain correspondence between the behaviour, in respect of convergence or divergence, of the series†

$$\sum_{\nu} \frac{1}{\chi(\nu)}, \quad \sum_{n} \frac{q_n}{\chi(q_n)};$$

and the latter series is

$$\sum \frac{1}{a_{n+1}}.$$

These rough considerations suggest that, if we compare the inequalities

$$a_n < \phi(n) \tag{11.11.1}$$

and

$$\left|\frac{p}{q} - \xi\right| < \frac{1}{q\chi(q)}, \tag{11.11.2}$$

there should be a certain correspondence between conditions on the two series

$$\sum \frac{1}{\phi(n)}, \quad \sum \frac{1}{\chi(q)}.$$

And the theorems of § 11.10 then suggest the two which follow.

THEOREM 198. *If*

$$\sum \frac{1}{\chi(q)}$$

is convergent, then the set of ξ *which satisfy* (11.11.2) *for an infinity of* q *is null.*

THEOREM 199*. *If* $\chi(q)/q$ *increases with* q, *and*

$$\sum \frac{1}{\chi(q)}$$

is divergent, then (11.11.2) *is true, for an infinity of* q, *for almost all* ξ.

† The idea is that underlying 'Cauchy's condensation test' for the convergence or divergence of a series of decreasing positive terms. See Hardy, *Pure mathematics,* 9th ed., 354.

Theorem 199 is difficult. But Theorem 198 is very easy, and can be proved without continued fractions. It shows, roughly, that most irrationals cannot be approximated by rationals with an error of order much less than q^{-2}, e.g. with an error

$$O\left\{\frac{1}{q^2(\log q)^2}\right\}.$$

The more difficult theorem shows that approximation to such orders as

$$O\left(\frac{1}{q^2 \log q}\right), \quad O\left(\frac{1}{q^2 \log q \log\log q}\right), \ldots$$

is usually possible.

We may suppose $0 < \xi < 1$. We enclose every p/q for which $q \geqslant N$ in an interval

$$\frac{p}{q} - \frac{1}{q\chi(q)}, \quad \frac{p}{q} + \frac{1}{q\chi(q)}.$$

There are less than q values of p corresponding to a given q, and the total length of the intervals is less (even without allowance for overlapping) than

$$2\sum_{N}^{\infty} \frac{1}{\chi(q)},$$

which tends to 0 when $N \to \infty$. Any ξ which has the property is included in an interval, whatever be N, and the set of ξ can therefore be included in a set of intervals whose total length is as small as we please.

11.12. Simultaneous approximation. So far we have been concerned with approximations to a single irrational ξ. Dirichlet's argument of § 11.3 has an important application to a multi-dimensional problem, that of the simultaneous approximation of k numbers

$$\xi_1, \ \xi_2, \ldots, \xi_k$$

by fractions

$$\frac{p_1}{q}, \frac{p_2}{q}, \ldots, \frac{p_k}{q}$$

with the same denominator q (but not necessarily irreducible).

THEOREM 200. *If $\xi_1, \xi_2, \ldots, \xi_k$ are any real numbers, then the system of inequalities*

$$\left|\frac{p_i}{q} - \xi_i\right| < \frac{1}{q^{1+\mu}} \qquad \left(\mu = \frac{1}{k}; \quad i = 1, 2, \ldots, k\right) \tag{11.12.1}$$

has at least one solution. If one ξ at least is irrational, then it has an infinity of solutions.

We may plainly suppose that $0 \leqslant \xi_i < 1$ for every i. We consider the k-dimensional 'cube' defined by $0 \leqslant x_i < 1$, and divide it into Q^k 'boxes' by drawing 'planes' parallel to its faces at distances $1/Q$. Of the Q^k+1 points

$$(l\xi_1),\ (l\xi_2), \ldots, (l\xi_k) \quad (l = 0, 1, 2, \ldots, Q^k),$$

some two, corresponding say to $l = q_1$ and $l = q_2 > q_1$, must lie in the same box. Hence, taking $q = q_2 - q_1$, as in § 11.3, there is a $q \leqslant Q^k$ such that

$$|\overline{q\xi_i}| < \frac{1}{Q} \leqslant \frac{1}{q^\mu}$$

for every i.

The proof may be completed as before; if a ξ, say ξ_i, is irrational, then ξ_i may be substituted for ξ in the final argument of § 11.3.

In particular we have

THEOREM 201. *Given ξ_1, ξ_2,..., ξ_k and any positive ϵ, we can find an integer q so that $q\xi_i$ differs from an integer, for every i, by less than ϵ.*

11.13. The transcendence of *e*. We conclude this chapter by proving that e and π are transcendental.

Our work will be considerably simplified by the introduction of a symbol h^r, which we define by

$$h^0 = 1, \quad h^r = r! \quad (r \geqslant 1).$$

If $f(x)$ is any polynomial in x of degree m, say

$$f(x) = \sum_{r=0}^{m} c_r x^r,$$

then we define $f(h)$ as

$$\sum_{r=0}^{m} c_r h^r = \sum_{r=0}^{m} c_r r!$$

(where 0! is to be interpreted as 1). Finally we define $f(x+h)$ in the manner suggested by Taylor's theorem, viz. as

$$\sum_{r=0}^{m} \frac{f^{(r)}(x)}{r!} h^r = \sum_{r=0}^{m} f^{(r)}(x).$$

If $f(x+y) = F(y)$, then $f(x+h) = F(h)$.

We define $u_r(x)$ and $\epsilon_r(x)$, for $r = 0, 1, 2, \ldots$, by

$$u_r(x) = \frac{x}{r+1} + \frac{x^2}{(r+1)(r+2)} + \cdots = e^{|x|}\epsilon_r(x).$$

It is obvious that $|u_r(x)| < e^{|x|}$, and so

$$|\epsilon_r(x)| < 1, \tag{11.13.1}$$

for all x.

We require two lemmas.

THEOREM 202. *If $\phi(x)$ is any polynomial and*

$$\phi(x) = \sum_{r=0}^{s} c_r x^r, \quad \psi(x) = \sum_{r=0}^{s} c_r \epsilon_r(x) x^r, \tag{11.13.2}$$

then

$$e^x \phi(h) = \phi(x+h) + \psi(x) e^{|x|}. \tag{11.13.3}$$

By our definitions above we have

$$\begin{aligned}
(x+h)^r &= h^r + rxh^{r-1} + \frac{r(r-1)}{1.2} x^2 h^{r-2} + \cdots + x^r \\
&= r! + r(r-1)!x + \frac{r(r-1)}{1.2}(r-2)!x^2 + \cdots + x^r \\
&= r!\left(1 + x + \frac{x^2}{2!} + \cdots + \frac{x^r}{r!}\right) \\
&= r!e^x - u_r(x)x^r - e^x h^r - u_r(x)x^r.
\end{aligned}$$

Hence

$$e^x h^r = (x+h)^r + u_r(x)x^r = (x+h)^r + e^{|x|}\epsilon_r(x)x^r.$$

Multiplying this throughout by c_r, and summing, we obtain (11.13.3).

As in § 7.2, we call a polynomial in x, or in $x, y, \ldots$, whose coefficients are integers, an integral polynomial in x, or $x, y, \ldots$.

THEOREM 203. *If $m \geqslant 2$, $f(x)$ is an integral polynomial in x, and*

$$F_1(x) = \frac{x^{m-1}}{(m-1)!}f(x), \quad F_2(x) = \frac{x^m}{(m-1)!}f(x),$$

then $F_1(h)$, $F_2(h)$ are integers and

$$F_1(h) \equiv f(0), \quad F_2(h) \equiv 0 \pmod{m}.$$

Suppose that

$$f(x) = \sum_{l=0}^{L} a_l x^l,$$

where $a_0, \ldots, a_L$ are integers. Then

$$F_1(x) = \sum_{l=0}^{L} a_l \frac{x^{l+m-1}}{(m-1)!},$$

and so

$$F_1(h) = \sum_{l=0}^{L} a_l \frac{(l+m-1)!}{(m-1)!}.$$

But

$$\frac{(l+m-1)!}{(m-1)!} = (l+m-1)(l+m-2)\cdots m$$

is an integral multiple of m if $l \geqslant 1$; and therefore

$$F_1(h) \equiv a_0 = f(0) \pmod{m}.$$

Similarly

$$F_2(x) = \sum_{l=0}^{L} a_l \frac{x^{l+m}}{(m-1)!},$$

$$F_2(h) = \sum_{l=0}^{L} a_l \frac{(l+m)!}{(m-1)!} \equiv 0 \pmod{m}.$$

We are now in a position to prove the first of our two main theorems, namely

Theorem 204. *e is transcendental.*

If the theorem is not true, then

$$\sum_{t=0}^{n} C_t e^t = 0, \tag{11.13.4}$$

where $n \geqslant 1$, $C_0, C_1, \ldots, C_n$ are integers, and $C_0 \neq 0$.

We suppose that p is a prime greater than $\max(n, |C_0|)$, and define $\phi(x)$ by

$$\phi(x) = \frac{x^{p-1}}{(p-1)!}\{(x-1)(x-2)\ldots(x-n)\}^p.$$

Ultimately, p will be large. If we multiply (11.13.4) by $\phi(h)$, and use (11.13.3), we obtain

$$\sum_{t=0}^{n} C_t \phi(t+h) + \sum_{t=0}^{n} C_t \psi(t) e^t = 0,$$

or

$$S_1 + S_2 = 0, \tag{11.13.5}$$

say.

By Theorem 203, with $m = p$, $\phi(h)$ is an integer and

$$\phi(h) \equiv (-1)^{pn}(n!)^p \pmod{p}.$$

Again, if $1 \leqslant t \leqslant n$,

$$\begin{aligned}\phi(t+x) &= \frac{(t+x)^{p-1}}{(p-1)!}\{(x+t-1)\ldots x(x-1)\ldots(x+t-n)\}^p \\ &= \frac{x^p}{(p-1)!}f(x),\end{aligned}$$

where $f(x)$ is an integral polynomial in x. It follows (again from Theorem 203) that $\phi(t+h)$ is an integer divisible by p. Hence

$$S_1 = \sum_{t=0}^{n} C_t\phi\,(t+h) \equiv (-1)^{pn}C_0\,(n!)^p \not\equiv 0\,(\mathrm{mod}\ p)\,,$$

since $C_0 \neq 0$ and $p > \max(n, |C_0|)$. Thus S_1 is an integer, not zero; and therefore

$$|S_1| \geqslant 1. \tag{11.13.6}$$

On the other hand, $|\epsilon_r(x)| < 1$, by (11.13.1), and so

$$\begin{aligned}|\psi(t)| &< \sum_{r=0}^{S}|c_r|\,t^r \\ &\leqslant \frac{t^{p-1}}{(p-1)!}\{(t+1)\,(t+2)\cdots(t+n)\}^p \to 0,\end{aligned}$$

when $p \to \infty$. Hence $S_2 \to 0$, and we can make

$$|S_2| < \frac{1}{2} \tag{11.13.7}$$

by choosing a sufficiently large value of p. The formulae (11.13.5), (11.13.6), and (11.13.7) are in contradiction. Hence (11.13.4) is impossible and e is transcendental.

The proof which precedes is a good deal more sophisticated than the simple proof of the irrationality of e given in § 4.7, but the ideas which underlie it are essentially the same. We use (i) the exponential series and (ii) the theorem that an integer whose modulus is less than 1 must be 0.

11.14. The transcendence of π. Finally we prove that π is transcendental. It is this theorem which settles the problem of the 'quadrature of the circle'.

THEOREM 205. *π is transcendental.*

The proof is very similar to that of Theorem 204, but there are one or two slight additional complications.

Suppose that $\beta_1, \beta_2, \ldots, \beta_m$ are the roots of an equation

$$dx^m + d_1x^{m-1} + \cdots + d_m = 0$$

with integral coefficients. Any symmetrical integral polynomial in

$$d\beta_1, d\beta_2, \ldots, d\beta_m$$

is an integral polynomial in

$$d_1, d_2, \ldots, d_m,$$

and is therefore an integer.

Now let us suppose that π is algebraic. Then $i\pi$ is algebraic,† and therefore the root of an equation

$$dx^m + d_1x^{m-1} + \cdots + d_m = 0,$$

where $m \geqslant 1$, $d, d_1, \ldots, d_m$ are integers, and $d \neq 0$. If the roots of this equation are

$$\omega_1, \ \omega_2, \ldots, \omega_m,$$

then $1+e^{\omega} = 1+e^{i\pi} = 0$ for some ω, and therefore

$$(1+e^{\omega_1})(1+e^{\omega_2})\ldots(1+e^{\omega_m}) = 0.$$

† If $a_0x^n + a_1x^{n-1} + \cdots + a_n = 0$ and $y = ix$, then

$$a_0y^n - a_2y^{n-2} + \cdots + i(a_1y^{n-1} - a_3y^{n-3} + \cdots) = 0$$

and so

$$(a_0y^n - a_2y^{n-2} + \cdots)^2 + (a_1y^{n-1} - a_3y^{n-3} + \cdots)^2 = 0.$$

Multiplying this out, we obtain

$$1+\sum_{t=1}^{2^m-1} e^{\alpha_t}=0, \tag{11.14.1}$$

where

$$\alpha_1,\alpha_2,\ldots,\alpha_{2^m-1} \tag{11.14.2}$$

are the 2^m-1 numbers

$$\omega_1,\ldots,\omega_m,\omega_1+\omega_2,\omega_1+\omega_3,\ldots,\omega_1+\omega_2+\cdots+\omega_m,$$

in some order.

Let us suppose that $C-1$ of the α are zero and that the remaining

$$n=2^m-1-(C-1)$$

are not zero; and that the non-zero α are arranged first, so that (11.14.2) reads

$$\alpha_1,\ldots,\alpha_n,0,0,\ldots,0.$$

Then it is clear that any symmetrical integral polynomial in

$$d\alpha 1,\ldots,\ d\alpha n \tag{11.14.3}$$

is a symmetrical integral polynomial in

$$d\alpha_1,\ldots,d\alpha_n,0,0,\ldots,0,$$

i.e. in

$$d\alpha_1,\ d\alpha_2,\ \ldots,\ d\alpha_{2^m-1}.$$

Hence any such function is a symmetrical integral polynomial in

$$d\omega_1,d\omega_2,\ldots,d\omega_m,$$

and so an integer.

We can write (11.14.1) as

$$C+\sum_{t=1}^{n} e^{\alpha_t}=0. \tag{11.14.4}$$

We choose a prime p such that

(11.14.5) $$p > \max(d,\ C, |d^n\alpha_1\cdots\alpha_n|)$$

and define $\phi(x)$ by

(11.14.6) $$\phi(x) = \frac{d^{np+p-1}x^{p-1}}{(p-1)!}\{(x-\alpha_1)(x-\alpha_2)\cdots(x-\alpha_n)\}^p.$$

Multiplying (11.14.4) by $\phi(h)$, and using (11.13.3), we obtain

(11.14.7) $$S_0 + S_1 + S_2 = 0,$$

where

(11.14.8) $$S_0 = C\phi(h),$$

(11.14.9) $$S_1 = \sum_{t=1}^{n}\phi(\alpha_t + h),$$

(11.14.10) $$S_2 = \sum_{t=1}^{n}\psi(\alpha_t)\,e^{|\alpha_t|}.$$

Now

$$\phi(x) = \frac{x^{p-1}}{(p-1)!}\sum_{l=0}^{np} g_l x^l,$$

where g_l is a symmetric integral polynomial in the numbers (11.14.3), and so an integer. It follows from Theorem 203 that $\phi(h)$ is an integer, and that

(11.14.11) $$\phi(h) \equiv g_0 = (-1)^{pn}\,d^{p-1}\,(d\alpha_1.d\alpha_2.\ \ldots\ .d\alpha_n)^p \pmod{p}.$$

Hence S_0 is an integer; and

(11.14.12) $$S_0 \equiv Cg_0 \not\equiv 0 \pmod{p},$$

because of (11.14.5).

Next, by substitution and rearrangement, we see that

$$\phi(\alpha_t + x) = \frac{x^p}{(p-1)!}\sum_{l=0}^{np-1} f_{l,t}x^l,$$

where

$$f_{l,t} = f_l(d\alpha_t; d\alpha_1, d\alpha_2, \ldots, d\alpha_{t-1}, d\alpha_{t+1}, \ldots, d\alpha_n)$$

is an integral polynomial in the numbers (11.14.3), symmetrical in all but $d\alpha_t$. Hence

$$\sum_{t=1}^{n} \phi(\alpha_l + x) = \frac{x^p}{(p-1)!} \sum_{l=0}^{np-1} F_l x^l,$$

where

$$F_l = \sum_{t=1}^{n} f_{l,t} = \sum_{t=1}^{n} f_l(d\alpha_t; d\alpha_1, \ldots, d\alpha_{t-1}, d\alpha_{t+1}, \ldots, d\alpha_n).$$

It follows that F_l is an integral polynomial symmetrical in all the numbers (11.14.3), and so an integer. Hence, by Theorem 203,

$$S_1 = \sum_{t=1}^{n} \phi(\alpha_t + h)$$

is an integer, and

(11.14.13) $$S_1 \equiv 0 \pmod{p}.$$

From (11.14.12) and (11.14.13) it follows that $S_0 + S_1$ is an integer not divisible by p, and so that

(11.14.14) $$|S_0 + S_1| \geqslant 1.$$

On the other hand,

$$|\psi(x)| < \frac{|d|^{np+p-1}|x|^{p-1}}{(p-1)!} \{(|x| + |\alpha_1|) \ldots (|x| + |\alpha_n|)\} p \to 0,$$

for any fixed x, when $p \to \infty$. It follows that

(11.14.15) $$|S_2| < \frac{1}{2}$$

for sufficiently large p. The three formulae (11.14.7), (11.14.14), and (11.14.15) are in contradiction, and therefore π is transcendental.

In particular π is not a 'Euclidean' number in the sense of § 11.5; and therefore it is impossible to construct, by Euclidean methods, a length equal to the circumference of a circle of unit diameter.

It may be proved by the methods of this section that

$$\alpha_1 e^{\beta_1} + \alpha_2 e^{\beta_2} + \cdots + \alpha_s e^{\beta_s} \neq 0$$

if the α and β are algebraic, the α are not all zero, and no two β are equal.

It has been proved more recently that α^β is transcendental if α and β are algebraic, α is not 0 or 1, and β is irrational. This shows, in particular, that $e^{-\pi}$, which is one of the values of i^{2i}, is transcendental. It also shows that

$$\theta = \frac{\log 3}{\log 2}$$

is transcendental, since $2^\theta = 3$ and θ is irrational.†

NOTES

§ 11.3. Dirichlet's argument depends upon the principle 'if there are n+1 objects in n boxes, there must be at least one box which contains two (or more) of the objects' (the *Schubfachprinzip* of German writers). That in § 11.12 is essentially the same.

§§ 11.6–7. A full account of Cantor's work in the theory of aggregates (*Mengenlehre*) will be found in Hobson's *Theory of functions of a real variable*, i.

Liouville's work was published in the *Journal de Math.* (1) 16 (1851), 133–42, over twenty years before Cantor's. See also the note on §§ 11.13–14.

Theorem 191 has been improved successively by Thue, Siegel, Dyson, and Gelfond. Finally Roth (*Mathematika*, 2 (1955), 1–20) showed that no irrational algebraic number is approximable to any order greater than 2. Roth's result can be re-phrased by saying that if one takes $\chi(q) = q^{1+\epsilon}$ in Theorem 198, with any fixed $\epsilon > 0$, then the resulting null set contains no irrational algebraic numbers. It is not known whether this remains true with any essentially smaller function $\chi(q)$. For an account of Schmidt's generalization of this to the simultaneous approximation to several algebraic numbers, see Baker, ch. 7, Th. 7.1. *et seq.* See also Bombieri and Gubler, *Heights in Diophantine geometry* (Cambridge University Press, Cambridge, 2006) for an account of the more general Subspace Theorem and its p-adic extensions. For stricter limitations on the degree of rational approximation possible to specific irrationals, e.g. $\sqrt[3]{2}$ see Baker, *Quart. J. Math. Oxford* (2) 15 (1964), 375–83. Curently (2007) it is known that

$$\left|\frac{p}{q} - \sqrt[3]{2}\right| > \frac{1}{4q^{2.4325}}$$

for all positive integers p, q (see Voutier *J. Théor. Nombres Bordeaux* 19 (2007), 265–90).

† See § 4.7.

§§ 11.8–9. Theorems 193 and 194 are due to Hurwitz, *Math. Ann.* 39 (1891), 279–84; and Theorem 195 to Borel, *Journal de Math.* (5), 9 (1903), 329–75. Our proofs follow Perron (*Kettenbrüche*, 49–52, and *Irrationalzahlen*, 129–31).

§ 11.10. The theorem with $2\sqrt{2}$ is also due to Hurwitz, *loc. cit. supra.* For fuller information see Koksma, 29 *et seq.*

Theorems 196 and 197 were proved by Borel, *Rendiconti del circolo mat. di Palermo,* 27 (1909), 247–71, and F. Bernstein, *Math. Ann.* 71 (1912), 417–39.

For further refinements see Khintchine, *Compositio Math.* 1 (1934), 361–83, and Dyson, *Journal London Math. Soc.* 18 (1943), 40–43.

§ 11.11. For Theorem 199 see Khintchine, *Math. Ann.* 92 (1924), 115–25.

§ 11.12. We lost nothing by supposing p/q irreducible throughout §§ 11.1–11. Suppose, for example, that p/q is a reducible solution of (11.1.1). Then if $(p,q) = d$ with $d > 1$, and we write $p = dp'$, $q = dq'$, we have $(p',q') = 1$ and

$$\left|\frac{p'}{q'} - \xi\right| = \left|\frac{p}{q} - \xi\right| < \frac{1}{q^2} < \frac{1}{q'^2},$$

so that p'/q' is an irreducible solution of (11.1.1).

This sort of reduction is no longer possible when we require a number of rational fractions with the same denominator, and some of our conclusions here would become false if we insisted on irreducibility. For example, in order that the system (11.12.1) should have an infinity of solutions, it would be necessary, after § 11.1 (1), that *every* ξ_i should be irrational.

We owe this remark to Dr. Wylie.

§§ 11.13–14. The transcendence of e was proved first by Hermite, *Comptes rendus*, 77 (1873), 18–24, etc. (*Œuvres,* iii. 150–81); and that of π by F. Lindemann, *Math. Ann.* 20 (1882), 213–25. The proofs were afterwards modified and simplified by Hilbert, Hurwitz, and other writers. The form in which we give them is in essentials the same as that in Landau, *Vorlesungen,* iii. 90–95, or Perron, *Irrationalzahlen*, 174–82.

Nesterenko (*Sb. Math.* 187 (1996), 1319–1348) showed that π and e^{π} are algebraically independent in the sense that there is no non-zero polynomial $P(x, y)$ with rational coefficients such that $P(\pi, e^{\pi}) = 0$. This result includes the transcendence of both numbers.

The problem of proving the transcendentality of α^{β}, under the conditions stated at the end of § 11.14, was propounded by Hilbert in 1900, and solved independently by Gelfond and Schneider, by different methods, in 1934. Fuller details, and references to the proofs of the transcendentality of the other numbers mentioned at the end of § 11.7, will be found in Koksma, ch. iv. and in Baker, ch. 2. Baker's book gives an up-to-date account of the whole subject of transcendental numbers, in which there have been important recent advances by him and others.

It is unknown whether log 2 and log 3 are algebraically independent, or indeed if there exist any two non-zero algebraic numbers α, β such that $\log \alpha$ and $\log \beta$ are algebraically independent.

XII

THE FUNDAMENTAL THEOREM OF ARITHMETIC IN $k(1)$, $k(i)$, AND $k(\rho)$

12.1. Algebraic numbers and integers. In this chapter we consider some simple generalizations of the notion of an integer.

We defined an algebraic number in § 11.5; ξ is an algebraic number if it is a root of an equation

$$c_0\xi^n + c_1\xi^{n-1} + \cdots + c_n = 0 \quad (c_0 \neq 0)$$

whose coefficients are rational integers.† If

$$c_0 = 1,$$

then ξ is said to be an *algebraic integer*. This is the natural definition, since a rational $\xi = a/b$ satisfies $b\xi - a = 0$, and is an integer when $b = 1$.

Thus

$$i = \surd(-1)$$

and

$$\rho = e^{\frac{2}{3}\pi i} = \tfrac{1}{2}(-1 + i\surd 3) \tag{12.1.1}$$

are algebraic integers, since

$$i^2 + 1 = 0$$

and

$$\rho^2 + \rho + 1 = 0.$$

When $n = 2$, ξ is said to be a *quadratic* number, or integer, as the case may be.

These definitions enable us to restate Theorem 45 in the form

THEOREM 206. *An algebraic integer, if rational, is a rational integer.*

† We defined the 'rational integers' in § 1.1. Since then we have described them simply as the 'integers', but now it becomes important to distinguish them explicitly from integers of other kinds.

12.2. The rational integers, the Gaussian integers, and the integers of $k(\rho)$. For the present we shall be concerned only with the three simplest classes of algebraic integers.

(1) The rational integers (defined in § 1.1) are the algebraic integers for which $n = 1$. For reasons which will appear later, we shall call the rational integers *the integers of* $k(1)$.†

(2) The complex or 'Gaussian' integers are the numbers

$$\xi = a + bi,$$

where a and b are rational integers. Since

$$\xi^2 - 2a\xi + a^2 + b^2 = 0,$$

a Gaussian integer is a quadratic integer. We call the Gaussian integers the *integers of* $k(i)$. In particular, any rational integer is a Gaussian integer.

Since

$$(a + bi) + (c + di) = (a + c) + (b + d)i,$$
$$(a + bi)(c + di) = ac - bd + (ad + bc)i,$$

sums and products of Gaussian integers are Gaussian integers. More generally, if $\alpha, \beta, \ldots, \kappa$ are Gaussian integers, and

$$\xi = P(\alpha, \beta, \ldots, \kappa),$$

where P is a polynomial whose coefficients are rational or Gaussian integers, then ξ is a Gaussian integer.

(3) If ρ is defined by (12.1.1), then

$$\rho^2 = e^{\frac{4}{3}\pi i} = \tfrac{1}{2}(-1 + i\surd 3),$$
$$\rho + \rho^2 = -1, \quad \rho\rho^2 = 1.$$

If

$$\xi = a + b\rho,$$

† We shall define $k(\theta)$ generally in § 14.1. $k(1)$ is in fact the class of rationals; we shall not use a special symbol for the sub-class of rational integers. $k(i)$ is the class of numbers $r+si$, where r and s are rational; and $k(\rho)$ is defined similarly.

where a and b are rational integers, then

$$(\xi - a - b\rho)(\xi - a - b\rho^2) = 0$$

or

$$\xi^2 - (2a - b)\xi + a^2 - ab + b^2 = 0,$$

so that ξ is a quadratic integer. We call the numbers ξ the *integers of* $k(\rho)$. Since

$$\rho^2 + \rho + 1 = 0,\ a + b\rho = a - b - b\rho^2,\ a + b\rho^2 = a - b - b\rho,$$

we might equally have defined the integers of $k(\rho)$ as the numbers $a+b\rho^2$.

The properties of the integers of $k(i)$ and $k(\rho)$ resemble in many ways those of the rational integers. Our object in this chapter is to study the simplest properties common to the three classes of numbers, and in particular the property of 'unique factorization'. This study is important for two reasons, first because it is interesting to see how far the properties of ordinary integers are susceptible to generalization, and secondly because many properties of the rational integers themselves follow most simply and most naturally from those of wider classes.

We shall use small Latin letters $a, b, \ldots$, as we have usually done, to denote rational integers, except that i will always be $\surd(-1)$. Integers of $k(i)$ or $k(\rho)$ will be denoted by Greek letters α, $\beta, \ldots$.

12.3. Euclid's algorithm. We have already proved the 'fundamental theorem of arithmetic', for the rational integers, by two different methods, in §§ 2.10 and 2.11. We shall now give a third proof which is important both logically and historically and will serve us as a model when extending it to other classes of numbers.†

Suppose that

$$a \geqslant b > 0.$$

Dividing a by b we obtain

$$a = q_1 b + r_1,$$

† The fundamental idea of the proof is the same as that of the proof of § 2.10: the numbers divisible by $d = (a, b)$ form a 'modulus'. But here we determine d by a direct construction.

where $0 \leqslant r_1 < b$. If $r_1 \neq 0$, we can repeat the process, and obtain

$$b = q_2 r_1 + r_2,$$

where $0 \leqslant r_2 < r_1$. If $r_2 \neq 0$,

$$r_1 = q_3 r_2 + r_3,$$

where $0 \leqslant r_3 < r_2$; and so on. The non-negative integers $b, r_1, r_2, \ldots,$ form a decreasing sequence, and so

$$r_{n+1} = 0,$$

for some n. The last two steps of the process will be

$$r_{n-2} = q_n r_{n-1} + r_n \quad (0 < r_n < r_{n-1}),$$
$$r_{n-1} = q_{n+1} r_n.$$

This system of equations for $r_1, r_2, \ldots$ is known as *Euclid's algorithm*. It is the same, except for notation, as that of § 10.6.

Euclid's algorithm embodies the ordinary process for finding the highest common divisor of a and b, as is shown by the next theorem.

THEOREM 207: $r_n = (a, b)$.

Let $d = (a, b)$. Then, using the successive steps of the algorithm, we have

$$d|a \,.\, d|b \to d|r_1 \to d|r_2 \to \cdots \to d|r_n,$$

so that $d \leqslant r_n$. Again, working backwards,

$$r_n|r_{n-1} \to r_n|r_{n-2} \to r_n|r_{n-3} \to \ldots \to r_n|b \to r_n|a.$$

Hence r_n divides both a and b. Since d is the greatest of the common divisors of a and b, it follows that $r_n \leqslant d$, and therefore that $r_n = d$.

12.4. Application of Euclid's algorithm to the fundamental theorem in *k*(1). We base the proof of the fundamental theorem on two preliminary theorems. The first is merely a repetition of Theorem 26, but it is convenient to restate it and deduce it from the algorithm. The second is substantially equivalent to Theorem 3.

THEOREM 208. *If $f|a$, $f|b$, then $f|(a, b)$.*

For

$$f|a \,.\, f|b \to f|r_1 \to f|r_2 \to \ldots \to f|r_n,$$

or $f|d$.

THEOREM 209. *If $(a,b) = 1$ and $b \mid ac$, then $b \mid c$.*

If we multiply each line of the algorithm by c, we obtain

$$ac = q_1bc + r_1c,$$

$$\cdot \quad \cdot \quad \cdot \quad \cdot \quad \cdot \quad \cdot$$

$$r_{n-2}c = q_nr_{n-1}c + r_nc,$$

$$r_{n-1}c = q_{n+1}r_nc,$$

which is the algorithm we should have obtained if we started with ac and bc instead of a and b. Here

$$r_n = (a,b) = 1$$

and so

$$(ac, bc) = r_nc = c.$$

Now $b|ac$, by hypothesis, and $b|bc$. Hence, by Theorem 208,

$$b|(ac, bc) = c,$$

which is what we had to prove.

If p is a prime, then either $p|a$ or $(a,p) = 1$. In the latter case, by Theorem 209, $p|ac$ implies $p|c$. Thus $p|ac$ implies $p|a$ or $p|c$. This is Theorem 3, and from Theorem 3 the fundamental theorem follows as in § 1.3.

It will be useful to restate the fundamental theorem in a slightly different form which extends more naturally to the integers of $k(i)$ and $k(\rho)$. We call the numbers

$$\epsilon = \pm 1,$$

the divisors of 1, the *unities* of $k(1)$. The two numbers

$$\epsilon m$$

we call *associates*. Finally we define a *prime* as an integer of $k(1)$ which is not 0 or a unity and is not divisible by any number except the unities and its associates. The primes are then

$$\pm 2, \quad \pm 3, \quad \pm 5, \ldots,$$

and the fundamental theorem takes the form: *any integer n of* $k(1)$, *not* 0 *or a unity, can be expressed as a product of primes, and the expression is unique except in regard to* (*a*) *the order of the factors,* (*b*) *the presence of unities as factors, and* (*c*) *ambiguities between associated primes.*

12.5. Historical remarks on Euclid's algorithm and the fundamental theorem. Euclid's algorithm is explained at length in Book vii of the *Elements* (Props. 1–3). Euclid deduces from the algorithm, effectively, that

$$f|a \,.\, f|b \to f|(a,b)$$

and

$$(ac, bc) = (a, b)c.$$

He has thus the weapons which were essential in our proof.

The actual theorem which he proves (vii. 24) is 'if two numbers be prime to any number, their product also will be prime to the same'; i.e.

$$(a,c) = 1 \,.\, (b,c) = 1 \to (ab,c) = 1. \tag{12.5.1}$$

Our Theorem 3 follows from this by taking c a prime p, and we can prove (12.5.1) by a slight change in the argument of § 12.4. But Euclid's method of proof, which depends on the notions of 'parts' and 'proportion', is essentially different.

It might seem strange at first that Euclid, having gone so far, could not prove the fundamental theorem itself; but this view would rest on a misconception. Euclid had no formal calculus of multiplication and exponentiation, and it would have been most difficult for him even to state the theorem. He had not even a *term* for the product of more than three factors. The omission of the fundamental theorem is in no way casual or accidental; Euclid knew very well that the theory of numbers turned upon his algorithm, and drew from it all the return he could.

12.6. Properties of the Gaussian integers. Throughout this and the next two sections the word 'integer' means Gaussian integer or integer of $k(i)$.

We define 'divisible' and 'divisor' in $k(i)$ in the same way as in $k(1)$; an integer ξ is said to be *divisible* by an integer η, not 0, if there exists an integer ζ such that

$$\xi = \eta\zeta;$$

and η is then said to be a *divisor* of ξ. We express this by $\eta|\xi$. Since $1, -1, i, -i$ are all integers, any ξ has the eight 'trivial' divisors

$$1, \xi, -1, -\xi, i, i\xi, -i, -i\xi.$$

Divisibility has the obvious properties expressed by

$$\alpha|\beta \,.\, \beta|\gamma \rightarrow \alpha|\gamma,$$
$$\alpha|\gamma_1 \,.\, \ldots \,.\, \alpha|\gamma_n \rightarrow \alpha|\beta_1\gamma_1 + \cdots + \beta_n\gamma_n.$$

The integer ϵ is said to be a *unity* of $k(i)$ if $\epsilon|\xi$ for every ξ of $k(i)$. Alternatively, we may define a unity as any integer which is a divisor of 1. The two definitions are equivalent, since 1 is a divisor of every integer of the field, and

$$\epsilon|1 \,.\, 1|\xi \rightarrow \epsilon|\xi.$$

The *norm* of an integer ξ is defined by

$$N\xi = N(a + bi) = a^2 + b^2.$$

If $\bar{\xi}$ is the conjugate of ξ, then

$$N\xi = \xi\bar{\xi} = |\xi|^2.$$

Since

$$(a^2 + b^2)(c^2 + d^2) = (ac - bd)^2 + (ad + bc)^2,$$

$N\xi$ has the properties

$$N\xi N\eta = N(\xi\eta), \quad N\xi N\eta \ldots = N(\xi\eta \ldots).$$

THEOREM 210. *The norm of a unity is* 1, *and any integer whose norm is* 1 *is a unity.*

If ϵ is a unity, then $\epsilon | 1$. Hence $1 = \epsilon\eta$, and so

$$1 = N\epsilon N\eta, \quad N\epsilon | 1, \quad N\epsilon = 1.$$

On the other hand, if $N(a+bi) = 1$, we have

$$1 = a^2 + b^2 = (a+bi)(a-bi), \quad a+bi \ | 1,$$

and so $a+bi$ is a unity.

THEOREM 211. *The unities of* $k(i)$ *are*

$$\epsilon = i^s \qquad (s = 0, 1, 2, 3).$$

The only solutions of $a^2 + b^2 = 1$ are

$$a = \pm 1, \quad b = 0; \qquad a = 0, \quad b = \pm 1,$$

so that the unities are $\pm 1, \pm i$.

If ϵ is any unity, then $\epsilon\xi$ is said to be *associated* with ξ. The associates of ξ are

$$\xi, i\xi, -\xi, -i\xi;$$

and the associates of 1 are the unities. It is clear that if $\xi | \eta$ then $\xi\epsilon_1 | \eta\epsilon_2$, where ϵ_1, ϵ_2 are any unities. Hence, if η is divisible by ξ, any associate of η is divisible by any associate of ξ.

12.7. Primes in $k(i)$. A *prime* is an integer, not 0 or a unity, divisible only by numbers associated with itself or with 1. We reserve the letter π for primes.† A prime π has no divisors except the eight trivial divisors

$$1, \pi, -1, -\pi, i, i\pi, -i, -i\pi.$$

The associates of a prime are clearly also primes.

THEOREM 212. *An integer whose norm is a rational prime is a prime.*

For suppose that $N\xi = p$, and that $\xi = \eta\zeta$. Then

$$p = N\xi = N\eta N\zeta.$$

Hence either $N\eta = 1$ or $N\zeta = 1$, and either η or ζ is a unity; and therefore ξ is a prime. Thus $N(2+i) = 5$, and $2+i$ is a prime.

† There will be no danger of confusion with the ordinary use of π.

The converse theorem is not true; thus $N3 = 9$, but 3 is a prime. For suppose that

$$3 = (a + bi)(c + di).$$

Then

$$9 = (a^2 + b^2)(c^2 + d^2).$$

It is impossible that

$$a^2 + b^2 = c^2 + d^2 = 3$$

(since 3 is not the sum of two squares), and therefore either $a^2 + b^2 = 1$ or $c^2 + d^2 = 1$, and either $a + bi$ or $c + di$ is a unity. It follows that 3 is a prime.

A rational integer, prime in $k(i)$, must be a rational prime; but not all rational primes are prime in $k(i)$. Thus

$$5 = (2 + i)(2 - i).$$

THEOREM 213. *Any integer, not* 0 *or a unity, is divisible by a prime.*

If γ is an integer, and not a prime, then

$$\gamma = \alpha_1\beta_1, \quad N\alpha_1 > 1, \quad N\beta_1 > 1, \quad N\gamma = N\alpha_1 N\beta_1,$$

and so

$$1 < N\alpha_1 < N\gamma.$$

If α_1 is not a prime, then

$$\alpha_1 = \alpha_2\beta_2, \quad N\alpha_2 > 1, \quad N\beta_2 > 1,$$
$$N\alpha_1 = N\alpha_2 N\beta_2, \quad 1 < N\alpha_2 < N\alpha_1.$$

We may continue this process so long as α_r is not prime. Since

$$N\gamma,\ N\alpha_1,\ N\alpha_2, \ldots$$

is a decreasing sequence of positive rational integers, we must sooner or later come to a prime α_r; and if α_r is the first prime in the sequence γ, α_1, α_2,..., then

$$\gamma = \beta_1\alpha_1 = \beta_1\beta_2\alpha_2 = \ldots = \beta_1\beta_2\beta_3\ldots\beta_r\alpha_r,$$

and so

$$\alpha_r | \gamma.$$

THEOREM 214. *Any integer, not* 0 *or a unity, is a product of primes.*

If γ is not 0 or a unity, it is divisible by a prime π_1. Hence

$$\gamma = \pi_1\gamma_1, \quad N\gamma_1 < N\gamma.$$

Either γ_1 is a unity or

$$\gamma_1 = \pi_2\gamma_2, \quad N\gamma_2 < N\gamma_1.$$

Continuing this process we obtain a decreasing sequence

$$N\gamma, N\gamma_1, N\gamma_2, \ldots,$$

of positive rational integers. Hence $N\gamma_r = 1$ for some r, and γ_r is a unity ϵ; and therefore

$$\gamma = \pi_1\pi_2 \ldots \pi_r\epsilon = \pi_1 \ldots \pi_{r-1}\pi_r',$$

where $\pi_r' = \pi_r\epsilon$ is an associate of π_r and so itself a prime.

12.8. The fundamental theorem of arithmetic in $k(i)$. Theorem 214 shows that every γ can be expressed in the form

$$\gamma = \pi_1\pi_2 \ldots \pi_r,$$

where every π is a prime. The fundamental theorem asserts that, apart from trivial variations, this representation is unique.

THEOREM 215 (THE FUNDAMENTAL THEOREM FOR GAUSSIAN INTEGERS). *The expression of an integer as a product of primes is unique, apart from the order of the primes, the presence of unities, and ambiguities between associated primes.*

We use a process, analogous to Euclid's algorithm, which depends upon

THEOREM 216. *Given any two integers γ, γ_1, of which $\gamma_1 \neq 0$, there is an integer κ such that*

$$\gamma = \kappa\gamma_1 + \gamma_2, \quad N\gamma_2 < N\gamma_1.$$

We shall actually prove more than this, viz. that

$$N\gamma_2 \leqslant \tfrac{1}{2}N\gamma_1,$$

but the essential point, on which the proof of the fundamental theorem depends, is what is stated in the theorem. If c and c_1 are positive rational integers, and $c_1 \neq 0$, there is a k such that

$$c = kc_1 + c_2, \quad 0 \leqslant c_2 < c_1.$$

It is on this that the construction of Euclid's algorithm depends, and Theorem 216 provides the basis for a similar construction in $k(i)$.

Since $\gamma_1 \neq 0$, we have

$$\frac{\gamma}{\gamma_1} = R + Si,$$

where R and S are real; in fact R and S are rational, but this is irrelevant. We can find two rational integers x and y such that

$$|R - x| \leqslant \tfrac{1}{2}, \quad |S - y| \leqslant \tfrac{1}{2};$$

and then

$$\left|\frac{\gamma}{\gamma_1} - (x + iy)\right| = |(R - x) + i(S - y)| = \{(R - x)^2 + (S - y)^2\}^{\frac{1}{2}} \leqslant \frac{1}{\sqrt{2}}.$$

If we take

$$\kappa = x + iy, \quad \gamma_2 = \gamma - \kappa\gamma_1,$$

we have

$$|\gamma - \kappa\gamma_1| \leqslant 2^{-\frac{1}{2}}|\gamma_1|,$$

and so, squaring,

$$N\gamma_2 = N(\gamma - \kappa\gamma_1) \leqslant \tfrac{1}{2}N\gamma_1.$$

We now apply Theorem 216 to obtain an analogue of Euclid's algorithm. If γ and γ_1 are given, and $\gamma_1 \neq 0$, we have

$$\gamma = \kappa\gamma_1 + \gamma_2 \quad (N\gamma_2 < N\gamma_1).$$

If $\gamma_2 \neq 0$, we have

$$\gamma_1 = \kappa_1\gamma_2 + \gamma_3 \quad (N\gamma_3 < N\gamma_2),$$

and so on. Since

$$N\gamma_1, N\gamma_2, \ldots$$

is a decreasing sequence of non-negative rational integers, there must be an n for which

$$N\gamma_{n+1} = 0, \quad \gamma_{n+1} = 0,$$

and the last steps of the algorithm will be

$$\gamma_{n-2} = \kappa_{n-2}\gamma_{n-1} + \gamma_n \quad (N\gamma n < N\gamma_{n-1}),$$
$$\gamma_{n-1} = \kappa_{n-1}\gamma_n.$$

It now follows, as in the proof of Theorem 207, that γ_n is a common divisor of γ and γ_1, and that every common divisor of γ and γ_1 is a divisor of γ_n.

We have nothing at this stage corresponding exactly to Theorem 207, since we have not yet defined 'highest common divisor'. If ζ is a common divisor of γ and γ_1, and every common divisor of γ and γ_1 is a divisor of ζ, we call ζ a *highest common divisor* of γ and γ_1, and write $\zeta = (\gamma, \gamma_1)$. Thus γ_n is a highest common divisor of γ and γ_1. The property of (γ, γ_1) corresponding to that proved in Theorem 208 is thus absorbed into its definition.

The highest common divisor is not unique, since any associate of a highest common divisor is also a highest common divisor. If η and ζ are each highest common divisors, then, by the definition,

$$\eta|\zeta, \quad \zeta|\eta,$$

and so

$$\zeta = \phi\eta, \quad \eta = \theta\zeta = \theta\phi\eta, \quad \theta\phi = 1.$$

Hence ϕ is a unity and ζ an associate of η, and *the highest common divisor is unique except for ambiguity between associates.*

It will be noticed that we defined the highest common divisor of two numbers of $k(1)$ differently, viz. as the greatest among the common divisors, and proved as a theorem that it possesses the property which we take as our definition here. We might define the highest common divisors of two integers of $k(i)$ as those whose norm is greatest, but the definition which we have adopted lends itself more naturally to generalization.

We now use the algorithm to prove the analogue of Theorem 209, viz.

THEOREM 217. *If* $(\gamma, \gamma_1) = 1$ *and* $\gamma_1 | \beta\gamma$, *then* $\gamma_1 | \beta$.

We multiply the algorithm throughout by β and find that

$$(\beta\gamma, \beta\gamma_1) = \beta\gamma_n.$$

Since $(\gamma, \gamma_1) = 1$, γ_n is a unity, and so

$$(\beta\gamma, \beta\gamma_1) = \beta.$$

Now $\gamma_1 | \beta\gamma$, by hypothesis, and $\gamma_1 | \beta\gamma_1$. Hence, by the definition of the highest common divisor,

$$\gamma_1 | (\beta\gamma, \beta\gamma_1)$$

or $\gamma_1 | \beta$.

If π is prime, and $(\pi, \gamma) = \mu$, then $\mu | \pi$ and $\mu | \gamma$. Since $\mu | \pi$, either (1) μ is a unity, and so $(\pi, \gamma) = 1$, or (2) μ is an associate of π, and so $\pi | \gamma$. Hence, if we take $\gamma_1 = \pi$ in Theorem 217, we obtain the analogue of Euclid's Theorem 3, viz.

THEOREM 218. *If* $\pi | \beta\gamma$, *then* $\pi | \beta$ *or* $\pi | \gamma$.

From this the fundamental theorem for $k(i)$ follows by the argument used for $k(1)$ in § 1.3.

12.9. The integers of $k(\rho)$. We conclude this chapter with a more summary discussion of the integers

$$\xi = a + b\rho$$

defined in § 12.2. Throughout this section 'integer' means 'integer of $k(\rho)$'.

We define divisor, unity, associate, and prime in $k(\rho)$ as in $k(i)$; but the *norm* of $\xi = a + b\rho$ is

$$N\xi = (a + b\rho)(a + b\rho^2) = a^2 - ab + b^2.$$

Since

$$a^2 - ab + b^2 = \left(a - \tfrac{1}{2}b\right)^2 + \tfrac{3}{4}b^2,$$

$N\xi$ is positive except when $\xi = 0$.

Since

$$|a+b\rho|^2 = a^2 - ab + b^2 = N(a+b\rho),$$

we have

$$N\alpha N\beta = N(\alpha\beta), \quad N\alpha N\beta\ldots = N(\alpha\beta\ldots),$$

as in $k(i)$.

Theorems 210, 212, 213, and 214 remain true in $k(\rho)$; and the proofs are the same except for the difference in the form of the norm.

The unities are given by

$$a^2 - ab + b^2 = 1,$$

or

$$(2a-b)^2 + 3b^2 = 4.$$

The only solutions of this equation are

$$a = \pm 1,\ b = 0;\ a = 0,\ b = \pm 1;\ a = 1,\ b = 1;\ a = -1,\ b = -1:$$

so that the unities are

$$\pm 1,\ \pm\rho, \pm(1+\rho)$$

or

$$\pm 1,\ \pm\rho,\ \pm\rho^2.$$

Any number whose norm is a rational prime is a prime; thus $1-\rho$ is a prime, since $N(1-\rho) = 3$. The converse is false; for example, 2 is a prime. For if

$$2 = (a+b\rho)(c+d\rho),$$

then

$$4 = (a^2 - ab + b^2)(c^2 - cd + d^2).$$

Hence either $a+b\rho$ or $c+d\rho$ is a unity, or

$$a^2 - ab + b^2 = \pm 2, \quad (2a-b)^2 + 3b^2 = \pm 8,$$

which is impossible.

The fundamental theorem is true in $k(\rho)$ also, and depends on a theorem verbally identical with Theorem 216.

THEOREM 219. *Given any two integers γ, γ_1, of which $\gamma_1 \neq 0$, there is an integer κ such that*

$$\gamma = \kappa\gamma_1 + \gamma_2, \quad N\gamma_2 < N\gamma_1.$$

For

$$\frac{\gamma}{\gamma_1} = \frac{a+b\rho}{c+d\rho} = \frac{(a+b\rho)(c+d\rho^2)}{(c+d\rho)(c+d\rho^2)}$$
$$= \frac{ac+bd-ad+(bc-ad)\rho}{c^2-cd+d^2} = R + S\rho,$$

say. We can find two rational integers x and y such that

$$|R-x| \leqslant \tfrac{1}{2}, \quad |S-y| \leqslant \tfrac{1}{2},$$

and then

$$\left|\frac{\gamma}{\gamma_1} - (x+y\rho)\right|^2 = (R-x)^2 - (R-x)(S-y) + (S-y)^2 \leqslant \tfrac{3}{4}.$$

Hence, if $\kappa = x + y\rho$, $\gamma_2 = \gamma - \kappa\gamma_1$, we have

$$N\gamma_2 = N(\gamma - \kappa\gamma_1) \leqslant \tfrac{3}{4}N\gamma_1 < N\gamma_1.$$

The fundamental theorem for $k(\rho)$ follows from Theorem 219 by the argument used in § 12.8.

THEOREM 220. [THE FUNDAMENTAL THEOREM FOR $k(\rho)$] *The expression of an integer of $k(\rho)$ as a product of primes is unique, apart from the order of the primes, the presence of unities, and ambiguities between associated primes.*

We conclude with a few trivial propositions about the integers of $k(\rho)$ which are of no intrinsic interest but will be required in Ch. XIII.

THEOREM 221. $\lambda = 1 - \rho$ *is a prime.*

This has been proved already.

THEOREM 222. *All integers of* $k(\rho)$ *fall into three classes* (mod λ), *typified by* 0, 1, *and* -1.

The definitions of a *congruence to modulus* λ, a *residue* (mod λ), and *a class of residues* (mod λ), are the same as in $k(1)$.

If γ is any integer of $k(\rho)$, we have

$$\gamma = a + b\rho = a + b - b\lambda \equiv a + b \pmod{\lambda}.$$

Since $3 = (1-\rho)(1-\rho^2)$, $\lambda|3$; and since $a+b$ has one of the three residues 0, 1, -1 (mod 3), γ has one of the same three residues (mod λ). These residues are incongruent, since neither $N1 = 1$ nor $N2 = 4$ is divisible by $N\lambda = 3$.

THEOREM 223. 3 *is associated with* λ^2.

For

$$\lambda^2 = 1 - 2\rho + \rho^2 = -3\rho.$$

THEOREM 224. *The numbers* $\pm(1-\rho)$, $\pm(1-\rho^2)$, $\pm\rho(1-\rho)$ *are all associated with* λ.

For

$$\pm(1-\rho) = \pm\lambda, \quad \pm(1-\rho^2) = \mp\lambda\rho^2, \quad \pm\rho(1-\rho) = \pm\lambda\rho.$$

NOTES

The terminology and notation of this chapter, and also of Chapters 14 and 15, has become out of date. In particular $k(1)$, $k(i)$, and $k(\rho)$ are alternatively denoted $\mathbb{Q}$, $\mathbb{Q}(i)$, and $\mathbb{Q}(\rho)$. Moreover 'unities' are alternatively referred to merely as 'units'.

§ 12.1. The Gaussian integers were used first by Gauss in his researches on biquadratic reciprocity. See in particular his memoirs entitled 'Theoria residuorum biquadraticorum', *Werke*, ii. 67–148. Gauss (here and in his memoirs on algebraic equations, *Werke*, iii. 3–64) was the first mathematician to use complex numbers in a really confident and scientific way.

The numbers $a + b\rho$ were introduced by Eisenstein and Jacobi in their work on cubic reciprocity. See Bachmann, *Allgemeine Arithmetik der Zahlkörper*, 142.

§ 12.5. We owe the substance of these remarks to Prof. S. Bochner.

Professor A. A. Mullin drew my attention to Euclid ix. 14, the theorem that, if n is the least number divisible by each of the primes $p_i, \ldots, p_j$, then n is not divisible by any other prime. This may perhaps be regarded as a further step on Euclid's part towards the Fundamental Theorem.

XIII

SOME DIOPHANTINE EQUATIONS

13.1. Fermat's last theorem. 'Fermat's last theorem' asserts that the equation

(13.1.1) $$x^n + y^n = z^n,$$

where n is an integer greater than 2, has no integral solutions, except the trivial solutions in which one of the variables is 0. The theorem has never been proved for all n,† or even in an infinity of genuinely distinct cases, but it is known to be true for $2 < n < 619$. In this chapter we shall be concerned only with the two simplest cases of the theorem, in which $n = 3$ and $n = 4$. The case $n = 4$ is easy, and the case $n = 3$ provides an excellent illustration of the use of the ideas of Ch. XII.

13.2. The equation $x^2 + y^2 = z^2$. The equation (13.1.1) is soluble when $n = 2$; the most familiar solutions are 3, 4, 5 and 5, 12, 13. We dispose of this problem first.

It is plain that we may suppose x, y, z positive, without loss of generality. Next

$$d \mid x \,.\, d \mid y \rightarrow d \mid z.$$

Hence, if x, y, z is a solution with $(x, y) = d$, then $x = dx', y = dy', z = dz'$, and x', y', z' is a solution with $(x', y') = 1$. We may therefore suppose that $(x, y) = 1$, the general solution being a multiple of a solution satisfying this condition. Finally

$$x \equiv 1 \pmod 2 \,.\, y \equiv 1 \pmod 2 \rightarrow z^2 \equiv 2 \pmod 4,$$

which is impossible; so that one of x and y must be odd and the other even.

It is therefore sufficient for our purpose to prove the theorem which follows.

THEOREM 225. *The most general solution of the equation*

(13.2.1) $$x^2 + y^2 = z^2,$$

satisfying the conditions

(13.2.2) $$x > 0, \quad y > 0, \quad z > 0, \quad (x, y) = 1, \; 2 \mid x,$$

† This has now been resolved. See the end of chapter notes.

is

$$x = 2ab, \quad y = a^2 - b^2, \quad z = a^2 + b^2, \tag{13.2.3}$$

where a, b are integers of opposite parity and

$$(a, b) = 1, \qquad a > b > 0. \tag{13.2.4}$$

There is a $(1, 1)$ *correspondence between different values of a, b and different values of x, y, z.*

First, let us assume (13.2.1) and (13.2.2). Since $2 \mid x$ and $(x, y) = 1$, y and z are odd and $(y, z) = 1$. Hence $\frac{1}{2}(z - y)$ and $\frac{1}{2}(z + y)$ are integral and

$$\left(\frac{z - y}{2}, \frac{z + y}{2}\right) = 1.$$

By (13.2.1),

$$\left(\frac{x}{2}\right)^2 = \left(\frac{z + y}{2}\right)\left(\frac{z - y}{2}\right),$$

and the two factors on the right, being coprime, must both be squares. Hence

$$\frac{z + y}{2} = a^2, \quad \frac{z - y}{2} = b^2,$$

where

$$a > 0, \quad b > 0, \quad a > b, \quad (a, b) = 1.$$

Also

$$a + b \equiv a^2 + b^2 = z \equiv 1 \pmod{2},$$

and a and b are of opposite parity. Hence any solution of (13.2.1), satisfying (13.2.2), is of the form (13.2.3); and a and b are of opposite parity and satisfy (13.2.4).

Next, let us assume that a and b are of opposite parity and satisfy (13.2.4). Then

$$x^2 + y^2 = 4a^2b^2 + (a^2 - b^2)^2 = (a^2 + b^2)^2 = z^2,$$
$$x > 0, \quad y > 0, \quad z > 0, \quad 2 \mid x.$$

If $(x,y) = d$, then $d \mid z$, and so

$$d \mid y = a^2 - b^2, \quad d \mid z = a^2 + b^2;$$

and therefore $d \mid 2a^2$, $d \mid 2b^2$. Since $(a,b) = 1$, d must be 1 or 2, and the second alternative is excluded because y is odd. Hence $(x,y) = 1$.

Finally, if y and z are given, a^2 and b^2, and consequently a and b, are uniquely determined, so that different values of x, y, and z correspond to different values of a and b.

13.3. The equation $x^4 + y^4 = z^4$. We now apply Theorem 225 to the proof of Fermat's theorem for $n = 4$. This is the only 'easy' case of the theorem. Actually we prove rather more.

THEOREM 226. *There are no positive integral solutions of*

$$x^4 + y^4 = z^2. \tag{13.3.1}$$

Suppose that u is the least number for which

$$x^4 + y^4 = u^2 \quad (x > 0,\ y > 0,\ u > 0) \tag{13.3.2}$$

has a solution. Then $(x,y) = 1$, for otherwise we can divide through by $(x,y)^4$ and so replace u by a smaller number. Hence at least one of x and y is odd, and

$$u^2 = x^4 + y^4 \equiv 1 \text{ or } 2 \pmod 4.$$

Since $u^2 \equiv 2 \pmod 4$ is impossible, u is odd, and just one of x and y is even.

If x, say, is even, then, by Theorem 225,

$$x^2 = 2ab, \quad y^2 = a^2 - b^2, \quad u = a^2 + b^2,$$
$$a > 0, \quad b > 0, \quad (a,b) = 1,$$

and a and b are of opposite parity. If a is even and b odd, then

$$y^2 \equiv -1 \pmod 4,$$

which is impossible; so that a is odd and b even, and say $b = 2c$.

Next

$$\left(\frac{1}{2}x\right)^2 = ac, \quad (a,c) = 1;$$

and so

$$a = d^2, \quad c = f^2, \quad d > 0, \quad f > 0, \quad (d,f) = 1,$$

and d is odd. Hence

$$y^2 = a^2 - b^2 = d^4 - 4f^4,$$
$$(2f^2)^2 + y^2 = (d^2)^2,$$

and no two of $2f^2, y, d^2$ have a common factor.

Applying Theorem 225 again, we obtain

$$2f^2 = 2lm, \ d^2 = l^2 + m^2, \ l > 0, \ m > 0, \ (l,m) = 1.$$

Since

$$f^2 = lm, \quad (l,m) = 1,$$

we have

$$l = r^2, \ m = s^2 \quad (r > 0, \ s > 0),$$

and so

$$r^4 + s^4 = d^2.$$

But

$$d \leqslant d^2 = a \leqslant a^2 < a^2 + b^2 = u,$$

and so u is not the least number for which (13.3.2) is possible. This contradiction proves the theorem.

The method of proof which we have used, and which was invented and applied to many problems by Fermat, is known as the 'method of descent'. If a proposition $P(n)$ is true for some positive integer n, there is a smallest such integer. If $P(n)$, for any positive n, implies $P(n')$ for some smaller positive n', then there is no such smallest integer; and the contradiction shows that $P(n)$ is false for every n.

13.4. The equation $x^3 + y^3 = z^3$. If Fermat's theorem is true for some n, it is true for any multiple of n, since $x^{ln} + y^{ln} = z^{ln}$ is

$$(x^l)^n + (y^l)^n = (z^l)^n.$$

The theorem is therefore true generally if it is true (*a*) when $n = 4$ (as we have shown) and (*b*) when n is an odd prime. The only case of (*b*) which we can discuss here is the case $n = 3$.

The natural method of attack, after Ch. XII, is to write Fermat's equation in the form

$$(x + y)(x + \rho y)(x + \rho^2 y) = z^3,$$

and consider the structure of the various factors in $k(\rho)$. As in § 13.3, we prove rather more than Fermat's theorem.

THEOREM 227. *There are no solutions of*

$$\xi^3 + \eta^3 + \zeta^3 = 0 \quad (\xi \neq 0,\ \eta \neq 0,\ \zeta \neq 0)$$

in integers of $k(\rho)$. *In particular, there are no solutions of*

$$x^3 + y^3 = z^3$$

in rational integers, except the trivial solutions in which one of x, y, z *is* 0.

In the proof that follows, Greek letters denote integers in $k(\rho)$, and λ is the prime $1 - \rho$.[†] We may plainly suppose that

$$(\eta, \zeta) = (\zeta, \xi) = (\xi, \eta) = 1. \tag{13.4.1}$$

We base the proof on four lemmas (Theorems 228–31).

THEOREM 228. *If* ω *is not divisible by* λ, *then*

$$\omega^3 \equiv \pm 1 \pmod{\lambda^4}.$$

Since ω is congruent to one of $0, 1, -1$, by Theorem 222, and $\lambda \nmid \omega$, we have

$$\omega \equiv \pm 1 \pmod{\lambda}.$$

We can therefore choose $\alpha = \pm\omega$ so that

$$\alpha \equiv 1 \pmod{\lambda}, \quad \alpha = 1 + \beta\lambda.$$

[†] See Theorem 221.

Then

$$\begin{aligned}\pm\left(\omega^3 \mp 1\right) &= \alpha^3 - 1 = (\alpha - 1)(\alpha - \rho)\left(\alpha - \rho^2\right) \\ &= \beta\lambda\,(\beta\lambda + 1 - \rho)\left(\beta\lambda + 1 - \rho^2\right) \\ &= \lambda^3\beta\,(\beta + 1)\left(\beta - \rho^2\right),\end{aligned}$$

since $1 - \rho^2 = \lambda(1 + \rho) = -\lambda\rho^2$. Also

$$\rho^2 \equiv 1 \pmod{\lambda},$$

so that

$$\beta(\beta + 1)(\beta - \rho^2) \equiv \beta(\beta + 1)(\beta - 1) \pmod{\lambda}.$$

But one of $\beta, \beta + 1, \beta - 1$ is divisible by λ, by Theorem 222; and so

$$\pm(\omega^3 \mp 1) \equiv 0 \pmod{\lambda^4}$$

or

$$\omega^3 \equiv \pm 1 \pmod{\lambda^4}.$$

THEOREM 229. *If $\xi^3 + \eta^3 + \zeta^3 = 0$, then one of ξ, η, ζ is divisible by λ.*

Let us suppose the contrary. Then

$$0 = \xi^3 + \eta^3 + \zeta^3 \equiv \pm 1 \pm 1 \pm 1 \pmod{\lambda^4},$$

and so $\pm 1 \equiv 0$ or $\pm 3 \equiv 0$, i.e. $\lambda^4 \mid 1$ or $\lambda^4 \mid 3$. The first hypothesis is untenable because λ is not a unity; and the second because 3 is an associate of λ^2† and therefore not divisible by λ^4. Hence one of ξ, η, ζ must be divisible by λ.

We may therefore suppose that $\lambda \mid \zeta$, and that

$$\zeta = \lambda^n \gamma,$$

where $\lambda \nmid \gamma$. Then $\lambda \nmid \xi$, $\lambda \nmid \eta$ by (13.4.1), and we have to prove the impossibility of

(13.4.2) $$\xi^3 + \eta^3 + \lambda^{3n}\gamma^3 = 0,$$

† Theorem 223.

where

$$(\xi, \eta) = 1, \quad n \geqslant 1, \quad \lambda \nmid \xi, \quad \lambda \nmid \eta, \quad \lambda \nmid \gamma. \tag{13.4.3}$$

It is convenient to prove more, viz. that

$$\xi^3 + \eta^3 + \epsilon\lambda^{3n}\gamma^3 = 0 \tag{13.4.4}$$

cannot be satisfied by any ξ, η, ζ, subject to (13.4.3) and any unity ϵ.

THEOREM 230. *If* ξ, η, *and* γ *satisfy* (13.4.3) *and* (13.4.4), *then* $n \geqslant 2$.

By Theorem 228,

$$-\epsilon\lambda^{3n}\gamma^3 = \xi^3 + \eta^3 \equiv \pm 1 \pm 1 \pmod{\lambda^4}.$$

If the signs are the same, then

$$-\epsilon\lambda^{3n}\gamma^3 \equiv \pm 2 \pmod{\lambda^4},$$

which is impossible because $\lambda \nmid 2$. Hence the signs are opposite, and

$$-\epsilon\lambda^{3n}\gamma^3 \equiv 0 \pmod{\lambda^4}.$$

Since $\lambda \nmid \gamma$, $n \geqslant 2$.

THEOREM 231. *If* (13.4.4) *is possible for* $n = m > 1$, *then it is possible for* $n = m - 1$.

Theorem 231 represents the critical stage in the proof of Theorem 227; when it is proved, Theorem 227 follows immediately. For if (13.4.4) is possible for any n, it is possible for $n = 1$, in contradiction to Theorem 230. The argument is another example of the 'method of descent'.

Our hypothesis is that

$$-\epsilon\lambda^{3m}\gamma^3 = (\xi + \eta)(\xi + \rho\eta)(\xi + \rho^2\eta). \tag{13.4.5}$$

The differences of the factors on the right are

$$\eta\lambda, \quad \rho\eta\lambda, \quad \rho^2\eta\lambda,$$

all associates of $\eta\lambda$. Each of them is divisible by λ but not by λ^2 (since $\lambda \nmid \eta$).

Since $m \geqslant 2$, $3m > 3$, and one of the three factors must be divisible by λ^2. The other two factors must be divisible by λ (since the differences are divisible), but not by λ^2 (since the differences are not). We may suppose that the factor divisible by λ^2 is $\xi + \eta$; if it were one of the other factors, we could replace η by one of its associates. We have then

(13.4.6) $$\xi + \eta = \lambda^{3m-2}\kappa_1, \quad \xi + \rho\eta = \lambda\kappa_2, \quad \xi + \rho^2\eta = \lambda\kappa_3,$$

where none of $\kappa_1, \kappa_2, \kappa_3$ is divisible by λ.

If $\delta \mid \kappa_2$ and $\delta \mid \kappa_3$, then δ also divides

$$\kappa_2 - \kappa_3 = \rho\eta$$

and

$$\rho\kappa_3 - \rho^2\kappa_2 = \rho\xi,$$

and therefore both ξ and η. Hence δ is a unity and $(\kappa_2, \kappa_3) = 1$. Similarly $(\kappa_3, \kappa_1) = 1$ and $(\kappa_1, \kappa_2) = 1$.

Substituting from (13.4.6) into (13.4.5), we obtain

$$-\epsilon\gamma^3 = \kappa_1\kappa_2\kappa_3.$$

Hence each of $\kappa_1, \kappa_2, \kappa_3$ is an associate of a cube, so that

$$\xi + \eta = \lambda^{3m-2}\kappa_1 = \epsilon_1\lambda^{3m-2}\theta^3, \ \xi + \rho\eta = \epsilon_2\lambda\phi^3, \ \xi + \rho^2\eta = \epsilon_3\lambda\psi^3,$$

where θ, ϕ, ψ have no common factor and are not divisible by λ, and ϵ_1, ϵ_2, ϵ_3 are unities. It follows that

$$\begin{aligned} 0 = (1 + \rho + \rho^2)(\xi + \eta) &= \xi + \eta + \rho(\xi + \rho\eta) + \rho^2(\xi + \rho^2\eta) \\ &= \epsilon_1\lambda^{3m-2}\theta^3 + \epsilon_2\rho\lambda\phi^3 + \epsilon_3\rho^2\lambda\psi^3; \end{aligned}$$

and so that

(13.4.7) $$\phi^3 + \epsilon_4\psi^3 + \epsilon_5\lambda^{3m-3}\theta^3 = 0,$$

where $\epsilon_4 = \epsilon_3\rho/\epsilon_2$ and $\epsilon_5 = \epsilon_1/\epsilon_2\rho$ are also unities.

Now $m \geqslant 2$ and so

$$\phi^3 + \epsilon_4\psi^3 \equiv 0 \pmod{\lambda^2}$$

(in fact, mod λ^3). But $\lambda \nmid \phi$ and $\lambda \nmid \psi$, and therefore, by Theorem 228,

$$\phi^3 \equiv \pm 1 \pmod{\lambda^2}, \qquad \psi^3 \equiv \pm 1 \pmod{\lambda^2}$$

(in fact, mod λ^4). Hence

$$\pm 1 \pm \epsilon_4 \equiv 0 \pmod{\lambda^2}.$$

Here ϵ_4 is ± 1, $\pm\rho$, or $\pm\rho^2$. But none of

$$\pm 1 \pm \rho, \quad \pm 1 \pm \rho^2$$

is divisible by λ^2, since each is an associate of 1 or of λ; and therefore $\epsilon_4 = \pm 1$.

If $\epsilon_4 = 1$, (13.4.7) is an equation of the type required. If $\epsilon_4 = -1$, we replace ψ by $-\psi$. In either case we have proved Theorem 231 and therefore Theorem 227.

13.5. The equation $x^3 + y^3 = 3z^3$. Almost the same reasoning will prove

THEOREM 232. *The equation*

$$x^3 + y^3 = 3z^3$$

has no solutions in integers, except the trivial solutions in which $z = 0$.

The proof is, as might be expected, substantially the same as that of Theorem 227, since 3 is an associate of λ^2. We again prove more, viz. that there are no solutions of

$$\xi^3 + \eta^3 + \epsilon\lambda^{3n+2}\gamma^3 = 0, \tag{13.5.1}$$

where

$$(\xi, \eta) = 1, \quad \lambda \nmid \gamma,$$

in integers of $k(\rho)$. And again we prove the theorem by proving two propositions, viz.

(*a*) if there is a solution, then $n > 0$;

(*b*) if there is a solution for $n = m \geqslant 1$, then there is a solution for $n = m - 1$;

which are contradictory if there is a solution for any n.

We have

$$(\xi + \eta)(\xi + \rho\eta)(\xi + \rho^2\eta) = -\epsilon\lambda^{3m+2}\gamma^3.$$

Hence at least one factor on the left, and therefore every factor, is divisible by λ; and hence $m > 0$. It then follows that $3m + 2 > 3$ and that one factor is divisible by λ^2, and (as in § 13.4) only one. We have therefore

$$\xi + \eta = \lambda^{3m}\kappa_1, \quad \xi + \rho\eta = \lambda\kappa_2, \quad \xi + \rho^2\eta = \lambda\kappa_3,$$

the κ being coprime in pairs and not divisible by λ.

Hence, as in § 13.4,

$$-\epsilon\gamma^3 = \kappa_1\kappa_2\kappa_3,$$

and κ_1, κ_2, κ_3 are the associates of cubes, so that

$$\xi + \eta = \epsilon\lambda^{3m}\theta^3, \quad \xi + \rho\eta = \epsilon_1\lambda\phi^3, \quad \xi + \rho^2\eta = \epsilon_1\lambda\psi^3.$$

It then follows that

$$\begin{aligned} 0 &= \xi + \eta + \rho(\xi + \rho\eta) + \rho^2(\xi + \rho^2\eta) \\ &= \epsilon_1\lambda^{3m}\theta^3 + \epsilon_2\rho\lambda\phi^3 + \epsilon_3\rho^2\lambda\psi^3, \\ &\phi^3 + \epsilon_4\psi^3 + \epsilon_5\lambda^{3m-1}\theta^3 = 0; \end{aligned}$$

and the remainder of the proof is the same as that of Theorem 227.

It is not possible to prove in this way that

$$(13.5.2) \qquad \xi^3 + \eta^3 + \epsilon\lambda^{3n+1}\gamma^3 \neq 0.$$

In fact

$$1^3 + 2^3 + 9(-1)^3 = 0,$$

and, since $9 = \rho\lambda^4$,† this equation is of the form (13.5.2). The reader will find it instructive to attempt the proof and observe where it fails.

13.6. The expression of a rational as a sum of rational cubes. Theorem 232 has a very interesting application to the 'additive' theory of numbers.

The typical problem of this theory is as follows. Suppose that x denotes an arbitrary member of a specified class of numbers, such as the class of positive integers or the class of rationals, and y is a member of some sub-class of the former class, such as the class of integral squares or rational cubes. Is it possible to express x in the form

$$x = y_1 + y_2 + \cdots + y_k;$$

† See the proof of Theorem 223.

and, if so, how economically, that is to say with how small a value of k?

For example, suppose x a positive integer and y an integral square. Lagrange's Theorem 369[†] shows that every positive integer is the sum of four squares, so that we may take $k = 4$. Since 7, for example, is not a sum of *three* squares, the value 4 of k is the least possible or the 'correct' one.

Here we shall suppose that x is a *positive rational*, and y a *non-negative rational cube*, and we shall show that the 'correct' value of k is 3.

In the first place we have, as a corollary of Theorem 232,

THEOREM 233. *There are positive rationals which are not sums of two non-negative rational cubes.*

For example, 3 is such a rational. For

$$\left(\frac{a}{b}\right)^3 + \left(\frac{c}{d}\right)^3 = 3$$

involves

$$(ad)^3 + (bc)^3 = 3(bd)^3,$$

in contradiction to Theorem 232.[‡]

In order to show that 3 is an admissible value of k, we require another theorem of a more elementary character.

THEOREM 234. *Any positive rational is the sum of three positive rational cubes.*

We have to solve

$$r = x^3 + y^3 + z^3, \tag{13.6.1}$$

where r is given, with positive rational x, y, z. It is easily verified that

$$x^3 + y^3 + z^3 = (x + y + z)^3 - 3(y + z)(z + x)(x + y)$$

and so (13.6.1) is equivalent to

$$(x + y + z)^3 - 3(y + z)(z + x)(x + y) = r.$$

[†] Proved in various ways in Ch. XX.

[‡] Theorem 227 shows that 1 is not the sum of two *positive* rational cubes, but it is of course expressible as $0^3 + 1^3$.

If we write $X = y + z$, $Y = z + x$, $Z = x + y$, this becomes

$$(X + Y + Z)^3 - 24XYZ = 8r. \tag{13.6.2}$$

If we put

$$u = \frac{X + Z}{Z}, \quad v = \frac{Y}{Z}, \tag{13.6.3}$$

(13.6.2) becomes

$$(u + v)^3 - 24v(u - 1) = 8rZ^{-3}. \tag{13.6.4}$$

Next we restrict Z and v to satisfy

$$r = 3Z^3 v, \tag{13.6.5}$$

so that (13.6.4) reduces to

$$(u + v)^3 = 24uv. \tag{13.6.6}$$

To solve (13.6.6), we put $u = vt$ and find that

$$u = \frac{24t^2}{(t + 1)^3}, \quad v = \frac{24t}{(t + 1)^3}. \tag{13.6.7}$$

This is a solution of (13.6.6) for every rational t. We have still to satisfy (13.6.5), which now becomes

$$r(t + 1)^3 = 72Z^3 t.$$

If we put $t = r/(72w^3)$, where w is any rational number, we have $Z = w(t + 1)$. Hence a solution of (13.6.2) is

$$X = (u - 1)Z, \quad Y = vZ, \quad Z = w(t + 1), \tag{13.6.8}$$

where u, v are given by (13.6.7) with $t = rw^{-3}/72$. We deduce the solution of (13.6.1) by using

$$2x = Y + Z - X, \quad 2y = Z + X - Y, \quad 2z = X + Y - Z. \tag{13.6.9}$$

To complete the proof of Theorem 234, we have to show that we can choose w so that x, y, z are all positive. If w is taken positive, then t and Z are positive. Now, by (13.6.8) and (13.6.9) we have

$$\frac{2x}{Z}=v+1-(u-1)=2+v-u, \quad \frac{2y}{Z}=u-v, \quad \frac{2z}{Z}=u+v-2.$$

These are all positive provided that

$$u > v \quad u-v < 2 < u+v,$$

that is

$$t > 1, \quad 12t(t-1) < (t+1)^3 < 12t(t+1).$$

These are certainly true if t is a little greater than 1, and we may choose w so that

$$t = \frac{r}{72w^3}$$

satisfies this requirement. (In fact, it is enough if $1 < t \leqslant 2$.)

Suppose for example that $r = \frac{2}{3}$. If we put $w = \frac{1}{6}$ so that $t = 2$, we have

$$\frac{2}{3} = \left(\frac{1}{18}\right)^3 + \left(\frac{4}{9}\right)^3 + \left(\frac{5}{6}\right)^3.$$

The equation

$$1 = \left(\frac{1}{2}\right)^3 + \left(\frac{2}{3}\right)^3 + \left(\frac{5}{6}\right)^3,$$

which is equivalent to

(13.6.10) $$6^3 = 3^3 + 4^3 + 5^3,$$

is even simpler, but is not obtainable by this method.

13.7. The equation $x^3 + y^3 + z^3 = t^3$. There are a number of other Diophantine equations which it would be natural to consider here; and the most interesting are

(13.7.1) $$x^3 + y^3 + z^3 = t^3$$

and

$$x^3 + y^3 = u^3 + v^3. \tag{13.7.2}$$

The second equation is derived from the first by writing $-u, v$ for z, t.

Each of the equations gives rise to a number of different problems, since we may look for solutions in (*a*) integers or (*b*) rationals, and we may or may not be interested in the signs of the solutions. The simplest problem (and the only one which has been solved completely) is that of the solution of the equations in *positive or negative rationals.* For this problem, the equations are equivalent, and we take the form (13.7.2). The complete solution was found by Euler and simplified by Binet.

If we put

$$x = X - Y, \quad y = X + Y, \quad u = U - V, \quad v = U + V,$$

(13.7.2) becomes

$$X(X^2 + 3Y^2) = U(U^2 + 3V^2). \tag{13.7.3}$$

We suppose that X and Y are not both 0. We may then write

$$\frac{U + V\surd(-3)}{X + Y\surd(-3)} = a + b\surd(-3), \quad \frac{U - V\surd(-3)}{X - Y\surd(-3)} = a - b\surd(-3),$$

where a, b are rational. From the first of these

$$U = aX - 3bY, \quad V = bX + aY, \tag{13.7.4}$$

while (13.7.3) becomes

$$X = U(a^2 + 3b^2).$$

This last, combined with the first of (13.7.4), gives us

$$cX = dY,$$

where

$$c = a(a^2 + 3b^2) - 1, \quad d = 3b(a^2 + 3b^2).$$

If $c = d = 0$, then $b = 0, a = 1, X = U, Y = V$. Otherwise

$$X = \lambda d = 3\lambda b(a^2 + 3b^2), \quad Y = \lambda c = \lambda\left\{a(a^2 + 3b^2) - 1\right\}, \tag{13.7.5}$$

where $\lambda \neq 0$. Using these in (13.7.4), we find that

$$U = 3\lambda b, \quad V = \lambda\left\{(a^2 + 3b^2)^2 - a\right\}. \tag{13.7.6}$$

Hence, apart from the two trivial solutions

$$X = Y = U = 0; \qquad X = U, \quad Y = V,$$

every rational solution of (13.7.3) takes the form given in (13.7.5) and (13.7.6) for appropriate rational λ, a, b.

Conversely, if λ, a, b are *any* rational numbers and X, Y, U, V are *defined* by (13.7.5) and (13.7.6), the formulae (13.7.4) follow at once and

$$\begin{aligned} U(U^2 + 3V^2) &= 3\lambda b\{(aX - 3bY)^2 + 3(bX + aY)^2\} \\ &= 3\lambda b(a^2 + 3b^2)(X^2 + 3Y^2) = X(X^2 + 3Y^2). \end{aligned}$$

We have thus proved

THEOREM 235. *Apart from the trivial solutions*

$$x = y = 0, \quad u = -v; \quad x = u, \quad y = v, \tag{13.7.7}$$

the general rational solution of (13.7.2) *is given by*

(13.7.8)

$$\begin{cases} x = \lambda\left\{1 - (a - 3b)(a^2 + 3b^2)\right\}, & y = \lambda\left\{(a + 3b)(a^2 + 3b^2) - 1\right\}, \\ u = \lambda\left\{(a + 3b) - (a^2 + 3b^2)^2\right\}, & v = \lambda\left\{(a^2 + 3b^2)^2 - (a - 3b)\right\}, \end{cases}$$

where λ, a, b are any rational numbers except that $\lambda \neq 0$.

The problem of finding all integral solutions of (13.7.2) is more difficult. Integral values of a, b, and λ in (13.7.8) give an integral solution, but there is no converse correspondence. The simplest solution of (13.7.2) in positive integers is

$$x = 1, \quad y = 12, \quad u = 9, \quad v = 10, \tag{13.7.9}$$

corresponding to

$$a = \tfrac{10}{19}, \quad b = -\tfrac{7}{19}, \quad \lambda = -\tfrac{361}{42}.$$

On the other hand, if we put $a = b = 1$, $\lambda = \frac{1}{3}$, we have

$$x = 3, \quad y = 5, \quad u = -4, \quad v = 6,$$

equivalent to (13.6.10).

Other simple solutions of (13.7.1) or (13.7.2) are

$$1^3 + 6^3 + 8^3 = 9^3, \quad 2^3 + 34^3 = 15^3 + 33^3, \quad 9^3 + 15^3 = 2^3 + 16^3.$$

Ramanujan gave

$$x = 3a^2 + 5ab - 5b^2, \quad y = 4a^2 - 4ab + 6b^2,$$
$$z = 5a^2 - 5ab - 3b^2, \quad t = 6a^2 - 4ab + 4b^2,$$

as a solution of (13.7.1). If we take $a = 2$, $b = 1$, we obtain the solution (17, 14, 7, 20). If we take $a = 1$, $b = -2$, we obtain a solution equivalent to (13.7.9). Other similar solutions are recorded in Dickson's *History*.

Much less is known about the equation

(13.7.10) $$x^4 + y^4 = u^4 + v^4,$$

first solved by Euler. The simplest parametric solution known is

(13.7.11) $$\begin{cases} x = a^7 + a^5b^2 - 2a^3b^4 + 3a^2b^5 + ab^6, \\ y = a^6b - 3a^5b^2 - 2a^4b^3 + a^2b^5 + b^7, \\ u = a^7 + a^5b^2 - 2a^3b^4 + 3a^2b^5 + ab^6, \\ v = a^6b + 3a^5b^2 - 2a^4b^3 + a^2b^5 + b^7, \end{cases}$$

but this solution is not in any sense complete. When $a = 1$, $b = 2$ it leads to

$$133^4 + 134^4 = 158^4 + 59^4,$$

and this is the smallest integral solution of (13.7.10).

To solve (13.7.10), we put

(13.7.12) $$x = aw + c, \quad y = bw - d, \quad u = aw + d, \quad v = bw + c.$$

We thus obtain a quartic equation for ω, in which the first and last coefficients are zero. The coefficient of ω^3 will also be zero if

$$c(a^3 - b^3) = d(a^3 + b^3),$$

in particular if $c = a^3 + b^3$, $d = a^3 - b^3$; and then, on dividing by ω, we find that

$$3\omega(a^2 - b^2)(c^2 - d^2) = 2(ad^3 - ac^3 + bc^3 + bd^3).$$

Finally, when we substitute these values of c, d, and ω in (13.7.12), and multiply throughout by $3a^2b^2$, we obtain (13.7.11).

We shall say something more about problems of this kind in Ch. XXI.

NOTES

§ 13.1. All this chapter, up to § 13.5, is modelled on Landau, *Vorlesungen,* iii. 201–17. See also Mordell, *Diophantine equations*, and the first pages of Cassels, *J. London Math. Soc.* 41 (1966), 193–291.

The phrase 'Diophantine equation' is derived from Diophantus of Alexandria (about A.D. 250), who was the first writer to make a systematic study of the solution of equations in integers. Diophantus proved the substance of Theorem 225. Particular solutions had been known to Greek mathematicians from Pythagoras onwards. Heath's *Diophantus of Alexandria* (Cambridge, 1910) includes translations of all the extant works of Diophantus, of Fermat's comments on them, and of many solutions of Diophantine problems by Euler.

There is a very large literature about 'Fermat's last theorem'. In particular we may refer to Bachmann, *Das Fermatproblem* (1919; reprinted Berlin, Springer, 1976); Dickson, *History*, ii, ch. xxvi; Landau, *Vorlesungen*, iii; Mordell, *Three lectures on Fermat's last theorem* (Cambridge, 1921); Vandiver, *Report of the committee on algebraic numbers*, ii (Washington, 1928), ch. ii, and *Amer. Math. Monthly,* 53 (1946), 556–78. An excellent account of the current state of knowledge about the theorem with full references is given by Ribenboim (*Canadian Math. Bull.* 20 (1977), 229–42). For a more detailed account of the subject and related theory, see Edwards, *Fermat's Last Theorem* (Berlin, Springer, 1977).

The theorem was enunciated by Fermat in 1637 in a marginal note in his copy of Bachet's edition of the works of Diophantus. Here he asserts definitely that he possessed a proof, but the later history of the subject seems to show that he must have been mistaken. A very large number of fallacious proofs have been published.

In view of the remark at the beginning of § 13.4, we can suppose that $n = p > 2$. Kummer (1850) proved the theorem for $n = p$, whenever the odd prime p is 'regular', i.e. when p does not divide the numerator of any of the numbers

$$B_1, B_2, \dots, B_{\frac{1}{2}(p-3)},$$

where B_k, is the kth Bernoulli number defined at the beginning of § 7.9. It is known, however, that there is an infinity of 'irregular' p. Various criteria have been developed (notably by Vandiver) for the truth of the theorem when p is irregular. The corresponding calculations have been carried out on a computer and, as a result, the theorem is now known to be true for all $p < 125000$. If, however, (13.1.1) is satisfied for any larger prime, then $\min(x, y)$ has more than 3 billion digits. See Ribenboim *loc. cit.* for references and Stewart, *Mathematika* 24 (1977), 130–2 for another result.

The problem is much simplified if it is assumed that no one of x, y, z is divisible by p. Wieferich proved in 1909 that there are no such solutions unless $2^{p-1} \equiv 1 \pmod{p^2}$, which is true for $p = 1093$ (§ 6.10) but for no other p less than 2000. Later writers have found further conditions of the same kind and by this means it has been shown that there are no solutions of this kind for $p < 3 \times 10^9$ or for p any Mersenne prime (and so for the largest known prime). See Ribenboim *loc. cit.*

Fermat's Last Theorem was finally settled in a pair of papers by Wiles, and by Wiles and Taylor, (*Ann. of Math.* (2) 141 (1995), 443–551 and 553–72). Unlike its predecessors described above, this work uses a connection between Fermat's equation and elliptic curves. Investigations by Hellegouarch, Frey, and Ribet had previously established that Fermat's

Last Theorem would follow from a standard conjecture on elliptic curves, namely the Taniyama–Shimura conjecture. Wiles was able to establish an important special case of the latter conjecture, which was sufficient to handle Fermat's Last Theorem. The paper by Wiles and Taylor provided the proof of a key step needed for Wiles' work.

§ 13.3. Theorem 226 was actually proved by Fermat. See Dickson, *History*, ii, ch. xxii.

§ 13.4. Theorem 227 was proved by Euler between 1753 and 1770. The proof was incomplete at one point, but the gap was filled by Legendre. See Dickson, *History*, ii, ch. xxi.

Our proof follows that given by Landau, but Landau presents it as a first exercise in the use of ideals, which we have to avoid.

§ 13.6. Theorem 234 is due to Richmond, *Proc. London Math. Soc.* (2) 21 (1923), 401–9. His proof is based on formulae given much earlier by Ryley [*The ladies' diary* (1825), 35].

Ryley's formulae have been reconsidered and generalized by Richmond [*Proc. Edinburgh Math. Soc.* (2) 2 (1930), 92–100, and *Journal London Math. Soc.* 17 (1942), 196–9] and Mordell [*Journal London Math. Soc.* 17 (1942), 194–6]. Richmond finds solutions not included in Ryley's; for example,

$$3(1-t+t^2)x = s(1+t^3), \quad 3(1-t+t^2)y = s(3t-1-t^3),$$
$$3(1-t+t^2)z = s(3t-3t^2),$$

where s is rational and $t = 3r/s^3$. Mordell solves the more general equation

$$(X+Y+Z)^3 - dXYZ = m,$$

of which (13.6.2) is a particular case. Our presentation of the proof is based on Mordell's. There are a number of other papers on cubic Diophantine equations in three variables, by Mordell and B. Segre, in later numbers of the *Journal.* Indeed Segre (*Math Notae*, 11 (1951), 1–68), has shown that if any non-degenerate cubic equation in three variables has a rational solution, it will have infinitely many solutions. This suffices to handle (13.6.1), which has a rational point 'at infinity'. A full account of much recent work on homogeneous equations of degree 3 and 4 variables is given by Manin (*Cubic forms*, Amsterdam, North Holland, 1974).

§ 13.7. The first results concerning 'equal sums of two cubes' were found by Vieta before 1591. See Dickson, *History*, ii. 550 *et seq*. Theorem 235 is due to Euler. Our method follows that of Hurwitz, *Math. Werke,* 2 (1933), 469–70.

The parameterization (13.7.8) has maximal degree 4 in a and b. There is an alternative parameterization of degree 3, namely

$$x = \lambda(A+B+C-D), \quad y = \lambda(A+B-C+D),$$
$$u = \lambda(A-B+C+D), \quad v = \lambda(A-B-C-D),$$

where

$$A = 9a^3+3ab^2+3b, \quad B = 6ab, \quad C = 9a^2b+3b^3+b, \quad D = 3a^2+3b^2+1,$$

see Hua, *Introduction to number theory*, (Springer, New York, 1982), 290–91.

Euler's solution of (13.7.10) is given in Dickson, *Introduction,* 60–62. His formulae, which are not quite so simple as (13.7.11), may be derived from the latter by writing $f+g$ and $f-g$ for a and b and dividing by 2. The formulae (13.7.11) themselves were first given by Gérardin, *L'Intermédiaire des mathématiciens,* 24 (1917), 51. The simple solution here is due to Swinnerton-Dyer, *Journal London Math. Soc.* 18 (1943), 2–4.

Leech (*Proc. Cambridge Phil. Soc.* 53 (1957), 778–80) lists numerical solutions of (13.7.2), of (13.7.10), and of several other Diophantine equations.

In 1844 Catalan conjectured that the only solution in integers p, q, x, y, each greater than 1, of the equation

$$x^p - y^q = 1$$

is $p = y = 2$, $q = x = 3$. This has been proved by Mihăilescu (*J. Reine Angew. Math.* 572 (2004), 167–195).

One of the most powerful results on Diophantine equations is due to Faltings (*Invent. Math.* 73 (1983), 349–66). A special case of this relates to equations of the form $f(x, y, z) = 0$, where f is a homogeneous polynomial of degree at least 4, with integral coefficients. One says that f is nonsingular if the partial derivatives of f cannot vanish simultaneously for any complex (x, y, z) apart from (0, 0, 0). For such an f, Falting's theorem asserts that the equation $f(x, y, z) = 0$ has at most finitely many distinct sloutions, up to multiplication by a constant. One may take $f(x, y, z) = ax^n + by^n - cz^n$ for $n \geqslant 4$, and deduce that the generalized Fermat equation has at most finitely many essentially distinct solutions for each n.

Many of the equations considered in this chapter take the form $a+b = c$, where a, b and c are constant multiples of powers. A very general conjecture about such equations, now known as the '*abc* conjecture' has been made by Oesterlé and by Masser in 1985. It states that if $\varepsilon > 0$ there is a constant $K(\varepsilon)$ with the following property. If a, b, c are any positive integers such that $a + b = c$, then $c \leqslant K(\varepsilon)r(abc)^{1+\varepsilon}$, where the function $r(m)$ is defined as the product of the distinct prime factors of m.

As an example of the potential applications of this conjecture, consider the Fermat equation (13.1.1). Taking $a = x^n$, $b = y^n$ and $c = z^n$, we observe that

$$r(abc) = r(x^n y^n z^n) \leqslant xyz \leqslant z^3$$

whence the conjecture would yield $z^n \leqslant K(\varepsilon)z^{3(1+\varepsilon)}$. Choosing $\varepsilon = 1/2$, and assuming that $n \geqslant 4$ we would then have

$$z^n \leqslant K(1/2)z^{7/2} \leqslant K(1/2)z^{7n/8}.$$

From this we can deduce that $z^n \leqslant K(1/2)^8$. Thus the *abc* conjecture immediately implies that Fermat's equation has at most finitely many solutions in x, y, z, n, for $n \geqslant 4$. In fact a whole host of other important results and conjectures are now known to follow from the *abc* conjecture.

XIV

QUADRATIC FIELDS (1)

14.1. Algebraic fields. In Ch. XII we considered the integers of $k(i)$ and $k(\rho)$, but did not develop the theory farther than was necessary for the purposes of Ch. XIII. In this and the next chapter we carry our investigation of the integers of quadratic fields a little farther.

An *algebraic field* is the aggregate of all numbers

$$R(\vartheta) = \frac{P(\vartheta)}{Q(\vartheta)},$$

where ϑ is a given algebraic number, $P(\vartheta)$ and $Q(\vartheta)$ are polynomials in ϑ with rational coefficients, and $Q(\vartheta) \neq 0$. We denote this field by $k(\vartheta)$. It is plain that sums and products of numbers of $k(\vartheta)$ belong to $k(\vartheta)$ and that α/β belongs to $k(\vartheta)$ if α and β belong to $k(\vartheta)$ and $\beta \neq 0$.

In § 11.5, we defined an *algebraic number* ξ as any root of an algebraic equation

$$a_0x^n + a_1x^{n-1} + \cdots + a_n = 0, \tag{14.1.1}$$

where $a_0, a_1, \ldots$ are rational integers, not all zero. If ξ satisfies an algebraic equation of degree n, but none of lower degree, we say that ξ is of degree n.

If $n = 1$, then ξ is rational and $k(\xi)$ is the aggregate of rationals. Hence, for every rational ξ, $k(\xi)$ denotes the same aggregate, the field of rationals, which we denote by $k(1)$. This field is part of every algebraic field.

If $n = 2$, we say that ξ is 'quadratic'. Then ξ is a root of a quadratic equation

$$a_0x^2 + a_1x + a_2 = 0,$$

and so

$$\xi = \frac{a + b\sqrt{m}}{c}, \quad \sqrt{m} = \frac{c\xi - a}{b}$$

for some rational integers a, b, c, m. Without loss of generality, we may take m to have no squared factor. It is then easily verified that the field $k(\xi)$ is the same aggregate as $k(\sqrt{m})$. Hence it will be enough for us to consider the quadratic fields $k(\sqrt{m})$ for every 'quadratfrei' rational integer m, positive or negative (apart from $m = 1$).

Any member ξ of $k(\sqrt{m})$ has the form

$$\xi = \frac{P(\sqrt{m})}{Q(\sqrt{m})} = \frac{t + u\sqrt{m}}{v + w\sqrt{m}} = \frac{(t + u\sqrt{m})(v - w\sqrt{m})}{v^2 - w^2 m} = \frac{a + b\sqrt{m}}{c}$$

for rational integers t, u, v, w, a, b, c. We have $(c\xi - a)^2 = mb^2$, and so ξ is a root of

$$c^2x^2 - 2acx + a^2 - mb^2 = 0. \tag{14.1.2}$$

Hence ξ is either rational or quadratic; i.e. every member of a quadratic field is either a rational or a quadratic number.

The field $k(\sqrt{m})$ includes a sub-class formed by all the algebraic integers of the field. In § 12.1 we defined an algebraic integer as any root of an equation

$$x^j + c_1x^{j-1} + \cdots + c_j = 0, \tag{14.1.3}$$

where $c_1, \ldots, c_j$ are rational integers. We appear then to have a choice in defining the integers of $k(\sqrt{m})$. We may say that a number ξ of $k(\sqrt{m})$ is an integer of $k(\sqrt{m})$ (i) if ξ satisfies an equation of the form (14.1.3) for *some* j, or (ii) if ξ satisfies an equation of the form (14.1.3) with $j = 2$. In the next section, however, we show that the set of integers of $k(\sqrt{m})$ is the same whichever definition we use.

14.2. Algebraic numbers and integers; primitive polynomials. We say that the integral polynomial

$$f(x) = a_0x^n + a_1x^{n-1} + \cdots + a_n \tag{14.2.1}$$

is a *primitive polynomial* if

$$a_0 > 0, \quad (a_0, a_1, \ldots, a_n) = 1$$

in the notation of p. 20. Under the same conditions, we call (14.1.1) a *primitive equation.* The equation (14.1.3) is obviously primitive.

THEOREM 236. *An algebraic number ξ of degree n satisfies a unique primitive equation of degree n. If ξ is an algebraic integer, the coefficient of x^n in this primitive equation is unity.*

For $n = 1$, the first part is trivial; the second part is equivalent to Theorem 206. Hence Theorem 236 is a generalization of Theorem 206. We shall deduce Theorem 236 from

THEOREM 237. *Let ξ be an algebraic number of degree n and let $f(x) = 0$ be a primitive equation of degree n satisfied by ξ. Let $g(x) = 0$ be any primitive equation satisfied by ξ. Then $g(x) = f(x)h(x)$ for some primitive polynomial $h(x)$ and all x.*

By the definition of ξ and n there must be at least one polynomial $f(x)$ of degree n such that $f(\xi) = 0$. We may clearly suppose $f(x)$ primitive. Again the degree of $g(x)$ cannot be less than n. Hence we can divide $g(x)$ by $f(x)$ by means of the division algorithm of elementary algebra and obtain a quotient $H(x)$ and a remainder $K(x)$, such that

$$g(x) = f(x)H(x) + K(x), \tag{14.2.2}$$

$H(x)$ and $K(x)$ are polynomials with rational coefficients, and $K(x)$ is of degree less than n.

If we put $x = \xi$ in (14.2.2), we have $K(\xi) = 0$. But this is impossible, since ξ is of degree n, unless $K(x)$ has all its coefficients zero. Hence

$$g(x) = f(x)H(x).$$

If we multiply this throughout by an appropriate rational integer, we obtain

$$cg(x) = f(x)h(x), \tag{14.2.3}$$

where c is a positive integer and $h(x)$ is an integral polynomial. Let d be the highest common divisor of the coefficients of $h(x)$. Since g is primitive, we must have $d|c$. Hence, if $d > 1$, we may remove the factor d; that is, we may take $h(x)$ primitive in (14.2.3). Now suppose that $p|c$, where p is prime. It follows that $f(x)h(x) \equiv 0 \pmod{p}$ and so, by Theorem 104 (i), either $f(x) \equiv 0$ or $h(x) \equiv 0 \pmod{p}$. Both are impossible for primitive f and h and so $c = 1$. This is Theorem 237.

The proof of Theorem 236 is now simple. If $g(x) = 0$ is a primitive equation of degree n satisfied by ξ, then $h(x)$ is a primitive polynomial of degree 0; i.e. $h(x) = 1$ and $g(x) = f(x)$ for all x. Hence $f(x)$ is unique.

If ξ is an algebraic integer, then ξ satisfies an equation of the form (14.1.3) for some $j \geqslant n$. We write $g(x)$ for the left-hand side of (14.1.3) and, by Theorem 237, we have

$$g(x) = f(x)h(x),$$

where $h(x)$ is of degree $j - n$. If $f(x) = a_0x^n + \cdots$ and $h(x) = h_0\, x^{j-n} + \cdots$, we have $1 = a_0h_0$, and so $a_0 = 1$. This completes the proof of Theorem 236.

14.3. The general quadratic field $k(\sqrt{m})$. We now define the *integers of* $k(\sqrt{m})$ as those algebraic integers which belong to $k(\sqrt{m})$. We use 'integer' throughout this chapter and Ch. XV for an integer of the particular field in which we are working.

With the notation of § 14.1, let

$$\xi = \frac{a + b\sqrt{m}}{c}$$

be an integer, where we may suppose that $c > 0$ and $(a, b, c) = 1$. If $b = 0$, then $\xi = a/c$ is rational, $c = 1$, and $\xi = a$, any rational integer.

If $b \neq 0$, ξ is quadratic. Hence, if we divide (14.1.2) through by c^2, we obtain a primitive equation whose leading coefficient is 1. Thus $c|2a$ and $c^2|(a^2 - mb^2)$. If $d = (a, c)$, we have

$$d^2|a^2,\ \ d^2|c^2,\ \ d^2|(a^2 - mb^2) \to d^2|mb^2 \to d|b,$$

since m has no squared factor. But $(a, b, c) = 1$ and so $d = 1$. Since $c|2a$, we have $c = 1$ or 2.

If $c = 2$, then a is odd and $mb^2 \equiv a^2 \equiv 1 \pmod 4$, so that b is odd and $m \equiv 1 \pmod 4$. We must therefore distinguish two cases.

(i) If $m \not\equiv 1 \pmod 4$, then $c = 1$ and the integers of $k(\sqrt{m})$ are

$$\xi = a + b\sqrt{m}$$

with rational integral a, b. In this case $m \equiv 2$ or $m \equiv 3 \pmod 4$.

(ii) If $m \equiv 1 \pmod 4$, one integer of $k(\sqrt{m})$ is $\tau = \frac{1}{2}(\sqrt{m} - 1)$ and all the integers can be expressed simply in terms of this τ. If $c = 2$, we have a and b odd and

$$\xi = \frac{a + b\sqrt{m}}{2} = \frac{a + b}{2} + b\tau = a_1 + (2b_1 + 1)\tau,$$

where a_1, b_1 are rational integers. If $c = 1$,

$$\xi = a + b\sqrt{m} = a + b + 2b\tau = a_1 + 2b_1\tau,$$

where a_1, b_1 are rational integers. Hence, if we change our notation a little, the integers of $k(\sqrt{m})$ are the numbers $a + b\tau$ with rational integral a, b.

THEOREM 238. *The integers of $k(\sqrt{m})$ are the numbers*

$$a + b\sqrt{m}$$

when $m \equiv 2$ *or* $m \equiv 3 \pmod 4$, *and the numbers*

$$a + b\tau = a + \tfrac{1}{2}b(\surd m - 1)$$

when $m \equiv 1 (\operatorname{mod} 4)$, *a and b being in either case rational integers.*

The field $k(i)$ is an example of the first case and the field $k\{\surd(-3)\}$ of the second. In the latter case

$$\tau = -\tfrac{1}{2} + \tfrac{1}{2}i\surd 3 = \rho$$

and the field is the same as $k(\rho)$. If the integers of $k(\vartheta)$ can be expressed as

$$a + b\phi,$$

where a and b run through the rational integers, then we say that $[1, \phi]$ is a *basis* of the integers of $k(\vartheta)$. Thus $[1, i]$ is a basis of the integers of $k(i)$, and $[1, \rho]$ of those of $k\{\surd(-3)\}$.

14.4. Unities and primes. The definitions of *divisibility*, *divisor*, *unity*, and *prime* in $k(\surd m)$ are the same as in $k(i)$; thus α is divisible by β, or $\beta | \alpha$, if there is an integer γ of $k(\surd m)$ such that $\alpha = \beta\gamma$.† A unity ϵ is a divisor of 1, and of every integer of the field. In particular 1 and -1 are unities. The numbers $\epsilon\xi$ are the *associates* of ξ, and a prime is a number divisible only by the unities and its associates.

THEOREM 239. *If* ϵ_1 *and* ϵ_2 *are unities, then* $\epsilon_1\epsilon_2$ *and* ϵ_1/ϵ_2 *are unities.*

There are a δ_1 and a δ_2 such that $\epsilon_1\delta_1 = 1, \epsilon_2\delta_2 = 1$, and

$$\epsilon_1\epsilon_2\delta_1\delta_2 = 1 \to \epsilon_1\epsilon_2 | 1.$$

Hence $\epsilon_1\epsilon_2$ is a unity. Also $\delta_2 = 1/\epsilon_2$ is a unity; and so, combining these results, ϵ_1/ϵ_2 is a unity.

We call $\bar{\xi} = r - s\surd m$ the *conjugate* of $\xi = r + s\surd m$. When $m < 0$, $\bar{\xi}$ is also the conjugate of ξ in the sense of analysis, ξ and $\bar{\xi}$ being conjugate complex numbers; but when $m > 0$ the meaning is different.

† If α and β are rational integers, then γ is rational, and so a rational integer, so that $\beta | \alpha$ then means the same in $k\{\surd(-m)\}$ as in $k(1)$.

The *norm* $N\xi$ of ξ is defined by

$$N\xi = \xi\bar{\xi} = (r + s\sqrt{m})(r - s\sqrt{m}) = r^2 - ms^2.$$

If ξ is an integer, then $N\xi$ is a rational integer. If $m \equiv 2$ or $3 \pmod 4$, and $\xi = a + b\sqrt{m}$, then

$$N\xi = a^2 - mb^2;$$

and if $m \equiv 1 \pmod 4$, and $\xi = a + b\omega$, then

$$N\xi = (a - \tfrac{1}{2}b)^2 - \tfrac{1}{4}mb^2.$$

Norms are positive in complex fields, but not necessarily in real fields. In any case $N(\xi\eta) = N\xi N\eta$.

THEOREM 240. *The norm of a unity is* ± 1, *and every number whose norm is* ± 1 *is a unity.*

For (*a*)

$$\epsilon | 1 \to \epsilon\delta = 1 \to N\epsilon N\delta = 1 \to N\epsilon = \pm 1,$$

and (*b*)

$$\xi\bar{\xi} = N\xi = \pm 1 \to \xi | 1.$$

If $m < 0$, $m = -\mu$, then the equations

$$a^2 + \mu b^2 = 1 \quad (m \equiv 2, 3 \pmod 4),$$
$$(a - \tfrac{1}{2}b)^2 + \tfrac{1}{4}\mu b^2 = 1 \quad (m \equiv 1 \pmod 4),$$

have only a finite number of solutions. This number is 4 in $k(i)$, 6 in $k(\rho)$, and 2 otherwise, since

$$a = \pm 1, b = 0$$

are the only solutions when $\mu > 3$.

There are an infinity of unities in a real field, as we shall see in a moment in $k(\sqrt{2})$.

$N\xi$ may be negative in a real field, but

$$M\xi = |N\xi|$$

is a positive integer, except when $\xi = 0$. Hence, repeating the arguments of § 12.7, with $M\xi$ in the place of $N\xi$ when the field is real, we obtain

THEOREM 241. *An integer whose norm is a rational prime is prime.*

THEOREM 242. *An integer, not* 0 *or a unity, can be expressed as a product of primes.*

The question of the uniqueness of the expression remains open.

14.5. The unities of $k(\sqrt{2})$. When $m = 2$,

$$N\xi = a^2 - 2b^2$$

and

$$a^2 - 2b^2 = -1$$

has the solutions 1, 1 and -1, 1. Hence

$$\omega = 1 + \sqrt{2}, \quad \omega^{-1} = -\bar{\omega} = -1 + \sqrt{2}$$

are unities. It follows, after Theorem 239, that all the numbers

$$\pm\omega^n, \pm\omega^{-n} \quad (n = 0, 1, 2, \ldots) \tag{14.5.1}$$

are unities. There are unities, of either sign, as large or as small as we please.

THEOREM 243. *The numbers* (14.5.1) *are the only unities of* $k(\sqrt{2})$.

(i) We prove first that there is no unity ϵ between 1 and ω. If there were, we should have

$$1 < x + y\sqrt{2} = \epsilon < 1 + \sqrt{2}$$

and

$$x^2 - 2y^2 = \pm 1;$$

so that

$$-1 < x - y\sqrt{2} < 1,$$
$$0 < 2x < 2 + \sqrt{2}.$$

Hence $x = 1$ and $1 < 1+y\sqrt{2} < 1+\sqrt{2}$, which is impossible for integral y.

(ii) If $\epsilon > 0$, then either $\epsilon = \omega^n$ or

$$\omega^n < \epsilon < \omega^{n+1}$$

for some integral n. In the latter case $\omega^{-n}\epsilon$ is a unity, by Theorem 239, and lies between 1 and ω. This contradicts (i); and therefore every positive ϵ is an ω^n. Since $-\epsilon$ is a unity if ϵ is a unity, this proves the theorem.

Since $N\omega = -1, N\omega^2 = 1$, we have proved incidentally

THEOREM 244. *All rational integral solutions of*

$$x^2 - 2y^2 = 1$$

are given by

$$x + y\sqrt{2} = \pm(1 + \sqrt{2})^{2n},$$

and all of

$$x^2 - 2y^2 = -1$$

by

$$x + y\sqrt{2} = \pm(1 + \sqrt{2})^{2n+1},$$

with n a rational integer.

The equation

$$x^2 - my^2 = 1,$$

where m is positive and not a square, has always an infinity of solutions, which may be found from the continued fraction for $\sqrt{m}$. In this case

$$\sqrt{2} = 1 + \frac{1}{2+}\,\frac{1}{2+\cdots},$$

the length of the period is 1, and the solution is particularly simple. If the convergents are

$$\frac{p_n}{q_n} = \frac{1}{1}, \frac{3}{2}, \frac{7}{5}, \ldots \quad (n = 0, 1, 2, \ldots)$$

then p_n, q_n, and

$$\phi_n = p_n + q_n\sqrt{2}, \quad \psi_n = p_n - q_n\sqrt{2}$$

are solutions of

$$x_n = 2x_{n-1} + x_{n-2}.$$

From

$$\phi_0 = \omega, \quad \phi_1 = \omega^2, \quad \psi_0 = -\omega^{-1}, \quad \psi_1 = \omega^{-2},$$

and

$$\omega^n = 2\omega^{n-1} + \omega^{n-2}, \quad (-\omega)^{-n} = 2(-\omega)^{-n+1} + (-\omega)^{-n+2},$$

it follows that

$$\phi_n = \omega^{n+1}, \quad \psi_n = (-\omega)^{-n-1}$$

for all n. Hence

$$p_n = \tfrac{1}{2}\left\{\omega^{n+1} + (-\omega)^{-n-1}\right\} = \tfrac{1}{2}\left\{(1+\sqrt{2})^{n+1} + (1-\sqrt{2})^{n+1}\right\},$$
$$q_n = \tfrac{1}{4}\sqrt{2}\left\{\omega^{n+1} - (-\omega)^{-n-1}\right\} = \tfrac{1}{4}\sqrt{2}\left\{(1+\sqrt{2})^{n+1} - (1-\sqrt{2})^{n+1}\right\},$$

and

$$p_n^2 - 2q_n^2 = \phi_n\psi_n = (-1)^{n+1}.$$

The convergents of odd rank give solutions of $x^2-2y^2 = 1$ and those of even rank solutions of $x^2-2y^2 = -1$.

If $x^2-2y^2 = 1$ and $x/y > 0$, then

$$0 < \frac{x}{y} - \sqrt{2} = \frac{1}{y(x+y\sqrt{2})} < \frac{1}{y.2y\sqrt{2}} < \frac{1}{2y^2}.$$

Hence, by Theorem 184, x/y is a convergent. The convergents also give all the solutions of the other equation, but this is not quite so easy to prove. In general, only some of the convergents to $\sqrt{m}$ yield unities of $k(\sqrt{m})$.

14.6. Fields in which the fundamental theorem is false. The fundamental theorem of arithmetic is true in $k(1)$, $k(i)$, $k(\rho)$, and (though we have not yet proved so) in $k(\surd 2)$. It is important to show by examples, before proceeding farther, that it is not true in every $k(\surd m)$. The simplest examples are $m = -5$ and (among real fields) $m = 10$.

(i) Since $-5 \equiv 3 \pmod 4$, the integers of $k\{\surd(-5)\}$ are $a + b\surd(-5)$. It is easy to verify that the four numbers

$$2,\ 3,\ 1 + \surd(-5),\ 1 - \surd(-5)$$

are prime. Thus

$$1 + \surd(-5) = \{a + b\surd(-5)\}\{c + d\surd(-5)\}$$

implies

$$6 = (a^2 + 5b^2)(c^2 + 5d^2);$$

and $a^2 + 5b^2$ must be 2 or 3, if neither factor is a unity. Since neither 2 nor 3 is of this form, $1 + \surd(-5)$ is prime; and the other numbers may be proved prime similarly. But

$$6 = 2 \,.\, 3 = \{1 + \surd(-5)\}\{1 - \surd(-5)\},$$

and 6 has two distinct decompositions into primes.

(ii) Since $10 \equiv 2 \pmod 4$, the integers of $k(\surd 10)$ are $a + b\surd 10$. In this case

$$6 = 2 \,.\, 3 = (4 + \surd 10)(4 - \surd 10),$$

and it is again easy to prove that all four factors are prime. Thus, for example,

$$2 = (a + b\surd 10)(c + d\surd 10)$$

implies

$$4 = (a^2 - 10b^2)(c^2 - 10d^2),$$

and $a^2 - 10b^2$ must be ± 2, if neither factor is a unity. This is impossible because neither of ± 2 is a quadratic residue of 10.†

† $1^2, 2^2, 3^2, 4^2, 5^2, 6^2, 7^2, 8^2, 9^2 \equiv 1, 4, 9, 6, 5, 6, 9, 4, 1 \pmod{10}$.

The falsity of the fundamental theorem in these fields involves the falsity of other theorems which are central in the arithmetic of $k(1)$. Thus, if α and β are integers of $k(1)$, without a common factor, there are integers λ and μ for which

$$\alpha\lambda + \beta\mu = 1.$$

This theorem is false in $k\{\surd(-5)\}$. Suppose, for example, that α and β are the primes 3 and $1 + \surd(-5)$. Then

$$3\{a + b\surd(-5)\} + \{1 + \surd(-5)\}\{c + d\surd(-5)\} = 1$$

involves

$$3a + c - 5d = 1, \qquad 3b + c + d = 0$$

and so

$$3a - 3b - 6d = 1,$$

which is impossible.

14.7. Complex Euclidean fields. A *simple* field is a field in which the fundamental theorem is true. The arithmetic of simple fields follows the lines of rational arithmetic, while in other cases a new foundation is required. The problem of determining all simple fields is very difficult, and no complete solution has been found, though Heilbronn has proved that, when m is *negative,* the number of simple fields is finite.

We proved the fundamental theorem in $k(i)$ and $k(\rho)$ by establishing an analogue of Euclid's algorithm in $k(1)$. Let us suppose, generally, that the proposition

(E) '*given integers γ and γ_1, with $\gamma_1 \neq 0$, then there is an integer κ such that*

$$\gamma = \kappa\gamma_1 + \gamma_2, \quad |N\gamma_2| < |N\gamma_1|\text{'}$$

is true in $k(\surd m)$. This is what we proved, for $k(i)$ and $k(\rho)$, in Theorems 216 and 219; but we have replaced $N\gamma$ by $|N\gamma|$ in order to include real fields. In these circumstances we say that *there is a Euclidean algorithm* in $k(\surd m)$, or that the field is *Euclidean.*

We can then repeat the arguments of §§ 12.8 and 12.9 (with the substitution of $|N\gamma|$ for $N\gamma$), and we conclude that

THEOREM 245. *The fundamental theorem is true in any Euclidean quadratic field.*

The conclusion is not confined to quadratic fields, but it is only in such fields that we have defined $N\gamma$ and are in a position to state it precisely.

(E) is plainly equivalent to

(E') '*given any δ (integral or not) of $k(\surd m)$, there is an integer κ such that*

$$|N(\delta-\kappa)|<1\text{'}. \tag{14.7.1}$$

Suppose now that

$$\delta = r + s\surd m,$$

where r and s are rational. If $m \not\equiv 1 \pmod 4$ then

$$\kappa = x + y\surd m,$$

where x and y are rational integers, and (14.7.1) is

$$\left|(r-x)^2 - m(s-y)^2\right| < 1. \tag{14.7.2}$$

If $m \equiv 1 \pmod 4$ then

$$\kappa = x + y + \tfrac{1}{2}y\left(\surd m - 1\right) = x + \tfrac{1}{2}y + \tfrac{1}{2}y\surd m,\dagger$$

where x and y are rational integers, and (14.7.1) is

$$\left|\left(r-x-\tfrac{1}{2}y\right)^2 - m\left(s-\tfrac{1}{2}y\right)^2\right| < 1. \tag{14.7.3}$$

When $m = -\mu < 0$, it is easy to determine all fields in which these inequalities can be satisfied for any r, s and appropriate x, y.

THEOREM 246. *There are just five complex Euclidean quadratic fields, viz. the fields in which*

$$m = -1, -2, -3, -7, -11.$$

† The form of § 14.3 with $x+y, y$ for a, b.

There are two cases.

(i) When $m \not\equiv 1 \pmod 4$, we take $r = \frac{1}{2}$, $s = \frac{1}{2}$ in (14.7.2); and we require

$$\tfrac{1}{4} + \tfrac{1}{4}\mu < 1,$$

or $\mu < 3$. Hence $\mu = 1$ and $\mu = 2$ are the only possible cases; and in these cases we can plainly satisfy (14.7.2), for any r and s, by taking x and y to be the integers nearest to r and s.

(ii) When $m \equiv 1 \pmod 4$ we take $r = \frac{1}{4}$, $s = \frac{1}{4}$ in (14.7.3). We require

$$\tfrac{1}{16} + \tfrac{1}{16}\mu < 1.$$

Since $\mu \equiv 3 \pmod 4$, the only possible values of μ are 3, 7, 11. Given s, there is a y for which

$$|2s - y| \leqslant \tfrac{1}{2},$$

and an x for which

$$\left|r - x - \tfrac{1}{2}y\right| \leqslant \tfrac{1}{2};$$

and then

$$\left|\left(r - x - \tfrac{1}{2}y\right)^2 - m\left(s - \tfrac{1}{2}y\right)^2\right| \leqslant \tfrac{1}{4} + \tfrac{11}{16} = \tfrac{15}{16} < 1.$$

Hence (14.7.3) can be satisfied when μ has one of the three values in question.

There are other simple fields, such as $k\{\sqrt{(-19)}\}$ and $k\{\sqrt{(-43)}\}$, which do not possess an algorithm; the condition is sufficient but not necessary for simplicity. There are just nine simple complex quadratic fields, viz. those corresponding to

$$m = -1, -2, -3, -7, -11, -19, -43, -67, -163.$$

14.8. Real Euclidean fields. The real fields with an algorithm are more numerous.

THEOREM 247.* *$k(\sqrt{m})$ is Euclidean when*

$$m = 2, 3, 5, 6, 7, 11, 13, 17, 19, 21, 29, 33, 37, 41, 57, 73$$

and for no other positive m.

We can plainly satisfy (14.7.2) when $m = 2$ or $m = 3$, since we can choose x and y so that $|r - x| \leqslant \frac{1}{2}$ and $|s - y| \leqslant \frac{1}{2}$. Hence $k(\sqrt{2})$ and

$k(\sqrt{3})$ are Euclidean, and therefore simple. We cannot prove Theorem 247 here, but we shall prove

THEOREM 248. *$k(\sqrt{m})$ is Euclidean when*

$$m = 2, 3, 5, 6, 7, 13, 17, 21, 29.$$

If we write

$$\lambda = 0, \qquad n = m \qquad (m \not\equiv 1 \pmod 4),$$

$$\lambda = \tfrac{1}{2}, \quad n = \tfrac{1}{4}m \quad (m \equiv 1 \pmod 4)),$$

and replace $2s$ by s when $m \equiv 1$, then we can combine (14.7.2) and (14.7.3) in the form

$$\left|(r - x - \lambda y)^2 - n(s - y)^2\right| < 1. \tag{14.8.1}$$

Let us assume that there is no algorithm in $k(\sqrt{m})$. Then (14.8.1) is false for some rational r, s and all integral x, y; and we may suppose that†

$$0 \leqslant r \leqslant \tfrac{1}{2}, \; 0 \leqslant s \leqslant \tfrac{1}{2}. \tag{14.8.2}$$

† This is very easy to see when $m \not\equiv 1 \pmod 4$ and the left-hand side of (14.8.1) is

$$|(r - x)^2 - m(s - y)^2|;$$

for this is unaltered if we write

$$\epsilon_1 r + u, \quad \epsilon_1 x + u, \quad \epsilon_2 s + v, \quad \epsilon_2 y + v,$$

where ϵ_1 and ϵ_2 are each 1 or -1, and u and v are integers, for

$$r, x, s, y;$$

and we can always choose $\epsilon_1, \epsilon_2, u, v$ so that $\epsilon_1 r + u$ and $\epsilon_2 s + v$ lie between 0 and $\frac{1}{2}$ inclusive. The situation is a little more complex when $m \equiv 1 \pmod 4$ and the left-hand side of (14.8.1) is

$$\left|\left(r - x - \tfrac{1}{2}y\right)^2 - \tfrac{1}{4}m(s - y)^2\right|.$$

This is unaltered by the substitution of any of

(1) $\epsilon_1 r + u, \epsilon_1 x + u, \epsilon_1 s, \epsilon_1 y$,

(2) $r, x - v, s + 2v, y + 2v$,

(3) $r, x + y, -s, -y$,

(4) $\frac{1}{2} - r, -x, 1 - s, 1 - y$,

for r, x, s, y. We first use (1) to make $0 \leqslant r \leqslant \frac{1}{2}$; then (2) to make $-1 \leqslant s \leqslant 1$; and then, if necessary, (3) to make $0 \leqslant s \leqslant 1$. If then $0 \leqslant s \leqslant \frac{1}{2}$, the reduction is completed. If $\frac{1}{2} \leqslant s \leqslant 1$, we end by using (4), as we can do because $\frac{1}{2} - r$ lies between 0 and $\frac{1}{2}$ if r does so.

There is therefore a pair r, s satisfying (14.8.2), such that one or other of

$$[P(x,y)] \quad (r-x-\lambda y)^2 \geqslant 1+n(s-y)^2,$$
$$[N(x,y)] \quad n(s-y)^2 \geqslant 1+(r-x-\lambda y)^2$$

is true for every x, y. The particular inequalities which we shall use are

$$\begin{array}{llll} [P(0,0)] & r^2 \geqslant 1+ns^2, & [N(0,0)] & ns^2 \geqslant 1+r^2, \\ [P(1,0)] & (1-r)^2 \geqslant 1+ns^2, & [N(1,0)] & ns^2 \geqslant 1+(1-r)^2, \\ [P(-1,0)] & (1+r)^2 \geqslant 1+ns^2, & [N(-1,0)] & ns^2 \geqslant 1+(1+r)^2. \end{array}$$

One at least of each of these pairs of inequalities is true for some r and s satisfying (14.8.2). If $r = s = 0$, $P(0, 0)$ and $N(0,0)$ are both false, so that this possibility is excluded.

Since r and s satisfy (14.8.2), and are not both 0, $P(0, 0)$ and $P(1, 0)$ are false; and therefore $N(0, 0)$ and $N(1, 0)$ are true. If $P(-1, 0)$ were true, then $N(1, 0)$ and $P(-1, 0)$ would give

$$(1+r)^2 \geqslant 1+ns^2 \geqslant 2+(1-r)^2$$

and so $4r \geqslant 2$. From this and (14.8.2) it would follow that $r = \frac{1}{2}$ and $ns^2 = \frac{5}{4}$, which is impossible.† Hence $P(-1, 0)$ is false, and therefore $N(-1, 0)$ is true. This gives

$$ns^2 \geqslant 1+(1+r)^2 \geqslant 2,$$

and this and (14.8.2) give $n \geqslant 8$.

It follows that there is an algorithm in all cases in which $n < 8$, and these are the cases enumerated in Theorem 248.

† Suppose that $s = p/q$, where $(p, q) = 1$. If $m \not\equiv 1 \pmod 4$, then $m = n$ and

$$4mp^2 = 5q^2.$$

Hence $p^2|5$, so that $p = 1$; and $q^2|4m$. But m has no squared factor, and $0 \leqslant s \leqslant \frac{1}{2}$. Hence $q = 2$, $s = \frac{1}{2}$ and $m = 5 \equiv 1 \pmod 4$, a contradiction.

If $m \equiv 1 \pmod 4$, then $m = 4n$ and

$$mp^2 = 5q^2.$$

From this we deduce $p = 1$, $q = 1$, $s = 1$, in contradiction to (14.8.2).

There is no algorithm when $m = 23$. Take $r = 0$, $s = \frac{7}{23}$. Then (14.8.1) is

$$|23x^2 - (23y - 7)^2| \leqslant 23.$$

Since

$$\xi = 23x^2 - (23y - 7)^2 \equiv -49 \equiv -3 \pmod{23},$$

ξ must be -3 or 20, and it is easy to see that each of these hypotheses is impossible. Suppose, for example, that

$$\xi = 23X^2 - Y^2 = -3.$$

Then neither X nor Y can be divisible by 3, and

$$X^2 \equiv 1, \quad Y^2 \equiv 1, \quad \xi \equiv 22 \equiv 1 \pmod{3},$$

a contradiction.

The field $k(\sqrt{23})$, though not Euclidean, is simple; but we cannot prove this here.

14.9. Real Euclidean fields. (*continued*). It is naturally more difficult to prove that $k(\sqrt{m})$ is not Euclidean for all positive m except those listed in Theorem 247, than to prove $k(\sqrt{m})$ Euclidean for particular values of m. In this direction we prove only

THEOREM 249. *The number of real Euclidean fields $k(\sqrt{m})$, where $m \equiv$ 2 or 3 $\pmod 4$, is finite.*

Let us suppose $k(\sqrt{m})$ Euclidean and $m \not\equiv 1 \pmod 4$. We take $r = 0$ and $s = t/m$ in (14.7.2), where t is an integer to be chosen later. Then there are rational integers x, y such that

$$\left|x^2 - m\left(y - \frac{t}{m}\right)^2\right| < 1, \quad |(my - t)^2 - mx^2| < m.$$

Since

$$(my - t)^2 - mx^2 \equiv t^2 \pmod{m},$$

there are rational integers x, z such that

$$z^2 - mx^2 \equiv t^2 (\text{mod } m), \quad |z^2 - mx^2| < m. \tag{14.9.1}$$

If $m \equiv 3$ (mod 4), we choose t an odd integer such that

$$5m < t^2 < 6m,$$

as we certainly can do if m is large enough. By (14.9.1), $z^2 - mx^2$ is equal to $t^2 - 5m$ or to $t^2 - 6m$, so that one of

$$t^2 - z^2 = m(5 - x^2), \quad t^2 - z^2 = m(6 - x^2) \tag{14.9.2}$$

is true. But, to modulus 8,

$$t^2 \equiv 1, \quad z^2, x^2 \equiv 0, 1, \text{ or } 4, \quad m \equiv 3 \text{ or } 7;$$

$$t^2 - z^2 \equiv 0, 1, \text{ or } 5,$$

$$5 - x^2 \equiv 1, 4, \text{ or } 5; \quad 6 - x^2 \equiv 2, 5, \text{ or } 6;$$

$$m(5 - x^2) \equiv 3, 4, \text{ or } 7; \quad m(6 - x^2) \equiv 2, 3, 6, \text{ or } 7;$$

and, however we choose the residues, each of (14.9.2) is impossible.

If $m \equiv 2$ (mod 4), we choose t odd and such that $2m < t^2 < 3m$, as we can if m is large enough. In this case, one of

$$t^2 - z^2 = m(2 - x^2), \quad t^2 - z^2 = m(3 - x)^2 \tag{14.9.3}$$

is true. But, to modulus 8, $m \equiv 2$ or 6:

$$2 - x^2 \equiv 1, 2, \text{ or } 6; \quad 3 - x^2 \equiv 2, 3, \text{ or } 7;$$

$$m(2 - x^2) \equiv 2, 4, \text{ or } 6; \quad m(3 - x^2) \equiv 2, 4, \text{ or } 6;$$

and each of (14.9.3) is impossible.

Hence, if $m \equiv 2$ or 3 (mod 4) and if m is large enough, $k(\sqrt{m})$ cannot be Euclidean. This is Theorem 249. The same is, of course, true for $m \equiv 1$, but the proof is distinctly more difficult.

NOTES

The terminology and notation of this chapter has become out of date since it was originally written. In particular it has become customary to write $\mathbb{Q}(\sqrt{m})$ rather than $k(\sqrt{m})$, and to refer to 'units' rather than 'unities'. Moreover, one usually says that the ring of integers of a field is a 'unique factorization domain', rather than calling the field 'simple'. The property (E) in §14.7 is generally referred to by saying that the field is 'Norm-Euclidean'. We say that the field (or its ring of integers) is 'Euclidean' if there is any function ϕ whatsoever, defined on the non-zero integers of the field and taking positive integer values, with the following two properties.

(i) If γ_1 and γ_2 are non-zero integers with $\gamma_1|\gamma_2$, then $\phi(\gamma_1) \leqslant \phi(\gamma_2)$.

(ii) If γ_1 and γ_2 are non-zero integers with $\gamma_1 \nmid \gamma_2$, then there is an integer κ such that $\phi(\gamma_1 - \kappa\gamma_2) < \phi(\gamma_2)$.

We shall follow this terminology for the two notions of Euclidean field for the remainder of the notes on this chapter.

§§ 14.1–6. The theory of quadratic fields is developed in detail in Bachmann's *Grundlehren der neueren Zahlentheorie* (Göschens Lehrbücherei, no. 3, ed. 2, 1931) and Sommer's *Vorlesungen über Zahlentheorie.* There is a French translation of Sommer's book, with the title *Introduction à la théorie des nombres algébriques* (Paris, 1911); and a more elementary account of the theory, with many numerical examples, in Reid's *The elements of the theory of algebraic numbers* (New York, 1910).

§ 14.5. The equation $x^2 - my^2 = 1$ is usually called Pell's equation, but this is the result of a misunderstanding. See Dickson, *History*, ii, ch. xii, especially pp. 341, 351, 354. There is a very full account of the history of the equation in Whitford's *The Pell equation* (New York, 1912).

§ 14.7. Theorem 245 is true for Euclidean fields in general, and not merely for Norm-Euclidean fields. This can be proved by the arguments of §§12.8 and 12.9. Theorem 246 refers to the Norm-Euclidean property, but in fact there are no further complex quadratic Euclidean fields, even with the wider definition given at the start of these notes, see Samuel (*J. Algebra*, 19 (1971), 282–301).

Heilbronn and Linfoot (*Quarterly Journal of Math.* (Oxford), 5 (1934), 150–60 and 293–301) proved that there was at most one simple complex quadratic field other than those listed at the end of § 14.7. Stark (*Michigan Math. J.* 14 (1967), 1–27) proved that this extra field did not exist. Baker (ch. 5) showed that the same result followed from his approach to transcendence.

An earlier approach to this problem by Heegner (*Math. Zeit.* 56 (1952), 227–53), had originally been supposed incomplete, but was later found to be essentially correct.

§ 14.8–9. Theorem 247, which refers to Norm-Euclidean fields, is essentially due to Chatland and Davenport [*Canadian Journal of Math.* 2 (1950), 289–96]. Davenport [*Proc. London Math. Soc.* (2) 53 (1951), 65–82] showed that $k(\sqrt{m})$ cannot be Norm-Euclidean if $m > 2^{14} = 16384$, which reduced the proof of Theorem 247 to the study of a finite number of values of m. Chatland [*Bulletin Amer. Math. Soc.* 55 (1949), 948–53] gives a list of references to previous results, including a mistaken announcement by another that $k(\sqrt{97})$ was Norm-Euclidean. Barnes and Swinnerton-Dyer [*Acta Math.* 87 (1952) 259–323] show that $k(\sqrt{97})$ is not, in fact, Norm-Euclidean.

Our proof of Theorem 249 is due to Oppenheim, *Math. Annalen* 109 (1934), 349–52, and that of Theorem 249 to E. Berg, *Fysiogr. Sällsk. Lund Förh.* 5 (1935), 1–6. Both theorems relate to the Norm-Euclidean property.

It has been shown by Harper, (*Canad. J. Math.* 56 (2004), 55–70), that the field $k(\sqrt{14})$ is Euclidean, and hence the integers satisfy the fundamental theorem, even though it is not Norm-Euclidean. It is conjectured that there are infinitely many real quadratic fields with the unique factorization property, and that they are all Euclidean, although only those listed in Theorem 247 can be Norm-Euclidean.

When p is a prime there appear to be a large number of fields $k(\sqrt{p})$ with the unique factorization property. Indeed Cohen and Lenstra (*Number theory, Noordwijkerhout 1983*,

Springer Lecture Notes in Math. 1068, 33–62), have given heuristics leading to a precise conjecture, which would show that $k(\sqrt{p})$ has the unique factorization property for asymptotically a positive proportion of primes.

We expect an infinity of real quadratic fields with the unique factorization property. However if we restrict attention to square-free integers m for which there is a small non-trivial unit, then the picture changes. Thus, for square-free numbers m of the form $m = 4r^2 + 1$, there is a 'small' unit $2m + \sqrt{r}$, and it has been shown by Biró (*Acta Arith.* 107 (2003), 179–94), that in this case one obtains a unique factorization domain if and only if $r = 1, 2, 3, 5, 7$ or 13.

XV

QUADRATIC FIELDS (2)

15.1. The primes of $k(i)$. We begin this chapter by determining the primes of $k(i)$ and a few other simple quadratic fields.

If π is a prime of $k(\sqrt{m})$, then

$$\pi | N\pi = \pi\bar{\pi}$$

and $\pi \,\big|\, |N\pi|$. There are therefore positive rational integers divisible by π. If z is the least such integer, $z = z_1 z_2$, and the field is simple, then

$$\pi | z_1 z_2 \to \pi | z_1 \text{ or } \pi | z_2,$$

a contradiction unless z_1 or z_2 is 1. Hence z is a rational prime. Thus π divides at least one rational prime p. If it divides two, say p and p', then

$$\pi | p \,.\, \pi | p' \to \pi | px - p'y = 1$$

for appropriate x and y, a contradiction.

THEOREM 250. *Any prime π of a simple field $k(\sqrt{m})$ is a divisor of just one positive rational prime.*

The primes of a simple field are therefore to be determined by the factorization, in the field, of rational primes.

We consider $k(i)$ first. If

$$\pi = a + bi | p, \quad \pi\lambda = p,$$

then

$$N\pi N\lambda = p^2.$$

Either $N\lambda = 1$, when λ is a unity and π an associate of p, or

$$N\pi = a^2 + b^2 = p. \tag{15.1.1}$$

(i) If $p = 2$, then

$$p = 1^2 + 1^2 = (1+i)(1-i) = i(1-i)^2.$$

The numbers $1+i$, $-1+i$, $-1-i$, $1-i$ (which are associates) are primes of $k(i)$.

(ii) If $p = 4n + 3$, (15.1.1) is impossible, since a square is congruent to 0 or 1 (mod 4). Hence the primes $4n + 3$ are primes of $k(i)$.

(iii) If $p = 4n + 1$, then

$$\left(\frac{-1}{p}\right) = 1,$$

by Theorem 82, and there is an x for which

$$p|x^2 + 1, \quad p|(x + i)(x - i).$$

If p were a prime of $k(i)$, it would divide $x + i$ or $x - i$, and this is false, since the numbers

$$\frac{x}{p} \pm \frac{i}{p}$$

are not integers. Hence p is not a prime. It follows that $p = \pi\lambda$, where $\pi = a + bi, \lambda = a - bi$, and

$$N\pi = a^2 + b^2 = p.$$

In this case p can be expressed as a sum of two squares.

The prime divisors of p are

$$\pi,\ i\pi,\ -\pi,\ -i\pi,\ \lambda,\ i\lambda,\ -\lambda,\ -i\lambda, \tag{15.1.2}$$

and any of these numbers may be substituted for π. The eight variations correspond to the eight equations

$$(\pm a)^2 + (\pm b)^2 = (\pm b)^2 + (\pm a)^2 = p. \tag{15.1.3}$$

And if $p = c^2 + d^2$ then $c + id|p$, so that $c + id$ is one of the numbers (15.1.2). Hence, apart from these variations, the expression of p as a sum of squares is unique.

THEOREM 251. *A rational prime $p = 4n + 1$ can be expressed as a sum $a^2 + b^2$ of two squares.*

THEOREM 252. *The primes of $k(i)$ are*

(1) *$1 + i$ and its associates,*

(2) *the rational primes $4n + 3$ and their associates,*

(3) *the factors $a + bi$ of the rational primes $4n + 1$.*

15.2. Fermat's theorem in $k(i)$. As an illustration of the arithmetic of $k(i)$, we select the analogue of Fermat's theorem. We consider only the analogue of Theorem 71 and not that of the more general Fermat–Euler theorem. It may be worth repeating that $\gamma\,|\,(\alpha-\beta)$ and

$$\alpha \equiv \beta \pmod{\gamma}$$

mean, when we are working in the field $k(\vartheta)$, that $\alpha-\beta=\kappa\gamma$, where κ is an integer of the field.

We denote rational primes $4n+1$ and $4n+3$ by p and q respectively, and a prime of $k(i)$ by π. We confine our attention to primes of the classes (2) and (3), i.e. primes whose norm is odd; thus π is a q or a divisor of a p. We write

$$\phi(\pi)=N\pi-1,$$

so that

$$\phi(\pi)=p-1 \quad (\pi\,|\,p), \quad \phi(\pi)=q^2-1 \ \ (\pi=q).$$

THEOREM 253. *If* $(\alpha,\pi)=1$, *then*

$$\alpha^{\phi(\pi)} \equiv 1 \pmod{\pi}.$$

Suppose that $\alpha=l+im$. Then, when $\pi\,|\,p$, $i^p=i$ and

$$\alpha^p=(l+im)^p \equiv l^p+(im)^p=l^p+im^p \pmod{p},$$

by Theorem 75; and so

$$\alpha^p \equiv l+im=\alpha \pmod{p},$$

by Theorem 70. The same congruence is true mod π, and we may remove the factor α.

When $\pi=q$, $i^q=-i$ and

$$\alpha^q=(l+im)^q \equiv l^q-im^q \equiv l-im=\bar{\alpha} \pmod{q}.$$

Similarly, $\bar{\alpha}^q \equiv \alpha$, so that

$$\alpha^{q^2} \equiv \alpha, \quad \alpha^{q^2-1} \equiv 1 \quad \pmod{q}.$$

The theorem can also be proved on lines corresponding to those of § 6.1. Suppose for example that $\pi = a + bi \mid p$. The number

$$(a + bi)(c + di) = ac - bd + i(ad + bc)$$

is a multiple of π and, since $(a, b) = 1$, we can choose c and d so that $ad + bc = 1$. Hence there is an s such that

$$\pi \mid s + i.$$

Now consider the numbers

$$r = 0, 1, 2, \ldots, N\pi - 1 = a^2 + b^2 - 1,$$

which are plainly incongruent (mod π). If $x + yi$ is any integer of $k(i)$, there is an r for which

$$x - sy \equiv r \pmod{N\pi};$$

and then

$$x + yi \equiv y(s + i) + r \equiv r \pmod{\pi}.$$

Hence the r form a 'complete system of residues' (mod π).

If α is prime to π, then, as in rational arithmetic, the numbers αr also form a complete system of residues.† Hence

$$\prod(\alpha r) \equiv \prod r \pmod{\pi},$$

and the theorem follows as in § 6.1.

The proof in the other case is similar, but the 'complete system' is constructed differently.

15.3. The primes of $k(\rho)$. The primes of $k(\rho)$ are also factors of rational primes, and there are again three cases.

(1) If $p = 3$, then

$$p = (1 - \rho)(1 - \rho^2) = (1 + \rho)(1 - \rho)^2 = -\rho^2(1 - \rho)^2.$$

By Theorem 221, $1 - \rho$ is a prime.

† Compare Theorem 58. The proof is essentially the same.

(2) If $p \equiv 2 \pmod 3$ then it is impossible that $N\pi = p$, since

$$4N\pi = (2a - b)^2 + 3b^2$$

is congruent to 0 or 1 (mod 3). Hence p is a prime in $k(\rho)$.

(3) If $p \equiv 1 \pmod 3$ then

$$\left(\frac{-3}{p}\right) = 1,$$

by Theorem 96, and $p|x^2 + 3$. It then follows as in § 15.1 that p is divisible by a prime $\pi = a + b\rho$, and that

$$p = N\pi = a^2 - ab + b^2.$$

THEOREM 254. *A rational prime* $3n + 1$ *is expressible in the form* $a^2 - ab + b^2$.

THEOREM 255. *The primes of* $k(\rho)$ *are*

(1) $1 - \rho$ *and its associates,*

(2) *the rational primes* $3n + 2$ *and their associates,*

(3) *the factors* $a + b\rho$ *of the rational primes* $3n + 1$.

15.4. The primes of $k(\surd 2)$ and $k(\surd 5)$. The discussion goes similarly in other simple fields. In $k(\surd 2)$, for example, either p is prime or

$$N\pi = a^2 - 2b^2 = \pm p. \tag{15.4.1}$$

Every square is congruent to 0, 1, or 4 (mod 8), and (15.4.1) is impossible when p is $8n \pm 3$. When p is $8n \pm 1$, 2 is a quadratic residue of p by Theorem 95, and we show as before that p is factorizable. Finally

$$2 = (\surd 2)^2,$$

and $\surd 2$ is prime.

THEOREM 256. *The primes of* $k(\surd 2)$ *are* (1) $\surd 2$, (2) *the rational primes* $8n \pm 3$, (3) *the factors* $a + b\surd 2$ *of rational primes* $8n \pm 1$ (*and the associates of these numbers*).

We consider one more example because we require the results in § 15.5. The integers of $k(\surd 5)$ are the numbers $a + b\omega$, where a and b are rational integers and

$$\omega = \tfrac{1}{2}\left(1 + \surd 5\right). \tag{15.4.2}$$

The norm of $a + b\omega$ is $a^2 + ab - b^2$. The numbers

$$\pm\omega^{\pm n} \quad (n = 0,\ 1,\ 2, \ldots) \tag{15.4.3}$$

are unities, and we can prove as in § 14.5 that there are no more.

The determination of the primes depends upon the equation

$$N\pi = a^2 + ab - b^2 = p,$$

or

$$(2a + b)^2 - 5b^2 = 4p.$$

If $p = 5n \pm 2$, then $(2a + b)^2 \equiv \pm 3 \pmod{5}$, which is impossible. Hence these primes are primes in $k(\surd 5)$.

If $p = 5n \pm 1$, then

$$\left(\frac{5}{p}\right) = 1,$$

by Theorem 97. Hence $p \mid (x^2 - 5)$ for some x, and we conclude as before that p is factorizable. Finally

$$5 = (\surd 5)^2 = (2\omega - 1)^2.$$

Theorem 257. *The unities of* $k(\surd 5)$ *are the numbers* (15.4.3). *The primes are* (1) $\surd 5$, (2) *the rational primes* $5n \pm 2$, (3) *the factors* $a + b\omega$ *of rational primes* $5n \pm 1$ (*and the associates of these numbers*).

We shall also need the analogue of Fermat's theorem.

Theorem 258. *If* p *and* q *are the rational primes* $5n \pm 1$ *and* $5n \pm 2$ *respectively;* $\phi(\pi) = |N\pi| - 1$, *so that*

$$\phi(\pi) = p - 1 \quad (\pi \mid p), \quad \phi(\pi) = q^2 - 1 \quad (\pi = q);$$

and $(\alpha, \pi) = 1$; *then*

$$\alpha^{\phi(\pi)} \equiv 1 \pmod{\pi}, \tag{15.4.4}$$

$$\alpha^{p-1} \equiv 1 \pmod{\pi}, \tag{15.4.5}$$

$$\alpha^{q+1} \equiv N\alpha \pmod{q}. \tag{15.4.6}$$

Further, if $\pi \mid p, \bar{\pi}$ *is the conjugate of* π, $(\alpha, \pi) = 1$ *and* $(\alpha, \bar{\pi}) = 1$, *then*

$$\alpha^{p-1} \equiv 1 \pmod{p}. \tag{15.4.7}$$

First, if

$$2\alpha = c + d\sqrt{5},$$

then

$$2\alpha^p \equiv (2\alpha)^p = \left(c + d\sqrt{5}\right)^p \equiv c^p + d^p 5^{\frac{1}{2}(p-1)}\sqrt{5} \pmod{p}.$$

But

$$5^{\frac{1}{2}(p-1)} \equiv \left(\frac{5}{p}\right) = 1 \pmod{p},$$

$c^p \equiv c$ and $d^p \equiv d$. Hence

$$2\alpha^p \equiv c + d\sqrt{5} = 2\alpha \pmod{p}, \tag{15.4.8}$$

and, *a fortiori*,

$$2\alpha^p \equiv 2\alpha \pmod{\pi}. \tag{15.4.9}$$

Since $(2, \pi) = 1$ and $(\alpha, \pi) = 1$, we may divide by 2α, and obtain (15.4.5). If also $(\alpha, \bar{\pi}) = 1$, so that $(\alpha, p) = 1$, then we may divide (15.4.8) by 2α, and obtain (15.4.7).

Similarly, if $q > 2$,

$$2\alpha^q \equiv c - d\sqrt{5} = 2\bar{\alpha}, \quad \alpha^q \equiv \bar{\alpha} \pmod{q}, \tag{15.4.10}$$

$$\alpha^{q+1} \equiv \alpha\bar{\alpha} = N\alpha \pmod{q}. \tag{15.4.11}$$

This proves (15.4.6). Also (15.4.10) involves

$$\alpha^{q^2} \equiv \bar{\alpha}^q \equiv \alpha \pmod{q},$$

$$\alpha^{q^2-1} \equiv 1 \pmod{q}. \tag{15.4.12}$$

Finally (15.4.5) and (15.4.12) together contain (15.4.4).

The proof fails if $q = 2$, but (15.4.4) and (15.4.6) are still true. If $\alpha = e + f\omega$ then one of e and f is odd, and therefore $N\alpha = e^2 + ef - f^2$ is odd. Also, to modulus 2,

$$\alpha^2 \equiv e^2 + f^2\omega^2 \equiv e + f\omega^2 = e + f(\omega + 1) \equiv e + f(1 - \omega)$$
$$= e + f\bar{\omega} = \bar{\alpha}$$

and

$$\alpha^3 \equiv \alpha\bar{\alpha} = N\alpha \equiv 1.$$

We note in passing that our results give incidentally another proof of Theorem 180.

The nth Fibonacci number is

$$u_n = \frac{\omega^n - \bar{\omega}^n}{\omega - \bar{\omega}} = \frac{\omega^n - \bar{\omega}^n}{\sqrt{5}},$$

where ω is the number (15.4.2) and $\bar{\omega} = -1/\omega$ is its conjugate.

If $n = p$, then

$$\omega^{p-1} \equiv 1 \pmod{p}, \quad \bar{\omega}^{p-1} \equiv 1 \pmod{p},$$
$$u_{p-1}\sqrt{5} = \omega^{p-1} - \bar{\omega}^{p-1} \equiv 0 \pmod{p},$$

and therefore $u_{p-1} \equiv 0 \pmod{p}$. If $n = q$, then

$$\omega^{q+1} \equiv N\omega, \quad \bar{\omega}^{q+1} \equiv N\omega \pmod{q},$$
$$u_{q+1}\sqrt{5} \equiv 0 \pmod{q}$$

and $u_{q+1} \equiv 0 \pmod{q}$.

15.5. Lucas's test for the primality of the Mersenne number M_{4n+3}. We are now in a position to prove a remarkable theorem which is due, in substance at any rate, to Lucas, and which contains a necessary and sufficient condition for the primality of M_{4n+3}. Many 'necessary and sufficient conditions' contain no more than a transformation of a problem, but this one gives a practical test which can be applied to otherwise inaccessible examples.

We define the sequence

$$r_1, r_2, r_3, \ldots = 3, 7, 47, \ldots$$

by

$$r_m = \omega^{2^m} + \bar{\omega}^{2^m},$$

where ω is the number (15.4.2) and $\bar{\omega} = -1/\omega$. Then

$$r_{m+1} = r_m^2 - 2.$$

In the notation of § 10.14,

$$r_m = v_{2^m}.$$

No two r_m have a common factor, since (i) they are all odd, and

(ii) $\qquad r_m \equiv 0 \to r_{m+1} \equiv -2 \to r_\nu \equiv 2\ (\nu > m+1),$

to any odd prime modulus.

THEOREM 259. *If p is a prime* $4n+3$, *and*

$$M = M_p = 2^p - 1$$

is the corresponding Mersenne number, then M is prime if

$$r_{p-1} \equiv 0 \pmod{M}, \tag{15.5.1}$$

and otherwise composite.

(1) Suppose M prime. Since

$$M \equiv 8.16^n - 1 \equiv 8 - 1 \equiv 2 \pmod{5},$$

we may take $\alpha = \omega, q = M$ in (15.4.6). Hence

$$\omega^{2^p} = \omega^{M+1} \equiv N\omega = -1 \pmod{M},$$
$$r_{p-1} = \bar{\omega}^{2^{p-1}}\left(\omega^{2^p} + 1\right) \equiv 0 \pmod{M},$$

which is (15.5.1).

(2) Suppose (15.5.1) true. Then

$$\omega^{2^p} + 1 = \omega^{2^{p-1}} r_{p-1} \equiv 0 \pmod{M},$$
$$\omega^{2^p} \equiv -1 \pmod{M}, \tag{15.5.2}$$
$$\omega^{2^{p+1}} \equiv 1 \pmod{M}. \tag{15.5.3}$$

The same congruences are true, *a fortiori,* to any modulus τ which divides M.

Suppose that

$$M = p_1 p_2 \dots q_1 q_2 \dots$$

is the expression of M as a product of rational primes, p_i being a prime $5n \pm 1$ (so that p_i is the product of two conjugate primes of the field) and q_i a prime $5n \pm 2$. Since $M \equiv 2 \pmod 5$, there is at least one q_i.

The congruence

$$\omega^x \equiv 1 (\operatorname{mod} \tau),$$

or $P(x)$, is true, after (15.5.3), when $x = 2^{p+1}$, and the smallest positive solution is, by Theorem 69, a divisor of 2^{p+1}. These divisors, apart from 2^{p+1}, are $2^p, 2^{p-1}, \dots$, and $P(x)$ is false for all of them, by (15.5.2). Hence 2^{p+1} is the smallest solution, and every solution is a multiple of this one.

But

$$\omega^{2^{p_i-1}} \equiv 1 \pmod{p_i},$$

$$\omega^{2^{(q_j+1)}} \equiv (N\omega)^2 \equiv 1 \pmod{q_j},$$

by (15.4.7) and (15.4.6). Hence $p_i - 1$ and $2(q_j + 1)$ are multiples of 2^{p+1}, and

$$p_i = 2^{p+1} h_i + 1,$$

$$q_j = 2^p k_j - 1,$$

for some h_i and k_j. The first hypothesis is impossible because the right-hand side is greater than M; and the second is impossible unless

$$k_j = 1, \qquad q_j = M.$$

Hence M is prime.

The test in Theorem 259 applies only when $p \equiv 3 \pmod 4$. The sequence

$$4, 14, 194, \dots$$

(constructed by the same rule) gives a test (verbally identical) for any p. In this case the relevant field is $k(\sqrt{3})$. We have selected the test in Theorem 259 because the proof is slightly simpler.

To take a trivial example, suppose $p = 7, M_p = 127$. The numbers r_m of Theorem 259, reduced (mod M), are

$$3,\ 7,\ 47,\ 2207 \equiv 48, \quad 2302 \equiv 16, \quad 254 \equiv 0,$$

and 127 is prime. If $p = 127$, for example, we must square 125 residues, which may contain as many as 39 digits (in the decimal scale). Such computations were, at one time, formidable, but quite practicable, and it was in this way that Lucas showed M_{127} to be prime. The construction of electronic digital computers enabled the tests to be applied to M_p with larger p. These computers usually work in the binary scale in which reduction to modulus $2^n - 1$ is particularly simple. But their great advantage is, of course, their speed. Thus M_{19937} was tested in about 35 minutes, in 1971, by Tuckerman on an IBM 360/91.

15.6. General remarks on the arithmetic of quadratic fields. The construction of an arithmetic in a field which is not simple, like $k\{\surd(-5)\}$ or $k(\surd 10)$, demands new ideas which (though they are not particularly difficult) we cannot develop systematically here. We add only some miscellaneous remarks which may be useful to a reader who wishes to study the subject more seriously.

We state below three properties, A, B, and C, common to the 'simple' fields which we have examined. These properties are all consequences of the Euclidean algorithm, when such an algorithm exists, and it was thus that we proved them in these fields. They are, however, true in any simple field, whether the field is Euclidean or not. We shall not prove so much as this; but a little consideration of the logical relations between them will be instructive.

A. *If α and β are integers of the field, then there is an integer δ with the properties*

(A i) $$\delta | \alpha, \quad \delta | \beta,$$

and

(A ii) $$\delta_1 | \alpha \,.\, \delta_1 | \beta \rightarrow \delta_1 | \delta.$$

Thus δ is the highest, or 'most comprehensive', common divisor (α, β) of α and β, as we defined it, in $k(i)$, in § 12.8.

B. *If α and β are integers of the field, then there is an integer δ with the properties*

(B i) $$\delta | \alpha, \quad \delta | \beta :$$

and

(B ii) δ *is a linear combination of* α *and* β; *there are integers* λ *and* μ *such that*

$$\lambda\alpha + \mu\beta = \delta.$$

It is obvious that B implies A; (B i) is the same as (A i), and a δ with the properties (B i) and (B ii) has the properties (A i) and (A ii). The converse, though true in the quadratic fields in which we are interested now, is less obvious, and depends upon the special properties of these fields.

There are 'fields' in which 'integers' possess a highest common divisor in sense A but not in sense B. Thus the aggregate of all rational functions

$$R(x,y) = \frac{P(x, y)}{Q(x, y)}$$

of two independent variables, with rational coefficients, is a field in the sense explained at the end of § 14.1. We may call the polynomials $P(x,y)$ of the field the 'integers', regarding two polynomials as the same when they differ only by a constant factor. Two polynomials have a greatest common divisor in sense A; thus x and y have the greatest common divisor 1. But there are no polynomials $P(x,y)$ and $Q(x,y)$ such that

$$xP(x,y) + yQ(x,y) = 1.$$

C. *Factorization in the field is unique: the field is simple.*

It is plain that B implies C; for (B i) and (B ii) imply

$$\delta\gamma|\alpha\gamma, \quad \delta\gamma|\beta\gamma, \quad \lambda\alpha\gamma + \mu\beta\gamma = \delta\gamma,$$

and so

$$(\alpha\gamma, \beta\gamma) = \delta\gamma; \tag{15.6.1}$$

and from this C follows as in § 12.8.

That A implies C is not quite so obvious, but may be proved as follows. It is enough to deduce (15.6.1) from A. Let

$$(\alpha\gamma, \beta\gamma) = \Delta.$$

Then

$$\delta|\alpha \,.\, \delta|\beta \rightarrow \delta\gamma|\alpha\gamma \,.\, \delta\gamma|\beta\gamma,$$

and so, by (A ii),

$$\delta\gamma|\Delta.$$

Hence

$$\Delta = \delta\gamma\rho,$$

say. But $\Delta|\alpha\gamma$, $\Delta|\beta\gamma$ and so

$$\delta\rho|\alpha, \quad \delta\rho|\beta;$$

and hence, again by (A ii), $\delta\rho|\delta$.

Hence ρ is a unity, and $\Delta = \delta\gamma$.

On the other hand, it is obvious that C implies A; for δ is the product of all prime factors common to α and β. That C implies B is again less immediate, and depends, like the inference from A to B, on the special properties of the fields in question.†

15.7. Ideals in a quadratic field. There is another property common to all simple quadratic fields. To fix our ideas, we consider the field $k(i)$, whose basis (§ 14.3) is $[1, i]$.

A lattice Λ is‡ the aggregate of all points‖

$$m\alpha + n\beta,$$

α and β being the points P and Q of § 3.5, and m and n running through the rational integers. We say that $[\alpha, \beta]$ is a basis of Λ, and write

$$\Lambda = [\alpha, \beta];$$

a lattice will, of course, have many different bases. The lattice is a modulus in the sense of § 2.9, and has the property

$$\rho \in \Lambda \,.\, \sigma \in \Lambda \rightarrow m\rho + n\sigma \in \Lambda \tag{15.7.1}$$

for any rational integral m and n.

Among lattices there is a sub-class of peculiar importance. Suppose that Λ has, in addition to (15.7.1), the property

$$\gamma \in \Lambda \rightarrow i\gamma \in \Lambda. \tag{15.7.2}$$

† In fact both inferences depend on just those arguments which are required in the elements of the theory of ideals in a quadratic field.

‡ See § 3.5. There, however, we reserved the symbol Λ for the principal lattice.

‖ We do not distinguish between a point and the number which is its affix in the Argand diagram.

Then plainly $m\gamma \in \Lambda$ and $ni\gamma \in \Lambda$, and so

$$\gamma \in \Lambda \rightarrow \mu\gamma \in \Lambda$$

for every integer μ of $k(i)$; *all multiples of points of* Λ *by integers of* $k(i)$ *are also points of* Λ. Such a lattice is called an *ideal.* If Λ is an ideal, and ρ and σ belong to Λ, then $\mu\rho + \nu\sigma$ belongs to Λ:

(15.7.3) $$\rho \in \Lambda \,.\, \sigma \in \Lambda \rightarrow \mu\rho + \nu\sigma \in \Lambda$$

for all integral μ and ν. This property includes, but states much more than, (15.7.1).

Suppose now that Λ is an ideal with basis $[\alpha, \beta]$, and that

$$(\alpha, \beta) = \delta.$$

Then every point of Λ is a multiple of δ. Also, since δ is a linear combination of α and β, δ and all its multiples are points of Λ. Thus Λ is the class of all multiples of δ; and it is plain that, conversely, the class of multiples of any δ is an ideal Λ. *Any ideal is the class of multiples of an integer of the field, and any such class is an ideal.*

If Λ is the class of multiples of ρ, we write

$$\Lambda = \{\rho\}.$$

In particular the fundamental lattice, formed by all the integers of the field, is $\{1\}$.

The properties of an integer ρ may be restated as properties of the ideal $\{\rho\}$. Thus $\sigma|\rho$ means that $\{\rho\}$ is a part of $\{\sigma\}$. We can then say that '$\{\rho\}$ is divisible by $\{\sigma\}$', and write

$$\{\sigma\}|\{\rho\}.$$

Or again we can write

$$\{\sigma\}|\rho, \rho \equiv 0 (\text{mod } \{\sigma\}),$$

these assertions meaning that the number ρ belongs to the ideal $\{\sigma\}$. In this way we can restate the whole of the arithmetic of the field in terms of ideals, though, in $k(i)$, we gain nothing substantial by such a restatement. An ideal being always the class of multiples of an integer, the new arithmetic is merely a verbal translation of the old one.

We can, however, define ideals in *any* quadratic field. We wish to use the geometrical imagery of the complex plane, and we shall therefore consider only complex fields.

Suppose that $k(\sqrt{m})$ is a complex field with basis $[1, \omega]$.† We may define a lattice as we defined it above in $k(i)$, and an ideal as a lattice which has the property

$$\gamma \in \Lambda \rightarrow \omega\gamma \in \Lambda, \tag{15.7.4}$$

analogous to (15.7.2). As in $k(i)$, such a lattice has also the property (15.7.3), and this property might be used as an alternative definition of an ideal.

Since two numbers α and β have not necessarily a 'greatest common divisor' we can no longer prove that an ideal $\mathbf{r}$ has necessarily the form $\{\rho\}$; any $\{\rho\}$ is an ideal, but the converse is not generally true. But the definitions above, which were logically independent of this reduction, are still available; we can define

$$\mathbf{s}|\mathbf{r}$$

as meaning that every number of $\mathbf{r}$ belongs to $\mathbf{s}$, and

$$\rho \equiv 0 \pmod{\mathbf{s}}$$

as meaning that ρ belongs to $\mathbf{s}$. We can thus define words like *divisible, divisor*, and *prime* with reference to ideals, and have the foundations for an arithmetic which is at any rate as extensive as the ordinary arithmetic of simple fields, and may perhaps be useful where such ordinary arithmetic fails. That this hope is justified, and that the notion of an ideal leads to a complete re-establishment of arithmetic in any field, is shown in systematic treatises on the theory of algebraic numbers. The reconstruction is as effective in real as in complex fields, though not all of our geometrical language is then appropriate.

An ideal of the special type $\{\rho\}$ is called a *principal ideal;* and the fourth characteristic property of simple quadratic fields, to which we referred at the beginning of this section, is

D. *Every ideal of a simple field is a principal ideal.*

This property may also be stated, when the field is complex, in a simple geometrical form. In $k(i)$ an ideal, that is to say a lattice with the property

† $\omega = \sqrt{m}$ when $m \not\equiv 1 \pmod 4$.

(15.7.2), is *square*; for it is of the form $\{\rho\}$, and may be regarded as the figure of lines based on the origin and the points ρ and $i\rho$. More generally

E. *If $m < 0$ and $k(\surd m)$ is simple, then every ideal of $k(\surd m)$ is a lattice similar in shape to the lattice formed by all the integers of the field.*

It is instructive to verify that this is not true in $k\{\surd(-5)\}$. The lattice

$$m\alpha + n\beta = m \,.\, 3 + n\{-1 + \surd(-5)\}$$

is an ideal, for $\omega = \surd(-5)$ and

$$\omega\alpha = \alpha + 3\beta, \quad \omega\beta = -2\alpha - \beta.$$

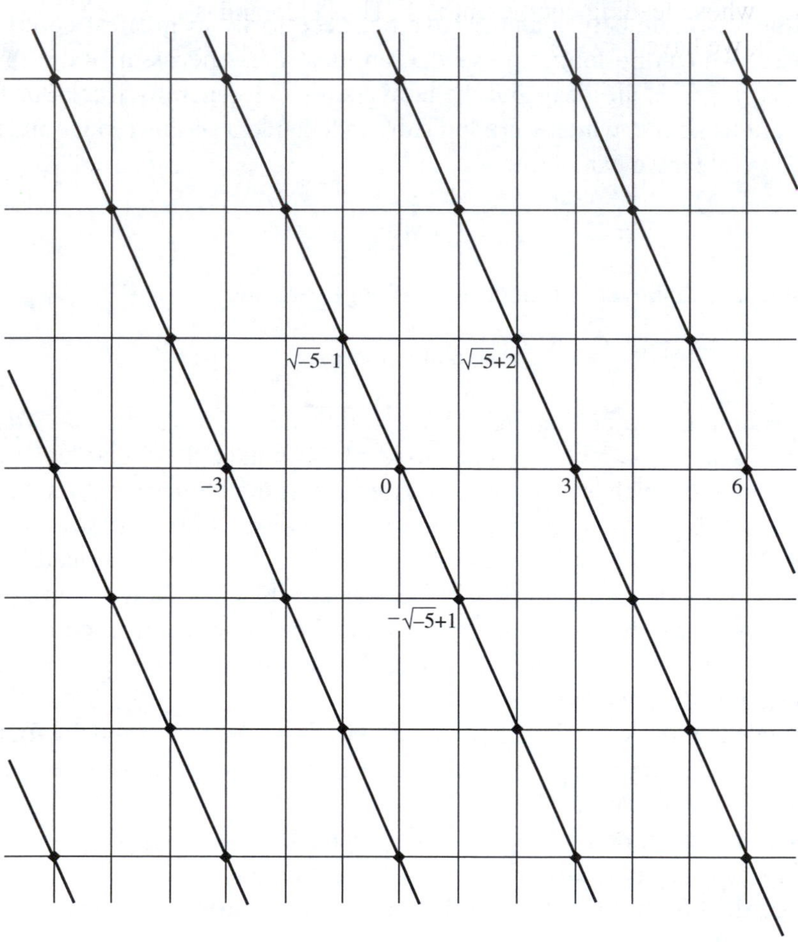

Fig. 7.

But, as is shown by Fig. 7 (and may, of course, be verified analytically), the lattice is not similar to the lattice of all integers of the field.

15.8. Other fields. We conclude this chapter with a few remarks about some non-quadratic fields of particularly interesting types. We leave the verification of most of our assertions to the reader.

(i) *The field* $k(\surd 2 + i)$. The number

$$\vartheta = \sqrt{2} + i$$

satisfies

$$\vartheta^4 - 2\vartheta^2 + 9 = 0,$$

and the number defines a field which we denote by $k(\surd 2 + i)$. The numbers of the field are

$$\xi = r + si + t\surd 2 + ui\surd 2, \tag{15.8.1}$$

where r, s, t, u are rational. The integers of the field are

$$\xi = a + bi + c\surd 2 + di\surd 2, \tag{15.8.2}$$

where a and b are integers and c and d are either both integers or both halves of odd integers.

The conjugates of ξ are the numbers ξ_1, ξ_2, ξ_3, formed by changing the sign of either or both of i and $\surd 2$ in (15.8.1) or (15.8.2), and the norm $N\xi$ of ξ is defined by

$$N\xi = \xi\xi_1\xi_2\xi_3.$$

Divisibility, and so forth, are defined as in the fields already considered. There is a Euclidean algorithm, and factorization is unique.†

(ii) *The field* $k(\surd 2 + \surd 3)$. The number

$$\vartheta = \surd 2 + \surd 3,$$

satisfies the equation

$$\vartheta^4 - 10\vartheta^2 + 1 = 0.$$

† Theorem 215 stands in the field as stated in §12.8. The proof demands some calculation.

The numbers of the field are

$$\xi = r + s\surd 2 + t\surd 3 + u\surd 6,$$

and the integers are the numbers

$$\xi = a + b\surd 2 + c\surd 3 + d\surd 6,$$

where a and c are integers and b and d are, either both integers or both halves of odd integers. There is again a Euclidean algorithm, and factorization is unique.

These fields are simple examples of 'biquadratic' fields.

(iii) *The field* $k(e^{\frac{2}{5}\pi i})$. The number $e^{\frac{2}{5}\pi i}$ satisfies the equation

$$\frac{\vartheta^5 - 1}{\vartheta - 1} = \vartheta^4 + \vartheta^3 + \vartheta^2 + \vartheta + 1 = 0.$$

The field is, after $k(i)$ and $k(\rho)$, the simplest 'cyclotomic' field.†

The numbers of the field are

$$\xi = r + s\vartheta + t\vartheta^2 + u\vartheta^3,$$

and the integers are the numbers in which r, s, t, u are integral. The conjugates of ξ are the numbers ξ_1, ξ_2, ξ_3, obtained by changing ϑ into $\vartheta^2, \vartheta^3, \vartheta^4$, and its norm is

$$N\xi = \xi\xi_1\xi_2\xi_3.$$

There is a Euclidean algorithm, and factorization is unique.

The number of unities in $k(i)$ and $k(\rho)$ is finite. In $k(e^{\frac{2}{5}\pi i})$ the number is infinite. Thus

$$(1 + \vartheta) \mid (\vartheta + \vartheta^2 + \vartheta^3 + \vartheta^4)$$

and $\vartheta + \vartheta^2 + \vartheta^3 + \vartheta^4 = -1$ so that $1 + \vartheta$ and all its powers are unities.

It is plainly this field which we must consider if we wish to prove 'Fermat's last theorem', when $n = 5$, by the method of § 13.4. The proof follows the same lines, but there are various complications of detail.

† The field $k(\vartheta)$ with ϑ a primitive nth root of unity, is called *cyclotomic* because ϑ and its powers are the complex coordinates of the vertices of a regular n-agon inscribed in the unit circle.

The field defined by a primitive nth root of unity is simple, in the sense of § 14.7, when†

$$n = 3, 4, 5, 8.$$

NOTES

§ 15.5. Lucas stated two tests for the primality of M_p, but his statements of his theorems vary, and he never published any complete proof of either. The argument in the text is due to Western, *Journal London Math. Soc.* 7 (1932), 130–7. The second theorem, not proved in the text, is that referred to in the penultimate paragraph of the section. Western proves this theorem by using the field $k(\surd 3)$. Other proofs, independent of the theory of algebraic numbers, have been given by D. H. Lehmer, *Annals of Math.* (2) 31 (1930), 419–48, and *Journal London Math. Soc.* 10 (1935), 162–5.

Professor Newman drew our attention to the following result, which can be proved by a simple extension of the argument of this section.

Let $h < 2^m$ be odd, $M = 2^m h - 1 \equiv \pm 2 \pmod 5$ and

$$R_1 = \omega^{2h} + \bar{\omega}^{2h},\ R_j = R_{j-1}^2 - 2 (j \geqslant 2).$$

Then a necessary and sufficient condition for M to be prime is that

$$R_{m-1} \equiv 0 \pmod M.$$

This result was stated by Lucas [*Amer. Journal of Math.* 1 (1878), 310], who gives a similar (but apparently erroneous) test for numbers of the form $N = h2^m + 1$. The primality of the latter can, however, be determined by the test of Theorem 102, which also requires about m squarings and reductions (mod N). The two tests would provide a practicable means of seeking large prime pairs $(p, p+2)$.

§§ 15.6–7. These sections have been much improved as a result of criticisms from Mr. Ingham, who read an earlier version. The remark about polynomials in § 15.6 is due to Bochner, *Journal London Math. Soc.* 9 (1934), 4.

§ 15.8. There is a proof that $k(e^{\frac{2}{5}\pi i})$ is Euclidean in Landau, *Vorlesungen*, iii. 228–31.

The list of fields $k(e^{2\pi i/m})$ with the unique factorization property has been completely determined by Masley and Montgomery (*J. Reine Angew. Math.* 286/287 (1976), 248–56). If m is odd, the values m and $2m$ lead to the same field. Bearing this in mind there are exactly 29 distinct fields for $m \geqslant 3$, corresponding to

$$m = 3, 4, 5, 7, 8, 9, 11, 12, 13, 15, 16, 17, 19, 20, 21, 24, 25, 27, 28,$$
$$32, 33, 35, 36, 40, 44, 45, 48, 60, 84.$$

† $e^{\frac{2}{8}\pi i} = e^{\frac{1}{4}\pi i} = \frac{1+i}{\surd 2}$ is a number of $k(\surd 2 + i)$.

XVI

THE ARITHMETICAL FUNCTIONS $\phi(n)$, $\mu(n)$, $d(n), \sigma(n)$, $r(n)$

16.1. The function $\phi(n)$. In this and the next two chapters we shall study the properties of certain 'arithmetical functions' of n, that is to say functions $f(n)$ of the positive integer n defined in a manner which expresses some arithmetical property of n.

The function $\phi(n)$ was defined in § 5.5, for $n > 1$, as the number of positive integers less than and prime to n. We proved (Theorem 62) that

$$\phi(n) = n \prod_{p|n} \left(1 - \frac{1}{p}\right). \tag{16.1.1}$$

This formula is also an immediate consequence of the general principle expressed by the theorem which follows.

THEOREM 260. *If there are N objects, of which N_α have the property α, N_β have $\beta, \ldots, N_{\alpha\beta}$ have both α and $\beta, \ldots, N_{\alpha\beta\gamma}$ have α, β, and $\gamma, \ldots$, and so on, then the number of the objects which have none of α, β, $\gamma, \ldots$ is*

$$N - N_\alpha - N_\beta - \cdots + N_{\alpha\beta} + \cdots - N_{\alpha\beta\gamma} - \cdots. \tag{16.1.2}$$

Suppose that O is an object which has just k of the properties $\alpha, \beta, \ldots$. Then O contributes 1 to N. If $k \geqslant 1$, O also contributes 1 to k of N_α, $N_\beta, \ldots$, to $\frac{1}{2}k(k-1)$ of $N_{\alpha\beta}, \ldots$, to

$$\frac{k(k-1)(k-2)}{1\,.\,2\,.\,3}$$

of $N_{\alpha\beta\gamma}, \ldots$, and so on. Hence, if $k \geqslant 1$, it contributes

$$1 - k\frac{k(k-1)}{1\,.\,2} - \frac{k(k-1)(k-2)}{1\,.\,2\,.\,3} + \cdots = (1-1)^k = 0$$

to the sum (16.1.2). On the other hand, if $k = 0$, it contributes 1. Hence (16.1.2) is the number of objects possessing none of the properties.

The number of integers not greater than n and divisible by a is

$$\left[\frac{n}{a}\right].$$

If a is prime to b, then the number of integers not greater than n, and divisible by both a and b, is

$$\left[\frac{n}{ab}\right];$$

and so on. Hence, taking α, β, $\gamma, \ldots$ to be divisibility by $a, b, c, \ldots$, we obtain

THEOREM 261. *The number of integers, less than or equal to n, and not divisible by any one of a coprime set of integers* $a, b, \ldots$, *is*

$$[n] - \sum \left[\frac{n}{a}\right] + \sum \left[\frac{n}{ab}\right] - \cdots.$$

If we take $a, b, \ldots$ to be the different prime factors $p, p', \ldots$ of n, we obtain

$$\phi(n) = n - \sum \frac{n}{p} + \sum \frac{n}{pp'} - \cdots = n \prod_{p|n} \left(1 - \frac{1}{p}\right), \tag{16.1.3}$$

which is Theorem 62.

16.2. A further proof of Theorem 63. Consider the set of n rational fractions

$$\frac{h}{n} \quad (1 \leqslant h \leqslant n). \tag{16.2.1}$$

We can express each of these fractions in 'irreducible' form in just one way, that is,

$$\frac{h}{n} = \frac{a}{d},$$

where $d|n$ and

$$1 \leqslant a \leqslant d, \quad (a, d) = 1, \tag{16.2.2}$$

and a and d are uniquely determined by h and n. Conversely, every fraction a/d, for which $d|n$ and (16.2.2) is satisfied, appears in the set (16.2.1), though in general not in reduced form. Hence, for any function $F(x)$, we have

$$\sum_{1 \leqslant h \leqslant n} F\left(\frac{h}{n}\right) = \sum_{d|n} \sum_{\substack{1 \leqslant a \leqslant d \\ (a,d)=1}} F\left(\frac{a}{d}\right). \tag{16.2.3}$$

Again, for a particular d, there are (by definition) just $\phi(d)$ values of a satisfying (16.2.2). Hence, if we put $F(x) = 1$ in (16.2.3), we have

$$n = \sum_{d|n} \phi(d).$$

16.3. The Möbius function. The Möbius function $\mu(n)$ is defined as follows:

(i) $\mu(1) = 1$;

(ii) $\mu(n) = 0$ if n has a squared factor;

(iii) $\mu(p_1 p_2 \ldots p_k) = (-1)^k$ if all the primes $p_1, p_2, \ldots, p_k$ are different.

Thus $\mu(2) = -1$, $\mu(4) = 0$, $\mu(6) = 1$.

THEOREM 262. *$\mu(n)$ is multiplicative.*[†]

This follows immediately from the definition of $\mu(n)$.

From (16.1.3) and the definition of $\mu(n)$ we obtain

$$\phi(n) = n \sum_{d|n} \frac{\mu(d)}{d} = \sum_{d|n} \frac{n}{d}\mu(d) = \sum_{d|n} d\mu\left(\frac{n}{d}\right) = \sum_{dd'=n} d'\mu(d).^{\ddagger}$$

Next, we prove

THEOREM 263:

$$\sum_{d|n} \mu(d) = 1 \quad (n = 1), \quad \sum_{d|n} \mu(d) = 0 \quad (n > 1).$$

THEOREM 264. *If $n > 1$, and k is the number of different prime factors of n, then*

$$\sum_{d|n} |\mu(d)| = 2^k.$$

In fact, if $k \geqslant 1$ and $n = p_1^{a_1} \ldots p_k^{a_k}$, we have

$$\sum_{d|n} \mu(d) = 1 + \sum_i \mu(p_i) + \sum_{i,j} \mu(p_i p_j) + \cdots$$

$$= 1 - k + \binom{k}{2} - \binom{k}{3} + \cdots = (1-1)^k = 0,$$

† See § 5.5.

‡ A sum extended over all pairs d, d' for which $dd' = n$.

while, if $n = 1, \mu(n) = 1$. This proves Theorem 263. The proof of Theorem 264 is similar. There is an alternative proof of Theorem 263 depending on an important general theorem.

THEOREM 265. *If $f(n)$ is a multiplicative function of n, then so is*

$$g(n) = \sum_{d|n} f(d).$$

If $(n, n') = 1, d|n$, and $d'|n'$, then $(d, d') = 1$ and $c = dd'$ runs through all divisors of nn'. Hence

$$g(nn') = \sum_{c|nn'} f(c) = \sum_{d|n, d'|n'} f(dd')$$

$$= \sum_{d|n} f(d) \sum_{d'|n'} f(d') = g(n)g(n').$$

To deduce Theorem 263 we write $f(n) = \mu(n)$, so that

$$g(n) = \sum_{d|n} \mu(d).$$

Then $g(1) = 1$, and

$$g(p^m) = 1 + \mu(p) = 0$$

when $m \geqslant 1$. Hence, when $n = p_1^{a_1} \ldots p_k^{a_h} > 1$,

$$g(n) = g(p_1^{a_1})g(p_2^{a_2}) \ldots = 0.$$

16.4. The Möbius inversion formula. In what follows we shall make frequent use of a general 'inversion' formula first proved by Möbius.

THEOREM 266. *If*

$$g(n) = \sum_{d|n} f(d),$$

then

$$f(n) = \sum_{d|n} \mu\left(\frac{n}{d}\right) g(d) = \sum_{d|n} \mu(d) g\left(\frac{n}{d}\right).$$

In fact

$$\sum_{d|n} \mu(d) g\left(\frac{n}{d}\right) = \sum_{d|n} \mu(d) \sum_{c|\frac{n}{d}} f(c) = \sum_{cd|n} \mu(d) f(c)$$

$$= \sum_{c|n} f(c) \sum_{d|\frac{n}{c}} \mu(d).$$

The inner sum here is 1 if $n/c = 1$, i.e. if $c = n$, and 0 otherwise, by Theorem 263, so that the repeated sum reduces to $f(n)$.

Theorem 266 has a converse expressed by

THEOREM 267:

$$f(n) = \sum_{d|n} \mu\left(\frac{n}{d}\right) g(d) \rightarrow g(n) = \sum_{d|n} f(d).$$

The proof is similar to that of Theorem 266. We have

$$\sum_{d|n} f(d) = \sum_{d|n} f\left(\frac{n}{d}\right) = \sum_{d|n} \sum_{c|\frac{n}{d}} \mu\left(\frac{n}{cd}\right) g(c)$$

$$= \sum_{cd|n} \mu\left(\frac{n}{cd}\right) g(c) = \sum_{c|n} g(c) \sum_{d|\frac{n}{c}} \mu\left(\frac{n}{cd}\right) = g(n).$$

If we put $g(n) = n$ in Theorem 267, and use (16.3.1), so that $f(n) = \phi(n)$, we obtain Theorem 63.

As an example of the use of Theorem 266, we give another proof of Theorem 110.

We suppose that $d \mid p - 1$ and $c|d$, and that $\chi(c)$ is the number of roots of the congruence $x^d \equiv 1 \pmod{p}$ which belong to c. Then (since the congruence has d roots in all)

$$\sum_{c|d} \chi(c) = d;$$

from which, by Theorem 266, it follows that

$$\chi(d) = \sum_{c|d} \mu(c) \frac{d}{c} = \phi(d).$$

16.5. Further inversion formulae. There are other inversion formulae involving $\mu(n)$, of a rather different type.

THEOREM 268. *If*

$$G(x) = \sum_{n=1}^{[x]} F\left(\frac{x}{n}\right)$$

for all positive x,[†] *then*

$$F(x) = \sum_{n=1}^{[x]} \mu(n)G\left(\frac{x}{n}\right).$$

For

$$\sum_{n=1}^{[x]} \mu(n)G\left(\frac{x}{n}\right) = \sum_{n=1}^{[x]} \mu(n) \sum_{m=1}^{[x/n]} F\left(\frac{x}{mn}\right)$$
$$= \sum_{1 \leqslant k \leqslant [x]} F\left(\tfrac{x}{k}\right) \sum_{n|k} \mu(n)^{\ddagger} = F(x),$$

by Theorem 263. There is a converse, viz.

THEOREM 269:

$$F(x) = \sum_{n=1}^{[x]} \mu(n)G\left(\frac{x}{n}\right) \to G(x) = \sum_{n=1}^{[x]} F\left(\frac{x}{n}\right).$$

This may be proved similarly.

Two further inversion formulae are contained in

THEOREM 270:

$$g(x) = \sum_{m=1}^{\infty} f(mx) \equiv f(x) = \sum_{n=1}^{\infty} \mu(n)g(nx).$$

[†] An empty sum is as usual to be interpreted as 0. Thus $G(x) = 0$ if $0 < x < 1$.

[‡] If $mn = k$ then $n|k$, and k runs through the numbers $1, 2, \ldots, [x]$.

The reader should have no difficulty in constructing a proof with the help of Theorem 263; but some care is required about convergence. A sufficient condition is that

$$\sum_{m,n} |f(mnx)| = \sum_{k} d(k)\,|f(kx)|$$

should be convergent. Here $d(k)$ is the number of divisors of k.†

16.6. Evaluation of Ramanujan's sum. Ramanujan's sum $c_n(m)$ was defined in § 5.6 by

$$c_n(m) = \sum_{\substack{1\leqslant h\leqslant n\\(h,n)=1}} e\left(\frac{hm}{n}\right). \tag{16.6.1}$$

We can now express $c_n(m)$ as a sum extended over the common divisors of m and n.

THEOREM 271:

$$c_n(m) = \sum_{d|m,d|n} \mu\left(\frac{n}{d}\right) d.$$

If we write

$$g(n) = \sum_{1\leqslant h\leqslant n} F\left(\frac{h}{n}\right),\quad f(n) = \sum_{\substack{1\leqslant h\leqslant n\\(h,n)=1}} F\left(\frac{h}{n}\right),$$

(16.2.3) becomes

$$g(n) = \sum_{d|n} f(d).$$

By Theorem 266, we have the inverse formula

$$f(n) = \sum_{d|n} \mu\left(\frac{n}{d}\right) g(d), \tag{16.6.2}$$

† See § 16.7.

that is

(16.6.3) $$\sum_{\substack{1\leqslant h\leqslant n\\(h,n)=1}} F\left(\frac{h}{n}\right) = \sum_{d|n}\mu\left(\frac{n}{d}\right)\sum_{1\leqslant a\leqslant d} F\left(\frac{a}{d}\right).$$

We now take $F(x) = e(mx)$. In this event,

$$f(n) = c_n(m)$$

by (16.6.1), while

$$g(n) = \sum_{1\leqslant h\leqslant n} e\left(\frac{hm}{n}\right),$$

which is n or 0 according as $n|m$ or $n \nmid m$. Hence (16.6.2) becomes

$$c_n(m) = \sum_{d|n,\,d|m} \mu\left(\frac{n}{d}\right)d.$$

Another simple expression for $c_n(m)$ is given by

THEOREM 272. *If $(n, m) = a$ and $n = aN$, then*

$$c_n(m) = \frac{\mu(N)\phi(n)}{\phi(N)}.$$

By Theorem 271,

$$c_n(m) = \sum_{d|a} d\mu\left(\frac{n}{d}\right) = \sum_{cd=a} d\mu(Nc) = \sum_{c|a}\frac{a}{c}\mu(Nc).$$

Now $\mu(Nc) = \mu(N)\mu(c)$ or 0 according as $(N, c) = 1$ or not. Hence

$$c_n(m) = a\mu(N)\sum_{\substack{c|a\\(c,N)=1}}\frac{\mu(c)}{c} = a\mu(N)\left(1 - \sum\frac{1}{p} + \sum\frac{1}{pp'} - \cdots\right),$$

where these sums run over those different p which divide a but do not divide N. Hence

$$c_n(m) = a\mu(N)\prod_{p|a,\,p\nmid N}\left(1 - \frac{1}{p}\right).$$

But, by Theorem 62,

$$\frac{\phi(n)}{\phi(N)} = \frac{n}{N} \prod_{p|n,\, p\nmid N} \left(1 - \frac{1}{p}\right) = a \prod_{p|n,\, p\nmid N} \left(1 - \frac{1}{p}\right)$$

and Theorem 272 follows at once.

When $m = 1$, we have $c_n(1) = \mu(n)$, that is

$$\mu(n) = \sum_{\substack{1 \leqslant h \leqslant n \\ (h,n)=1}} e\left(\frac{h}{n}\right). \tag{16.6.4}$$

16.7. The functions $d(n)$ and $\sigma_k(n)$. The function $d(n)$ is the number of divisors of n, including 1 and n, while $\sigma_k(n)$ is the sum of the kth powers of the divisors of n. Thus

$$\sigma_k(n) = \sum_{d|n} d^k, \quad d(n) = \sum_{d|n} 1,$$

and $d(n) = \sigma_0(n)$. We write $\sigma(n)$ for $\sigma_1(n)$, the sum of the divisors of n.

If

$$n = p_1^{a_1} p_2^{a_2} \dots p_l^{a_l},$$

then the divisors of n are the numbers

$$p_1^{b_1} p_2^{b_2} \dots p_l^{b_l},$$

where

$$0 \leqslant b_1 \leqslant a_1, \quad 0 \leqslant b_2 \leqslant a_2, \quad \dots, \quad 0 \leqslant b_l \leqslant a_l.$$

There are

$$(a_1 + 1)(a_2 + 1) \dots (a_l + 1)$$

of these numbers. Hence

THEOREM 273:

$$d(n) = \prod_{i=1}^{l} (a_i + 1).$$

More generally, if $k > 0$,

$$\sigma_k(n) = \sum_{b_1=0}^{a_1} \sum_{b_2=0}^{a_2} \cdots \sum_{b_l=0}^{a_l} p_1^{b_1 k} p_2^{b_2 k} \cdots p_l^{b_l k}$$
$$= \prod_{i=1}^{l} \left(1 + p_i^k + p_i^{2k} + \cdots + p_i^{a_i k}\right).$$

Hence

THEOREM 274:

$$\sigma_k(n) = \prod_{i=1}^{l} \left(\frac{p_i^{(a_i+1)k} - 1}{p_i^k - 1}\right).$$

In particular,

THEOREM 275:

$$\sigma(n) = \prod_{i=1}^{l} \left(\frac{p_i^{a_i+1} - 1}{p_i - 1}\right).$$

16.8. Perfect numbers. *A perfect* number is a number n such that $\sigma(n) = 2n$. In other words a number is perfect if it is the sum of its divisors other than itself. Since $1 + 2 + 3 = 6$, and

$$1 + 2 + 4 + 7 + 14 = 28,$$

6 and 28 are perfect numbers.

The only general class of perfect numbers known occurs in Euclid.

THEOREM 276. *If* $2^{n+1} - 1$ *is prime, then* $2^n(2^{n+1} - 1)$ *is perfect.*

Write $2^{n+1} - 1 = p, N = 2^n p$. Then, by Theorem 275,

$$\sigma(N) = (2^{n+1} - 1)(p + 1) = 2^{n+1}(2^{n+1} - 1) = 2N,$$

so that N is perfect.

Theorem 276 shows that to every Mersenne prime there corresponds a perfect number. On the other hand, if $N = 2^n p$ is perfect, we have

$$\sigma(N) = (2^{n+1} - 1)(p + 1) = 2^{n+1} p$$

and so

$$p = 2^{n+1} - 1.$$

Hence there is a Mersenne prime corresponding to any perfect number of the form $2^n p$. But we can prove more than this.

THEOREM 277. *Any even perfect number is a Euclid number, that is to say of the form* $2^n(2^{n+1} - 1)$, *where* $2^{n+1} - 1$ *is prime.*

We can write any such number in the form $N = 2^n b$, where $n > 0$ and b is odd. By Theorem 275, $\sigma(n)$ is multiplicative, and therefore

$$\sigma(N) = \sigma(2^n)\sigma(b) = (2^{n+1} - 1)\sigma(b).$$

Since N is perfect,

$$\sigma(N) = 2N = 2^{n+1}b;$$

and so

$$\frac{b}{\sigma(b)} = \frac{2^{n+1} - 1}{2^{n+1}}.$$

The fraction on the right-hand side is in its lowest terms, and therefore

$$b = (2^{n+1} - 1)c, \quad \sigma(b) = 2^{n+1}c,$$

where c is an integer.

If $c > 1$, b has at least the divisors $b, c, 1$, so that

$$\sigma(b) \geqslant b + c + 1 = 2^{n+1}c + 1 > 2^{n+1}c = \sigma(b),$$

a contradiction. Hence $c = 1$, $N = 2^n(2^{n+1} - 1)$, and

$$\sigma(2^{n+1} - 1) = 2^{n+1}.$$

But, if $2^{n+1} - 1$ is not prime, it has divisors other than itself and 1, and

$$\sigma(2^{n+1} - 1) > 2^{n+1}.$$

Hence $2^{n+1} - 1$ is prime, and the theorem is proved.

The Euclid numbers corresponding to the Mersenne primes are the only perfect numbers known. It seems probable that there are no odd perfect numbers, but this has not been proved. The most that is known in this direction is that any odd perfect number must be greater than 10^{200}, that it must have at least 8 different prime factors and that its largest prime factor must be greater than 100110.†

16.9. The function $r(n)$. We define $r(n)$ as the number of representations of n in the form

$$n = A^2 + B^2,$$

where A and B are rational integers. We count representations as distinct even when they differ only 'trivially', i.e. in respect of the sign or order of A and B. Thus

$$0 = 0^2 + 0^2, \quad r(0) = 1;$$

$$1 = (\pm 1)^2 + 0^2 = 0^2 + (\pm 1)^2, \quad r(1) = 4;$$

$$5 = (\pm 2)^2 + (\pm 1)^2 = (\pm 1)^2 + (\pm 2)^2, \quad r(5) = 8.$$

We know already (§ 15.1) that $r(n) = 8$ when n is a prime $4m + 1$; the representation is unique apart from its eight trivial variations. On the other hand, $r(n) = 0$ when n is of the form $4m + 3$.

We define $\chi(n)$, for $n > 0$, by

$$\chi(n) = 0 \ \ (2 \mid n), \quad \chi(n) = (-1)^{\frac{1}{2}(n-1)} \ \ (2 \nmid n).$$

Thus $\chi(n)$ assumes the values $1, 0, -1, 0, 1, \ldots$ for $n = 1, 2, 3, \ldots$. Since

$$\tfrac{1}{2}(nn' - 1) - \tfrac{1}{2}(n - 1) - \tfrac{1}{2}(n' - 1) = \tfrac{1}{2}(n - 1)(n' - 1) \equiv 0 \pmod 2$$

When n and n' are odd, $\chi(n)$ satisfies

$$\chi(nn') = \chi(n)\chi(n')$$

for all n and n'. In particular $\chi(n)$ is multiplicative in the sense of § 5.5.

It is plain that, if we write

$$\delta(n) = \sum_{d|n} \chi(d), \tag{16.9.1}$$

then

$$\delta(n) = d_1(n) - d_3(n), \tag{16.9.2}$$

† See end of chapter notes.

where $d_1(n)$ and $d_3(n)$ are the numbers of divisors of n of the forms $4m+1$ and $4m+3$ respectively.

Suppose now that

$$n = 2^{\alpha}N = 2^{\alpha}\mu\nu = 2^{\alpha}\prod p^{r}\prod q^{s}, \tag{16.9.3}$$

where p and q are primes $4m+1$ and $4m+3$ respectively. If there are no factors q, so that Πq^s is 'empty', then we define ν as 1. Plainly

$$\delta(n) = \delta(N).$$

The divisors of N are the terms in the product

$$\prod(1+p+\cdots+p^{r})\prod(1+q+\cdots+q^{s}). \tag{16.9.4}$$

A divisor is $4m+1$ if it contains an even number of factors q, and $4m+3$ in the contrary case. Hence $\delta(N)$ is obtained by writing 1 for p and -1 for q in (16.9.4); and

$$\delta(N) = \prod(r+1)\prod\left(\frac{1+(-1)^{s}}{2}\right). \tag{16.9.5}$$

If any s is odd, i.e. if ν is not a square, then

$$\delta(n) = \delta(N) = 0;$$

while

$$\delta(n) = \delta(N) = \prod(r+1) = d(\mu)$$

if ν is a square.

Our object is to prove

THEOREM 278. *If $n \geqslant 1$, then*

$$r(n) = 4\delta(n).$$

We have therefore to show that $r(n)$ is $4d(\mu)$ when ν is a square, and zero otherwise.

16.10. Proof of the formula for $r(n)$. We write (16.9.3) in the form

$$n = \{(1+i)(1-i)\}^{\alpha} \prod \{(a+bi)(a-bi)\}^{r} \prod q^{s},$$

where a and b are positive and unequal and

$$p = a^2 + b^2.$$

This expression of p is unique (after § 15.1) except for the order of a and b. The factors

$$1 \pm i, \quad a \pm bi, \quad q$$

are primes of $k(i)$.

If

$$n = A^2 + B^2 = (A+Bi)(A-Bi),$$

then

$$A + Bi = i^{t}(1+i)^{\alpha_1}(1-i)^{\alpha_2} \prod \{(a+bi)^{r_1}(a-bi)^{r_2}\} \prod q^{s_1},$$
$$A - Bi = i^{-t}(1+i)^{\alpha_1}(1-i)^{\alpha_2} \prod \{(a-bi)^{r_1}(a+bi)^{r_2}\} \prod q^{s_2},$$

where

$$t = 0, 1, 2, \text{ or } 3, \quad \alpha_1 + \alpha_2 = \alpha, \quad r_1 + r_2 = r, \quad s_1 + s_2 = s.$$

Plainly $s_1 = s_2$, so that every s is even, and v is a square. Unless this is so, there is no representation.

We suppose then that

$$v = \prod q^{s} = \prod q^{2s_1}$$

is a square. There is no choice in the division of the factors q between $A + Bi$ and $A - Bi$. There are

$$4(\alpha + 1) \prod (r + 1)$$

choices in the division of the other factors. But

$$\frac{1-i}{1+i} = -i$$

is a unity, so that a change in α_1 and α_2 produces no variation in A and B beyond that produced by variation of t. We are thus left with

$$4\prod(r+1) = 4d(\mu)$$

possibly effective choices, i.e. choices which may produce variation in A and B.

The trivial variations in a representation $n = A^2 + B^2$ correspond (i) to multiplication of $A + Bi$ by a unity and (ii) to exchange of $A + Bi$ with its conjugate. Thus

$$1(A+Bi) = A+Bi, \qquad i(A+Bi) = -B+Ai,$$
$$i^2(A+Bi) = -A-Bi, \qquad i^3(A+Bi) = B-Ai,$$

and $A - Bi, -B - Ai, -A + Bi, B + Ai$ are the conjugates of these four numbers. Any change in t varies the representation. Any change in the r_1 and r_2 also varies the representation, and in a manner not accounted for by any change in t; for

$$i^t(1+i)^{a_1}(1-i)^{a_2}\prod\{(a+bi)^{r_1}(a-bi)^{r_2}\}$$
$$= i^\theta i^{t'}(1+i)^{\alpha_1'}(1-i)^{\alpha_2'}\prod\{(a+bi)^{r_1'}(a-bi)^{r_2'}\}$$

is impossible, after Theorem 215, unless $r_1 = r_1'$ and $r_2 = r_2'$.† There are therefore $4d(\mu)$ different sets of values of A and B, or of representations of n; and this proves Theorem 278.

NOTES

§ 16.1. The argument follows Pólya and Szegő, Nos. 21, 25. Theorem 260 is widely known as the Inclusion–Exclusion Theorem.

§§ 16.3–5. The function $\mu(n)$ occurs implicitly in the work of Euler as early as 1748, but Möbius, in 1832, was the first to investigate its properties systematically. See Landau, *Handbuch*, 567–87 and 901.

§ 16.6. Ramanujan, *Collected papers*, 180. Our method of proof of Theorem 271 was suggested by Professor van der Pol. Theorem 272 is due to Holder, *Prace Mat. Fiz.* 43 (1936), 13–23. See also Zuckerman, *American Math. Monthly*, 59 (1952), 230 and Anderson and Apostol, *Duke Math. Journ.* 20 (1953), 211–16.

§§ 16.7–8. There is a very full account of the history of the theorems of these sections in Dickson, *History*, i, chs. i–ii. References to the theorems referred to at the end of § 16.8 are given by Kishore (*Math. Comp.* 31 (1977), 274–9).

† Change of r_1 into r_2, and r_2 into r_1 (together with corresponding changes in t, α_1, α_2) changes $A + Bi$ into its conjugate.

Euler showed that any odd perfect number must take the form $p^a q_1^{2e_1} \dots q_r^{2e_r}$ with primes $p, q_1, \dots, q_r$, and with $a \equiv p \equiv 1 \pmod 4$. It is now (2007) known that an odd perfect number would have to exceed 10^{300} (Brent, Cohen, and te Riele, *Math. Comp.* 57 (1991), 857–68). Moreover, Nielsen has announced (http://arxiv.org/pdf/math/0602485) that an odd perfect number must have at least 9 distinct prime factors. It is known that the largest prime factor must exceed 10^7 (Jenkins, *Math. Comp.* 72 (2003), no. 243, 1549–1554 (electronic)). Indeed Goto and Ohno have announced that this bound can be increased to 10^8. Neilsen (*Integers* 3 (2003), A14, (electronic)) has also shown that an odd perfect number n with k distinct prime factors must satisfy $n < 2^{4^k}$.

§ 16.9. Theorem 278 was first proved by Jacobi by means of the theory of elliptic functions. It is, however, equivalent to one stated by Gauss, *D.A.*, § 182; and there had been many incomplete proofs or statements published before. See Dickson, *History*, ii, ch. vi, and Bachmann, *Niedere Zahlentheorie*, ii, ch. vii.

XVII
GENERATING FUNCTIONS OF ARITHMETICAL FUNCTIONS

17.1. The generation of arithmetical functions by means of Dirichlet series. A *Dirichlet series* is a series of the form

$$F(s) = \sum_{n=1}^{\infty} \frac{\alpha_n}{n^s}. \tag{17.1.1}$$

The variable s may be real or complex, but here we shall be concerned with real values only. $F(s)$, the sum of the series, is called the *generating function* of α_n.

The theory of Dirichlet series, when studied seriously for its own sake, involves many delicate questions of convergence. These are mostly irrelevant here, since we are concerned primarily with the formal side of the theory; and most of our results could be proved (as we explain later in § 17.6) without the use of any theorem of analysis or even the notion of the sum of an infinite series. There are, however, some theorems which must be considered as theorems of analysis; and, even when this is not so, the reader will probably find it easier to think of the series which occur as sums in the ordinary analytical sense.

We shall use the four theorems which follow. These are special cases of more general theorems which, when they occur in their proper places in the general theory, can be proved better by different methods. We confine ourselves here to what is essential for our immediate purpose.

(1) If $\sum \alpha_n n^{-s}$ is absolutely convergent for a given s, then it is absolutely convergent for all greater s. This is obvious because

$$\left|\alpha_n n^{-s_2}\right| \leqslant \left|\alpha_n n^{-s_1}\right|$$

when $n \geqslant 1$ and $s_2 > s_1$.

(2) If $\sum \alpha_n n^{-s}$ is absolutely convergent for $s > s_0$ then the equation (17.1.1) may be differentiated term by term, so that

$$F'(s) = -\sum \frac{\alpha_n \log n}{n^s} \tag{17.1.2}$$

for $s > s_0$. To prove this, suppose that

$$s_0 < s_0 + \delta = s_1 \leqslant s \leqslant s_2.$$

Then $\log n < K(\delta)n^{\frac{1}{2}\delta}$, where $K(\delta)$ depends only on δ, and

$$\left|\frac{\alpha_n \log n}{n^s}\right| \leqslant K(\delta)\left|\frac{\alpha_n}{n^{s_0+\frac{1}{2}\delta}}\right|$$

for all s of the interval (s_1, s_2). Since

$$\sum\left|\frac{\alpha_n}{n^{s_0+\frac{1}{2}\delta}}\right|$$

is convergent, the series on the right of (17.1.2) is uniformly convergent in (s_1, s_2), and the differentiation is justifiable.

(3) If

$$F(s) = \sum \alpha_n n^{-s} = 0$$

for $s > s_0$, then $\alpha_n = 0$ for all n. To prove this, suppose that α_m is the first non-zero coefficient. Then

$$(17.1.3)\quad 0 = F(s) = \alpha_m m^{-s}\left\{1 + \frac{\alpha_{m+1}}{\alpha_m}\left(\frac{m+1}{m}\right)^{-s} + \frac{\alpha_{m+2}}{\alpha_m}\left(\frac{m+2}{m}\right)^{-s} + \cdots\right\} = \alpha_m m^{-s}\{1 + G(s)\},$$

say. If $s_0 < s_1 < s$, then

$$\left(\frac{m+k}{m}\right)^{-s} \leqslant \left(\frac{m+1}{m}\right)^{-(s-s_1)}\left(\frac{m+k}{m}\right)^{-s_1}$$

and

$$|G(s)| \leqslant \frac{1}{|\alpha_m|}\left(\frac{m+1}{m}\right)^{-(s-s_1)} m^{s_1}\sum_{k=1}^{\infty}\frac{|\alpha_{m+k}|}{(m+k)^{s_1}},$$

which tends to 0 when $s \to \infty$. Hence

$$|1 + G(s)| > \tfrac{1}{2}$$

for sufficiently large s; and (17.1.3) implies $\alpha_m = 0$, a contradiction.

It follows that if

$$\sum \alpha_n n^{-s} = \sum \beta_n n^{-s}$$

for $s > s_1$, then $\alpha_n = \beta_n$ for all n. We refer to this theorem as the 'uniqueness theorem'.

(4) Two absolutely convergent Dirichlet series may be multiplied in a manner explained in § 17.4.

17.2. The zeta function. The simplest infinite Dirichlet series is

$$\zeta(s) = \sum_{n=1}^{\infty} \frac{1}{n^s}. \tag{17.2.1}$$

It is convergent for $s > 1$, and its sum $\zeta(s)$ is called the Riemann zeta function. In particular[†]

$$\zeta(2) = \sum_{n=1}^{\infty} \frac{1}{n^2} = \frac{\pi^2}{6}. \tag{17.2.2}$$

If we differentiate (17.2.1) term by term with respect to s, we obtain

THEOREM 279:

$$\zeta'(s) = -\sum_{1}^{\infty} \frac{\log n}{n^s} \quad (s > 1).$$

The zeta function is fundamental in the theory of prime numbers. Its importance depends on a remarkable identity discovered by Euler, which expresses the function as a product extended over prime numbers only.

THEOREM 280: *If* $s > 1$ *then*

$$\zeta(s) = \prod_p \frac{1}{1 - p^{-s}}.$$

† $\zeta(2n)$ is a rational multiple of π^{2n} for all positive integral n. Thus $\zeta(4) = \frac{1}{90}\pi^4$, and generally

$$\zeta(2n) = \frac{2^{2n-1} B_n}{(2n)!} \pi^{2n},$$

where B_n is Bernoulli's number.

Since $p \geqslant 2$, we have

$$\frac{1}{1-p^{-s}} = 1 + p^{-s} + p^{-2s} + \cdots \tag{17.2.3}$$

for $s > 1$ (indeed for $s > 0$). If we take $p = 2, 3, \ldots, P$, and multiply the series together, the general term resulting is of the type

$$2^{-a_2 s} 3^{-a_3 s} \ldots P^{-a_P s} = n^{-s},$$

where

$$n = 2^{a_2} 3^{a_3} \ldots P^{a_P} \quad (a_2 \geqslant 0, a_3 \geqslant 0, \ldots, a_P \geqslant 0).$$

A number n will occur if and only if it has no prime factors greater than P, and then, by Theorem 2, once only. Hence

$$\prod_{p \leqslant P} \frac{1}{1-p^{-s}} = \sum_{(P)} n^{-s},$$

the summation on the right-hand side extending over numbers formed from the primes up to P.

These numbers include all numbers up to P, so that

$$0 < \sum_{n=1}^{\infty} n^{-s} - \sum_{(P)} n^{-s} < \sum_{P+1}^{\infty} n^{-s},$$

and the last sum tends to 0 when $P \to \infty$. Hence

$$\sum_{n=1}^{\infty} n^{-s} = \lim_{P \to \infty} \sum_{(P)} n^{-s} = \lim_{P \to \infty} \prod_{p \leqslant P} \frac{1}{1-p^{-s}},$$

the result of Theorem 280.

Theorem 280 may be regarded as an analytical expression of the fundamental theorem of arithmetic.

17.3. The behaviour of $\zeta(s)$ when $s \to 1$. We shall require later to know how $\zeta(s)$ and $\zeta'(s)$ behave when s tends to 1 through values greater than 1.

We can write $\zeta(s)$ in the form

$$\zeta(s) = \sum_1^\infty n^{-s} = \int_1^\infty x^{-s}dx + \sum_1^\infty \int_n^{n+1} (n^{-s} - x^{-s})\,dx. \tag{17.3.1}$$

Here

$$\int_1^\infty x^{-s}dx = \frac{1}{s-1},$$

since $s > 1$. Also

$$0 < n^{-s} - x^{-s} = \int_n^x st^{-s-1}dt < \frac{s}{n^2},$$

if $n < x < n+1$, and so

$$0 < \int_n^{n+1} (n^{-s} - x^{-s})\,dx < \frac{s}{n^2};$$

and the last term in (17.3.1) is positive and numerically less than $s\sum n^{-2}$. Hence

THEOREM 281:

$$\zeta(s) = \frac{1}{s-1} + O(1).$$

Also

$$\log\zeta(s) = \log\frac{1}{s-1} + \log\{1 + O(s-1)\},$$

and so

THEOREM 282:

$$\log\zeta(s) = \log\frac{1}{s-1} + O(s-1).$$

We may also argue with

$$-\zeta'(s) = \sum_1^\infty n^{-s} \log n$$

$$= \int_1^\infty x^{-s} \log x \, dx + \sum_1^\infty \int_n^{n+1} (n^{-s} \log n - x^{-s} \log x) \, dx$$

much as with $\zeta(s)$, and deduce

THEOREM 283:

$$\zeta'(s) = -\frac{1}{(s-1)^2} + O(1).$$

In particular,

$$\zeta(s) \sim \frac{1}{s-1}.$$

This may also be proved by observing that, if $s > 1$,

$$(1 - 2^{1-s})\zeta(s) = 1^{-s} + 2^s + 3^{-s} + \cdots - 2(2^{-s} + 4^{-s} + 6^{-s} + \cdots)$$
$$= 1^{-s} - 2^{-s} + 3^{-s} - \cdots,$$

and that the last series converges to $\log 2$ for $s = 1$. Hence†

$$(s-1)\zeta(s) = (1 - 2^{1-s})\zeta(s) \frac{s-1}{1 - 2^{1-s}} \to \log 2 \frac{1}{\log 2} = 1.$$

17.4. Multiplication of Dirichlet series. Suppose that we are given a finite set of Dirichlet series

(17.4.1) $$\sum \alpha_n n^{-s}, \quad \sum \beta_n n^{-s}, \quad \sum \gamma_n n^{-s}, \quad \ldots,$$

† We assume here that

$$\lim_{s \to 1} \sum \frac{a_n}{n^s} = \sum \frac{a_n}{n}$$

whenever the series on the right is convergent, a theorem not included in those of § 17.1. We do not prove this theorem because we require it only for an alternative proof.

and that we multiply them together in the sense of forming all possible products with one factor selected from each series. The general term resulting is

$$\alpha_u u^{-s} . \beta_v v^{-s} . \gamma_\omega \omega^{-s} \ldots = \alpha_u \beta_v \gamma_\omega \ldots n^{-s},$$

where $n = uvw\ldots$. If now we add together all terms for which n has a given value, we obtain a single term $\chi_n n^{-s}$ where

$$\chi_n = \sum_{uvw\ldots=n} \alpha_u \beta_v \gamma_\omega \ldots . \tag{17.4.2}$$

The series $\sum \chi_n n^{-s}$, with χ_n defined by (17.4.2), is called the *formal product* of the series (17.4.1).

The simplest case is that in which there are only two series (17.4.1), $\sum \alpha_u u^{-s}$ and $\sum \beta_v v^{-s}$. If (changing our notation a little) we denote their formal product by $\sum \gamma_n n^{-s}$, then

$$\gamma_n = \sum_{uv=n} \alpha_u \beta_v = \sum_{d|n} \alpha_d \beta_{n/d} = \sum_{d|n} \alpha_{n/d} \beta_d, \tag{17.4.3}$$

a sum of a type which occurred frequently in Ch. XVI. And if the two given series are absolutely convergent, and their sums are $F(s)$ and $G(s)$, then

$$F(s)G(s) = \sum_u \alpha_u u^{-s} \sum_v \beta_v v^{-s} = \sum_{u,v} \alpha_u \beta_v (uv)^{-s},$$

$$= \sum_n n^{-s} \sum_{uv=n} \alpha_u \beta_v = \sum \gamma_n n^{-s},$$

since we may multiply two absolutely convergent series and arrange the terms of the product in any order that we please.

Theorem 284. *If the series*

$$F(s) = \sum \alpha_u u^{-s}, \quad G(s) = \sum \beta_v v^{-s}$$

are absolutely convergent, then

$$F(s)G(s) = \sum \gamma_n n^{-s},$$

where γ_n *is defined by* (17.4.3).

Conversely, if

$$H(s) = \sum \delta_n n^{-s} = F(s)G(s)$$

then it follows from the uniqueness theorem of § 17.1 that $\delta_n = \gamma_n$.

Our definition of the formal product may be extended, with proper precautions, to an infinite set of series. It is convenient to suppose that

$$\alpha_1 = \beta_1 = \gamma_1 = \ldots = 1.$$

Then the term

$$\alpha_u \beta_v \gamma_w \ldots$$

in (17.4.2) contains only a finite number of factors which are not 1, and we may define χ_n by (17.4.2) whenever the series is absolutely convergent.[†]

The most important case is that in which $f(1) = 1$, $f(n)$ is multiplicative, and the series (17.4.1) are

$$1 + f(p)p^{-s} + f(p^2)p^{-2s} + \cdots + f(p^a)p^{-as} + \cdots \tag{17.4.4}$$

for $p = 2, 3, 5, \ldots$; so that, for example, α_u is $f(2^a)$ when $u = 2^a$ and 0 otherwise. Then, after Theorem 2, every n occurs just once as a product $uvw\ldots$ with a non-zero coefficient, and

$$\chi_n = f(p_1^{a_1})f(p_2^{a_2})\ldots = f(n)$$

when $n = p_1^{a_1}p_2^{a_2}\ldots$. It will be observed that the series (17.4.2) reduces to a single term, so that no question of convergence arises.

Hence

THEOREM 285. *If $f(1) = 1$ and $f(n)$ is multiplicative, then*

$$\sum f(n)n^{-s}$$

is the formal product of the series (17.4.4).

In particular, $\sum n^{-s}$ is the formal product of the series

$$1 + p^{-s} + p^{-2s} + \ldots.$$

[†] We must assume *absolute* convergence because we have not specified the order in which the terms are to be taken.

Theorem 280 says in some ways more than this, namely that $\zeta(s)$, the sum of the series $\sum n^{-s}$ when $s > 1$, is equal to the product of the sums of the series $1+p^{-s}+p^{-2s}\ldots$. The proof can be generalized to cover the more general case considered here.

THEOREM 286. *If $f(n)$ satisfies the conditions of Theorem 285, and*

$$\sum |f(n)|n^{-s} \tag{17.4.5}$$

is convergent, then

$$F(s) = \sum f(n)n^{-s} = \prod_p \{1+f(p)p^{-s}+f(p^2)p^{-2s}+\cdots\}.$$

We write

$$F_p(s) = 1+f(p)p^{-s}+f(p^2)p^{-2s}+\cdots;$$

the absolute convergence of the series is a corollary of the convergence of (17.4.5). Hence, arguing as in § 17.2, and using the multiplicative property of $f(n)$, we obtain

$$\prod_{p\leqslant P} F_p(s) = \sum_{(P)} f(n)\,n^{-s}.$$

Since

$$\left|\sum_{n=1}^{\infty} f(n)n^{-s} - \sum_{(P)} f(n)n^{-s}\right| \leqslant \sum_{P+1}^{\infty} |f(n)|n^{-s} \to 0$$

the result follows as in § 17.2.

17.5. The generating functions of some special arithmetical functions. The generating functions of most of the arithmetical functions which we have considered are simple combinations of zeta functions. In this section we work out some of the most important examples.

THEOREM 287:

$$\frac{1}{\zeta(s)} = \sum_{n=1}^{\infty} \frac{\mu(n)}{n^s} \qquad (s > 1).$$

This follows at once from Theorems 280, 262, and 286, since

$$\frac{1}{\zeta(s)}=\prod_p(1-p^{-s})=\prod\{1+\mu(p)p^{-s}+\mu(p^2)p^{-2s}+\ldots\}=\sum_{n=1}^{\infty}\mu(n)n^{-s}.$$

THEOREM 288:

$$\frac{\zeta(s-1)}{\zeta(s)}=\sum_{n=1}^{\infty}\frac{\phi(n)}{n^s}\quad(s>2).$$

By Theorem 287, Theorem 284, and (16.3.1)

$$\frac{\zeta(s-1)}{\zeta(s)}=\sum_{n=1}^{\infty}\frac{n}{n^s}\sum_{n=1}^{\infty}\frac{\mu(n)}{n^s}=\sum_{n=1}^{\infty}\frac{1}{n^s}\sum_{d|n}d\mu\left(\frac{n}{d}\right)=\sum_{n=1}^{\infty}\frac{\phi(n)}{n^s}.$$

THEOREM 289:

$$\zeta^2(s)=\sum_{n=1}^{\infty}\frac{d(n)}{n^s}\quad(s>1).$$

THEOREM 290:

$$\zeta(s)\zeta(s-1)=\sum_{n=1}^{\infty}\frac{\sigma(n)}{n^s}\quad(s>2).$$

These are special cases of the theorem

THEOREM 291:

$$\zeta(s)\zeta(s-k)=\sum_{n=1}^{\infty}\frac{\sigma_k(n)}{n^s}\quad(s>1,\ s>k+1).$$

In fact

$$\zeta(s)\zeta(s-k)=\sum_{n=1}^{\infty}\frac{1}{n^s}\sum_{r=1}^{\infty}\frac{n^k}{n^s}=\sum_{n=1}^{\infty}\frac{1}{n^s}\sum_{d|n}d^k=\sum_{n=1}^{\infty}\frac{\sigma_k(n)}{n^s},$$

by Theorem 284.

THEOREM 292:

$$\frac{\sigma_{s-1}(m)}{m^{s-1}\zeta(s)} = \sum_{n=1}^{\infty} \frac{c_n(m)}{n^s} \quad (s > 1).$$

By Theorem 271,

$$c_n(m) = \sum_{d|m,\, d|n} \mu\left(\frac{n}{d}\right) d = \sum_{d|m,\, dd'=n} \mu(d')d;$$

and so

$$\sum_{n=1}^{\infty} \frac{c_n(m)}{n^s} = \sum_{n=1}^{\infty} \sum_{d|m,\, dd'=n} \frac{\mu(d')d}{d'^s d^s}$$

$$= \sum_{d'=1}^{\infty} \frac{\mu(d')}{d'^s} \sum_{d|m} \frac{1}{d^{s-1}} = \frac{1}{\zeta(s)} \sum_{d|m} \frac{1}{d^{s-1}}.$$

Finally

$$\sum_{d|m} d^{1-s} = m^{1-s} \sum_{d|m} d^{s-1} = m^{1-s}\sigma_{s-1}(m).$$

In particular,

THEOREM 293:

$$\sum_{n} \frac{c_n(m)}{n^2} = \frac{6}{\pi^2} \frac{\sigma(m)}{m}.$$

17.6. The analytical interpretation of the Möbius formula. Suppose that

$$g(n) = \sum_{d|n} f(d),$$

and that $F(s)$ and $G(s)$ are the generating functions of $f(n)$ and $g(n)$. Then, if the series are absolutely convergent, we have

$$F(s)\zeta(s) = \sum_{n=1}^{\infty} \frac{f(n)}{n^s} \sum_{n=1}^{\infty} \frac{1}{n^s} = \sum_{n=1}^{\infty} \frac{1}{n^s} \sum_{d|n} f(d) = \sum_{n=1}^{\infty} \frac{g(n)}{n^s} = G(s);$$

and therefore

$$F(s) = \frac{G(s)}{\zeta(s)} = \sum_{n=1}^{\infty} \frac{g(n)}{n^s} \sum_{n=1}^{\infty} \frac{\mu(n)}{n^s} = \sum_{n=1}^{\infty} \frac{h(n)}{n^s},$$

where

$$h(n) = \sum_{d|n} g(d)\mu\left(\frac{n}{d}\right).$$

It then follows from the uniqueness theorem of § 17.1 (3) that

$$h(n) = f(n),$$

which is the inversion formula of Möbius (Theorem 266). This formula then appears as an arithmetical expression of the equivalence of the equations

$$G(s) = \zeta(s)F(s), \qquad F(s) = \frac{G(s)}{\zeta(s)}.$$

We cannot regard this argument, as it stands, as a proof of the Möbius formula, since it depends upon the convergence of the series for $F(s)$. This hypothesis involves a limitation on the order of magnitude of $f(n)$, and it is obvious that such limitations are irrelevant. The 'real' proof of the Möbius formula is that given in § 16.4.

We may, however, take this opportunity of expanding some remarks which we made in § 17.1. We could construct a formal theory of Dirichlet series in which 'analysis' played no part. This theory would include all identities of the 'Möbius' type, but the notions of the sum of an infinite series, or the value of an infinite product, would never occur. We shall not attempt to construct such a theory in detail, but it is interesting to consider how it would begin.

We denote the formal series $\sum a_n n^{-s}$ by A, and write

$$A = \sum a_n n^{-s}.$$

In particular we write

$$I = 1 \,.\, 1^{-s} + 0 \,.\, 2^{-s} + 0 \,.\, 3^{-s} + \cdots,$$
$$Z = 1 \,.\, 1^{-s} + 1 \,.\, 2^{-s} + 1 \,.\, 3^{-s} + \cdots,$$
$$M = \mu(1)1^{-s} + \mu(2)2^{-s} + \mu(3)3^{-s} + \cdots.$$

By

$$A = B$$

we mean that $a_n = b_n$ for all values of n.

The equation

$$A \times B = C$$

means that C is the formal product of A and B, in the sense of § 17.4. The definition may be extended, as in § 17.4, to the product of any finite number of series, or, with proper precautions, of an infinity. It is plain from the definition that

$$A \times B = B \times A, \quad A \times B \times C = (A \times B) \times C = A \times (B \times C),$$

and so on and that

$$A \times I = A.$$

The equation

$$A \times Z = B$$

means that

$$b_n = \sum_{d|n} a_d.$$

Let us suppose that there is a series L such that

$$Z \times L = I.$$

Then

$$A = A \times I = A \times (Z \times L) = (A \times Z) \times L = B \times L,$$

i.e.

$$a_n = \sum_{d|n} b_d l_{n/d}.$$

The Möbius formula asserts that $l_n = \mu(n)$, or that $L = M$, or that

(17.6.1) $$Z \times M = I;$$

and this means that

$$\sum_{d|n} \mu(d)$$

is 1 when $n = 1$ and 0 when $n > 1$ (Theorem 263).

We may prove this as in § 16.3, or we may continue as follows. We write

$$P_p = 1 - p^{-s}, \quad Q_p = 1 + p^{-s} + p^{-2s} + \cdots,$$

where p is a prime (so that P_p, for example, is the series A in which $a_1 = 1$, $a_p = -1$, and the remaining coefficients are 0); and calculate the coefficient of n^{-s} in the formal product of P_p and Q_p. This coefficient is 1 if $n = 1$, $1 - 1 = 0$ if n is a positive power of p, and 0 in all other cases; so that

$$P_p \times Q_p = I$$

for every p.

The series P_p, Q_p, and I are of the special type considered in § 17.4; and

$$Z = \prod Q_p, \quad M = \prod P_p,$$
$$Z \times M = \prod Q_p \times \prod P_p,$$

while

$$\prod (Q_p \times P_p) = \prod I = I.$$

But the coefficient of n^{-s} in

$$(Q_2 \times Q_3 \times Q_5 \times \ldots) \times (P_2 \times P_3 \times P_5 \times \ldots)$$

(a product of two series of the general type) is the same as in

$$Q_2 \times P_2 \times Q_3 \times P_3 \times Q_5 \times P_5 \times \ldots$$

or in

$$(Q_2 \times P_2) \times (Q_3 \times P_3) \times (Q_5 \times P_5) \times \ldots$$

(which are each products of an infinity of series of the special type); in each case the χ_n of § 17.4 contains only a finite number of terms. Hence

$$Z \times M = \prod Q_p \times \prod P_p = \prod (Q_p \times P_p) = \prod I = I.$$

It is plain that this proof of (17.6.1) is, at bottom, merely a translation into a different language of that of § 16.3; and that, in a simple case like this, we gain nothing by the translation. More complicated formulae become much easier to grasp and prove when stated in the language of infinite series and products, and it is important to realize that we can use it without analytical assumptions. In what follows, however, we continue to use the language of ordinary analysis.

17.7. The function $\Lambda(n)$. The function $\Lambda(n)$, which is particularly important in the analytical theory of primes, is defined by

$$\Lambda(n) = \log p \quad (n = p^m),$$
$$\Lambda(n) = 0 \quad (n \neq p^m),$$

i.e. as being $\log p$ when n is a prime p or one of its powers, and 0 otherwise.

From Theorem 280, we have

$$\log \zeta(s) = \sum_p \log\left(\frac{1}{1-p^{-s}}\right).$$

Differentiating with respect to s, and observing that

$$\frac{d}{ds}\log\frac{1}{1-p^{-s}} = -\frac{\log p}{p^s-1},$$

we obtain

$$-\frac{\zeta'(s)}{\zeta(s)} = \sum_p \frac{\log p}{p^s-1}. \tag{17.7.1}$$

The differentiation is legitimate because the derived series is uniformly convergent for $s \geqslant 1+\delta > 1$.†

We may write (17.7.1) in the form

$$-\frac{\zeta'(s)}{\zeta(s)} = \sum_p \log p \sum_{m=1}^{\infty} p^{-ms}$$

and the double series $\sum\sum p^{-ms}\log p$ is absolutely convergent when $s > 1$. Hence it may be written as

$$\sum_{p,m} p^{-ms}\log p = \sum \Lambda(n) n^{-s},$$

by the definition of $\Lambda(n)$.

THEOREM 294:

$$-\frac{\zeta'(s)}{\zeta(s)} = \sum \Lambda(n) n^{-s} \qquad (s > 1).$$

Since

$$-\zeta'(s) = \sum_{n=1}^{\infty} \frac{\log n}{n^s},$$

† The nth prime p_n is greater than n, and the series may be compared with $\sum n^{-s}\log n$.

by Theorem 279, it follows that

$$\sum_{n=1}^{\infty} \frac{\Lambda(n)}{n^s} = \frac{1}{\zeta(s)} \sum_{n=1}^{\infty} \frac{\log n}{n^s} = \sum_{n=1}^{\infty} \frac{\mu(n)}{n^s} \sum_{n=1}^{\infty} \frac{\log n}{n^s},$$

and

$$\sum_{n=1}^{\infty} \frac{\log n}{n^s} = \zeta(s) \sum_{n=1}^{\infty} \frac{\Lambda(n)}{n^s} = \sum_{n=1}^{\infty} \frac{1}{n^s} \sum_{n=1}^{\infty} \frac{\Lambda(n)}{n^s}.$$

From these equations, and the uniqueness theorem of § 17.1, we deduce†

THEOREM 295:

$$\Lambda(n) = \sum_{d|n} \mu\left(\frac{n}{d}\right) \log d.$$

THEOREM 296:

$$\log n = \sum_{d|n} \Lambda(d).$$

We may also prove these theorems directly. If $n = \prod p^a$, then

$$\sum_{d|n} \Lambda(d) = \sum_{p^a|n} \log p.$$

The summation extends over all values of p, and all positive values of a for which $p^a|n$, so that $\log p$ occurs a times. Hence

$$\sum_{p^a|n} \log p = \sum a \log p = \log \prod p^a = \log n.$$

This proves Theorem 296, and Theorem 295 follows by Theorem 266.

Again

$$-\frac{d}{ds}\left\{\frac{1}{\zeta(s)}\right\} = \frac{\zeta'(s)}{\zeta^2(s)} = -\frac{1}{\zeta(s)}\left\{-\frac{\zeta'(s)}{\zeta(s)}\right\},$$

† Compare § 17.6.

so that

$$\sum_{n=1}^{\infty} \frac{\mu(n)\log n}{n^s} = -\sum_{n=1}^{\infty} \frac{\mu(n)}{n^s} \sum_{n=1}^{\infty} \frac{\Lambda(n)}{n^s}.$$

Hence, as before, we deduce

THEOREM 297:

$$-\mu(n)\log n = \sum_{d|n} \mu\left(\frac{n}{d}\right)\Lambda(d).$$

Similarly

$$-\frac{\zeta'(s)}{\zeta(s)} = \zeta(s)\frac{d}{ds}\left\{\frac{1}{\zeta(s)}\right\},$$

and from this (or from Theorems 297 and 267) we deduce

THEOREM 298:

$$\Lambda(n) = -\sum_{d|n} \mu(d)\log d.$$

17.8. Further examples of generating functions. We add a few examples of a more miscellaneous character. We define $d_k(n)$ as the number of ways of expressing n as the product of k positive factors (of which any number may be unity), expressions in which only the order of the factors being different is regarded as distinct. In particular, $d_2(n) = d(n)$. Then

THEOREM 299:

$$\zeta^k(s) = \sum \frac{d_k(n)}{n^s} \quad (s > 1).$$

Theorem 289 is a particular case of this theorem.

Again

$$\frac{\zeta(2s)}{\zeta(s)} = \prod_p \left(\frac{1-p^{-s}}{1-p^{-2s}}\right) = \prod_p \left(1+\frac{1}{p^s}\right)^{-1}$$
$$= \prod_p \left(1 - \frac{1}{p^s} + \frac{1}{p^{2s}} - \cdots\right)$$
$$= \sum_{n=1}^{\infty} \frac{\lambda(n)}{n^s},$$

where $\lambda(n) = (-1)^{\rho}$, ρ being the total number of prime factors of n, when multiple factors are counted multiply. Thus

THEOREM 300:

$$\frac{\zeta(2s)}{\zeta(s)} = \sum \frac{\lambda(n)}{n^s} \quad (s > 1).$$

Similarly we can prove

THEOREM 301:

$$\frac{\zeta^2(s)}{\zeta(2s)} = \sum_{n=1}^{\infty} \frac{2^{\omega(n)}}{n^s} \quad (s > 1),$$

where $\omega(n)$ is the number of different prime factors of n.

A number n is said to be *squarefree*† if it has no squared factor. If we write $q(n) = 1$ when n is *squarefree*, and $q(n) = 0$ when n has a squared factor, so that $q(n) = |\mu(n)|$, then

$$\frac{\zeta(s)}{\zeta(2s)} = \prod_p \left(\frac{1-p^{-2s}}{1-p^{-s}}\right) = \prod_p \left(1+p^{-s}\right) = \sum_{n=1}^{\infty} \frac{q(n)}{n^s} \quad (s > 1),$$

by Theorems 280 and 286. Thus

THEOREM 302:

$$\frac{\zeta(s)}{\zeta(2s)} = \sum_{n=1}^{\infty} \frac{q(n)}{n^s} = \sum_{n=1}^{\infty} \frac{|\mu(n)|}{n^s} \quad (s > 1).$$

† Some writers (in English) use the German word 'quadratfrei'.

More generally, if $q_k(n) = 0$ or 1 according as n has or has not a kth power as a factor, then

THEOREM 303:

$$\frac{\zeta(s)}{\zeta(ks)} = \sum_{n=1}^{\infty} \frac{q_k(n)}{n^s} \quad (s > 1).$$

Another example, due to Ramanujan, is

THEOREM 304:

$$\frac{\zeta^4(s)}{\zeta(2s)} = \sum_{n=1}^{\infty} \frac{\{d(n)\}^2}{n^s} \quad (s > 1).$$

This may be proved as follows. We have

$$\frac{\zeta^4(s)}{\zeta(2s)} = \prod_p \frac{1 - p^{-2s}}{(1 - p^{-s})^4} = \prod_p \frac{1 + p^{-s}}{(1 - p^{-s})^3}.$$

Now

$$\frac{1 + x}{(1 - x)^3} = (1 + x)(1 + 3x + 6x^2 + \cdots)$$

$$= 1 + 4x + 9x^2 + \cdots = \sum_{l=0}^{\infty} (l + 1)^2 x^l.$$

Hence

$$\frac{\zeta^4(s)}{\zeta(2s)} = \prod_p \left\{ \sum_{l=0}^{\infty} (l + 1)^2 p^{-ls} \right\}.$$

The coefficient of n^{-s}, when $n = p_1^{l_1} p_2^{l_2} \ldots$, is

$$(l_1 + 1)^2 (l_2 + 1)^2 \ldots = \{d(n)\}^2,$$

by Theorem 273.

More generally we can prove, by similar reasoning,

THEOREM 305. *If s, $s-a, s-b$, and $s-a-b$ are all greater than* 1, *then*

$$\frac{\zeta(s)\zeta(s-a)\zeta(s-b)\zeta(s-a-b)}{\zeta(2s-a-b)} = \sum_{n=1}^{\infty} \frac{\sigma_a(n)\sigma_b(n)}{n^s}.$$

17.9. The generating function of $r(n)$. We saw in § 16.10 that

$$r(n) = 4\sum_{d|n} \chi(d),$$

where $\chi(n)$ is 0 when n is even and $(-1)^{\frac{1}{2}(n-1)}$ when n is odd. Hence

$$\sum \frac{r(n)}{n^s} = 4\sum \frac{1}{n^s} \sum \frac{\chi(n)}{n^s} = 4\zeta(s)L(s),$$

where

$$L(s) = 1^{-s} - 3^{-s} + 5^{-s} - \cdots,$$

if $s > 1$.

THEOREM 306:

$$\sum \frac{r(n)}{n^s} = 4\zeta(s)L(s) \quad (s > 1).$$

The function

$$\eta(s) = 1^{-s} - 2^{-s} + 3^{-s} - \cdots$$

is expressible in terms of $\zeta(s)$ by the formula

$$\eta(s) = (1 - 2^{1-s})\zeta(s);$$

but $L(s)$, which can also be expressed in the form

$$L(s) = \prod_p \left(\frac{1}{1 - \chi(p)p^{-s}}\right),$$

is an independent function. It is the basis of the analytical theory of the distribution of primes in the progressions $4m+1$ and $4m+3$.

17.10. Generating functions of other types. The generating functions discussed in this chapter have been defined by Dirichlet series; but any function

$$F(s) = \sum \alpha_n u_n(s)$$

may be regarded as a generating function of α_n. The most usual form of $u_n(s)$ is

$$u_n(s) = e^{-\lambda_n s},$$

where λ_n is a sequence of positive numbers which increases steadily to infinity. The most important cases are the cases $\lambda_n = \log n$ and $\lambda_n = n$. When $\lambda_n = \log n, u_n(s) = n^{-s}$ and the series is a Dirichlet series. When $\lambda_n = n$, it is a power series in

$$x = e^{-s}.$$

Since

$$m^{-s}.n^{-s} = (mn)^{-s},$$

and

$$x^m.x^m = x^{m+n},$$

the first type of series is more important in the 'multiplicative' side of the theory of numbers (and in particular in the theory of primes). Such functions as

$$\sum \mu(n)x^n, \quad \sum \phi(n)x^n, \quad \sum \Lambda(n)x^n$$

are extremely difficult to handle. But generating functions defined by power series are dominant in the 'additive' theory.†

Another interesting type of series is obtained by taking

$$u_n(s) = \frac{e^{-ns}}{1 - e^{-ns}} = \frac{x^n}{1 - x^n}.$$

† See Chs. XIX–XXI.

We write

$$F(x) = \sum_{n=1}^{\infty} a_n \frac{x^n}{1-x^n},$$

and disregard questions of convergence, which are not interesting here.[†] A series of this type is called a 'Lambert series'. Then

$$F(x) = \sum_{n=1}^{\infty} a_n \sum_{m=1}^{\infty} x^{mn} = \sum_{N=1}^{\infty} b_N x^N,$$

where

$$b_N = \sum_{n|N}^{\infty} a_n.$$

This relation between the a and b is that considered in §§ 16.4 and 17.6, and it is equivalent to

$$\zeta(s)f(s) = g(s),$$

where $f(s)$ and $g(s)$ are the Dirichlet series associated with a_n and b_n.

THEOREM 307. *If*

$$f(s) = \sum a_n n^{-s}, \quad g(s) = \sum b_n n^{-s},$$

then

$$F(x) = \sum a_n \frac{x^n}{1-x^n} = \sum b_n x^n$$

if and only if

$$\zeta(s)f(s) = g(s).$$

If $f(s) = \sum \mu(n)n^{-s}$, $g(s) = 1$, by Theorem 287. If $f(s) = \sum \phi(n)n^{-s}$,

$$g(s) = \zeta(s-1) = \sum \frac{n}{n^s},$$

by Theorem 288. Hence we derive

† All the series of this kind which we consider are absolutely convergent when $0 \leqslant x < 1$.

THEOREM 308:

$$\sum_1^\infty \frac{\mu(n)x^n}{1-x^n} = x.$$

THEOREM 309:

$$\sum_1^\infty \frac{\phi(n)x^n}{1-x^n} = \frac{x}{(1-x)^2}.$$

Similarly, from Theorems 289 and 306, we deduce

THEOREM 310:

$$\sum_{n=1}^\infty d(n)x^n = \frac{x}{1-x} + \frac{x^2}{1-x^2} + \frac{x^3}{1-x^3} + \cdots.$$

THEOREM 311:

$$\sum_{n=1}^\infty r(n)x^n = 4\left(\frac{x}{1-x} - \frac{x^3}{1-x^3} + \frac{x^5}{1-x^5} - \cdots\right).$$

Theorem 311 is equivalent to a famous identity in the theory of elliptic functions, viz.

THEOREM 312:

$$(1+2x+2x^4+2x^9+\cdots)^2$$
$$= 1 + 4\left(\frac{x}{1-x} - \frac{x^3}{1-x^3} + \frac{x^5}{1-x^5} - \cdots\right).$$

In fact, if we square the series

$$1+2x+2x^4+2x^9+\cdots = \sum_{-\infty}^\infty x^{m^2},$$

the coefficient of x^n is $r(n)$, since every pair (m_1, m_2) for which $m_1^2+m_2^2 = n$ contributes a unit to it.†

NOTES

§ 17.1. There is a short account of the analytical theory of Dirichlet series in Titchmarsh, *Theory of functions*, ch. ix; and fuller accounts, including the theory of series of the more general type

$$\sum a_n e^{-\lambda_n s}$$

(referred to in § 17.10) in Hardy and Riesz, *The general theory of Dirichlet's series* (Cambridge Math. Tracts, no. 18, 1915), and Landau, *Handbuch*, 103–24, 723–75.

§ 17.2. There is a large literature concerned with the zeta function and its application to the theory of primes. See in particular the books of Ingham and Landau, Titchmarsh, *The Riemann zeta-function* (Oxford, 1951) and Edwards, *Riemann's zeta-function* (New York, Academic Press, 1974), the last especially from the historical point of view.

For the value of $\zeta(2n)$ see Bromwich, *Infinite series*, ed. 2, 298.

§ 17.3. The proof of Theorem 283 depends on the formulae

$$0 < n^{-s}\log n - x^{-s}\log x = \int_n^x t^{-s-1}(s\log t - 1)\,dt < \frac{s}{n^2}\log(n+1),$$

valid for $3 \leqslant n \leqslant x \leqslant n+1$ and $s > 1$.

There are proofs of the theorem referred to in the footnote to p. 247 in Landau, *Handbuch*, 106–7, and Titchmarsh, *Theory of functions*, 289–90.

§§ 17.5–10. Many of the identities in these sections, and others of similar character, occur in Pólya and Szegő, Nos. 38–83. Some of them go back to Euler. We do not attempt to assign them systematically to their discoverers, but Theorems 304 and 305 were first stated by Ramanujan in the *Messenger of Math.* 45 (1916), 81–84 (*Collected papers*, 133–5 and 185).

§ 17.6. The discussion in small print was the result of conversation with Professor Harald Bohr.

§ 17.10. Theorem 312 is due to Jacobi, *Fundamenta nova* (1829), § 40 (4) and § 65 (6).

† Thus 5 arises from 8 pairs, viz. (2, 1), (1, 2), and those derived by changes of sign.

XVIII

THE ORDER OF MAGNITUDE OF ARITHMETICAL FUNCTIONS

18.1. The order of $d(n)$. In the last chapter we discussed formal relations satisfied by certain arithmetical functions, such as $d(n)$, $\sigma(n)$, and $\phi(n)$. We now consider the behaviour of these functions for large values of n, beginning with $d(n)$. It is obvious that $d(n) \geqslant 2$ when $n > 1$, while $d(n) = 2$ if n is a prime. Hence

THEOREM 313. *The lower limit of $d(n)$ as $n \to \infty$ is* 2:

$$\varliminf_{n\to\infty} d(n) = 2.$$

It is less trivial to find any upper bound for the order of magnitude of $d(n)$. We first prove a negative theorem.

THEOREM 314. *The order of magnitude of $d(n)$ is sometimes larger than that of any power of* $\log n$*: the equation*

$$d(n) = O\{(\log n)^{\Delta}\} \tag{18.1.1}$$

is false for every Δ.[†]

If $n = 2^m$, then

$$d(n) = m + 1 \sim \frac{\log n}{\log 2}.$$

If $n = (2\,.\,3)^m$, then

$$d(n) = (m+1)^2 \sim \left(\frac{\log n}{\log 6}\right)^2;$$

and so on. If

$$l \leqslant \Delta < l + 1$$

and

$$n = (2\,.\,3 \ldots p_{l+1})^m,$$

[†] The symbols $O, o, \sim$ were defined in § 1.6.

then

$$d(n) = (m+1)^{l+1} \sim \left\{ \frac{\log n}{\log(2 \, . \, 3 \dots p_{l+1})} \right\}^{l+1} > K(\log n)^{l+1},$$

where K is independent of n. Hence (18.1.1) is false for an infinite sequence of values of n.

On the other hand we can prove

THEOREM 315:

$$d(n) = O(n^{\delta})$$

for all positive δ.

The assertions that $d(n) = O(n^{\delta})$, for all positive δ, and that $d(n) = o(n^{\delta})$, for all positive δ, are equivalent, since $n^{\delta'} = o(n^{\delta})$ when $0 < \delta' < \delta$.

We require the lemma

THEOREM 316. *If $f(n)$ is multiplicative, and $f(p^m) \to 0$ as $p^m \to \infty$, then $f(n) \to 0$ as $n \to \infty$.*

Given any positive ϵ, we have

(i) $|f(p^m)| < A$ for all p and m,
(ii) $|f(p^m)| < 1 \quad$ if $\quad p^m > B$,
(iii) $|f(p^m)| < \epsilon \quad$ if $\quad p^m > N(\epsilon)$,

where A and B are independent of p, m, and ϵ, and $N(\epsilon)$ depends on ϵ only. If

$$n = p_1^{a_1} p_2^{a_2} \dots p_r^{a_r},$$

then

$$f(n) = f(p_1^{a_1}) f(p_2^{a_2}) \dots f(p_r^{a_r}).$$

Of the factors $p_1^{a_1}, p_2^{a_2}, \dots$, not more than C are less than or equal to B, C being independent of n and ϵ. The product of the corresponding factors $f(p^a)$ is numerically less than A^C, and the rest of the factors of $f(n)$ are numerically less than 1.

The number of integers which can be formed by the multiplication of factors $p^a \leqslant N(\epsilon)$ is $M(\epsilon)$, and every such number is less than $P(\epsilon)$, $M(\epsilon)$

and $P(\epsilon)$ depending only on ϵ. Hence, if $n > P(\epsilon)$ there is at least one factor p^a of n such that $p^a > N(\epsilon)$ and then, by (iii),

$$|f(p^a)| < \epsilon.$$

It follows that

$$|f(n)| < A^C \epsilon.$$

when $n > P(\epsilon)$, and therefore that $f(n) \to 0$.

To deduce Theorem 315, we take $f(n) = n^{-\delta} d(n)$. Then $f(n)$ is multiplicative, by Theorem 273, and

$$f(p^m) = \frac{m+1}{p^{m\delta}} \leqslant \frac{2m}{p^{m\delta}} = \frac{2}{p^{m\delta}} \frac{\log p^m}{\log p} \leqslant \frac{2}{\log 2} \frac{\log p^m}{(p^m)^\delta} \to 0$$

when $p^m \to \infty$. Hence $f(n) \to 0$ when $n \to \infty$, and this is Theorem 315 (with o for O).

We can also prove Theorem 315 directly. By Theorem 273,

$$\frac{d(n)}{n^\delta} = \prod_{i=1}^{r} \left(\frac{a_i + 1}{p_i^{ai\delta}} \right). \tag{18.1.2}$$

Since

$$a\delta \log 2 \leqslant e^{a\delta \log 2} = 2^{a\delta} \leqslant p^{a\delta},$$

we have

$$\frac{a+1}{p^{a\delta}} \leqslant 1 + \frac{a}{p^{a\delta}} \leqslant 1 + \frac{1}{\delta \log 2} \leqslant \exp\left(\frac{1}{\delta \log 2} \right).$$

We use this in (18.1.2) for those p which are less than $2^{1/\delta}$; there are less than $2^{1/\delta}$ such primes. If $p \geqslant 2^{1/\delta}$, we have

$$p^\delta \geqslant 2, \quad \frac{a+1}{p^{a\delta}} \leqslant \frac{a+1}{2^a} \leqslant 1.$$

Hence

$$\frac{d(n)}{n^\delta} \leqslant \prod_{p \leqslant 2^{1/\delta}} \exp\left(\frac{1}{\delta \log 2} \right) < \exp\left(\frac{2^{1/\delta}}{\delta \log 2} \right) = O(1). \tag{18.1.3}$$

This is Theorem 315.

We can use this type of argument to improve on Theorem 315. We suppose $\epsilon > 0$ and replace δ in the last paragraph by

$$\alpha = \frac{(1 + \frac{1}{2}\epsilon) \log 2}{\log \log n}.$$

Nothing is changed until we reach the final step in (18.1.3) since it is here that, for the first time, we use the fact that δ is independent of n. This time we have

$$\log\left(\frac{d(n)}{n^{\alpha}}\right) < \frac{2^{1/\alpha}}{\alpha \log 2} = \frac{(\log n)^{1/(1+\frac{1}{2}\epsilon)} \log \log n}{\left(1 + \frac{1}{2}\epsilon\right) \log^2 2} \leqslant \frac{\epsilon \log 2 \log n}{2 \log \log n}$$

for all $n > n_0(\epsilon)$ (by the remark at the top of p. 9). Hence

$$\log d(n) \leqslant \alpha \log n + \frac{\epsilon \log 2 \log n}{2 \log \log n} = \frac{(1 + \epsilon) \log 2 \log n}{\log \log n}.$$

We have thus proved part of

THEOREM 317: $$\varlimsup \frac{\log d(n) \log \log n}{\log n} = \log 2;$$

that is, if $\epsilon > 0$ *then*

$$d(n) < 2^{(1+\epsilon) \log n / \log \log n}$$

for all $n > n_0(\epsilon)$ *and*

$$d(n) > 2^{(1-\epsilon) \log n / \log \log n} \tag{18.1.4}$$

for an infinity of values of n.

Thus the true 'maximum order' of $d(n)$ is about

$$2^{\log n / \log \log n}.$$

It follows from Theorem 315 that

$$\frac{\log d(n)}{\log n} \to 0$$

and so

$$d(n) = n^{\log d(n) / \log n} = n^{\epsilon_n},$$

where $\epsilon_n \to 0$ as $n \to \infty$. On the other hand, since

$$2^{\log n/\log\log n} = n^{\log 2/\log\log n}$$

and loglog n tends very slowly to infinity, ϵ_n tends very slowly to 0. To put it roughly, $d(n)$ is, for some n, much more like a power of n than a power of log n. But this happens only very rarely† and, as Theorem 313 shows, $d(n)$ is sometimes quite small.

To complete the proof of Theorem 317, we have to prove (18.1.4) for a suitable sequence of n. We take n to be the product of the first r primes, so that

$$n = 2.\,3.\,5.\,7\ldots P, \qquad d(n) = 2^r = 2^{\pi(P)},$$

where P is the rth prime. It is reasonable to expect that such a choice of n will give us a large value of $d(n)$. The function

$$\vartheta(x) = \sum_{p \leqslant x} \log p$$

is discussed in Ch. XXII, where we shall prove (Theorem 414) that

$$\vartheta(x) > Ax$$

for some fixed positive A and all $x \geqslant 2$.‡ We have then

$$AP < \vartheta(P) = \sum_{p \leqslant P} \log p = \log n,$$

$$\pi(P) \log P = \log P \sum_{p \leqslant P} 1 \geqslant \vartheta(P) = \log n,$$

and so

$$\log d(n) = \pi(P) \log 2 \geqslant \frac{\log n \log 2}{\log P} > \frac{\log n \log 2}{\log\log n - \log A}$$
$$> \frac{(1-\epsilon)\log n \log 2}{\log\log n}$$

for $n > n_0(\epsilon)$.

† See § 22.13.

‡ In fact, we prove (Theorem 6 and 420) that $\vartheta(x) \sim x$, but it is of interest that the much simpler Theorem 414 suffices here.

18.2. The average order of $d(n)$. If $f(n)$ is an arithmetical function and $g(n)$ is any simple function of n such that

$$f(1)+f(2)+\cdots+f(n) \sim g(1)+\cdots+g(n), \tag{18.2.1}$$

we say that $f(n)$ is of the *average order* of $g(n)$. For many arithmetical functions, the sum of the left-hand side of (18.2.1) behaves much more regularly for large n than does $f(n)$ itself. For $d(n)$, in particular, this is true and we can prove very precise results about it.

THEOREM 318: $$d(1)+d(2)+\cdots+d(n) \sim n\log n.$$

Since $$\log 1+\log 2+\cdots+\log n \sim \int_1^n \log t\,dt \sim n\log n,$$

the result of Theorem 318 is equivalent to

$$d(1)+d(2)+\cdots+d(n) \sim \log 1+\log 2+\cdots+\log n.$$

We may express this by saying

THEOREM 319. *The average order of $d(n)$ is* $\log n$.

Both theorems are included in a more precise theorem, viz.

THEOREM 320:

$$d(1)+d(2)+\cdots+d(n) = n\log n+(2\gamma-1)n+O(\surd n),$$

where γ is Euler's constant.†

We prove these theorems by use of the lattice L of Ch. III, whose vertices are the points in the (x, y)-plane with integral coordinates. We denote by **D** the region in the upper right-hand quadrant contained between the axes and the rectangular hyperbola $xy = n$. We count the lattice points in **D**, including those on the hyperbola but not those on the axes. Every lattice point in **D** appears on a hyperbola

$$xy = s \qquad (1 \leqslant s \leqslant n);$$

† In Theorem 422 we prove that

$$1+\frac{1}{2}+\cdots+\frac{1}{n}-\log n = \gamma+O\left(\frac{1}{n}\right),$$

where γ is a constant, known as Euler's constant.

and the number on such a hyperbola is $d(s)$. Hence the number of lattice points in **D** is

$$d(1) + d(2) + \cdots + d(n).$$

Of these points, $n = [n]$ have the x-coordinate 1, $\left[\frac{1}{2}n\right]$ have the x-coordinate 2, and so on. Hence their number is

$$[n] + \left[\frac{n}{2}\right] + \left[\frac{n}{3}\right] + \cdots + \left[\frac{n}{n}\right] = n\left(1 + \frac{1}{2} + \cdots + \frac{1}{n}\right) + O(n)$$
$$= n \log n + O(n),$$

since the error involved in the removal of any square bracket is less than 1. This result includes Theorem 318.

Theorem 320 requires a refinement of the method. We write

$$u = [\sqrt{n}],$$

so that

$$u^2 = n + O(\sqrt{n}) = n + O(u)$$

and

$$\log u = \log\left\{\sqrt{n} + O(1)\right\} = \tfrac{1}{2}\log n + O\left(\frac{1}{\sqrt{n}}\right).$$

In Fig. 8 the curve $GEFH$ is the rectangular hyperbola $xy = n$, and the coordinates of A, B, C, D are $(0, 0)$, $(0, u)$, (u, u), $(u, 0)$. Since $(u+1)^2 > n$, there is no lattice point inside the small triangle ECF; and the figure is symmetrical as between x and y. Hence the number of lattice points in **D** is equal to twice the number in the strip between AY and DF, counting those on DF and the curve but not those on AY, less the number in the square $ADCB$, counting those on BC and CD but not those on AB and AD; and therefore

$$\sum_{l=1}^{n} d(l) = 2\left(\left[\frac{n}{1}\right] + \left[\frac{n}{2}\right] + \cdots + \left[\frac{n}{u}\right]\right) - u^2$$
$$= 2n\left(1 + \frac{1}{2} + \cdots + \frac{1}{u}\right) - n + O(u).$$

Now

$$2\left(1 + \frac{1}{2} + \cdots + \frac{1}{u}\right) = 2\log u + 2\gamma + O\left(\frac{1}{u}\right),$$

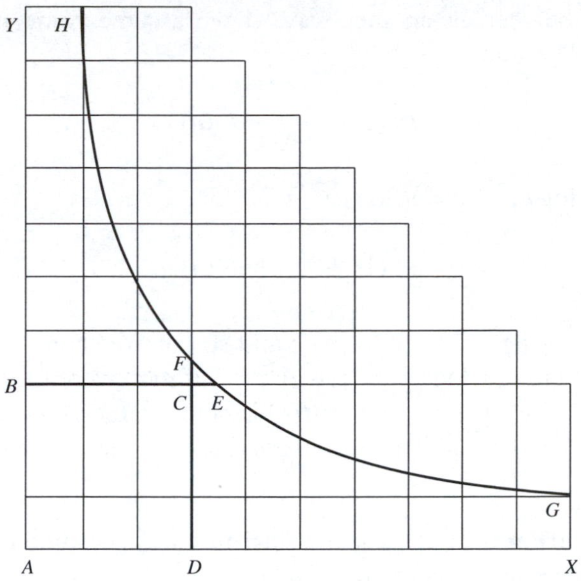

FIG. 8.

so that

$$\sum_{l=1}^{n} d(l) = 2n\log u + (2\gamma - 1)n + O(u) + O\left(\frac{n}{u}\right)$$

$$= n\log n + (2\gamma - 1)n + O(\surd n).$$

Although

$$\frac{1}{n}\sum_{l=1}^{n} d(l) \sim \log n,$$

it is not true that 'most' numbers n have about $\log n$ divisors. Actually 'almost all' numbers have about

$$(\log n)^{\log 2} = (\log n)^{\cdot 6\ldots}$$

divisors. The average $\log n$ is produced by the contributions of the small proportion of numbers with abnormally large $d(n)$.†

† 'Almost all' is used in the sense of § 1.6. The theorem is proved in § 22.13.

This may be seen in another way, if we assume some theorems of Ramanujan. The sum

$$d^2(1) + \cdots + d^2(n)$$

is of order $n(\log n)^{2^2-1} = n(\log n)^3$;

$$d^3(1) + \cdots + d^3(n)$$

is of order $n(\log n)^{2^3-1} = n(\log n)^7$; and so on. We should expect these sums to be of order $n(\log n)^2, n(\log n)^3, \ldots$, if $d(n)$ were generally of the order of $\log n$. But, as the power of $d(n)$ becomes larger, the numbers with an abnormally large number of divisors dominate the average more and more.

18.3. The order of $\sigma(n)$. The irregularities in the behaviour of $\sigma(n)$ are much less pronounced than those of $d(n)$.

Since $1|n$ and $n|n$, we have first

THEOREM 321:

$$\sigma(n) > n.$$

On the other hand,

THEOREM 322:

$$\sigma(n) = O(n^{1+\delta})$$

for every positive δ.

More precisely,

THEOREM 323:

$$\overline{\lim} \frac{\sigma(n)}{n \log \log n} = e^{\gamma}.$$

We shall prove Theorem 322 in the next section, but must postpone the proof of Theorem 323, which, with Theorem 321, shows that the order of $\sigma(n)$ is always 'very nearly n', to § 22.9.

As regards the average order, we have

THEOREM 324. *The average order of* $\sigma(n)$ *is* $\frac{1}{6}\pi^2 n$. *More precisely,*

$$\sigma(1)+\sigma(2)+\cdots+\sigma(n)=\tfrac{1}{12}\pi^2 n^2+O(n\log n).$$

For

$$\sigma(1)+\cdots+\sigma(n)=\sum y,$$

where the summation extends over all the lattice points in the region **D** of § 18.2. Hence

$$\sum_{l=1}^{n}\sigma(l)=\sum_{x=1}^{n}\sum_{y\leqslant n/x}y=\sum_{x=1}^{n}\frac{1}{2}\left[\frac{n}{x}\right]\left(\left[\frac{n}{x}\right]+1\right)$$

$$=\tfrac{1}{2}\sum_{x=1}^{n}\left(\frac{n}{x}+O(1)\right)\left(\frac{n}{x}+O(1)\right)=\tfrac{1}{2}n^2\sum_{x=1}^{n}\frac{1}{x^2}+O\left(n\sum_{x=1}^{n}\frac{1}{x}\right)+O(n).$$

Now

$$\sum_{x=1}^{n}\frac{1}{x^2}=\sum_{x=1}^{\infty}\frac{1}{x^2}+O\left(\frac{1}{n}\right)=\tfrac{1}{6}\pi^2+O\left(\frac{1}{n}\right),$$

by (17.2.2), and

$$\sum_{x=1}^{n}\frac{1}{x}=O(\log n).$$

Hence

$$\sum_{l=1}^{n}\sigma(l)=\tfrac{1}{12}\pi^2 n^2+O(n\log n).$$

In particular, the average order of $\sigma(n)$ is $\frac{1}{6}\pi^2 n$.[†]

[†] Since $\sum_{1}^{n} m\sim\frac{1}{2}n^2$.

18.4. The order of $\phi(n)$. The function $\phi(n)$ is also comparatively regular, and its order is also always 'nearly n'. In the first place

THEOREM 325: *$\phi(n) < n$ if $n > 1$.*

Next, if $n = p^m$, and $p > 1/\epsilon$ then

$$\phi(n) = n\left(1 - \frac{1}{p}\right) > n(1 - \epsilon).$$

Hence

THEOREM 326: $$\overline{\lim}\frac{\phi(n)}{n} = 1.$$

There are also two theorems for $\phi(n)$ corresponding to Theorems 322 and 323.

THEOREM 327:

$$\frac{\phi(n)}{n^{1-\delta}} \to \infty$$

for every positive δ.

THEOREM 328:

$$\underline{\lim}\frac{\phi(n)\log\log n}{n} = e^{-\gamma}.$$

Theorem 327 is equivalent to Theorem 322, in virtue of

THEOREM 329:

$$A < \frac{\sigma(n)\phi(n)}{n^2} < 1$$

(*for a positive constant A*).

To prove the last theorem we observe that, if $n = \prod p^a$, then

$$\sigma(n) = \prod_{p|n}\frac{p^{a+1} - 1}{p - 1} = n\prod_{p|n}\frac{1 - p^{-a-1}}{1 - p - 1}$$

and

$$\phi(n) = n\prod_{p|n}(1 - p^{-1}).$$

Hence

$$\frac{\sigma(n)\phi(n)}{n^2} = \prod_{p|n}(1 - p^{-a-1}),$$

which lies between 1 and $\prod(1 - p^{-2})$.† It follows that $\sigma(n)/n$ and $n/\phi(n)$ have the same order of magnitude, so that Theorem 327 is equivalent to Theorem 322.

To prove Theorem 327 (and so Theorem 322) we write

$$f(n) = \frac{n^{1-\delta}}{\phi(n)}.$$

Then $f(n)$ is multiplicative, and so, by Theorem 316, it is sufficient to prove that

$$f(p^m) \to 0$$

when $p^m \to \infty$. But

$$\frac{1}{f(p^m)} = \frac{\phi(p^m)}{p^{m(1-\delta)}} = p^{m\delta}\left(1 - \frac{1}{p}\right) \geqslant \tfrac{1}{2}p^{m\delta} \to \infty.$$

We defer the proof of Theorem 328 to Ch. XXII.

18.5. The average order of $\phi(n)$. The average order of $\phi(n)$ is $6n/\pi^2$. More precisely

THEOREM 330:

$$\Phi(n) = \phi(1) + \cdots + \phi(n) = \frac{3n^2}{\pi^2} + O(n \log n).$$

For, by (16.3.1),

$$\Phi(n) = \sum_{m=1}^{n} m \sum_{d|m} \frac{\mu(d)}{d} = \sum_{dd' \leqslant n} d'\mu(d)$$

$$= \sum_{d=1}^{n} \mu(d) \sum_{d'=1}^{[n/d]} d' = \frac{1}{2}\sum_{d=1}^{n} \mu(d)\left(\left[\frac{n}{d}\right]^2 + \left[\frac{n}{d}\right]\right)$$

† By Theorem 280 and (17.2.2), we see that the A of Theorem 329 is in fact

$$\{\zeta(2)\}^{-1} = 6\pi^{-2}.$$

$$= \tfrac{1}{2}\sum_{d=1}^{n}\mu(d)\left\{\frac{n^2}{d^2}+O\left(\frac{n}{d}\right)\right\}$$

$$= \tfrac{1}{2}n^2\sum_{d=1}^{n}\frac{\mu(d)}{d^2}+O\left(n\sum_{d=1}^{n}\frac{1}{d}\right)$$

$$= \tfrac{1}{2}n^2\sum_{d=1}^{\infty}\frac{\mu(d)}{d^2}+O\left(n^2\sum_{n+1}^{\infty}\frac{1}{d^2}\right)+O(n\log n)$$

$$= \frac{n^2}{2\zeta(2)}+O(n)+O(n\log n)=\frac{3n^2}{\pi^2}+O(n\log n),$$

by Theorem 287 and (17.2.2).

The number of terms in the Farey series $\mathfrak{F}_n$ is $\Phi(n)+1$, so that an alternative form of Theorem 330 is

THEOREM 331. *The number of terms in the Farey series of order n is approximately* $3n^2/\pi^2$.

Theorems 330 and 331 may be stated more picturesquely in the language of probability. Suppose that n is given, and consider all pairs of integers (p, q) for which

$$q > 0, \qquad 1 \leqslant p \leqslant q \leqslant n,$$

and the corresponding fractions p/q. There are

$$\psi_n = \tfrac{1}{2}n(n+1) \sim \tfrac{1}{2}n^2$$

such fractions, and χ_n, the number of them which are in their lowest terms, is $\Phi(n)$. If, as is natural, we define 'the probability that p and q are prime to one another' as

$$\lim_{n\to\infty}\frac{\chi_n}{\psi_n},$$

we obtain

THEOREM 332. *The probability that two integers should be prime to one another is* $6/\pi^2$.

18.6. The number of squarefree numbers. An allied problem is that of finding the probability that a number should be 'squarefree',† i.e. of determining approximately the number $Q(x)$ of squarefree numbers not exceeding x.

We can arrange all the positive integers $n \leqslant y^2$ in sets $S_1, S_2, \ldots$, such that S_d contains just those n whose largest square factor is d^2. Thus S_1 is the set of all squarefree $n \leqslant y^2$ The number of n belonging to S_d is

$$Q\left(\frac{y^2}{d^2}\right)$$

and, when $d > y, S_d$ is empty. Hence

$$[y^2] = \sum_{d \leqslant y} Q\left(\frac{y^2}{d^2}\right)$$

and so, by Theorem 268,

$$\begin{aligned} Q(y^2) &= \sum_{d \leqslant y} \mu(d) \left[\frac{y^2}{d^2}\right] = \sum_{d \leqslant y} \mu(d) \left(\frac{y^2}{d^2} + O(1)\right) \\ &= y^2 \sum_{d \leqslant y} \frac{\mu(d)}{d^2} + O(y) \\ &= y^2 \sum_{d=1}^{\infty} \frac{\mu(d)}{d^2} + O\left(y^2 \sum_{d>y} \frac{1}{d^2}\right) + O(y) \\ &= \frac{y^2}{\zeta(2)} + O(y) = \frac{6y^2}{\pi^2} + O(y). \end{aligned}$$

Replacing y^2 by x, we obtain

THEOREM 333. *The probability that a number should be squarefree is* $6/\pi^2$: *more precisely*

$$Q(x) = \frac{6x}{\pi^2} + O(\sqrt{x}).$$

† Without square factors, a product of different primes: see § 17.8.

A number n is squarefree if $\mu(n) = \pm 1$, or $|\mu(n)| = 1$. Hence an alternative statement of Theorem 333 is

THEOREM 334:

$$\sum_{n=1}^{x} |\mu(n)| = \frac{6x}{\pi^2} + O(\sqrt{x}).$$

It is natural to ask whether, among the squarefree numbers, those for which $\mu(n) = 1$ and those for which $\mu(n) = -1$ occur with about the same frequency. If they do so, then the sum

$$M(x) = \sum_{n=1}^{x} \mu(n)$$

should be of lower order than x; i.e.

THEOREM 335:

$$M(x) = o(x).$$

This is true, but we must defer the proof until § 22.17.

18.7. The order of $r(n)$. The function $r(n)$ behaves in some ways rather like $d(n)$, as is to be expected after Theorem 278 and (16.9.2). If $n \equiv 3 \pmod 4$, then $r(n) = 0$. If $n = (p_1 p_2 \ldots p_{l+1})^m$, and every p is $4k+1$, then $r(n) = 4d(n)$. In any case $r(n) \leqslant 4d(n)$. Hence we obtain the analogues of Theorems 313, 314, and 315, viz.

THEOREM 336:

$$\underline{\lim}\, r(n) = 0.$$

THEOREM 337:

$$r(n) = O\{(\log n)^{\Delta}\}$$

is false for every Δ.

THEOREM 338:

$$r(n) = O(n^{\delta})$$

for every positive δ.

There is also a theorem corresponding to Theorem 317; the maximum order of $r(n)$ is

$$2^{\frac{\log n}{\log\log n}}.$$

A difference appears when we consider the average order.

THEOREM 339. *The average order of* $r(n)$ *is* π; *i.e.*

$$\lim_{n\to\infty} \frac{r(1)+r(2)+\cdots+r(n)}{n} = \pi.$$

More precisely

$$r(1)+r(2)+\cdots+r(n) = \pi n + O(\sqrt{n}). \tag{18.7.1}$$

We can deduce this from Theorem 278, or prove it directly. The direct proof is simpler. Since $r(m)$, the number of solutions of $x^2+y^2=m$, is the number of lattice points of L on the circle $x^2+y^2=m$, the sum (18.7.1) is one less than the number of lattice points inside or on the circle $x^2+y^2=n$. If we associate with each such lattice point the lattice square of which it is the south-west corner, we obtain an area which is included in the circle

$$x^2+y^2 = (\sqrt{n}+\sqrt{2})^2$$

and includes the circle

$$x^2+y^2 = (\sqrt{n}-\sqrt{2})^2;$$

and each of these circles has an area $\pi n + O(\sqrt{n})$.

This geometrical argument may be extended to space of any number of dimensions. Suppose, for example, that $r_3(n)$ is the number of integral solutions of

$$x^2+y^2+z^2=n$$

(solutions differing only in sign or order being again regarded as distinct). Then we can prove

THEOREM 340:

$$r_3(1)+r_3(2)+\cdots+r_3(n) = \tfrac{4}{3}\pi n^{\frac{3}{2}} + O(n).$$

If we use Theorem 278, we have

$$\sum_{1\leqslant v\leqslant x} r(v) = 4\sum_{1}^{[x]}\sum_{d|v}\chi(d) = 4\sum_{1\leqslant uv\leqslant x}\chi(u),$$

the sum being extended over all the lattice points of the region **D** of § 18.2. If we write this in the form

$$4\sum_{1\leqslant u\leqslant x}\chi(u)\sum_{1\leqslant v\leqslant x/u} 1 = 4\sum_{1\leqslant u\leqslant x}\chi(u)\left[\frac{x}{u}\right],$$

we obtain

THEOREM 341:

$$\sum_{1\leqslant v\leqslant x} r(v) = 4\left(\left[\frac{x}{1}\right]-\left[\frac{x}{3}\right]+\left[\frac{x}{5}\right]-\cdots\right).$$

This formula is true whether x is an integer or not. If we sum separately over the regions *ADFY* and *DFX* of § 18.2, and calculate the second part of the sum by summing first along the horizontal lines of Fig. 8, we obtain

$$4\sum_{u\leqslant\surd x}\chi(u)\left[\frac{x}{u}\right]+4\sum_{v\leqslant\surd x}\sum_{\surd x<u\leqslant x/v}\chi(u).$$

The second sum is $O(\surd x)$, since $\sum\chi(u)$, between any limits, is 0 or ± 1, and

$$\sum_{u\leqslant\surd x}\chi(u)\left[\frac{x}{u}\right] = \sum_{u\leqslant\surd x}\chi(u)\frac{x}{u} + O(\surd x)$$

$$= x\left(1-\frac{1}{3}+\frac{1}{5}-\cdots+\frac{\chi([\surd x])}{[\surd x]}+O(\surd x)\right)$$

$$= x\left\{\tfrac{1}{4}\pi + O\left(\frac{1}{\surd x}\right)\right\} + O(\surd x) = \tfrac{1}{4}\pi x + O(\surd x).$$

This gives the result of Theorem 339.

NOTES

§ 18.1. For the proof of Theorem 315 see Pólya and Szegő, No. 264.

Theorem 317 is due to Wigert, *Arkiv för matematik*, 3, no. 18 (1907), 1–9 (Landau, *Handbuch*, 219–22). Wigert's proof depends upon the 'prime number theorem' (Theorem 6), but Ramanujan (*Collected papers*, 85–86) showed that it is possible to prove it in a more elementary way. Our proof is essentially Wigert's, modified so as not to require Theorem 6.

§ 18.2. Theorem 320 was proved by Dirichlet, *Abhandl. Akad. Berlin* (1849), 69–83 (*Werke*, ii. 49–66).

A great deal of work has been done since on the very difficult problem ('Dirichlet's divisor problem') of finding better bounds for the error in the approximation. Suppose that θ is the lower bound of numbers β such that

$$d(1) + d(2) + \cdots + d(n) = n \log n + (2\gamma - 1)n + O(n^{\beta}).$$

Theorem 320 shows that $\theta \leqslant \frac{1}{2}$. Voronöi proved in 1903 that $\theta \leqslant \frac{1}{3}$, and van der Corput in 1922 that $\theta < \frac{33}{100}$, and these numbers have been improved further by later writers. The current (2007) record is due to Huxley (*Proc. London Math. Soc.* (3) 87 (2003), 591–609) and states that $\theta \leqslant \frac{131}{416}$. On the other hand, Hardy and Landau proved independently in 1915 that $\theta \geqslant \frac{1}{4}$. The true value of θ is still unknown. See also the note on § 18.7.

As regards the sums $d^2(1) + \cdots + d^2(n)$, etc., see Ramanujan, *Collected papers*, 133–5, and B. M. Wilson, *Proc. London Math. Soc.* (2) 21 (1922), 235–55.

§ 18.3. Theorem 323 is due to Gronwall, *Trans. American Math. Soc.* 14 (1913), 113–22. Theorem 324 stands as stated here in Bachmann, *Analytische Zahlentheorie*, 402. The substance of it is contained in the memoir of Dirichlet referred to under § 18.2. The error term has been improved slightly to $O(n(\log n)^{2/3})$ by Walfisz, *Weylsche Exponentialsummen in der neueren Zahlentheorie* (Berlin, 1963). He similarly improved the error term in Theorem 330 to $O(n(\log n)^{2/3}(\log\log n)^{4/3})$.

§§ 18.4–5. Theorem 328 was proved by Landau, *Archiv d. Math. u. Phys.* (3) 5 (1903), 86–91 (*Handbuch*, 216–19); and Theorem 330 by Mertens, *Journal für Math.* 77 (1874), 289–338 (Landau, *Handbuch*, 578–9). Dirichlet (1849) proved a slightly weaker form of Theorem 330, i.e. with error $O(n^{1+\varepsilon})$ for any $\epsilon > 0$ (Dickson, *History*, i, 119).

§ 18.6. Theorem 333 is due to Gegenbauer, *Denkschriften Akad. Wien*, 49, Abt. 1 (1885), 37–80 (Landau, *Handbuch*, 580–2). The error term has been improved by various authors, the current (2007) record being $O(x^{\theta})$, for any $\theta > \frac{17}{54}$, due to Jia (*Sci. China Ser.* A 36 (1993), 154–169).

Landau [*Handbuch*, ii. 588–90] showed that Theorem 335 follows simply from the 'prime number theorem' (Theorem 6) and later [*Sitzungsberichte Akad. Wien*, 120, Abt. 2 (1911), 973–88] that Theorem 6 follows readily from Theorem 335. Mertens conjectured that $|M(x)| \leqslant x^{1/2}$ for all $x > 1$. However this was disproved by Odlyzko and te Riele (*J. Reine Angew. Math.* 357 (1985), 138–160), who showed in fact that there are infinitely many integral x for which $M(x) > \sqrt{x}$, and similarly for which $M(x) < -\sqrt{x}$. No specific example of such an $x > 1$ is known, and Odlyzko and te Riele suggest that there is no example below 10^{20}, or even 10^{30}.

§ 18.7. For Theorem 339 See Gauss, *Werke*, ii. 272–5.

This theorem, like Theorem 320, has been the starting-point of a great deal of modern work, the aim being the determination of the number θ corresponding to the θ of the note on § 18.2. The problem is very similar to the divisor problem, and the numbers $\frac{1}{2}, \frac{1}{3}, \frac{1}{4}$ occur in the same kind of way; but the analysis required is in some ways a little simpler. See

Landau, *Vorlesungen,* ii. 183–308. As with Theorem 320 the current (2007) record is due to Huxley (*Proc. London Math. Soc.* (3) 87 (2003), 591–609) and states again that $\theta \leqslant \frac{131}{416}$.

The error term in Theorem 340 has been investigated by a number of authors. The best known result up to 2007 is due to Health-Brown (*Number theory in progress*, Vol. 2, 883–92, (Berlin, 1999)), and states that the error is $O(n^{\theta})$ for any $\theta > \frac{21}{32}$.

Atkinson and Cherwell (*Quart. J. Math. Oxford,* 20 (1949), 65–79) give a general method of calculating the 'average order' of arithmetical functions belonging to a wide class. For deeper methods, see Wirsing (*Acta Math. Acad. Sci. Hungaricae* 18 (1967), 411–67) and Halász (*ibid.* 19 (1968), 365–403).

XIX

PARTITIONS

19.1. The general problem of additive arithmetic. In this and the next two chapters we shall be occupied with the additive theory of numbers. The general problem of the theory may be stated as follows.

Suppose that A or

$$a_1, a_2, a_3, \ldots$$

is a given system of integers. Thus A might contain all the positive integers, or the squares, or the primes. We consider all possible representations of an arbitrary positive integer n in the form

$$n = a_{i_1} + a_{i_2} + \cdots + a_{i_s},$$

where s may be fixed or unrestricted, the a may or may not be necessarily different, and order may or may not be relevant, according to the particular problem considered. We denote by $r(n)$ the number of such representations. Then what can we say about $r(n)$? For example, is $r(n)$ always positive? Is there always at any rate one representation of every n?

19.2. Partitions of numbers. We take first the case in which A is the set $1, 2, 3, \ldots$ of all positive integers, s is unrestricted, repetitions are allowed, and order is irrelevant. This is the problem of 'unrestricted partitions'.

A *partition* of a number n is a representation of n as the sum of any number of positive integral parts. Thus

$$\begin{aligned} 5 &= 4+1 = 3+2 = 3+1+1 = 2+2+1 \\ &= 2+1+1+1 = 1+1+1+1+1 \end{aligned}$$

has 7 partitions.† The order of the parts is irrelevant, so that we may, when we please, suppose the parts to be arranged in descending order of magnitude. We denote by $p(n)$ the number of partitions of n; thus $p(5) = 7$.

We can represent a partition graphically by an array of dots or 'nodes' such as

† We have, of course, to count the representation by one part only.

```
. . . . . . .
. . . .
. . .
. . .
.            A
```

the dots in a row corresponding to a part. Thus A represents the partition

$$7 + 4 + 3 + 3 + 1$$

of 18.

We might also read A by columns, in which case it would represent the partition

$$5 + 4 + 4 + 2 + 1 + 1 + 1$$

of 18. Partitions related in this manner are said to be *conjugate*.

A number of theorems about partitions follow immediately from this graphical representation. A graph with m rows, read horizontally, represents a partition into m parts; read vertically, it represents a partition into parts the largest of which is m. Hence

THEOREM 342. *The number of partitions of n into m parts is equal to the number of partitions of n into parts the largest of which is m.*

Similarly,

THEOREM 343. *The number of partitions of n into at most m parts is equal to the number of partitions of n into parts which do not exceed m.*

We shall make further use of 'graphical' arguments of this character, but usually we shall need the more powerful weapons provided by the theory of generating functions.

19.3. The generating function of $p(n)$. The generating functions which are useful here are power series[†]

$$F(x) = \sum f(n)x^n.$$

The sum of the series whose general coefficient is $f(n)$ is called the *generating function* of $f(n)$, and is said to *enumerate* $f(n)$.

[†] Compare § 17.10.

The generating function of $p(n)$ was found by Euler, and is

$$F(x) = \frac{1}{(1-x)(1-x^2)(1-x^3)\dots} = 1 + \sum_1^\infty p(n)x^n. \tag{19.3.1}$$

We can see this by writing the infinite product as

$$\begin{gathered}(1 + x + x^2 + \cdots)\\ (1 + x^2 + x^4 + \cdots)\\ (1 + x^3 + x^6 + \cdots)\\ \cdot\quad\cdot\quad\cdot\quad\cdot\end{gathered}$$

and multiplying the series together. Every partition of n contributes just 1 to the coefficient of x^n. Thus the partition

$$10 = 3 + 2 + 2 + 2 + 1$$

corresponds to the product of x^3 in the third row, $x^6 = x^{2+2+2}$ in the second, and x in the first; and this product contributes a unit to the coefficient of x^{10}.

This makes (19.3.1) intuitive, but (since we have to multiply an infinity of infinite series) some development of the argument is necessary.

Suppose that $0 < x < 1$, so that the product which defines $F(x)$ is convergent. The series

$$1 + x + x^2 + \cdots, \quad 1 + x^2 + x^4 + \cdots, \quad \cdots, \quad 1 + x^m + x^{2m} + \cdots,$$

are absolutely convergent, and we can multiply them together and arrange the result as we please. The coefficient of x^n in the product is

$$p_m(n),$$

the number of partitions of n into parts not exceeding m. Hence

$$F_m(x) = \frac{1}{(1-x)(1-x^2)\dots(1-x^m)} = 1 + \sum_{n=1}^\infty p_m(n)x^n. \tag{19.3.2}$$

It is plain that

$$p_m(n) \leqslant p(n), \tag{19.3.3}$$

that

(19.3.4) $$p_m(n) = p(n)$$

for $n \leqslant m$, and that

(19.3.5) $$p_m(n) \to p(n),$$

when $m \to \infty$, for every n. And

(19.3.6) $$F_m(x) = 1 + \sum_{n=1}^{m} p(n)x^n + \sum_{m+1}^{\infty} p_m(n)x^n.$$

The left-hand side is less than $F(x)$ and tends to $F(x)$ when $m \to \infty$.

Thus

$$1 + \sum_{n=1}^{m} p(n)x^n < F_m(x) < F(x),$$

which is independent of m. Hence $\sum p(n)x^n$ is convergent, and so, after (19.3.3), $\sum p_m(n)x^n$ converges, for any fixed x of the range $0 < x < 1$, uniformly for all values of m. Finally, it follows from (19.3.5) that

$$1 + \sum_{n=1}^{\infty} p(n)x^n = \lim_{m\to\infty} \left(1 + \sum_{n=1}^{\infty} p_m(n)x^n\right) = \lim_{m\to\infty} F_m(x) = F(x).$$

Incidentally, we have proved that

(19.3.7) $$\frac{1}{(1-x)(1-x^2)\dots(1-x^m)}$$

enumerates the partitions of n into parts which do not exceed m or (what is the same thing, after Theorem 343) into at most m parts.

We have written out the proof of the fundamental formula (19.3.1) in detail. We have proved it for $0 < x < 1$, and its truth for $|x| < 1$ follows at once from familiar theorems of analysis. In what follows we shall pay no attention to such 'convergence theorems',† since the interest of the subject-matter is essentially formal. The series and products with which we deal are all absolutely convergent for small x (and usually, as here, for $|x| < 1$).

† Except once in § 19.8, where again we are concerned with a fundamental identity, and once in § 19.9, where the limit process involved is less obvious.

The questions of convergence, identity, and so on, which arise are trivial, and can be settled at once by any reader who knows the elements of the theory of functions.

19.4. Other generating functions. It is equally easy to find the generating functions which enumerate the partitions of n into parts restricted in various ways. Thus

$$\frac{1}{(1-x)(1-x^3)(1-x^5)\dots} \tag{19.4.1}$$

enumerates partitions into *odd* parts;

$$\frac{1}{(1-x^2)(1-x^4)(1-x^6)\dots} \tag{19.4.2}$$

partitions into *even* parts;

$$(1+x)(1+x^2)(1+x^3)\dots \tag{19.4.3}$$

partitions into *unequal* parts;

$$(1+x)(1+x^3)(1+x^5)\dots \tag{19.4.4}$$

partitions into parts which are *both odd and unequal*; and

$$\frac{1}{(1-x)(1-x^4)(1-x^6)(1-x^9)\dots}, \tag{19.4.5}$$

where the indices are the numbers $5m+1$ and $5m+4$, partitions into parts each of which is of one of these forms.

Another function which will occur later is

$$\frac{x^N}{(1-x^2)(1-x^4)\dots(1-x^{2m})}. \tag{19.4.6}$$

This enumerates the partitions of $n-N$ into even parts not exceeding $2m$, or of $\frac{1}{2}(n-N)$ into parts not exceeding m; or again, after Theorem 343, the partitions of $\frac{1}{2}(n-N)$ into at most m parts.

Some properties of partitions may be deduced at once from the forms of these generating functions. Thus

$$(19.4.7)\qquad (1+x)(1+x^2)(1+x^3)\ldots = \frac{1-x^2}{1-x}\,\frac{1-x^4}{1-x^2}\,\frac{1-x^6}{1-x^3}\cdots$$
$$= \frac{1}{(1-x)(1-x^3)(1-x^5)\ldots}$$

Hence

THEOREM 344. *The number of partitions of n into unequal parts is equal to the number of its partitions into odd parts.*

It is interesting to prove this without the use of generating functions. Any number l can be expressed uniquely in the binary scale, i.e. as

$$l = 2^a + 2^b + 2^c + \cdots \qquad (0 \leqslant a < b < c \ldots).^{\dagger}$$

Hence a partition of n into odd parts can be written as

$$\begin{aligned} n &= l_1.1 + l_2.3 + l_3.5 + \cdots \\ &= (2^{a_1} + 2^{b_1} + \cdots)1 + (2^{a_2} + 2^{b_2} + \cdots)3 + (2^{a_3} + \cdots)5 + \cdots ; \end{aligned}$$

and there is a (1,1) correspondence between this partition and the partition into the unequal parts

$$2^{a_1}, 2^{b_1}, \ldots, 2^{a_2}.3, 2^{b_2}.3, \ldots, 2^{a_3}.5, 2^{b_3}.5, \ldots, \ldots .$$

19.5. Two theorems of Euler. There are two identities due to Euler which give instructive illustrations of different methods of proof used frequently in this theory.

THEOREM 345:

$$\begin{aligned} &(1+x)(1+x^3)(1+x^5)\ldots \\ &\quad = 1 + \frac{x}{1-x^2} + \frac{x^4}{(1-x^2)(1-x^4)} + \frac{x^9}{(1-x^2)(1-x^4)(1-x^6)} + \cdots . \end{aligned}$$

† This is the arithmetic equivalent of the identity

$$(1+x)(1+x^2)(1+x^4)(1+x^8)\ldots = \frac{1}{1-x}.$$

THEOREM 346:

$$(1+x^2)(1+x^4)(1+x^6)\dots$$
$$= 1 + \frac{x^2}{1-x^2} + \frac{x^6}{(1-x^2)(1-x^4)} + \frac{x^{12}}{(1-x^2)(1-x^4)(1-x^6)} + \cdots.$$

In Theorem 346 the indices in the numerators are 1.2, 2.3, 3.4, ...

(i) We first prove these theorems by Euler's device of the introduction of a second parameter a.

Let

$$K(a) = K(a,x) = (1+ax)(1+ax^3)(1+ax^5)\dots$$
$$= 1 + c_1 a + c_2 a^2 + \dots,$$

where $c_n = c_n(x)$ is independent of a. Plainly

$$K(a) = (1+ax)K(ax^2)$$

or

$$1 + c_1 a + c_2 a^2 + \cdots = (1+ax)(1 + c_1 ax^2 + c_2 a^2 x^4 + \cdots).$$

Hence, equating coefficients, we obtain

$$c_1 = x + c_1 x^2, c_2 = c_1 x^3 + c_2 x^4, \dots, c_m = c_{m-1}x^{2m-1} + c_m x^{2m}, \dots,$$

and so

$$c_m = \frac{x^{2m-1}}{1-x^{2m}} c_{m-1} = \frac{x^{1+3+\cdots+(2m-1)}}{(1-x^2)(1-x^4)\dots(1-x^{2m})}$$
$$= \frac{x^{m^2}}{(1-x^2)(1-x^4)\dots(1-x^{2m})}.$$

It follows that

$$(1+ax)(1+ax^3)(1+ax^5)\dots \tag{19.5.1}$$
$$= 1 + \frac{ax}{1-x^2} + \frac{a^2x^4}{(1-x^2)(1-x^4)} + \dots,$$

and Theorems 345 and 346 are the special cases $a = 1$ and $a = x$.

(ii) The theorems can also be proved by arguments independent of the theory of infinite series. Such proofs are sometimes described as 'combinatorial'. We select Theorem 345.

We have seen that the left-hand side of the identity enumerates partitions into odd and unequal parts: thus

$$15 = 11 + 3 + 1 = 9 + 5 + 1 = 7 + 5 + 3$$

has 4 such partitions. Let us take, for example, the partition 11+3+1, and represent it graphically as in B, the points on one bent line corresponding to a part of the partition.

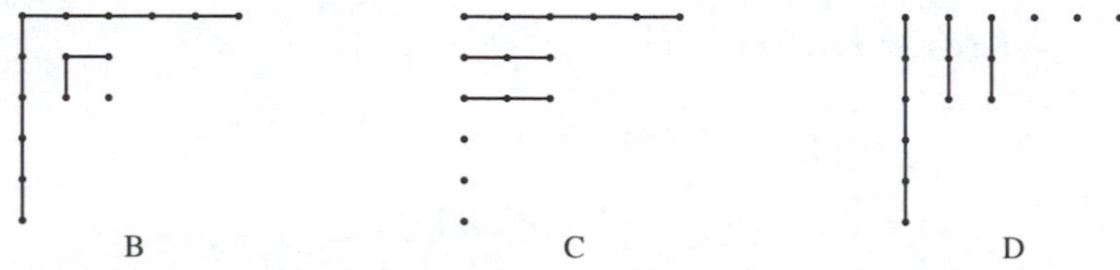

We can also read the graph (considered as an array of points) as in C or D, along a series of horizontal or vertical lines. The graphs C and D differ only in orientation, and each of them corresponds to another partition of 15, viz. 6+3+3+1+1+1. A partition like this, symmetrical about the south-easterly direction, is called by Macmahon a *self-conjugate* partition, and the graphs establish a (1,1) correspondence between self-conjugate partitions and partitions into odd and unequal parts. The left-hand side of the identity enumerates odd and unequal partitions, and therefore the identity will be proved if we can show that its right-hand side enumerates self-conjugate partitions.

Now our array of points may be read in a fourth way, viz. as in E.

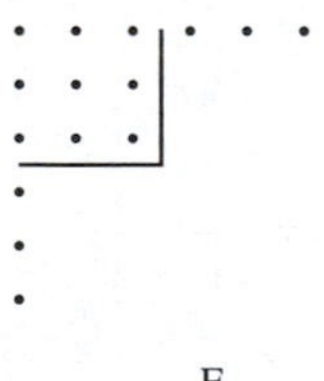

E

Here we have a square of 3^2 points, and two 'tails', each representing a partition of $\frac{1}{2}(15 - 3^2) = 3$ into 3 parts at most (and in this particular case all 1's). Generally, a self-conjugate partition of n can be read as a square of m^2 points, and two tails representing partitions of

$$\tfrac{1}{2}(n - m^2)$$

into m parts at most. Given the (self-conjugate) partition, then m and the reading of the partition are fixed; conversely, given n, and given any square m^2 not exceeding n, there is a group of self-conjugate partitions of n based upon a square of m^2 points.

Now

$$\frac{x^{m^2}}{(1-x^2)(1-x^4)\dots(1-x^{2m})}$$

is a special case of (19.4.6), and enumerates the number of partitions of $\frac{1}{2}(n-m^2)$ into at most m parts, and each of these corresponds as we have seen to a self-conjugate partition of n based upon a square of m^2 points. Hence, summing with respect to m,

$$1+\sum_{1}^{\infty}\frac{x^{m^2}}{(1-x^2)(1-x^4)\dots(1-x^{2m})}$$

enumerates all self-conjugate partitions of n, and this proves the theorem.

Incidentally, we have proved

THEOREM 346. *The number of partitions of n into odd and unequal parts is equal to the number of its self-conjugate partitions.*

Our argument suffices to prove the more general identity (19.5.1), and show its combinatorial meaning. The number of partitions of n into just m odd and unequal parts is equal to the number of self-conjugate partitions of n based upon a square of m^2 points. The effect of putting $a = 1$ is to obliterate the distinction between different values of m.

The reader will find it instructive to give a combinatorial proof of Theorem 346. It is best to begin by replacing x^2 by x, and to use the decomposition $1+2+3+\cdots+m$ of $\frac{1}{2}m(m+1)$. The square of (ii) is replaced by an isosceles right-angled triangle.

19.6. Further algebraical identities. We can use the method (i) of § 19.5 to prove a large number of algebraical identities. Suppose, for example, that

$$K_j(a) = K_j(a,x) = (1+ax)(1+ax^2)\dots(1+ax^j) = \sum_{m=0}^{j} c_m a^m.$$

Then

$$\left(1+ax^{j+1}\right)K_j(a)=(1+ax)K_j(ax).$$

Inserting the power series, and equating the coefficients of a^m, we obtain

$$c_m+c_{m-1}x^{j+1}=(c_m+c_{m-1})x^m$$

or

$$(1-x^m)c_m=(x^m-x^{j+1})c_{m-1}=x^m(1-x^{j-m+1})c_{m-1},$$

for $1\leqslant m\leqslant j$. Hence

THEOREM 348:

$$(1+ax)(1+ax^2)\dots(1+ax^j)=1+ax\frac{1-x^j}{1-x}+a^2x^3\frac{(1-x^j)(1-x^{j-1})}{(1-x)(1-x^2)}+$$
$$+\cdots+a^mx^{\frac{1}{2}m(m+1)}\frac{(1-x^j)\dots(1-x^{j-m+1})}{(1-x)\dots(1-x^m)}+\cdots+a^jx^{\frac{1}{2}j(j+1)}.$$

If we write x^2 for x, $1/x$ for a, and make $j\to\infty$, we obtain Theorem 345. Similarly we can prove

THEOREM 349:

$$\frac{1}{(1-ax)(1-ax^2)\dots(1-ax^j)}=1+ax\frac{1-x^j}{1-x}$$
$$+a^2x^2\frac{(1-x^j)(1-x^{j+1})}{(1-x)(1-x^2)}+\cdots.$$

In particular, if we put $a=1$, and make $j\to\infty$, we obtain

THEOREM 350:

$$\frac{1}{(1-x)(1-x^2)\dots}=1+\frac{x}{1-x}+\frac{x^2}{(1-x)(1-x^2)}+\cdots.$$

19.7. Another formula for $F(x)$. As a further example of 'combinatorial' reasoning we prove another theorem of Euler, viz.

THEOREM 351:

$$\frac{1}{(1-x)(1-x^2)(1-x^3)\ldots} = 1 + \frac{x}{(1-x)^2} + \frac{x^4}{(1-x)^2(1-x^2)^2} + \frac{x^9}{(1-x)^2(1-x^2)^2(1-x^3)^2} + \ldots.$$

The graphical representation of any partition, say

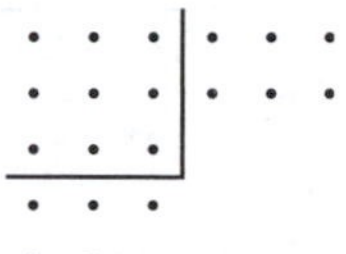

F

contains a square of nodes in the north-west corner. If we take the largest such square, called the 'Durfee square' (here a square of 9 nodes), then the graph consists of a square containing i^2 nodes and two tails; one of these tails represents the partition of a number, say l, into not more than i parts, the other the partition of a number, say m, into parts not exceeding i; and

$$n = i^2 + l + m.$$

In the figure $n = 20$, $i = 3$, $l = 6$, $m = 5$.

The number of partitions of l (into at most i parts) is, after § 19.3, the coefficient of x^l in

$$\frac{1}{(1-x)(1-x^2)\ldots(1-x^i)},$$

and the number of partitions of m (into parts not exceeding i) is the coefficient of x^m in the same expansion. Hence the coefficient of x^{n-i^2} in

$$\left\{\frac{1}{(1-x)(1-x^2)\ldots(1-x^i)}\right\}^2,$$

or of x^n in

$$\frac{x^{i^2}}{(1-x)^2(1-x^2)^2\dots(1-x^i)^2},$$

is the number of possible pairs of tails in a partition of n in which the Durfee square is i^2. And hence the total number of partitions of n is the coefficient of x^n in the expansion of

$$1+\frac{x}{(1-x)^2}+\frac{x^4}{(1-x)^2(1-x^2)^2}+\cdots$$
$$+\frac{x^{i^2}}{(1-x)^2(1-x^2)^2\dots(1-x^i)^2}+\cdots.$$

This proves the theorem.

There are also simple algebraical† proofs.

19.8. A theorem of Jacobi. We shall require later certain special cases of a famous identity which belongs properly to the theory of elliptic functions.

THEOREM 352. *If* $|x|<1$, *then*

$$\prod_{n=1}^{\infty}\{(1-x^{2n})(1+x^{2n-1}z)(1+x^{2n-1}z^{-1})\} \tag{19.8.1}$$
$$=1+\sum_{n=1}^{\infty}x^{n^2}(z^n+z^{-n})=\sum_{-\infty}^{\infty}x^{n^2}z^n$$

for all z *except* $z=0$.

The two forms of the series are obviously equivalent.

Let us write

$$P(x,z)=Q(x)R(x,z^{-1}),$$

† We use the word 'algebraical' in its old-fashioned sense, in which it includes elementary manipulation of power series or infinite products. Such proofs involve (though sometimes only superficially) the use of limiting processes, and are, in the strict sense of the word, 'analytical'; but the word 'analytical' is usually reserved, in the theory of numbers, for proofs which depend upon analysis of a deeper kind (usually upon the theory of functions of a complex variable).

where

$$Q(x) = \prod_{n=1}^{\infty}(1 - x^{2n}), \quad R(x,z) = \prod_{n=1}^{\infty}(1 + x^{2n-1}z).$$

When $|x| < 1$ and $z \neq 0$, the infinite products

$$\prod_{n=1}^{\infty}(1 + |x|^{2n}), \quad \prod_{n=1}^{\infty}(1 + |x^{2n-1}z|), \quad \prod_{n=1}^{\infty}(1 + |x^{2n-1}z^{-1}|)$$

are all convergent. Hence the products $Q(x)$, $R(x,z)$, $R(x,z^{-1})$ and the product $P(x,z)$ may be formally multiplied out and the resulting terms collected and arranged in any way we please; the resulting series is absolutely convergent and its sum is equal to $P(x,z)$. In particular,

$$P(x,z) = \sum_{n=-\infty}^{\infty} a_n(x)z^n,$$

where $a_n(x)$ does not depend on z and

(19.8.2) $$a_{-n}(x) = a_n(x).$$

Provided $x \neq 0$, we can easily verify that

$$(1 + xz)R(x, zx^2) = R(x,z), \quad R(x,z^{-1}x^{-2}) = (1 + z^{-1}x^{-1})R(x,z^{-1}),$$

so that $xzP(x, zx^2) = P(x,z)$. Hence

$$\sum_{n=-\infty}^{\infty} x^{2n+1}a_n(x)z^{n+1} = \sum_{n=-\infty}^{\infty} a_n(x)z^n.$$

Since this is true for all values of z (except $z = 0$) we can equate the coefficients of z^n and find that $a_{n+1}(x) = x^{2n+1}a_n(x)$. Thus, for $n \geqslant 0$, we have

$$a_{n+1}(x) = x^{(2n+1)+(2n-1)+\cdots+1}a_0(x) = x^{(n+1)^2}a_0(x).$$

By (19.8.2) the same is true when $n+1 < 0$ and so $a_n(x) = x^{n^2}a_0(x)$ for all n, provided $x \neq 0$. But, when $x = 0$, the result is trivial. Hence

(19.8.3) $$P(x,z) = a_0(x)S(x,z),$$

where

$$S(x,z) = \sum_{n=-\infty}^{\infty} x^{n^2} z^n.$$

To complete the proof of the theorem, we have to show that $a_0(x) = 1$.

If z has any fixed value other than zero and if $|x| < \frac{1}{2}$ (say), the products $Q(x)$, $R(x,z)$, $R(x,z^{-1})$ and the series $S(x,z)$ are all uniformly convergent with respect to x. Hence $P(x,z)$ and $S(x,z)$ represent continuous functions of x and, as $x \to 0$,

$$P(x,z) \to P(0,z) = 1, \quad S(x,z) \to S(0,z) = 1.$$

It follows from (19.8.3) that $a_0(x) \to 1$ as $x \to 0$.

Putting $z = i$, we have

(19.8.4) $$S(x,i) = 1 + 2\sum_{n=1}^{\infty} (-1)^n x^{4n^2} = S(x^4, -1).$$

Again

$$R(x,i)R(x,i^{-1}) = \prod_{n=1}^{\infty} \left\{(1 + ix^{2n-1})(1 - ix^{2n-1})\right\} = \prod_{n=1}^{\infty} (1 + x^{4n-2}),$$

$$Q(x) = \prod_{n=1}^{\infty} (1 - x^{2n}) = \prod_{n=1}^{\infty} \left\{(1 - x^{4n})(1 - x^{4n-2})\right\},$$

and so

(19.8.5) $$P(x,i) = \prod_{n=1}^{\infty} \left\{(1 - x^{4n})(1 - x^{8n-4})\right\}$$
$$= \prod_{n=1}^{\infty} \left\{(1 - x^{8n})(1 - x^{8n-4})^2\right\} = P(x^4, -1).$$

Clearly $P(x^4, -1) \neq 0$, and so it follows from (19.8.3), (19.8.4), and (19.8.5) that $a_0(x) = a_0(x^4)$. Using this repeatedly with $x^4, x^{4^2}, x^{4^3}, \ldots$ replacing x, we have

$$a_0(x) = a_0(x^4) = \ldots = a_0(x^{4^k})$$

for any positive integer k. But $|x| < 1$ and so $x^{4^k} \to 0$ as $k \to \infty$. Hence

$$a_0(x) = \lim_{x \to 0} a_0(x) = 1.$$

This completes the proof of Theorem 352.

19.9. Special cases of Jacobi's identity. If we write x^k for x, $-x^l$ and x^l for z, and replace n by $n+1$ on the left-hand side of (19.8.1), we obtain

(19.9.1)
$$\prod_{n=0}^{\infty} \{(1 - x^{2kn+k-l})(1 - x^{2kn+k+l})(1 - x^{2kn+2k})\} = \sum_{n=-\infty}^{\infty} (-1)^n x^{kn^2+ln},$$

(19.9.2)
$$\prod_{n=0}^{\infty} \{(1 + x^{2kn+k-l})(1 - x^{2kn+k+l})(1 - x^{2kn+2k})\} = \sum_{n=-\infty}^{\infty} x^{kn^2+ln}.$$

Some special cases are particularly interesting.

(i) $k = 1, l = 0$ gives

$$\prod_{n=0}^{\infty} \{(1 - x^{2n+1})^2(1 - x^{2n+2})\} = \sum_{n=-\infty}^{\infty} (-1)^n x^{n^2},$$

$$\prod_{n=0}^{\infty} \{(1 + x^{2n+1})^2(1 - x^{2n+2})\} = \sum_{n=-\infty}^{\infty} x^{n^2},$$

two standard formulae from the theory of elliptic functions.

(ii) $k = \frac{3}{2}, l = \frac{1}{2}$ in (19.9.1) gives

$$\prod_{n=0}^{\infty} \{(1 - x^{3n+1})(1 - x^{3n+2})(1 - x^{3n+3})\} = \sum_{n=-\infty}^{\infty} (-1)^n x^{\frac{1}{2}n(3n+1)}$$

or

THEOREM 353:

$$(1-x)(1-x^2)(1-x^3)\ldots = \sum_{n=-\infty}^{\infty} (-1)^n x^{\frac{1}{2}n(3n+1)}.$$

This famous identity of Euler may also be written in the form

(19.9.3)
$$(1-x)(1-x^2)(1-x^3)\ldots$$
$$= 1 + \sum_{n=1}^{\infty} (-1)^n \left\{ x^{\frac{1}{2}n(3n-1)} + x^{\frac{1}{2}n(3n+1)} \right\}$$
$$= 1 - x - x^2 + x^5 + x^7 - x^{12} - x^{15} + \ldots.$$

(iii) $k = l = \frac{1}{2}$ in (19.9.2) gives

$$\prod_{n=0}^{\infty} \{(1+x^n)(1-x^{2n+2})\} = \sum_{n=-\infty}^{\infty} x^{\frac{1}{2}n(n+1)},$$

which may be transformed, by use of (19.4.7), into

THEOREM 354:

$$\frac{(1-x^2)(1-x^4)(1-x^6)\ldots}{(1-x)(1-x^3)(1-x^5)\ldots} = 1 + x + x^3 + x^6 + x^{10} + \ldots.$$

Here the indices on the right are the triangular numbers.†

(iv) $k = \frac{5}{2}, l = \frac{3}{2}$ and $k = \frac{5}{2}, l = \frac{1}{2}$ in (19.9.1) give

THEOREM 355:

$$\prod_{n=0}^{\infty} \{(1-x^{5n+1})(1-x^{5n+4})(1-x^{5n+5})\} = \sum_{n=-\infty}^{\infty} (-1)^n x^{\frac{1}{2}n(5n+3)}.$$

THEOREM 356:

$$\prod_{n=0}^{\infty} \{(1-x^{5n+2})(1-x^{5n+3})(1-x^{5n+5})\} = \sum_{n=-\infty}^{\infty} (-1)^n x^{\frac{1}{2}n(5n+1)}.$$

† The numbers $\frac{1}{2}n(n+1)$.

We shall require these formulae later.

As a final application, we replace x by $x^{\frac{1}{2}}$ and z by $x^{\frac{1}{2}}\zeta$ in (19.8.1). This gives

$$\prod_{n=1}^{\infty}\{(1-x^n)(1+x^n\zeta)(1+x^{n-1}\zeta^{-1})\} = \sum_{n=-\infty}^{\infty} x^{\frac{1}{2}n(n+1)}\zeta^n$$

or

$$(1+\zeta^{-1})\prod_{n=1}^{\infty}\{(1-x^n)(1+x^n\zeta)(1+x^{n-1}\zeta^{-1})\}$$
$$= \sum_{n=0}^{\infty}(\zeta^m+\zeta^{-m-1})x^{\frac{1}{2}m(m+1)},$$

where on the right-hand side we have combined the terms which correspond to $n = m$ and $n = -m-1$. We deduce that

$$\prod_{n=1}^{\infty}\{(1-x^n)(1+x^n\zeta)(1+x^n\zeta^{-1})\} \tag{19.9.4}$$
$$= \sum_{m=0}^{\infty}\zeta^{-m}\left(\frac{1+\zeta^{2m+1}}{1+\zeta}\right)x^{\frac{1}{2}m(m+1)}$$
$$= \sum_{m=0}^{\infty}x^{\frac{1}{2}m(m+1)}\zeta^{-m}(1-\zeta+\zeta^2-\cdots+\zeta^{2m})$$

for all ζ except $\zeta = 0$ and $\zeta = -1$. We now suppose the value of x fixed and that ζ lies in the closed interval $-\frac{3}{2} \leqslant \zeta \leqslant -\frac{1}{2}$. The infinite product on the left and the infinite series on the right of (19.9.4) are then uniformly convergent with respect to ζ. Hence each represents a continuous function of ζ in this interval and we may let $\zeta \to -1$.
We have then

THEOREM 357:

$$\prod_{n=1}^{\infty}(1-x^n)^3 = \sum_{m=0}^{\infty}(-1)^m(2m+1)x^{\frac{1}{2}m(m+1)}.$$

This is another famous theorem of Jacobi.

19.10. Applications of Theorem 353. Euler's identity (19.9.3) has a striking combinatorial interpretation. The coefficient of x^n in

$$(1-x)(1-x^2)(1-x^3)\dots$$

is

$$\sum (-1)^{\nu}, \tag{19.10.1}$$

where the summation is extended over all partitions of n into unequal parts, and ν is the number of parts in such a partition. Thus the partition 3+2+1 of 6 contributes $(-1)^3$ to the coefficient of x^6. But (19.10.1) is $E(n)-U(n)$, where $E(n)$ is the number of partitions of n into an even number of unequal parts, and $U(n)$ that into an odd number. Hence Theorem 353 may be restated as

THEOREM 358. *$E(n)=U(n)$ except when $n=\frac{1}{2}k(3k\pm 1)$, when*

$$E(n)-U(n)=(-1)^k.$$

Thus

$$7=6+1=5+2=4+3=4+2+1,$$

$$E(7)=3,\quad U(7)=2,\quad E(7)-U(7)=1,$$

and

$$7=\tfrac{1}{2}\,.\,2\,.\,(3\,.\,2+1),\quad k=2.$$

The identity may be used effectively for the calculation of $p(n)$. For

$$(1-x-x^2+x^5+x^7-\dots)\left\{1+\sum_{1}^{\infty}p(n)x^n\right\}$$

$$=\frac{1-x-x^2+x^5+x^7-\dots}{(1-x)(1-x^2)(1-x^3)\dots}=1.$$

Hence, equating coefficients,

(19.10.2)

$$p(n)-p(n-1)-p(n-2)+p(n-5)+\dots$$
$$+(-1)^k p\{n-\tfrac{1}{2}k(3k-1)\}+(-1)^k p\{n-\tfrac{1}{2}k(3k+1)\}+\dots=0.$$

The number of terms on the left is about $2\surd(\frac{2}{3}n)$ for large n.

Macmahon used (19.10.2) to calculate $p(n)$ up to $n = 200$, and found that

$$p(200) = 3972999029388.$$

19.11. Elementary proof of Theorem 358. There is a very beautiful proof of Theorem 358, due to Franklin, which uses no algebraical machinery.

We try to establish a (1,1) correspondence between partitions of the two sorts considered in § 19.10. Such a correspondence naturally cannot be exact, since an exact correspondence would prove that $E(n) = U(n)$ for all n.

We take a graph G representing a partition of n into any number of unequal parts, in descending order. We call the lowest line AB

(which may contain one point only) the 'base' β of the graph. From C, the extreme north-east node, we draw the longest south-westerly line possible in the graph; this also may contain one node only. This line CDE we call the 'slope' σ of the graph. We write $\beta < \sigma$ when, as in graph G, there are more nodes in σ than in β, and use a similar notation in other cases. Then there are three possibilities.

(*a*) $\beta < \sigma$. We move β into a position parallel to and outside σ, as shown in graph H. This gives a new partition into decreasing unequal parts, and into a number of such parts whose parity is opposite to that of the number in G. We call this operation O, and the converse operation (removing σ and placing it below β) Ω. It is plain that Ω is not possible, when $\beta < \sigma$, without violating the conditions of the graph.

(*b*) $\beta = \sigma$. In this case O is possible (as in graph I) unless β meets σ (as in graph J), when it is impossible. Ω is not possible in either case.

(*c*) $\beta > \sigma$. In this case O is always impossible. Ω is possible (as in graph K) unless β meets σ and $\beta = \sigma+1$ (as in graph L). Ω is impossible in the last case because it would lead to a partition with two equal parts.

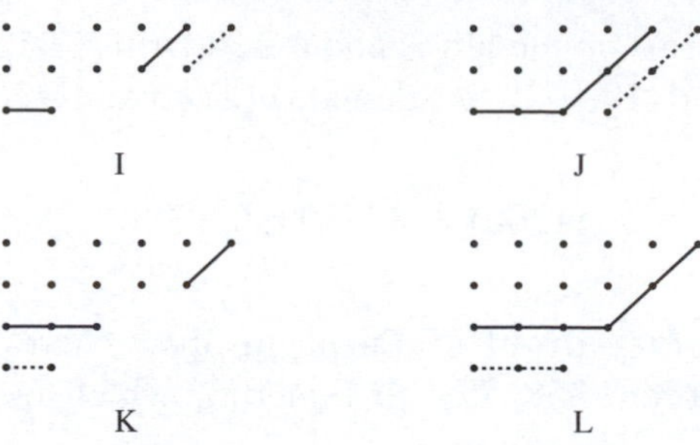

To sum up, there is a (1, 1) correspondence between the two types of partitions except in the cases exemplified by J and L. In the first of these exceptional cases n is of the form

$$k + (k+1) + \cdots + (2k-1) = \tfrac{1}{2}(3k^2 - k),$$

and in this case there is an excess of one partition into an even number of parts, or one into an odd number, according as k is even or odd. In the second case n is of the form

$$(k+1) + (k+2) + \cdots + 2k = \tfrac{1}{2}(3k^2 + k),$$

and the excess is the same. Hence $E(n) - U(n)$ is 0 unless $n = \frac{1}{2}(3k^2 \pm k)$, when $E(n) - U(n) = (-1)^k$. This is Euler's theorem.

19.12. Congruence properties of $p(n)$. In spite of the simplicity of the definition of $p(n)$, not very much is known about its arithmetic properties.

The simplest arithmetic properties known were found by Ramanujan. Examining Macmahon's table of $p(n)$, he was led first to conjecture, and then to prove, three striking arithmetic properties associated with the moduli 5, 7, and 11. No analogous results are known to modulus 2 or 3, although Newman has found some further results to modulus 13.

THEOREM 359:

$$p(5m+4) \equiv 0 \pmod{5}.$$

THEOREM 360:

$$p(7m+5) \equiv 0 \pmod{7}.$$

THEOREM 361*:

$$p(11m+6) \equiv 0 \pmod{11}.$$

We give here a proof of Theorem 359. Theorem 360 may be proved in the same kind of way, but Theorem 361 is more difficult.

By Theorems 353 and 357,

$$\begin{aligned} x\{(1-x)(1-x^2)\ldots\}^4 &= x(1-x)(1-x^2)\ldots\{(1-x)(1-x^2)\ldots\}^3 \\ &= x(1-x-x^2+x^5+\ldots) \\ &\quad\times(1-3x+5x^3-7x^6+\ldots) \\ &= \sum_{r=-\infty}^{\infty}\sum_{s=0}^{\infty}(-1)^{r+s}(2s+1)x^k, \end{aligned}$$

where

$$k = k(r,s) = 1+\tfrac{1}{2}r(3r+1)+\tfrac{1}{2}s(s+1).$$

We consider in what circumstances k is divisible by 5.

Now

$$2(r+1)^2+(2s+1)^2 = 8k-10r^2-5 \equiv 8k \pmod{5}.$$

Hence $k \equiv 0 \pmod{5}$ implies

$$2(r+1)^2+(2s+1)^2 \equiv 0 \pmod{5}.$$

Also

$$2(r+1)^2 \equiv 0, 2, \text{ or } 3, \quad (2s+1)^2 \equiv 0, 1, \text{ or } 4 \pmod{5},$$

and we get 0 on addition only if $2(r+1)^2$ and $(2s+1)^2$ are each divisible by 5. Hence k can be divisible by 5 only if $2s+1$ is divisible by 5, and thus *the coefficient of* x^{5m+5} *in*

$$x\{(1-x)(1-x^2)\ldots\}^4$$

is divisible by 5.

Next, in the binomial expansion of $(1-x)^{-5}$, all the coefficients are divisible by 5, except those of $1, x^5, x^{10},\ldots$, which have the remainder 1.[†] We may express this by writing

$$\frac{1}{(1-x)^5} \equiv \frac{1}{1-x^5} \pmod{5};$$

[†] Theorem 76 of Ch. VI.

the notation, which is an extension of that used for polynomials in § 7.2, implying that the coefficients of every power of x are congruent. It follows that

$$\frac{1-x^5}{(1-x)^5} \equiv 1 \pmod{5}$$

and

$$\frac{(1-x^5)(1-x^{10})(1-x^{15})\dots}{\{(1-x)(1-x^2)(1-x^3)\dots\}^5} \equiv 1 \pmod{5}.$$

Hence the coefficient of x^{5m+5} in

$$x\frac{(1-x^5)(1-x^{10})\dots}{(1-x)(1-x^2)\dots} = x\{(1-x)(1-x^2)\dots\}^4 \frac{(1-x^5)(1-x^{10})\dots}{\{(1-x)(1-x^2)\dots\}^5}$$

is a multiple of 5. Finally, since

$$\frac{x}{(1-x)(1-x^2)\dots} = x\frac{(1-x^5)(1-x^{10})\dots}{(1-x)(1-x^2)\dots} \times (1+x^5+x^{10}+\dots)(1+x^{10}+x^{20}+\dots)\dots,$$

the coefficient of x^{5m+5} in

$$\frac{x}{(1-x)(1-x^2)(1-x^3)\dots} = x + \sum_2^\infty p(n-1)x^n$$

is a multiple of 5; and this is Theorem 359.

The proof of Theorem 360 is similar. We use the square of Jacobi's series $1-3x+5x^3-7x^6+\dots$ instead of the product of Euler's and Jacobi's series.

There are also congruences to moduli 5^2, 7^2, and 11^2, such as

$$p(25m+24) \equiv 0 \pmod{5^2}.$$

Ramanujan made the general conjecture that *if*

$$\delta = 5^a 7^b 11^c,$$

and

$$24n \equiv 1 \pmod{\delta},$$

then

$$p(n) \equiv 0 \pmod{\delta}.$$

It is only necessary to consider the cases $\delta = 5^a$, 7^b, 11^c, since all others would follows as corollaries.

Ramanujan proved the congruences for 5^2, 7^2, 11^2, Krečmar that for 5^3, and Watson that for general 5^a. But Gupta, in extending Macmahon's table up to 300, found that

$$p(243) = 133978259344888$$

is not divisible by $7^3 = 343$; and, since $24 \,.\, 243 \equiv 1 \pmod{343}$, this contradicts the conjecture for 7^3. The conjecture for 7^b had therefore to be modified, and Watson found and proved the appropriate modification, viz. that $p(n) \equiv 0 \pmod{7^b}$ if $b > 1$ and $24n \equiv 1 \pmod{7^{2b-2}}$.

D. H. Lehmer used a quite different method based upon the analytic theory of Hardy and Ramanujan and of Rademacher to calculate $p(n)$ for particular n. By this means he verified the truth of the conjecture for the first values of n associated with the moduli 11^3 and 11^4. Subsequently Lehner proved the conjecture for 11^3 and Atkin for general 11^c.

Dyson conjectured and Atkin and Swinnerton-Dyer proved certain remarkable results from which Theorems 359 and 360, but not 361, are immediate corollaries. Thus, let us define the *rank* of a partition as the largest part minus the number of parts, so that, for example, the rank of a partition and that of the conjugate partition differ only in sign. Next we arrange the partitions of a number in five classes, each class containing the partitions whose rank has the same residue (mod 5). Then, if $n \equiv 4$ (mod 5), the number of partitions in each of the five classes is the same and Theorem 359 is an immediate corollary. There is a similar result leading to Theorem 360.

19.13. The Rogers–Ramanujan identities. We end this chapter with two theorems which resemble Theorems 345 and 346 superficially, but are much more difficult to prove. These are

THEOREM 362:

$$1 + \frac{x}{1-x} + \frac{x^4}{(1-x)(1-x^2)} + \frac{x^9}{(1-x)(1-x^2)(1-x^3)} + \cdots$$
$$= \frac{1}{(1-x)(1-x^6)\ldots(1-x^4)(1-x^9)\ldots},$$

i.e.

$$1+\sum_{1}^{\infty}\frac{x^{m^2}}{(1-x)(1-x^2)\dots(1-x^m)} = \prod_{0}^{\infty}\frac{1}{(1-x^{5m+1})(1-x^{5m+4})}. \tag{19.13.1}$$

THEOREM 363:

$$1+\frac{x^2}{1-x}+\frac{x^6}{(1-x)(1-x^2)}+\frac{x^{12}}{(1-x^2)(1-x^3)}+\dots = \frac{1}{(1-x^2)(1-x^7)\dots(1-x^3)(1-x^8)\dots},$$

i.e.

$$1+\sum_{1}^{\infty}\frac{x^{m(m+1)}}{(1-x)(1-x^2)\dots(1-x^m)} = \prod_{0}^{\infty}\frac{1}{(1-x^{5m+2})(1-x^{5m+3})}. \tag{19.13.2}$$

The series here differ from those in Theorems 345 and 346 only in that x^2 is replaced by x in the denominators. The peculiar interest of the formulae lies in the unexpected part played by the number 5.

We observe first that the theorems have, like Theorems 345 and 346, a combinatorial interpretation. Consider Theorem 362, for example. We can exhibit any square m^2 as

$$m^2 = 1+3+5+\dots+(2m-1)$$

or as shown by the black dots in the graph M, in which $m=4$. If we now take any partition of $n-m^2$ into m parts at most, with the parts in descending order, and add it to the graph, as shown by the circles of M, where $m=4$ and $n=4^2+11=27$, we obtain a partition of n (here $27=11+8+6+2$) into parts without repetitions or sequences, or *parts whose minimal difference is* 2. The left-hand side of (19.13.1) enumerates this type of partition of n.

```
• • • • • • • ○ ○ ○ ○
• • • • • ○ ○ ○
• • • ○ ○ ○
• ○
```

M

On the other hand, the right-hand side enumerates partitions into numbers of the forms $5m + 1$ and $5m + 4$. Hence Theorem 362 may be restated as a purely 'combinatorial' theorem, viz.

THEOREM 364. *The number of partitions of n with minimal difference* 2 *is equal to the number of partitions into parts of the forms* $5m + 1$ *and* $5m + 4$.

Thus, when $n = 9$, there are 5 partitions of each type,

$$9, \quad 8+1, \quad 7+2, \quad 6+3, \quad 5+3+1$$

of the first kind, and

$$9, \quad 6+1+1+1, \quad 4+4+1, \quad 4+1+1+1+1+1, \\ 1+1+1+1+1+1+1+1+1$$

of the second.

Similarly, the combinatorial equivalent of Theorem 363 is

THEOREM 365. *The number of partitions of n into parts not less than* 2, *and with minimal difference* 2, *is equal to the number of partitions of n into parts of the forms* $5m + 2$ *and* $5m + 3$.

We can prove this equivalence in the same way, starting from the identity

$$m(m+1) = 2 + 4 + 6 + \cdots + 2m.$$

The proof which we give of these theorems in the next section was found independently by Rogers and Ramanujan. We state it in the form given by Rogers. It is fairly straightforward, but unilluminating, since it depends on writing down an auxiliary function whose genesis remains obscure. It is natural to ask for an elementary proof on some such lines as those of § 19.11, and such a proof was found by Schur; but Schur's proof is too elaborate for insertion here. There are other proofs by Rogers and Schur, and one by Watson based on different ideas. No proof is really easy (and it would perhaps be unreasonable to expect an easy proof).

19.14. Proof of Theorems 362 and 363. We write

$$P_0 = 1, P_r = \prod_{s=1}^{r} \frac{1}{1 - x^s}, \quad Q_r = Q_r(a) = \prod_{s=r}^{\infty} \frac{1}{1 - ax^s},$$

$$\lambda(r) = \tfrac{1}{2}r(5r + 1),$$

and define the operator η by

$$\eta f(a) = f(ax).$$

We introduce the auxiliary function

$$H_m = H_m(a) = \sum_{r=0}^{\infty} (-1)^r a^{2r} x^{\lambda(r)-mr} (1 - a^m x^{2mr}) P_r Q_r, \tag{19.14.1}$$

where $m = 0$, 1, or 2. Our object is to expand H_1 and H_2 in powers of a. We prove first that

$$H_m - H_{m-1} = a^{m-1} \eta H_{3-m} \quad (m = 1, 2). \tag{19.14.2}$$

We have

$$H_m - H_{m-1} = \sum_{r=0}^{\infty} (-1)^r a^{2r} x^{\lambda(r)} C_{mr} P_r Q_r,$$

where

$$\begin{aligned} C_{mr} &= x^{-mr} - a^m x^{mr} - x^{(1-m)r} + a^{m-1} x^{r(m-1)} \\ &= a^{m-1} x^{r(m-1)} (1 - ax^r) + x^{-mr}(1 - x^r). \end{aligned}$$

Now

$$(1 - ax^r) Q_r = Q_{r+1}, \quad (1 - x^r) P_r = P_{r-1}, \quad 1 - x^0 = 0,$$

and so

$$\begin{aligned} H_m - H_{m-1} = &\sum_{r=0}^{\infty} (-1)^r a^{2r+m-1} x^{\lambda(r)+r(m-1)} P_r Q_{r+1} \\ &+ \sum_{r=1}^{\infty} (-1)^r a^{2r} x^{\lambda(r)-mr} P_{r-1} Q_r. \end{aligned}$$

In the second sum on the right-hand side of this identity we change r into $r+1$. Thus

$$H_m - H_{m-1} = \sum_{r=0}^{\infty} (-1)^r D_{mr} P_r Q_{r+1},$$

where

$$\begin{aligned} D_{mr} &= a^{2r+m-1} x^{\lambda(r)+r(m-1)} - a^{2(r+1)} x^{\lambda(r+1)-m(r+1)} \\ &= a^{m-1+2r} x^{\lambda(r)+r(m-1)} (1 - a^{3-m} x^{(2r+1)(3-m)}) \\ &= a^{m-1} \eta \left\{ a^{2r} x^{\lambda(r)-r(3-m)} (1 - a^{3-m} x^{2r(3-m)}) \right\}, \end{aligned}$$

since $\lambda(r+1) - \lambda(r) = 5r + 3$. Also $Q_{r+1} = \eta Q_r$ and so

$$\begin{aligned} &H_m - H_{m-1} \\ &\quad = a^{m-1} \eta \sum_{r=0}^{\infty} (-1)^r a^{2r} x^{\lambda(r)-r(3-m)} (1 - a^{3-m} x^{2r(3-m)}) P_r Q_r \\ &\quad = a^{m-1} \eta H_{3-m}, \end{aligned}$$

which is (19.14.2).

If we put $m = 1$ and $m = 2$ in (19.14.2) and remember that $H_0 = 0$, we have

(19.14.3)
$$H_1 = \eta H_2,$$
$$H_2 - H_1 = a\eta H_1,$$

so that

(19.14.4)
$$H_2 = \eta H_2 + a\eta^2 H_2.$$

We use this to expand H_2 in powers of a. If

$$H_2 = c_0 + c_1 a + \cdots = \sum c_s a^s,$$

where the c_s are independent of a, then $c_0 = 1$ and (19.14.4) gives

$$\sum c_s a^s = \sum c_s x^s a^s + \sum c_s x^{2s} a^{s+1},$$

Hence, equating the coefficients of a^s, we have

$$c_1 = \frac{1}{1-x}, \quad c_s = \frac{x^{2s-2}}{1-x^s} c_{s-1} = \frac{x^{2+4+\cdots+2(s-1)}}{(1-x)\ldots(1-x^s)} = x^{s(s-1)} P_s.$$

Hence

$$H_2(a) = \sum_{s=0}^{\infty} a^s x^{s(s-1)} P_s.$$

If we put $a = x$, the right-hand side of this is the series in (19.13.1). Also $P_r Q_r(x) = P_\infty$ and so, by (19.14.1),

$$\begin{aligned} H_2(x) &= P_\infty \sum_{r=0}^{\infty} (-1)^r x^{\lambda(r)} (1 - x^{2(2r+1)}) \\ &= P_\infty \left\{ \sum_{r=0}^{\infty} (-1)^r x^{\lambda(r)} + \sum_{r=1}^{\infty} (-1)^r x^{\lambda(r-1)+2(2r-1)} \right\} \\ &= P_\infty \left\{ 1 + \sum_{r=1}^{\infty} (-1)^r (x^{\frac{1}{2}r(5r+1)} + x^{\frac{1}{2}r(5r-1)}) \right\}. \end{aligned}$$

Hence, by Theorem 356,

$$\begin{aligned} H_2(x) &= P_\infty \prod_{n=0}^{\infty} \{(1 - x^{5n+2})(1 - x^{5n+3})(1 - x^{5n+5})\} \\ &= \prod_{n=0}^{\infty} \frac{1}{(1 - x^{5n+1})(1 - x^{5n+4})}. \end{aligned}$$

This completes the proof of Theorem 362.

Again, by (19.14.3),

$$H_1(a) = \eta H_2(a) = H_2(ax) = \sum_{s=0}^{\infty} a^s x^{s^2} P_s$$

and, for $a = x$, the right-hand side becomes the series in (19.13.2). Using (19.14.1) and Theorem 355, we complete the proof of Theorem 363 in the same way as we did that of Theorem 362.

19.15. Ramanujan's continued fraction. We can write (19.14.14) in the form

$$H_2(a,x) = H_2(ax, x) + aH_2(ax^2, x)$$

so that

$$H_2(ax, x) = H_2(ax^2, x) + axH_2(ax^3, x).$$

Hence, if we define $F(a)$ by

$$\begin{aligned} F(a) = F(a,x) = H_1(a,x) = \eta H_2(a,x) = H_2(ax,x) \\ = 1 + \frac{ax}{1-x} + \frac{a^2x^4}{(1-x)(1-x^2)} + \cdots, \end{aligned}$$

then $F(a)$ satisfies

$$F(ax^n) = F(ax^{n+1}) + ax^{n+1}F(ax^{n+2}).$$

Hence, if

$$u_n = \frac{F(ax^n)}{F(ax^{n+1})},$$

we have

$$u_n = 1 + \frac{ax^{n+1}}{u_{n+1}};$$

and hence $u_0 = F(a)/F(ax)$ may be developed formally as

(19.15.1) $$\frac{F(a)}{F(ax)} = 1 + \frac{ax}{1+}\,\frac{ax^2}{1+}\,\frac{ax^3}{1+\ldots},$$

a 'continued fraction' of a different type from those which we considered in Ch. X.

We have no space to construct a theory of such fractions here. It is not difficult to show that, when $|x| < 1$,

$$1 + \frac{ax}{1+}\,\frac{ax^2}{1+\cdots}\cdots\frac{ax^n}{1}$$

tends to a limit by means of which we can define the right-hand side of (19.15.1). If we take this for granted, we have, in particular,

$$\frac{F(1)}{F(x)} = 1 + \frac{x}{1+}\frac{x^2}{1+}\frac{x^3}{1+\cdots},$$

and so

$$\begin{aligned} 1 + \frac{x}{1+}\frac{x^2}{1+\cdots} &= \frac{1 - x^2 - x^3 + x^9 + \ldots}{1 - x - x^4 + x^7 + \cdots} \\ &= \frac{(1-x^2)(1-x^7)\ldots(1-x^3)(1-x^8)\ldots}{(1-x)(1-x^6)\ldots(1-x^4)(1-x^9)\cdots}. \end{aligned}$$

It is known from the theory of elliptic functions that these products and series can be calculated for certain special values of x, and in particular when $x = e^{-2\pi\sqrt{h}}$ and h is rational. In this way Ramanujan proved that, for example,

$$1 + \frac{e^{-2\pi}}{1+}\frac{e^{-4\pi}}{1+}\frac{e^{-6\pi}}{1+\cdots} = \left\{\sqrt{\left(\frac{5+\sqrt{5}}{2}\right)} - \frac{\sqrt{5}+1}{2}\right\} e^{\frac{2}{5}\pi}.$$

NOTES

§19.1. There are general accounts of the earlier theory of partitions in Bachmann, *Niedere Zahlentheorie,* ii, ch. 3; Netto, *Combinatorik* (second ed. by Brun and Skolem, 1927); and MacMahon, *Combinatory analysis*, ii. For references to later work, see the survey by Gupta (*J. Res. Nat. Bur. Standards* B74 (1970), 1–29); Andrews, *Partitions*; Andrews and Eriksson, *Integer Partitions*; Ono and Ahlgren (*Notices Amer. Math. Soc.*, 48 (2001), 978–84); Ono, *The Web of Modularity.*

§§19.3–5. All of the formulas of these sections are Euler's. More extensive developments of these methods can be found in Andrews, *Partitions*, ch. 2 and Andrews and Eriksson, *Integer Partitions*, ch. 5. For historical references, see Dickson, *History*, ii, ch.3.

§19.6. Theorem 348 (the q-binomial theorem) and Theorem 349 (the q-binomial series) are not in Euler's works. Cauchy studied them, but probably they predate him. Further applications of these results appear in Andrews, *Partitions*, ch. 3, and Andrews and Eriksson, ch. 7.

§19.7. While this formula is often attributed to Euler, its first published appearance is by Jacobi, *Fundamenta nova*, §64. Indeed, Jacobi needed a generalization of Theorem 351 for his original proof of Theorem 352.

§19.8. Theorem 352 is often referred to as Jacobi's triple product identity, (Jacobi, *Fundamenta nova*, §64). The theorem was known to Gauss. The proof given here is ascribed to Jacobi by Enneper; Mr. R. F. Whitehead drew our attention to it. Wright (*J. London Math. Soc.* 40 (1965), 55–57) gives a simple combinatorial proof of Theorem 352, using arrays of points as in §§19.5, 19.6, and 19.11. A full history of the method used by Wright and an extensive application of it are given by Andrews (*Memoirs of the Amer. Math. Soc.*

49 (1984)). Alternative proofs appear in Andrews, *Partitions*, ch. 2, and in Andrews and Eriksson, *Integer partitions*, ch. 8.

§19.9. Theorem 353 is due to Euler; for references see Bachmann, *Niedere Zahlentheorie* ii, 163, or Dickson, *History*, ii. 103. Theorem 354 was proved by Gauss in 1808 (*Werke*, ii. 20), and Theorem 357 by Jacobi (*Fundamenta nova*, §66). Professor D. H. Lehmer suggested the proof of Theorem 357 given here.

§19.10. MacMahon's table is printed in (*Proc. London Math Soc.* (2) 17 (1918), 114–15), and has subsequently been extended to 600 (Gupta, *ibid.* 39 (1935), 142–9, and 42 (1937), 546–9), and to 1000 (Gupta, Gwyther, and Miller, *Roy. Soc. Math. Tables* 4 (Cambridge, 1958)). Recently Sun Tae Soh has prepared a program for computing $p(n)$ for $n \leqslant 22{,}000{,}000$ (cf. http://trinitas.mju.ac.kr/intro2numbpart.html).

§19.11 F. Franklin, (*Comptes rendus*, 92 (1881), 448–50). We observe that, if we use this method to prove Theorem 358, i.e. Theorem 353, we can shorten the proof of Theorem 352 in §19.8. We proceed as before up to (19.8.3). We then put $x = y^{3/2}\ z = -y^{1/2}$ and have

$$P(x,z) = \prod_{n=1}^{\infty} \left\{ \left(1-y^{3n}\right)\left(1-y^{3n-1}\right)\left(1-y^{3n-2}\right) \right\} = \prod_{m=1}^{\infty} \left(1-y^{m}\right)$$

and

$$S(x,z) = \sum_{n=-\infty}^{\infty} (-1)^n y^{\frac{1}{2}n(3n+1)} = P(x,z)$$

by Theorem 353, so that $a_0(x) = 1$.

§19.12. See Ramanujan, *Collected Papers*, nos. 25, 28, 30. These papers contain complete proofs of the congruences to moduli 5, 7, and 11 only. On p. 213 he states identities which involve the congruences to moduli 5^2 and 7^2 as corollaries, and these identities were proved later by Darling (*Proc. London Math. Soc.* (2) 19 (1921), 350–72) and Mordell (*ibid.* 20 (1922), 408–16). An unpublished manuscript of Ramanujan dealt with many instances of his conjecture; this document has been retrieved by Berndt and Ono (*The Andrews Festschrift*, Springer, 2001, pp. 39–110).

The papers referred to at the end of the section are Gupta's mentioned in the Note to §19.10; Krečmar (*Bulletin de l'acad. des sciences de l'URSS* (7) 6 (1933), 763–800); Lehmer (*Journal London Math. Soc.* 11 (1936), 114–18 and *Bull. Amer. Math. Soc.* 44 (1938), 84–90); Watson (*Journal für Math.* 179 (1938), 97–128); Lehner (*Proc. Amer. Math. Soc.* 1 (1950), 172–81); Dyson (*Eureka* 8 (1994) 10–15); Atkin and Swinnerton-Dyer (*Proc. London Math. Soc.* (3) 4 (1954), 84–106). Atkin (*Glasgow Math. J.* 8 (1967), 14–32) proved the 11^c result for general c and has also found a number of other congruences of a more complicated character.

More recently Ono, *The Web of Modularity*, and his colleagues have vastly expanded our knowledge of partition function congruences. Andrews and Garvan (*Bull. Amer. Math. Soc.* 18 (1998), 167–71) found the 'crank' conjectured by Dyson; Mahlburg (*Proc. Nat. Acad. Sci.* 102 (2005), 15373–76) has related the crank to the cornucopia of congruences discovered by Ono.

§§ 19.13–14. For the history of the Rogers–Ramanujan identities, first found by Rogers in 1894, see the note by Hardy reprinted on pp. 344–5 of Ramanujan's *Collected papers*, and Hardy, *Ramanujan*, ch. 6. Schur's proofs appeared in the *Berliner Sitzungsberichte* (1917), 302–21, and Watson's in the *Journal London Math. Soc.* 4 (1929), 4–9. Hardy, *Ramanujan*, 95–99 and 107–11, gives other variations of the proofs.

Selberg, *Avhandlinger Norske Akad.* (1936), no. 8, has generalized the argument of Rogers and Ramanujan, and found similar, but less simple, formulae associated with the number 7. Dyson, *Journal London Math. Soc.* 18 (1943), 35–39, has pointed out that these also may be found in Rogers's work, and has simplified the proofs considerably.

More recently, development of the theory and extension of the Rogers–Ramanujan identities has been very active. Accounts of these discoveries can be found in surveys by Alder (*Amer. Math. Monthly*, 76 (1969), 733–46); Alladi (*Number Theory, Paris* 1992–93, Cambridge University Press (1995), 1–36); Andrews (*Advances in Math.*, 9 (1972), 10–51; *Bull. Amer. Math. Soc.*, 80 (1974), 1033–52; *Memoirs Amer. Math. Soc.*, 152 (1974) I+86 pp.; *Pac. J. Math.* 114 (1984), 267–83). Applications in physics are surveyed by Berkovich and McCoy (*Proc. ICM* 1998, III, 163–72). See also Andrews, *Partitions*.

Mr. C. Sudler suggested a substantial improvement in the presentation of the proof in § 19.14.

§19.15. Recent discoveries concerning the Rogers–Ramanujan continued fraction are discussed in Andrews and Berndt, *Ramanujan's Lost Notebook, Part I*, chs. 1–8.

XX

THE REPRESENTATION OF A NUMBER BY TWO OR FOUR SQUARES

20.1. Waring's problem: the numbers $g(k)$ and $G(k)$. Waring's problem is that of the representation of positive integers as sums of a fixed number s of non-negative kth powers. It is the particular case of the general problem of § 19.1 in which the a are

$$0^k, 1^k, 2^k, 3^k, \ldots$$

and s is fixed. When $k = 1$, the problem is that of partitions into s parts of unrestricted form; such partitions are enumerated, as we saw in Ch. XIX, by the function

$$\frac{1}{(1-x)\left(1-x^2\right)\ldots(1-x^s)}.$$

Hence we take $k \geqslant 2$.

It is plainly impossible to represent all integers if s is too small, for example if $s = 1$. Indeed it is impossible if $s < k$. For the number of values of x_1 for which $x_1^k \leqslant n$ does not exceed $n^{1/k} + 1$; and so the number of sets of values $x_1, x_2, \ldots, x_{k-1}$ for which

$$x_1^k + \cdots + x_{k-1}^k \leqslant n$$

does not exceed

$$(n^{1/k} + 1)^{k-1} = n^{(k-1)/k} + O(n^{(k-2)/k}).$$

Hence most numbers are not representable by $k - 1$ or fewer kth powers.

The first question that arises is whether, for a given k, there is any fixed $s = s(k)$ such that

$$n = x_1^k + x_2^k + \cdots + x_s^k \tag{20.1.1}$$

is soluble for every n.

The answer is by no means obvious. For example, if the a of § 19.1 are the numbers

$$1, 2, 2^2, \ldots, 2^m, \ldots,$$

then the number

$$2^{m+1} - 1 = 1 + 2 + 2^2 + \cdots + 2^m$$

is not representable by less than $m + 1$ numbers a, and we have $m + 1 \to \infty$ when $n = 2^{m+1} - 1 \to \infty$. Hence it is not true that all numbers are representable by a fixed number of powers of 2.

Waring stated without proof that every number is the sum of 4 squares, of 9 cubes, of 19 biquadrates, 'and so on'. His language implies that he believed that the answer to our question is affirmative, that (20.1.1) is soluble for each fixed k, any positive n, and an $s = s(k)$ depending only on k. It is very improbable that Waring had any sufficient grounds for his assertion, and it was not until more than 100 years later that Hilbert first proved it true.

A number representable by s kth powers is plainly representable by any larger number. Hence, if all numbers are representable by s kth powers, there is a least value of s for which this is true. This least value of s is denoted by $g(k)$. We shall prove in this chapter that $g(2) = 4$, that is to say that any number is representable by four squares and that four is the least number of squares by which all numbers are representable. In Ch. XXI we shall prove that $g(3)$ and $g(4)$ exist, but without determining their values.

There is another number in some ways still more interesting than $g(k)$. Let us suppose, to fix our ideas, that $k = 3$. It is known that $g(3) = 9$; every number is representable by 9 or fewer cubes, and every number, except $23 = 2 \, . \, 2^3 + 7 \, . \, 1^3$ and

$$239 = 2 \, . \, 4^3 + 4 \, . \, 3^3 + 3 \, . \, 1^3,$$

can be represented by 8 or fewer cubes. In fact, all sufficiently large numbers are representable by 7 or fewer. Numerical evidence indicates that only 15 other numbers, of which the largest is 454, require so many cubes as 8, and that 7 suffice from 455 onwards.

It is plain, if this be so, that 9 is not the number which is really most significant in the problem. The facts that just two numbers require 9 cubes, and, if it is a fact, that just 15 more require 8, are, so to say, arithmetical flukes, depending on comparatively trivial idiosyncrasies of special numbers. The most fundamental and most difficult problem is that of deciding, not how many cubes are required for the representation of *all* numbers, but how many are required for the representation of *all large* numbers, i.e. of all numbers with some finite number of exceptions.

We define $G(k)$ as the least value of s for which it is true that all sufficiently large numbers, i.e. all numbers with at most a finite number of exceptions, are representable by s kth powers. Thus $G(3) \leqslant 7$. On the other hand, as we shall see in the next chapter, $G(3) \geqslant 4$; there are infinitely many numbers not representable by three cubes. Thus $G(3)$ is 4, 5, 6, or 7; it is still not known which.

It is plain that

$$G(k) \leqslant g(k)$$

for every k. In general, $G(k)$ is much smaller than $g(k)$, the value of $g(k)$ being swollen by the difficulty of representing certain comparatively small numbers.

20.2. Squares. In this chapter we confine ourselves to the case $k = 2$. Our main theorem is Theorem 369, which, combined with the trivial result[†] that no number of the form $8m + 7$ can be the sum of three squares, shows that

$$g(2) = G(2) = 4.$$

We give three proofs of this fundamental theorem. The first (§ 20.5) is elementary and depends on the 'method of descent', due in principle to Fermat. The second (§§ 20.6–9) depends on the arithmetic of quaternions. The third (§ 20.11–12) depends on an identity which belongs properly to the theory of elliptic functions (though we prove it by elementary algebra),[‡] and gives a formula for the number of representations.

But before we do this, we return for a time to the problem of the representation of a number by two squares.

THEOREM 366. *A number n is the sum of two squares if and only if all prime factors of n of the form* $4m + 3$ *have even exponents in the standard form of n.*

This theorem is an immediate consequence of (16.9.5) and Theorem 278. There are, however, other proofs of Theorem 366, some independent of the arithmetic of $k(i)$, which involve interesting and important ideas.

20.3. Second proof of Theorem 366. We have to prove that n is of the form of $x^2 + y^2$ if and only if

$$n = n_1^2 n_2, \tag{20.3.1}$$

where n_2 has no prime factors of the form $4m + 3$.

We say that

$$n = x^2 + y^2$$

is a *primitive* representation of n if $(x, y) = 1$, and otherwise an *imprimitive* representation.

[†] See § 20.10.

[‡] See the footnote to p. 372.

THEOREM 367. *If $p = 4m + 3$ and $p|n$, then n has no primitive representations.*

If n has a primitive representation, then

$$p|(x^2 + y^2), \quad (x, y) = 1,$$

and so $p \nmid x, p \nmid y$. Hence, by Theorem 57, there is a number l such that $y \equiv lx \pmod{p}$ and so

$$x^2(1 + l^2) \equiv x^2 + y^2 \equiv 0 \pmod{p}.$$

It follows that

$$1 + l^2 \equiv 0 \pmod{p}$$

and therefore that -1 is a quadratic residue of p, which contradicts Theorem 82.

THEOREM 368. *If $p = 4m + 3$, $p^c|n$, $p^{c+1} \nmid n$, and c is odd, then n has no representations (primitive or imprimitive).*

Suppose that $n = x^2 + y^2$, $(x, y) = d$; and let p^γ be the highest power of p which divides d. Then

$$x = dX, \quad y = dY, \quad (X, Y) = 1,$$

$$n = d^2(X^2 + Y^2) = d^2N,$$

say. The index of the highest power of p which divides N is $c - 2\gamma$, which is positive because c is odd. Hence

$$N = X^2 + Y^2, \quad (X, Y) = 1, \quad p|N;$$

which contradicts Theorem 367.

It remains to prove that n is representable when n is of the form (20.3.1), and it is plainly enough to prove n_2 representable. Also

$$\left(x_1^2 + y_1^2\right)\left(x_2^2 + y_2^2\right) = (x_1x_2 + y_1y_2)^2 + (x_1y_2 - x_2y_1)^2,$$

so that the product of two representable numbers is itself representable. Since $2 = 1^2 + 1^2$ is representable, the problem is reduced to that of proving Theorem 251, i.e. of proving that *if $p = 4m + 1$, then p is representable.*

Since -1 is a quadratic residue of such a p, there is an l for which

$$l^2 \equiv -1 \pmod{p}.$$

Taking $n = [\sqrt{p}]$ in Theorem 36, we see that there are integers a and b such that

$$0 < b < \sqrt{p}, \qquad \left| -\frac{l}{p} - \frac{a}{b} \right| < \frac{1}{b\sqrt{p}}.$$

If we write

$$c = lb + pa,$$

then

$$|c| < \sqrt{p}, \quad 0 < b^2 + c^2 < 2p.$$

But $c \equiv lb \pmod{p}$, and so

$$b^2 + c^2 \equiv b^2 + l^2 b^2 \equiv b^2(1 + l^2) \equiv 0 \pmod{p};$$

and therefore

$$b^2 + c^2 = p.$$

20.4. Third and fourth proofs of Theorem 366. (1) Another proof of Theorem 366, due (in principle at any rate) to Fermat, is based on the 'method of descent'. To prove that $p = 4m+1$ is representable, we prove (i) that *some* multiple of p is representable, and (ii) that the *least* representable multiple of p must be p itself. The rest of the proof is the same.

By Theorem 86, there are numbers x, y such that

$$x^2 + y^2 = mp, \qquad p \nmid x, \qquad p \nmid y, \tag{20.4.1}$$

and $0 < m < p$. Let m_0 be the least value of m for which (20.4.1) is soluble, and write m_0 for m in (20.4.1). If $m_0 = 1$, our theorem is proved.

If $m_0 > 1$, then $1 < m_0 < p$. Now m_0 cannot divide both x and y, since this would involve

$$m_0^2 \,|\, (x^2 + y^2) \rightarrow m_0^2 \,|\, m_0 p \rightarrow m_0 \,|\, p.$$

Hence we can choose c and d so that

$$x_1 = x - cm_0, \quad y_1 = y - dm_0,$$
$$|x_1| \leqslant \tfrac{1}{2}m_0, \quad |y_1| \leqslant \tfrac{1}{2}m_0, \quad x_1^2 + y_1^2 > 0,$$

and therefore

$$0 < x_1^2 + y_1^2 \leqslant 2\left(\tfrac{1}{2}m_0\right)^2 < m_0^2. \tag{20.4.2}$$

Now

$$x_1^2 + y_1^2 \equiv x^2 + y^2 \equiv 0 \pmod{m_0}$$

or

$$x_1^2 + y_1^2 = m_1 m_0, \tag{20.4.3}$$

where $0 < m_1 < m_0$, by (20.4.2). Multiplying (20.4.3) by (20.4.1), with $m = m_0$, we obtain

$$m_0^2 m_1 p = \left(x^2 + y^2\right)\left(x_1^2 + y_1^2\right) = (xx_1 + yy_1)^2 + (xy_1 - x_1y)^2 .$$

But

$$xx_1 + yy_1 = x\,(x - cm_0) + y\,(y - dm_0) = m_0 X,$$
$$xy_1 - x_1y = x\,(y - dm_0) - y\,(x - cm_0) = m_0 Y,$$

where $X = p - cx - dy$, $Y = cy - dx$. Hence

$$m_1 p = X^2 + Y^2 \quad (0 < m_1 < m_0),$$

which contradicts the definition of m_0. It follows that m_0 must be 1.

(2) A fourth proof, due to Grace, depends on the ideas of Ch. III.

By Theorem 82, there is a number l for which

$$l^2 + 1 \equiv 0 \pmod{p}.$$

We consider the points (x, y) of the fundamental lattice Λ which satisfy

$$y \equiv lx \pmod{p}.$$

These points define a lattice M.† It is easy to see that the proportion of points of Λ, in a large circle round the origin, which belong to M is asymptotically $1/p$, and that the area of a fundamental parallelogram of M is therefore p.

Suppose that A or (ξ, η) is one of the points of M nearest to the origin. Then $\eta \equiv l\xi$ and so

$$-\xi \equiv l^2\xi \equiv l\eta \pmod{p},$$

and therefore B or $(-\eta, \xi)$ is also a point of M. There is no point of M inside the triangle OAB, and therefore none within the square with sides OA, OB.

† We state the proof shortly, leaving some details to the reader.

Hence this square is a fundamental parallelogram of M, and therefore its area is p. It follows that

$$\xi^2 + \eta^2 = p.$$

20.5. The four-square theorem. We pass now to the principal theorem of this chapter.

THEOREM 369 (LAGRANGE'S THEOREM). *Every positive integer is the sum of four squares.*

Since

(20.5.1)

$$\begin{aligned}&\left(x_1^2+x_2^2+x_3^2+x_4^2\right)\left(y_1^2+y_2^2+y_3^2+y_4^2\right)\\&\quad=(x_1y_1+x_2y_2+x_3y_3+x_4y_4)^2+(x_1y_2-x_2y_1+x_3y_4+x_4y_3)^2\\&\qquad+(x_1y_3-x_3y_1+x_4y_2-x_2y_4)^2+(x_1y_4-x_4y_1+x_2y_3-x_3y_2)^2,\end{aligned}$$

the product of two representable numbers is itself representable. Also $1 = 1^2 + 0^2 + 0^2 + 0^2$. Hence Theorem 369 will follow from

THEOREM 370. *Any prime p is the sum of four squares.*

Our first proof proceeds on the same lines as the proof of Theorem 366 in § 20.4 (1). Since $2 = 1^2 + 1^2 + 0^2 + 0^2$, we can take $p > 2$.

It follows from Theorem 87 that there is a multiple of p, say mp, such that

$$mp = x_1^2 + x_2^2 + x_3^2 + x_4^2,$$

with x_1, x_2, x_3, x_4 not all divisible by p; and we have to prove that the least such multiple of p is p itself.

Let m_0p be the least such multiple. If $m_0 = 1$, there is nothing more to prove; we suppose therefore that $m_0 > 1$. By Theorem 87, $m_0 < p$.

If m_0 is even, then $x_1 + x_2 + x_3 + x_4$ is even and so either (i) x_1, x_2, x_3, x_4, are all even, or (ii) they are all odd, or (iii) two are even and two are odd. In the last case, let us suppose that x_1, x_2 are even and x_3, x_4 are odd. Then in all three cases

$$x_1 + x_2, \quad x_1 - x_2, \quad x_3 + x_4, \quad x_3 - x_4$$

are all even, and so

$$\tfrac{1}{2}m_0p = \left(\frac{x_1+x_2}{2}\right)^2 + \left(\frac{x_1-x_2}{2}\right)^2 + \left(\frac{x_3+x_4}{2}\right)^2 + \left(\frac{x_3-x_4}{2}\right)^2$$

is the sum of four integral squares. These squares are not all divisible by p, since x_1, x_2, x_3, x_4 are not all divisible by p. But this contradicts our definition of m_0. Hence m_0 must be odd.

Next, x_1, x_2, x_3, x_4, are not all divisible by m_0, since this would imply

$$m_0^2 \,|m_0\, p \to m_0|\, p,$$

which is impossible. Also m_0 is odd, and therefore at least 3. We can therefore choose b_1, b_2, b_3, b_4 so that

$$y_i = x_i - b_i m_0 \quad (i = 1, 2, 3, 4)$$

satisfy

$$|y_i| < \tfrac{1}{2} m_0, \qquad y_1^2 + y_2^2 + y_3^2 + y_4^2 > 0.$$

Then

$$0 < y_1^2 + y_2^2 + y_3^2 + y_4^2 < 4\left(\tfrac{1}{2} m_0\right)^2 = m_0^2,$$

and

$$y_1^2 + y_2^2 + y_3^2 + y_4^2 \equiv 0 \pmod{m_0}.$$

It follows that

$$x_1^2 + x_2^2 + x_3^2 + x_4^2 = m_0 p \qquad (m_0 < p),$$
$$y_1^2 + y_2^2 + y_3^2 + y_4^2 = m_0 m_1 \quad (0 < m_1 < m_0);$$

and so, by (20.5.1),

(20.5.2) $$m_0^2 m_1 p = z_1^2 + z_2^2 + z_3^2 + z_4^2,$$

where z_1, z_2, z_3, z_4 are the four numbers which occur on the right-hand side of (20.5.1). But

$$z_1 = \sum x_i y_i = \sum x_i (x_i - b_i m_0) \equiv \sum x_i^2 \equiv 0 \pmod{m_0};$$

and similarly z_2, z_3, z_4 are divisible by m_0. We may therefore write

$$z_i = m_0 t_i \quad (i = 1, 2, 3, 4);$$

and then (20.5.2) becomes

$$m_1 p = t_1^2 + t_2^2 + t_3^2 + t_4^2,$$

which contradicts the definition of m_0 because $m_1 < m_0$.

It follows that $m_0 = 1$.

20.6. Quaternions. In Ch. XV we deduced Theorem 251 from the arithmetic of the Gaussian integers, a subclass of the complex numbers of ordinary analysis. There is a proof of Theorem 370 based on ideas which are similar, but more sophisticated because we use numbers which do not obey all the laws of ordinary algebra.

Quaternions† are 'hyper-complex' numbers of a special kind. The numbers of the system are of the form

$$\alpha = a_0 + a_1 i_1 + a_2 i_2 + a_3 i_3, \tag{20.6.1}$$

where a_0, a_1, a_2, a_3 are real numbers (the *coordinates* of α), and i_1, i_2, i_3 elements characteristic of the system. Two quaternions are *equal* if their coordinates are equal.

These numbers are combined according to rules which resemble those of ordinary algebra in all respects but one. There are, as in ordinary algebra, operations of addition and multiplication. The laws of addition are the same as in ordinary algebra; thus

$$\begin{aligned}\alpha + \beta &= (a_0 + a_1 i_1 + a_2 i_2 + a_3 i_3) + (b_0 + b_1 i_1 + b_2 i_2 + b_3 i_3)\\ &= (a_0 + b_0) + (a_1 + b_1) i_1 + (a_2 + b_2) i_2 + (a_3 + b_3) i_3.\end{aligned}$$

Multiplication is associative and distributive, but not generally commutative. It is commutative for the coordinates, and between the coordinates and i_1, i_2, i_3; but

$$\begin{cases} i_1^2 = i_2^2 = i_3^2 = -1, \\ i_2 i_3 = i_1 = -i_3 i_2, \; i_3 i_1 = i_2 = -i_1 i_3, \; i_1 i_2 = i_3 = -i_2 i_1. \end{cases} \tag{20.6.2}$$

Generally,

$$\begin{aligned}\alpha\beta &= (a_0 + a_1 i_1 + a_2 i_2 + a_3 i_3)(b_0 + b_1 i_1 + b_2 i_2 + b_3 i_3)\\ &= c_0 + c_1 i_1 + c_2 i_2 + c_3 i_3,\end{aligned} \tag{20.6.3}$$

where

$$\begin{cases} c_0 = a_0 b_0 - a_1 b_1 - a_2 b_2 - a_3 b_3, \\ c_1 = a_0 b_1 + a_1 b_0 + a_2 b_3 - a_3 b_2, \\ c_2 = a_0 b_2 - a_1 b_3 + a_2 b_0 + a_3 b_1, \\ c_3 = a_0 b_3 + a_1 b_2 - a_2 b_1 + a_3 b_0. \end{cases} \tag{20.6.4}$$

† We take the elements of the algebra of quaternions for granted. A reader who knows nothing of quaternions, but accepts what is stated here, will be able to follow §§ 20.7–9.

In particular,

(20.6.5)
$$(a_0 + a_1 i_1 + a_2 i_2 + a_3 i_3)(a_0 - a_1 i_1 - a_2 i_2 - a_3 i_3) = a_0^2 + a_1^2 + a_2^2 + a_3^2,$$

the coefficients of i_1, i_2, i_3 in the product being zero.

We shall say that the quaternion α is *integral* if a_0, a_1, a_2, a_3 are either (i) all rational integers or (ii) all halves of odd rational integers. We are interested only in integral quaternions; and henceforth we use 'quaternion' to mean 'integral quaternion'. We shall use Greek letters for quaternions, except that, when $a_1 = a_2 = a_3 = 0$ and so $\alpha = a_0$, we shall use a_0 both for the quaternion

$$a_0 + 0\,.\,i_1 + 0\,.\,i_2 + 0\,.\,i_3$$

and for the rational integer a_0.

The quaternion

(20.6.6) $$\bar{\alpha} = a_0 - a_1 i_1 - a_2 i_2 - a_3 i_3$$

is called the *conjugate* of $\alpha = a_0 + a_1 i_1 + a_2 i_2 + a_3 i_3$, and

(20.6.7) $$N\alpha = \alpha\bar{\alpha} = \bar{\alpha}\alpha = a_0^2 + a_1^2 + a_2^2 + a_3^2$$

the *norm* of α. The norm of an integral quaternion is a rational integer. We shall say that α is *odd* or *even* according as $N\alpha$ is odd or even.

It follows from (20.6.3), (20.6.4), and (20.6.6) that

$$\overline{\alpha\beta} = \bar{\beta}\bar{\alpha},$$

and so

(20.6.8) $$N(\alpha\beta) = \alpha\beta\,.\,\overline{\alpha\beta} = \alpha\beta\,.\,\bar{\beta}\bar{\alpha} = \alpha\,.\,N\beta\,.\,\bar{\alpha} = \alpha\bar{\alpha}\,.\,N\beta = N\alpha N\beta.$$

We define α^{-1}, when $\alpha \neq 0$, by

(20.6.9) $$\alpha^{-1} = \frac{\bar{\alpha}}{N\alpha},$$

so that

(20.6.10) $$\alpha\alpha^{-1} = \alpha^{-1}\alpha = 1.$$

If α and α^{-1} are both integral, then we say that α is a *unity,* and write $\alpha = \epsilon$. Since $\epsilon\epsilon^{-1} = 1$, $N\epsilon N\epsilon^{-1} = 1$ and so $N\epsilon = 1$. Conversely, if α is integral and $N\alpha = 1$, then $\alpha^{-1} = \bar{\alpha}$ is also integral, so that α is a unity. Thus a unity may be defined alternatively as an integral quaternion whose norm is 1.

If a_0, a_1, a_2, a_3 are all integral, and $a_0^2 + a_1^2 + a_2^2 + a_3^2 = 1$, then one of $a_0^2, \ldots$ must be 1 and the rest 0. If they are all halves of odd integers, then each of $a_0^2, \ldots$ must be $\frac{1}{4}$. Hence there are just 24 unities, viz.

$$\pm 1, \quad \pm i_1, \quad \pm i_2, \quad \pm i_3, \quad \tfrac{1}{2}(\pm 1 \pm i_1 \pm i_2 \pm i_3). \tag{20.6.11}$$

If we write

$$\rho = \tfrac{1}{2}(1 + i_1 + i_2 + i_3), \tag{20.6.12}$$

then any integral quaternion may be expressed in the form

$$k_0\rho + k_1 i_1 + k_2 i_2 + k_3 i_3, \tag{20.6.13}$$

where k_0, k_1, k_2, k_3 are rational integers; and any quaternion of this form is integral. It is plain that the sum of any two integral quaternions is integral. Also, after (20.6.3) and (20.6.4),

$$\rho^2 = \tfrac{1}{2}(-1 + i_1 + i_2 + i_3) = \rho - 1,$$
$$\rho i_1 = \tfrac{1}{2}(-1 + i_1 + i_2 - i_3) = -\rho + i_1 + i_2,$$
$$i_1\rho = \tfrac{1}{2}(-1 + i_1 - i_2 + i_3) = -\rho + i_1 + i_3,$$

with similar expressions for ρi_2, etc. Hence all these products are integral, and therefore the product of any two integral quaternions is integral.

If ϵ is any unity, then $\epsilon\alpha$ and $\alpha\epsilon$ are said to be *associates* of α. Associates have equal norms, and the associates of an integral quaternion are integral.

If $\gamma = \alpha\beta$, then γ is said to have α as a *left-hand divisor* and β as a *right-hand divisor.* If $\alpha = a_0$ or $\beta = b_0$, then $\alpha\beta = \beta\alpha$ and the distinction of right and left is unnecessary.

20.7. Preliminary theorems about integral quaternions. Our second proof of Theorem 370 is similar in principle to that of Theorem 251 contained in §§ 12.8 and 15.1. We need some preliminary theorems.

THEOREM 371. *If α is an integral quaternion, then one at least of its associates has integral coordinates; and if α is odd, then one at least of its associates has non-integral coordinates.*

(1) If the coordinates of α itself are not integral, then we can choose the signs so that

$$\alpha = (b_0 + b_1 i_1 + b_2 i_2 + b_3 i_3) + \tfrac{1}{2}(\pm 1 \pm i_1 \pm i_2 \pm i_3) = \beta + \gamma,$$

say, where b_0, b_1, b_2, b_3 are even. Any associate of β has integral coordinates, and $\gamma\bar{\gamma}$; an associate of γ, is 1. Hence $\alpha\bar{\gamma}$, an associate of α, has integral coordinates.

(2) If α is odd, and has integral coordinates, then

$$\alpha = (b_0 + b_1 i_1 + b_2 i_2 + b_3 i_3) + (c_0 + c_1 i_1 + c_2 i_2 + c_3 i_3) = \beta + \gamma,$$

say, where b_0, b_1, b_2, b_3 are even, each of c_0, c_1, c_2, c_3 is 0 or 1, and (since $N\alpha$ is odd) either one is 1 or three are. Any associate of β has integral coordinates. It is therefore sufficient to prove that each of the quaternions

$$1, \quad i_1, \quad i_2, \quad i_3, \quad 1+i_2+i_3, \quad 1+i_1+i_3, \quad 1+i_1+i_2, \quad i_1+i_2+i_3$$

has an associate with non-integral coordinates, and this is easily verified. Thus, if $\gamma = i_1$ then $\gamma\rho$ has non-integral coordinates. If

$$\gamma = 1 + i_2 + i_3 = (1 + i_1 + i_2 + i_3) - i_1 = \lambda + \mu$$

or

$$\gamma = i_1 + i_2 + i_3 = (1 + i_1 + i_2 + i_3) - 1 = \lambda + \mu,$$

then

$$\lambda\epsilon = \lambda \,.\, \tfrac{1}{2}(1 - i_1 - i_2 - i_3) = 2$$

and the coordinates of $\mu\epsilon$ are non-integral.

Theorem 372. *If κ is an integral quaternion, and m a positive integer, then there is an integral quaternion λ such that*

$$N(\kappa - m\lambda) < m^2.$$

The case $m = 1$ is trivial, and we may suppose $m > 1$. We use the form (20.6.13) of an integral quaternion, and write

$$\kappa = k_0\rho + k_1 i_1 + k_2 i_2 + k_3 i_3, \quad \lambda = l_0\rho + l_1 i_1 + l_2 i_2 + l_3 i_3,$$

where $k_0, \ldots, l_0, \ldots$ are integers. The coordinates of $\kappa - m\lambda$ are

$$\tfrac{1}{2}(k_0 - ml_0),\ \tfrac{1}{2}\{k_0 + 2k_1 - m(l_0 + 2l_1)\},\ \tfrac{1}{2}\{k_0 + 2k_2 - m(l_0 + 2l_2)\},$$
$$\tfrac{1}{2}\{k_0 + 2k_2 - m(l_0 + 2l_3)\}.$$

We can choose l_0, l_1, l_2, l_3 in succession so that these have absolute values not exceeding $\frac{1}{4}m, \frac{1}{2}m, \frac{1}{2}m, \frac{1}{2}m$; and then

$$N(\kappa - m\lambda) \leqslant \tfrac{1}{16}m^2 + 3 \,.\, \tfrac{1}{4}m^2 < m^2.$$

THEOREM 373. *If α and β are integral quaternions, and $\beta \neq 0$, then there are integral quaternions λ and γ such that*

$$\alpha = \lambda\beta + \gamma, \quad N\gamma < N\beta.$$

We take

$$\kappa = \alpha\bar{\beta}, \quad m = \beta\bar{\beta} = N\beta,$$

and determine λ as in Theorem 372. Then

$$(\alpha - \lambda\beta)\bar{\beta} = \kappa - \lambda m = \kappa - m\lambda,$$

$$N(\alpha - \lambda\beta)N\bar{\beta} = N(\kappa - m\lambda) < m^2,$$

$$N\gamma = N(\alpha - \lambda\beta) < m = N\beta.$$

20.8. The highest common right-hand divisor of two quaternions. We shall say that two integral quaternions α and β have a *highest common right-hand divisor* δ if (i) δ is a right-hand divisor of α and β, and (ii) every right-hand divisor of α and β is a right-hand divisor of δ; and we shall prove that any two integral quaternions, not both 0, have a highest common right-hand divisor which is effectively unique. We could use Theorem 373 for the construction of a 'Euclidean algorithm' similar to those of §§ 12.3 and 12.8, but it is simpler to use ideas like those of §§ 2.9 and 15.7.

We call a system S of integral quaternions, one of which is not 0, a *right-ideal* if it has the properties

(i) $\alpha \in S \,.\, \beta \epsilon S \rightarrow \alpha \pm \beta \in S$,

(ii) $\alpha \in S \rightarrow \lambda\alpha \in S$ for all integral quaternions λ:

the latter property corresponds to the characteristic property of the ideals of § 15.7. If δ is any integral quaternion, and S is the set $(\lambda\delta)$ of all left-hand multiples of δ by integral quaternions λ, then it is plain that S is a right-ideal. We call such a right-ideal a *principal right-ideal.*

THEOREM 374. *Every right-ideal is a principal right-ideal.*

Among the members of S, not 0, there are some with minimum norm: we call one of these δ. If $\gamma \epsilon S, N\gamma < N\delta$ then $\gamma = 0$.

If $\alpha \in S$ then $\alpha - \lambda\delta \in S$, for every integral λ, by (i) and (ii). By Theorem 373, we can choose λ so that $N\gamma = N(\alpha - \lambda\delta) < N\delta$. But then $\gamma = 0, \alpha = \lambda\delta$, and so S is the principal right-ideal $(\lambda\delta)$.

We can now prove

THEOREM 375. *Any two integral quaternions α and β, not both 0, have a highest common right-hand divisor δ, which is unique except for a left-hand unit factor, and can be expressed in the form*

$$\delta = \mu\alpha + \nu\beta, \tag{20.8.1}$$

where μ and ν are integral.

The set S of all quaternions $\mu\alpha + \nu\beta$ is plainly a right-ideal which, by Theorem 374, is the principal right-ideal formed by all integral multiples $\lambda\delta$ of a certain δ. Since S includes δ, δ can be expressed in the form (20.8.1). Since S includes α and β, δ is a common right-hand divisor of α and β; and any such divisor is a right-hand divisor of every member of S, and therefore of δ. Hence δ is a highest common right-hand divisor of α and β.

Finally, if both δ and δ' satisfy the conditions, $\delta' = \lambda\delta$ and $\delta = \lambda'\delta'$, where λ and λ' are integral. Hence $\delta = \lambda'\lambda\delta, 1 = \lambda'\lambda$, and λ and λ' are unities.

If δ is a unity ϵ, then all highest common right-hand divisors of α and β are unities. In this case

$$\mu'\alpha + \nu'\beta = \epsilon,$$

for some integral μ', ν'; and

$$(\epsilon^{-1}\mu')\alpha + (\epsilon^{-1}\nu')\beta = 1;$$

so that

$$\mu\alpha + \nu\beta = 1, \tag{20.8.2}$$

for some integral μ, ν. We then write

$$(\alpha, \beta)_r = 1. \tag{20.8.3}$$

We could of course establish a similar theory of the highest common left-hand divisor.

If α and β have a common right-hand divisor δ, not a unity, then $N\alpha$ and $N\beta$ have the common right-hand divisor $N\delta > 1$. There is one important case in which the converse is true.

THEOREM 376. *If α is integral and $\beta = m$, a positive rational integer, then a necessary and sufficient condition that* $(\alpha, \beta)_r = 1$ *is that* $(N\alpha, N\beta) = 1$, *or (what is the same thing) that* $(N\alpha, m) = 1$.

For if $(\alpha, \beta)_r = 1$ then (20.8.2) is true for appropriate μ, ν. Hence

$$N(\mu\alpha) = N(1 - \nu\beta) = (1 - m\nu)(1 - m\bar{\nu}),$$

$$N\mu N\alpha = 1 - m\nu - m\bar{\nu} + m^2 N\nu,$$

and $(N\alpha, m)$ divides every term in this equation except 1. Hence $(N\alpha, m) = 1$. Since $N\beta = m^2$, the two forms of the condition are equivalent.

20.9. Prime quaternions and the proof of Theorem 370. An integral quaternion π, not a unity, is said to be *prime* if its only divisors are the unities and its associates, i.e. if $\pi = \alpha\beta$ implies that either α or β is a unity. It is plain that all associates of a prime are prime. If $\pi = \alpha\beta$, then $N\pi = N\alpha N\beta$, so that π is certainly prime if $N\pi$ is a rational prime. We shall prove that the converse is also true.

THEOREM 377. *An integral quaternion π is prime if and only if its norm $N\pi$ is a rational prime.*

Since $Np = p^2$, a particular case of Theorem 377 is

THEOREM 378. *A rational prime p cannot be a prime quaternion.*

We begin by proving Theorem 378 (which is all that we shall actually need).

Since

$$2 = (1 + i_1)(1 - i_1),$$

2 is not a prime quaternion. We may therefore suppose p odd.

By Theorem 87, there are integers r and s such that

$$0 < r < p, \quad 0 < s < p, \quad 1 + r^2 + s^2 \equiv 0 \pmod{p}.$$

If

$$\alpha = 1 + si_2 - ri_3,$$

then

$$N\alpha = 1 + r^2 + s^2 \equiv 0 \pmod{p},$$

and $(N\alpha, p) > 1$. It follows, by Theorem 376, that α and p have a common right-hand divisor δ which is not a unity. If

$$\alpha = \delta_1\delta, \quad p = \delta_2\delta,$$

then δ_2 is not a unity; for if it were then δ would be an associate of p, in which case p would divide all the coordinates of

$$\alpha = \delta_1\delta = \delta_1\delta_2^{-1}p,$$

and in particular 1. Hence $p = \delta_2\delta$, where neither δ nor δ_2 is a unity, and so p is not prime.

To complete the proof of Theorem 377, suppose that π is prime and p a rational prime divisor of $N\pi$. By Theorem 376, π and p have a common right-hand divisor π' which is not a unity. Since π is prime, π' is an associate of π and $N\pi' = N\pi$. Also $p = \lambda\pi'$, where λ is integral; and $p^2 = N\lambda N\pi' = N\lambda N\pi$, so that $N\lambda$ is 1 or p. If $N\lambda$ were 1, p would be an associate of π' and π, and so a prime quaternion, which we have seen to be impossible. Hence $N\pi = p$, a rational prime.

It is now easy to prove Theorem 370. If p is any rational prime, $p = \lambda\pi$, where $N\lambda = N\pi = p$. If π has integral coordinates a_0, a_1, a_2, a_3, then

$$p = N\pi = a_0^2 + a_1^2 + a_2^2 + a_3^2.$$

If not then, by Theorem 371, there is an associate π' of π which has integral coordinates. Since

$$p = N\pi = N\pi',$$

the conclusion follows as before.

The analysis of the preceding sections may be developed so as to lead to a complete theory of the factorization of integral quaternions and of the representation of rational integers by sums of four squares. In particular it leads to formulae for the number of representations, analogous to those of §§ 16.9–10. We shall prove these formulae by a different method in § 20.12, and shall not pursue the arithmetic of quaternions further here. There is however one other interesting theorem which is an immediate consequence of our analysis. If we suppose p odd, and select an associate π' of π whose coordinates are halves of odd integers (as we may by Theorem 371), then

$$p = N\pi = N\pi' = (b_0 + \tfrac{1}{2})^2 + (b_1 + \tfrac{1}{2})^2 + (b_2 + \tfrac{1}{2})^2 + (b_3 + \tfrac{1}{2})^2,$$

where $b_0, \ldots$ are integers, and

$$4p = (2b_0 + 1)^2 + (2b_1 + 1)^2 + (2b_2 + 1)^2 + (2b_3 + 1)^2.$$

Hence we obtain

THEOREM 379. *If p is an odd prime, then $4p$ is the sum of four odd integral squares.*

Thus $4 \,.\, 3 = 12 = 1^2 + 1^2 + 1^2 + 3^2$ (but $4 \,.\, 2 = 8$ is not the sum of four odd integral squares).

20.10. The values of $g(2)$ and $G(2)$. Theorem 369 shows that

$$G(2) \leqslant g(2) \leqslant 4.$$

On the other hand,

$$(2m)^2 \equiv 0 \pmod 4, \quad (2m+1)^2 \equiv 1 \pmod 8,$$

so that

$$x^2 \equiv 0, 1, \text{ or } 4 \pmod 8$$

and

$$x^2 + y^2 + z^2 \not\equiv 7 \pmod 8.$$

Hence no number $8m + 7$ is representable by three squares, and we obtain

THEOREM 380:

$$g(2) = G(2) = 4.$$

If $x^2 + y^2 + z^2 \equiv 0 \pmod 4$, then all of x, y, z are even, and

$$\tfrac{1}{4}\left(x^2 + y^2 + z^2\right) = \left(\tfrac{1}{2}x\right)^2 + (\tfrac{1}{2}y)^2 + (\tfrac{1}{2}z)^2$$

is representable by three squares. It follows that no number $4^a(8m+7)$ is the sum of three squares. It can be proved that any number not of this form is the sum of three squares, so that

$$n \neq 4^a(8m+7)$$

is a necessary and sufficient condition for n to be representable by three squares; but the proof depends upon the theory of ternary quadratic forms and cannot be included here.

20.11. Lemmas for the third proof of Theorem 369. Our third proof of Theorem 369 is of a quite different kind and, although 'elementary', belongs properly to the theory of elliptic functions.

The coefficient $r_4(n)$ of x^n in

$$(1+2x+2x^4+\cdots)^4=\left(\sum_{m=-\infty}^{\infty}x^{m^2}\right)^4$$

is the number of solutions of

$$n=m_1^2+m_2^2+m_3^2+m_4^2$$

in rational integers, solutions differing only in the sign or order of the m being reckoned as distinct. We have to prove that this coefficient is positive for every n.

By Theorem 312

$$(1+2x+2x^4+\cdots)^2=1+4\left(\frac{x}{1-x}-\frac{x^3}{1-x^3}+\cdots\right),$$

and we proceed to find a transformation of the square of the right-hand side.

In what follows x is any number, real or complex, for which $|x|<1$. The series which we use, whether simple or multiple, are absolutely convergent for $|x|<1$. The rearrangements to which we subject them are all justified by the theorem that any absolutely convergent series, simple or multiple, may be summed in any manner we please.

We write

$$u_r=\frac{x^r}{1-x^r},$$

so that

$$\frac{x^r}{(1-x^r)^2}=u_r(1+u_r).$$

We require two preliminary lemmas.

THEOREM 381:

$$\sum_{m=1}^{\infty}u_m(1+u_m)=\sum_{n=1}^{\infty}nu_n.$$

For

$$\sum_{m=1}^{\infty} \frac{x^m}{(1-x^m)^2} = \sum_{m=1}^{\infty}\sum_{n=1}^{\infty} nx^{mn} = \sum_{n=1}^{\infty} n \sum_{m=1}^{\infty} x^{mn} = \sum_{n=1}^{\infty} \frac{nx^n}{1-x^n}.$$

THEOREM 382:

$$\sum_{m=1}^{\infty} (-1)^{m-1} u_{2m}(1+u_{2m}) = \sum_{n=1}^{\infty} (2n-1)u_{4n-2}.$$

For

$$\begin{aligned}\sum_{m=1}^{\infty} \frac{(-1)^{m-1}x^{2m}}{\left(1-x^{2m}\right)^2} &= \sum_{m=1}^{\infty}(-1)^{m-1}\sum_{r=1}^{\infty} rx^{2mr} \\ &= \sum_{r=1}^{\infty} r \sum_{m=1}^{\infty}(-1)^{m-1}x^{2mr} = \sum_{r=1}^{\infty} \frac{rx^{2r}}{1+x^{2r}} \\ &= \sum_{r=1}^{\infty}\left(\frac{rx^{2r}}{1-x^{2r}} - \frac{2rx^{4r}}{1-x^{4r}}\right) = \sum_{n=1}^{\infty} \frac{(2n-1)x^{4n-2}}{1-x^{4n-2}}.\end{aligned}$$

20.12. Third proof of Theorem 369: the number of representations. We begin by proving an identity more general than the actual one we need.

THEOREM 383. *If θ is real and not an even multiple of π, and if*

$$\begin{aligned} L = L(x,\theta) &= \tfrac{1}{4}\cot\tfrac{1}{2}\theta + u_1 \sin\theta + u_2 \sin 2\theta + \cdots, \\ T_1 = T_1(x,\theta) &= \left(\tfrac{1}{4}\cot\tfrac{1}{2}\theta\right)^2 + u_1(1+u_1)\cos\theta \\ &\qquad + u_2(1+u_2)\cos 2\theta + \cdots, \\ T_2 = T_2(x,\theta) &= \tfrac{1}{2}\{u_1(1-\cos\theta) + 2u_2(1-\cos 2\theta) \\ &\qquad + 3u_2(1-\cos 3\theta) + \cdots\}, \end{aligned}$$

then

$$L^2 = T_1 + T_2.$$

We have

$$L^2 = \left\{ \tfrac{1}{4}\cot\tfrac{1}{2}\theta + \sum_{n=1}^{\infty} u_n \sin n\theta \right\}^2$$

$$= \left(\tfrac{1}{4}\cot\tfrac{1}{2}\theta\right)^2 + \tfrac{1}{2}\sum_{n=1}^{\infty} u_n \cot\tfrac{1}{2}\theta \sin n\theta + \sum_{m=1}^{\infty}\sum_{n=1}^{\infty} u_m u_n \sin m\theta \sin n\theta$$

$$= \left(\tfrac{1}{4}\cot\tfrac{1}{2}\theta\right)^2 + S_1 + S_2,$$

say. We now use the identities

$$\tfrac{1}{2}\cot\tfrac{1}{2}\theta \sin n\theta = \tfrac{1}{2} + \cos\theta + \cos 2\theta + \cdots + \cos(n-1)\theta + \tfrac{1}{2}\cos n\theta,$$

$$2\sin m\theta \sin n\theta = \cos(m-n)\theta - \cos(m+n)\theta,$$

which give

$$S_1 = \sum_{n=1}^{\infty} u_n \left\{\tfrac{1}{2} + \cos\theta + \cos 2\theta + \cdots + \cos(n-1)\theta + \tfrac{1}{2}\cos n\theta\right\},$$

$$S_2 = \tfrac{1}{2}\sum_{m=1}^{\infty}\sum_{n=1}^{\infty} u_m u_n\{\cos(m-n)\theta - \cos(m+n)\theta\}.$$

and

$$L^2 = \left(\tfrac{1}{4}\cot\tfrac{1}{2}\theta\right)^2 + C_0 + \sum_{k=1}^{\infty} C_k \cos k\theta,$$

say, on rearranging S_1 and S_2 as series of cosines of multiples of θ.†

† To justify this rearrangement we have to prove that

$$\sum_{n=1}^{\infty} |u_n| \left(\tfrac{1}{2} + |\cos\theta| + \cdots + \tfrac{1}{2}|\cos n\theta|\right)$$

and

$$\sum_{m=1}^{\infty}\sum_{n=1}^{\infty} |u_m||u_n|(|\cos(m+n)\theta| + |\cos(m-n)\theta|)$$

are convergent. But this is an immediate consequence of the absolute convergence of

$$\sum_{n=1}^{\infty} nu_n, \quad \sum_{m=1}^{\infty}\sum_{n=1}^{\infty} u_m u_n.$$

We consider C_0 first. This coefficient includes a contribution $\frac{1}{2}\sum_1^{\infty} u_n$ from S_1, and a contribution $\frac{1}{2}\sum_1^{\infty} u_n^2$ from the terms of S_2 for which $m = n$. Hence

$$C_0 = \tfrac{1}{2}\sum_{n=1}^{\infty}(u_n + u_n^2) = \tfrac{1}{2}\sum_{n=1}^{\infty} nu_n,$$

by Theorem 381.

Now suppose $k > 0$. Then S_1 contributes

$$\tfrac{1}{2}u_k + \sum_{n=k+1}^{\infty} u_n = \tfrac{1}{2}u_k + \sum_{l=1}^{\infty} u_{k+l}$$

to C_k, while S_2 contributes

$$\tfrac{1}{2}\sum_{m-n=k} u_m u_n + \tfrac{1}{2}\sum_{n-m=k} u_m u_n - \tfrac{1}{2}\sum_{m+n=k} u_m u_n,$$

where $m \geqslant 1, n \geqslant 1$ in each summation. Hence

$$C_k = \tfrac{1}{2}u_k + \sum_{l=1}^{\infty} u_{k+l} + \sum_{l=1}^{\infty} u_l u_{k+l} - \tfrac{1}{2}\sum_{l=1}^{k-1} u_l u_{k-1}.$$

The reader will easily verify that

$$u_l u_{k-l} = u_k(1 + u_l + u_{k-1})$$

and

$$u_{k+1} + u_l u_{k+l} = u_k(u_l - u_{k+l}).$$

Hence

$$\begin{aligned} C_k &= u_k\left\{\tfrac{1}{2} + \sum_{l=1}^{\infty}(u_l - u_{k+1}) - \tfrac{1}{2}\sum_{l=1}^{k-1}(1 + u_1 + u_{k-l})\right\} \\ &= u_k\left\{\tfrac{1}{2} + u_1 + u_2 + \cdots + u_k - \tfrac{1}{2}(k-1) - (u_1 + u_2 + \cdots + u_{k-1})\right\} \\ &= u_k\left(1 + u_k - \tfrac{1}{2}k\right), \end{aligned}$$

and so

$$L^2 = \left(\tfrac{1}{4}\cot\tfrac{1}{2}\theta\right)^2 + \tfrac{1}{2}\sum_{n=1}^{\infty} nu_n + \sum_{k=1}^{\infty} u_k\left(1 + u_k - \tfrac{1}{2}k\right)\cos k\theta$$

$$= \left(\tfrac{1}{4}\cot\tfrac{1}{2}\theta\right)^2 + \sum_{k=1}^{\infty} u_k(1+u_k)\cos k\theta + \tfrac{1}{2}\sum_{k=1}^{\infty} ku_k(1-\cos k\theta)$$

$$= T_1(x,\theta) + T_2(x,\theta).$$

THEOREM 384:

$$\left(\tfrac{1}{4} + u_1 - u_3 + u_5 - u_7 + \cdots\right)^2$$
$$= \tfrac{1}{16} + \tfrac{1}{2}(u_1 + 2u_2 + 3u_3 + 5u_5 + 6u_6 + 7u_7 + 9u_9 + \cdots),$$

where in the last series there are no terms in u_4, u_8, $u_{12}, \ldots$.

We put $\theta = \frac{1}{2}\pi$ in Theorem 383. Then we have

$$T_1 = \tfrac{1}{16} - \sum_{m=1}^{\infty} (-1)^{m-1} u_{2m}(1 + u_{2m}),$$

$$T_2 = \tfrac{1}{2}\sum_{m=1}^{\infty} (2m-1)u_{2m-1} + 2\sum_{m=1}^{\infty} (2m-1)u_{4m-2}.$$

Now, by Theorem 382,

$$T_1 = \tfrac{1}{16} - \sum_{m=1}^{\infty} (2m-1)u_{4m-2},$$

and so

$$T_1 + T_2 = \tfrac{1}{16} + \tfrac{1}{2}(u_1 + 2u_2 + 3u_3 + 5u_5 + \cdots).$$

From Theorems 312 and 384 we deduce

THEOREM 385:

$$(1 + 2x + 2x^4 + 2x^9 + \cdots)^4 = 1 + 8\sum{}' mu_m,$$

where m runs through all positive integral values which are not multiples of 4.

Finally,

$$8\sum' mu_m = 8\sum' \frac{mx^m}{1-x^m} = 8\sum' m\sum_{r=1}^{\infty} x^{mr} = 8\sum_{n=1}^{\infty} c_n x^n,$$

where

$$c_n = \sum_{m|n,4\nmid m} m$$

is the sum of the divisors of n which are not multiples of 4.

It is plain that $c_n > 0$ for all $n > 0$, and so $r_4(n) > 0$. This provides us with another proof of Theorem 369; and we have also proved

THEOREM 386. *The number of representations of a positive integer n as the sum of four squares, representations which differ only in order or sign being counted as distinct, is* 8 *times the sum of the divisors of n which are not multiples of* 4.

20.13. Representations by a larger number of squares. There are similar formulae for the numbers of representations of n by 6 or 8 squares. Thus

$$r_6(n) = 16\sum_{d|n} \chi(d')d^2 - 4\sum_{d|n} \chi(d)d^2,$$

where $dd' = n$ and $\chi(d)$, as in § 16.9, is 1, -1, or 0 according as d is $4k+1$, $4k-1$, or $2k$; and

$$r_8(n) = 16(-1)^n \sum_{d|n} (-1)^d d^3.$$

These formulae are the arithmetical equivalents of the identities

$$(1+2x+2x^4+\cdots)^6 = 1+16\left(\frac{1^2x}{1+x^2}+\frac{2^2x^2}{1+x^4}+\frac{3^2x^3}{1+x^6}+\cdots\right)$$
$$-4\left(\frac{1^2x}{1-x}-\frac{3^2x^3}{1-x^3}+\frac{5^2x^5}{1-x^5}-\cdots\right),$$

and

$$(1+2x+2x^4+\cdots)^8 = 1+16\left(\frac{1^3x}{1+x}+\frac{2^3x^2}{1-x^2}+\frac{3^3x^3}{1+x^3}+\cdots\right).$$

These identities also can be proved in an elementary manner, but have their roots in the theory of the elliptic modular functions. That $r_6(n)$ and $r_8(n)$ are positive for all n is trivial after Theorem 369.

The formulae for $r_s(n)$, where $s = 10, 12, \ldots$, involve other arithmetical functions of a more recondite type. Thus $r_{10}(n)$ involves sums of powers of the complex divisors of n.

The corresponding problems for representations of n by sums of an *odd* number of squares are more difficult, as may be inferred from § 20.10. When s is 3, 5, or 7 the number of representations is expressible as a finite sum involving the symbol $\left(\frac{m}{n}\right)$ of Legendre and Jacobi.

NOTES

§ 20.1. Waring made his assertion in *Meditationes algebraicae* (1770), 204–5, and Lagrange proved that $g(2) = 4$ later in the same year. There is an exhaustive account of the history of the four-square theorem in Dickson, *History*, ii, ch. viii.

Hilbert's proof of the existence of $g(k)$ for every k was published in *Göttinger Nachrichten* (1909), 17–36, and *Math. Annalen*, 67 (1909), 281–305. Previous writers had proved its existence when $k = 3, 4, 5, 6, 7, 8$, and 10, but its value had been determined only for $k = 3$. The value of $g(k)$ is now known for all k: that of $G(k)$ for $k = 2$ and $k = 4$ only. The determinations of $g(k)$ rest on a previous determination of an upper bound for $G(k)$.

See also Dickson, *History*, ii, ch. 25, and our notes on Ch. XXI.

Lord Saltoun drew my attention to an error on p. 394.

§ 20.3. This proof is due to Hermite, *Journal de math.* (1), 13 (1848), 15 (*Œuvres*, i. 264).

§ 20.4. The fourth proof is due to Grace, *Journal London Math. Soc.* 2 (1927), 3–8. Grace also gives a proof of Theorem 369 based on simple properties of four-dimensional lattices.

§ 20.5. Bachet enunciated Theorem 369 in 1621, though he did not profess to have proved it. The proof in this section is substantially Euler's.

§§ 20.6–9. These sections are based on Hurwitz, *Vorlesungen über die Zahlentheorie der Quaternionen* (Berlin, 1919). Hurwitz develops the theory in much greater detail, and uses it to find the formulae of § 20.12. We go so far only as is necessary for the proof of Theorem 370; we do not, for example, prove any general theorem concerning uniqueness of factorization. There is another account of Hurwitz's theory, with generalizations, in Dickson, *Algebren und ihre Zahlentheorie* (Zürich, 1927), ch. 9.

Lipschitz (*Untersuchungen über die Summen von Quadrat,* Bonn, 1886) was the first to develop and publish an arithmetic of quaternions, though Hamilton, the inventor of quaternions, gave the same method in an unpublished letter in 1856 (see *The Mathematical papers of Sir. Wm. R. Hamilton* (ed. Halberstam and Ingram), xviii and Appendix 4). Lipschitz (like Hamilton) defines an integral quaternion in the most obvious manner, viz.

as one with integral coordinates, but his theory is much more complicated than Hurwitz's. Later, Dickson [*Proc. London Math. Soc.* (2) 20 (1922), 225–32] worked out an alternative and much simpler theory based on Lipschitz's definition. We followed this theory in our first edition, but it is less satisfactory than Hurwitz's: it is not true, for example, in Dickson's theory, that any two integral quaternions have a highest common right-hand divisor.

§ 20.10. The 'three-square theorem', which we do not prove, is due to Legendre, *Essai sur la théorie des nombres* (1798), 202, 398–9, and Gauss, *D.A.*, § 291. Gauss determined the number of representations. See Landau, *Vorlesungen,* i. 114–25. There is a proof, depending on the methods of Liouville, referred to in the note on § 20.13 below, in Uspensky and Heaslet, 465–74 and another proof, due to Ankeny (*Proc. American Math. Soc.* 8 (1957), 316–19) depending only on Minkowski's theorem (our Theorem 447) and Dirichlet's theorem (our Theorem 15).

§§ 20.11–12. Ramanujan, *Collected papers*, 138 *et seq*.

§ 20.13. The results for 6 and 8 squares are due to Jacobi, and are contained implicitly in the formulae of §§ 40–42 of the *Fundamenta nova.* They are stated explicitly in Smith's *Report on the theory of numbers* (*Collected papers,* i. 306–7). Liouville gave formulae for 12 and 10 squares in the *Journal de math.* (2) 9 (1864), 296–8, and 11 (1866), 1–8. Glaisher, *Proc. London Math. Soc.* (2) 5 (1907), 479–90, gave a systematic table of formulae for $r_{2s}(n)$ up to $2s = 18$, based on previous work published in vols. 36–39 of the *Quarterly Journal of Math.* The formulae for 14 and 18 squares contain functions defined only as the coefficients in certain modular functions and not arithmetically. Ramanujan (*Collected papers*, no. 18) continues Glaisher's table up to $2s = 24$.

Boulyguine, in 1914, found general formulae for $r_{2s}(n)$ in which every function which occurs has an arithmetical definition. Thus the formula for $r_{2s}(n)$ contains functions $\sum \phi(x_1, x_2, \ldots, x_t)$, where ϕ is a polynomial, t has one of the values $2s-8, 2s-16, \ldots$, and the summation is over all solutions of $x_1^2 + x_2^2 + \cdots + x_t^2 = n$. There are references to Boulyguine's work in Dickson's *History*, ii. 317.

Uspensky developed the elementary methods which seem to have been used by Liouville in a series of papers published in Russian: references will be found in a later paper in *Trans. Amer. Math. Soc.* 30 (1928), 385–404. He carries his analysis up to $2s = 12$, and states that his methods enable him to prove Boulyguine's general formulae.

A more analytic method, applicable also to representations by an odd number of squares, has been developed by Hardy, Mordell, and Ramanujan. See Hardy, *Trans. Amer. Math. Soc.* 21 (1920), 255–84, and *Ramanujan*, ch. 9; Mordell, *Quarterly Journal of Math.* 48 (1920), 93–104, and *Trans. Camb. Phil. Soc.* 22 (1923), 361–72; Estermann, *Acta arithmetica*, 2 (1936), 47–79; and nos. 18 and 21 of Ramanujan's *Collected papers.*

We defined Legendre's symbol in § 6.5. Jacobi's generalization is defined in the more systematic treatises, e.g. in Landau, *Vorlesungen*, i. 47.

Self-contained formulae for the number of representations of a positive integer as the sum of squares are nowadays seen to be explained by the theory of modular forms (see, for example, Chapter 11 of H. Iwaniec, *Topics in classical automorphic forms,* Amer. Math. Soc., 1997). Indeed one may consider positive-definite quadratic forms

$$Q(x_1, \ldots, x_n) = \sum_{i,j=1}^{n} a_{ij} x_i x_j \quad (a_{ij} = a_{ji} \text{ integers})$$

in complete generality by such methods.

An elegant result for such forms has been proved by Conway and Schneeberger (unpublished). This states that if Q represents every positive integer up to and including 15, then it represent all positive inttegers. One cannot reduce the number 15, since in fact

$x_1^2 + 2x_2^2 + 5x_3^2 + 5x_4^2$ represents all positive integers except 15. A more difficult version of this result has been established by Bhargava (*Quadratic forms and their applications (Dublin, 1999)*, 27–37, Contemp. Math., 272, Amer. Math. Soc., Providence, RI, 2000), referring to forms

$$Q(x_1, \ldots, x_n) = \sum_{1 \leqslant i \leqslant j \leqslant n} a_{ij} x_i x_j \quad (a_{ij} \text{ integers}).$$

In this case, if every integer up to 290 is represented then all integers are represented.

XXI

REPRESENTATION BY CUBES AND HIGHER POWERS

21.1. Biquadrates. We defined 'Waring's problem' in § 20.1 as the problem of determining $g(k)$ and $G(k)$, and solved it completely when $k = 2$. The general problem is much more difficult. Even the proof of the existence of $g(k)$ and $G(k)$ requires quite elaborate analysis; and the value of $G(k)$ is not known for any k but 2 and 4. We give a summary of the present state of knowledge at the end of the chapter, but we shall prove only a few special theorems, and these usually not the best of their kind that are known.

It is easy to prove the existence of $g(4)$.

THEOREM 387. *$g(4)$ exists, and does not exceed* 50.

The proof depends on Theorem 369 and the identity

$$6(a^2+b^2+c^2+d^2)^2 = (a+b)^4+(a-b)^4+(c+d)^4 \\ +(c-d)^4+(a+c)^4+(a-c)^4 \\ +(b+d)^4+(b-d)^4+(a+d)^4 \\ +(a-d)^4+(b+c)^4+(b-c)^4. \tag{21.1.1}$$

We denote by B_s a number which is the sum of s or fewer biquadrates. Thus (21.1.1) shows that

$$6\left(a^2+b^2+c^2+d^2\right)^2 = B_{12},$$

and therefore, after Theorem 369, that

$$6x^2 = B_{12}, \tag{21.1.2}$$

for every x.

Now any positive integer n is of the form

$$n = 6N + r,$$

where $N \geqslant 0$ and r is 0, 1, 2, 3, 4, or 5. Hence (again by Theorem 369)

$$n = 6\left(x_1^2+x_2^2+x_3^2+x_4^2\right)+r;$$

and therefore, by (21.1.2),

$$n = B_{12} + B_{12} + B_{12} + B_{12} + r = B_{48} + r = B_{53}$$

(since r is expressible by at most 5 1's). Hence $g(4)$ exists and is at most 53.

It is easy to improve this result a little. Any $n \geqslant 81$ is expressible as

$$n = 6N + t,$$

where $N \geqslant 0$, and $t = 0, 1, 2, 81, 16$, or 17, according as $n \equiv 0, 1, 2, 3, 4$, or 5 (mod 6). But

$$1 = 1^4, \quad 2 = 1^4 + 1^4, \quad 81 = 3^4, \quad 16 = 2^4, \quad 17 = 2^4 + 1^4.$$

Hence $t = B_2$, and therefore

$$n = B_{48} + B_2 = B_{50},$$

so that any $n \geqslant 81$ is B_{50}.

On the other hand it is easily verified that $n = B_{19}$ if $1 \leqslant n \leqslant 80$. In fact only

$$79 = 4 \,.\, 2^4 + 15 \,.\, 1^4$$

requires 19 biquadrates.

21.2. Cubes: the existence of $G(3)$ and $g(3)$. The proof of the existence of $g(3)$ is more sophisticated (as is natural because a cube may be negative). We prove first

Theorem 388:

$$G(3) \leqslant 13.$$

We denote by C_s a number which is the sum of s non-negative cubes.

We suppose that z runs through the values $7, 13, 19, \ldots$ congruent to 1 (mod 6), and that I_z is the interval

$$\phi(z) = 11z^9 + (z^3 + 1)^3 + 125z^3 \leqslant n \leqslant 14z^9 = \psi(z).$$

It is plain that $\phi(z + 6) < \psi(z)$ for large z, so that the intervals I_z ultimately overlap, and every large n lies in some I_z. It is therefore sufficient to prove that every n of I_z is the sum of 13 non-negative cubes.

We prove that any n of I_z can be expressed in the form

$$n = N + 8z^9 + 6mz^3, \tag{21.2.1}$$

where

$$N = C_5, \quad 0 < m < z^6. \tag{21.2.2}$$

We shall then have

$$m = x_1^2 + x_2^2 + x_3^2 + x_4^2,$$

where $0 \leqslant x_i < z^3$; and so

$$\begin{aligned} n &= N + 8z^9 + 6z^3(x_1^2 + x_2^2 + x_3^2 + x_4^2) \\ &= N + \sum_{i=1}^{4} \{(z^3 + x_i)^3 + (z^3 - x_i)^3\} \\ &= C_5 + C_8 = C_{13}. \end{aligned}$$

It remains to prove (21.2.1). We define r, s, and N by

$$\begin{aligned} &n \equiv 6r \pmod{z^3} \quad (1 \leqslant r \leqslant z^3), \\ &n \equiv s + 4 \pmod{6} \quad (0 \leqslant s \leqslant 5), \\ &N = (r+1)^3 + (r-1)^3 + 2(z^3 - r)^3 + (sz)^3. \end{aligned}$$

Then $N = C_5$ and

$$0 < N < (z^3+1)^3 + 3z^9 + 125z^3 = \phi(z) - 8z^9 \leqslant n - 8z^9,$$

so that

$$8z^9 < n - N < 14z^9. \tag{21.2.3}$$

Now

$$N \equiv (r+1)^3 + (r-1)^3 - 2r^3 = 6r \equiv n \equiv n - 8z^9 \pmod{z^3}.$$

Also $x^3 \equiv x \pmod{6}$ for every x, and so

$$\begin{aligned} N &\equiv r + 1 + r - 1 + 2(z^3 - r) + sz = 2z^3 + sz \\ &\equiv (2+s)z \equiv 2 + s \equiv n - 2 \\ &\equiv n - 8 \equiv n - 8z^9 \pmod{6}. \end{aligned}$$

Hence $n - N - 8z^9$ is a multiple of $6z^3$. This proves (21.2.1), and the inequality in (21.2.2) follows from (21.2.3).

The existence of $g(3)$ is a corollary of Theorem 388. It is however interesting to show that the bound for $G(3)$ stated in the theorem is also a bound for $g(3)$.

21.3. A bound for $g(3)$. We must begin by proving a sharpened form of Theorem 388, with a definite limit beyond which all numbers are C_{13}.

THEOREM 389. *If $n \geqslant 10^{25}$, then $n = C_{13}$.*

We prove first that $\phi(z+6) \leqslant \psi(z)$ if $z \geqslant 373$, or that

$$11t^9 + (t^3+1)^3 + 125t^3 \leqslant 14(t-6)^9,$$

i.e.

$$14\left(1 - \frac{6}{t}\right)^9 \geqslant 12 + \frac{3}{t^3} + \frac{128}{t^6} + \frac{128}{t^6} + \frac{1}{t^9}, \tag{21.3.1}$$

if $t \geqslant 379$. Now

$$(1-\delta)^m > 1 - m\delta$$

if $0 < \delta < 1$. Hence

$$\left(1 - \frac{6}{t}\right)^9 > 1 - \frac{54}{t}$$

if $t > 6$; and so (21.3.1) is satisfied if

$$14\left(1 - \frac{54}{t}\right) \geqslant 12 + \frac{3}{t^3} + \frac{128}{t^6} + \frac{1}{t^9},$$

or if

$$2(t - 7\,.\,54) \geqslant \frac{3}{t^2} + \frac{128}{t^5} + \frac{1}{t^8}.$$

This is clearly true if $t \geqslant 7\,.\,54 + 1 = 379$.

It follows that the intervals I_z overlap from $z = 373$ onwards, and n certainly lies in an I_z if

$$n \geqslant 14(373)^9,$$

which is less than 10^{25}.

We have now to consider representations of numbers less than 10^{25}. It is known from tables that all numbers up to 40000 are C_9, and that, among these numbers, only 23 and 239 require as many cubes as 9.
Hence

$$n = C_9 \quad (1 \leqslant n \leqslant 239), \quad n = C_8 \quad (240 \leqslant n \leqslant 40000).$$

Next, if $N \geqslant 1$ and $m = \left[N^{\frac{1}{3}}\right]$, we have

$$N - m^3 = \left(N^{\frac{1}{3}}\right)^3 - m^3 \leqslant 3N^{\frac{2}{3}}\left(N^{\frac{1}{3}} - m\right) < 3N^{\frac{2}{3}}.$$

Now let us suppose that

$$240 \leqslant n \leqslant 10^{25}$$

and put $\quad n = 240 + N, \quad 0 \leqslant N < 10^{25}.$
Then

$$N = m^3 + N_1, \quad m = \left[N^{\frac{1}{3}}\right], \quad 0 \leqslant N_1 < 3N^{\frac{2}{3}},$$

$$N_1 = m_1^3 + N_2, \quad m_1 = \left[N_1^{\frac{1}{3}}\right], \quad 0 \leqslant N_2 < 3N_1^{\frac{2}{3}},$$

$$\cdots\cdots\cdots\cdots$$

$$N_4 = m_4^3 + N_5, \quad m_4 = \left[N_4^{\frac{1}{3}}\right], \quad 0 \leqslant N_5 < 3N_4^{\frac{2}{3}}.$$

Hence

$$n = 240 + N = 240 + N_5 + m^3 + m_1^3 + m_2^3 + m_3^3 + m_4^3. \tag{21.3.2}$$

Here

$$\begin{aligned} 0 &\leqslant N_5 \leqslant 3N_4^{\frac{2}{3}} \leqslant 3(3N_3^{\frac{2}{3}})^{\frac{2}{3}} \leqslant \dots \\ &< 3 \,.\, 3^{\frac{2}{3}} 3^{\left(\frac{2}{3}\right)^2} 3^{\left(\frac{2}{3}\right)^3} 3^{\left(\frac{2}{3}\right)^4} N^{\left(\frac{2}{3}\right)^5} \\ &= 27\left(\frac{N}{27}\right)^{\left(\frac{2}{3}\right)^5} < 27\left(\frac{10^{25}}{27}\right)^{\left(\frac{2}{3}\right)^5} < 35000. \end{aligned}$$

Hence

$$240 \leqslant 240 + N_5 < 35240 < 40000,$$

and so $240 + N_5$ is C_8; and therefore, by (21.3.2), n is C_{13}. Hence all positive integers are sums of 13 cubes.

THEOREM 390:

$$g(3) \leqslant 13.$$

The true value of $g(3)$ is 9, but the proof of this demands Legendre's theorem (§ 20.10) on the representation of numbers by sums of *three* squares. We have not proved this theorem and are compelled to use Theorem 369 instead, and it is this which accounts for the imperfection of our result.

21.4. Higher powers. In § 21.1 we used the identity (21.1.1) to deduce the existence of $g(4)$ from that of $g(2)$. There are similar identities which enable us to deduce the existence of $g(6)$ and $g(8)$ from that of $g(3)$ and $g(4)$. Thus

$$60(a^2+b^2+c^2+d^2)^3 = \sum (a \pm b \pm c)^6 + 2\sum (a \pm b)^6 + 36 \sum a^6. \tag{21.4.1}$$

On the right there are

$$16 + 2\,.\,12 + 36\,.\,4 = 184$$

sixth powers. Now any n is of the form

$$60N + r \quad (0 \leqslant r \leqslant 59);$$

and

$$60N = 60 \sum_{i=1}^{g(3)} X_i^3 = 60 \sum_{i=1}^{g(3)} \left(a_i^2 + b_i^2 + c_i^2 + d_i^2\right)^3,$$

which, by (21.4.1), is the sum of $184g(3)$ sixth powers. Hence n is the sum of

$$184g(3) + r \leqslant 184g(3) + 59$$

sixth powers; and so, by Theorem 390,

THEOREM 391:

$$g(6) \leqslant 184g(3) + 59 \leqslant 2451.$$

Again, the identity

$$(21.4.2) \qquad \begin{aligned} &5040(a^2 + b^2 + c^2 + d^2)^4 \\ &= 6\sum(2a)^8 + 60\sum(a \pm b)^8 \\ &\quad + \sum(2a \pm b \pm c)^8 + 6\sum(a \pm b \pm c \pm d)^8 \end{aligned}$$

has

$$6 \,.\, 4 + 60 \,.\, 12 + 48 + 6 \,.\, 8 = 840$$

eighth powers on its right-hand side. Hence, as above, any number $5040N$ is the sum of $840g(4)$ eighth powers. Now any number up to 5039 is the sum of at most 273 eighth powers of 1 or 2.† Hence, by Theorem 387,

THEOREM 392:

$$g(8) \leqslant 540g(4) + 273 \leqslant 42273.$$

The results of Theorems 391 and 392 arc, numcrically, very poor; and the theorems are really interesting only as existence theorems. It is known that $g(6) = 73$ and that $g(8) = 279$.

21.5. A lower bound for $g(k)$. We have found upper bounds for $g(k)$, and *a fortiori* for $G(k)$, for $k = 3$, 4, 6, and 8, but they are a good deal larger than those given by deeper methods. There is also the problem of finding lower bounds, and here elementary methods are relatively much more effective. It is indeed quite easy to prove all that is known at present.

We begin with $g(k)$. Let us write $q = \left[\left(\frac{3}{2}\right)^k\right]$. The number

$$n = 2^k q - 1 < 3^k$$

can only be represented by the powers 1^k and 2^k. In fact

$$n = (q-1)2^k + (2^k - 1)1^k,$$

† The worst number is $4863 = 18 \,.\, 2^8 + 255 \,.\, 1^8$.

and so n requires just

$$q - 1 + 2^k - 1 = 2^k + q - 2$$

kth powers. Hence

THEOREM 393:

$$g(k) \geqslant 2^k + q - 2.$$

In particular $g(2) \geqslant 4$, $g(3) \geqslant 9$, $g(4) \geqslant 19$, $g(5) \geqslant 37, \ldots$. It is known that $g(k) = 2^k + q - 2$ for all values of k up to 400 except perhaps 4 and 5, and it is quite likely that this is true for every k.

21.6. Lower bounds for $G(k)$. Passing to $G(k)$, we prove first a general theorem for every k.

THEOREM 394:

$$G(k) \geqslant k + 1 \textit{ for } k \geqslant 2.$$

Let $A(N)$ be the number of numbers $n \leqslant N$ which are representable in the form

$$n = x_1^k + x_2^k + \cdots + x_k^k, \tag{21.6.1}$$

where $x_i \geqslant 0$. We may suppose the x_i arranged in ascending order of magnitude, so that

$$0 \leqslant x_1 \leqslant x_2 \leqslant \cdots \leqslant x_k \leqslant N^{1/k}. \tag{21.6.2}$$

Hence $A(N)$ does not exceed the number of solutions of the inequalities (21.6.2), which is

$$B(N) = \sum_{x_k=0}^{[N^{1/k}]} \sum_{x_{k-1}=0}^{x_k} \sum_{x_{k-2}=0}^{x_{k-1}} \cdots \sum_{x_1=0}^{x_2} 1.$$

The summation with respect to x_1 gives $x_2 + 1$, that with respect to x_2 gives

$$\sum_{x_2=0}^{x_3} (x_2 + 1) = \frac{(x_3 + 1)(x_3 + 2)}{2!},$$

that with respect to x_3 gives

$$\sum_{x_3=0}^{x_4} \frac{(x_3+1)(x_3+2)}{2!} = \frac{(x_4+1)(x_4+2)(x_4+3)}{3!},$$

and so on; so that

(21.6.3) $$B(N) = \frac{1}{k!}\prod_{r=1}^{k}\left(\left[N^{1/k}\right]+r\right) \sim \frac{N}{k!}$$

for large N.

On the other hand, if $G(k) \leqslant k$, all but a finite number of n are representable in the form (21.6.1), and

$$A(N) > N - C,$$

where C is independent of N. Hence

$$N - C < A(N) \leqslant B(N) \sim \frac{N}{k!},$$

which is plainly impossible when $k > 1$. It follows that $G(k) > k$.

Theorem 394 gives the best known universal lower bound for $G(k)$. There are arguments based on congruences which give equivalent, or better, results for special forms of k. Thus

$$x^3 \equiv 0, 1, \text{ or } -1 \pmod 9,$$

and so at least 4 cubes are required to represent a number $N = 9m \pm 4$. This proves that $G(3) \geqslant 4$, a special case of Theorem 394.

Again

(21.6.4) $$x^4 \equiv 0 \text{ or } 1 \pmod{16},$$

and so all numbers $16m+15$ require at least 15 biquadrates. It follows that $G(4) \geqslant 15$. This is a much better result than that given by Theorem 394, and we can improve it slightly.

It follows from (21.6.4) that, if $16n$ is the sum of 15 or fewer biquadrates, each of these biquadrates must be a multiple of 16. Hence

$$16n = \sum_{i=1}^{15} x_i^4 = \sum_{i=1}^{15} (2y_i)^4$$

and so

$$n = \sum_{i=1}^{15} y_i^4.$$

Hence, if $16n$ is the sum of 15 or fewer biquadrates, so is n. But 31 is not the sum of 15 or fewer biquadrates; and so $16^m . 31$ is not, for any m. Hence

THEOREM 395:

$$G(4) \geqslant 16.$$

More generally

THEOREM 396:

$$G(2^\theta) \geqslant 2^{\theta+2} \textit{ if } \theta \geqslant 2.$$

The case $\theta = 2$ has been dealt with already. If $\theta > 2$, then

$$k = 2^\theta > \theta + 2.$$

Hence, if x is even,

$$x^{2^\theta} \equiv 0 \pmod{2^{\theta+2}},$$

while if x is odd then

$$x^{2^\theta} = (1+2m)^{2^\theta} \equiv 1 + 2^{\theta+1}m + 2^{\theta+1}(2^\theta - 1)m^2$$
$$\equiv 1 - 2^{\theta+1}m(m-1) \equiv 1 \pmod{2^{\theta+2}}.$$

Thus

$$x^{2^\theta} \equiv 0 \text{ or } 1 \pmod{2^{\theta+2}}. \tag{21.6.5}$$

Now let n be any odd number and suppose that $2^{\theta+2}n$ is the sum of $2^{\theta+2} - 1$ or fewer kth powers. Then each of these powers must be even, by (21.6.5), and so divisible by 2^k. Hence $2^{k-\theta-2} | n$, and so n is even; a contradiction which proves Theorem 396.

It will be observed that the last stage in the proof fails for $\theta = 2$, when a special device is needed.

There are three more theorems which, when they are applicable, give better results than Theorem 394.

THEOREM 397. *If $p > 2$ and $\theta \geqslant 0$, then $G\{p^{\theta}(p-1)\} \geqslant p^{\theta+1}$.*

For example,

$$G(6) \geqslant 9.$$

If $k = p^{\theta}(p-1)$, then $\theta + 1 \leqslant 3^{\theta} < k$. Hence

$$x^k \equiv 0 \pmod{p^{\theta+1}}$$

if $p|x$. On the other hand, if $p \nmid x$, we have

$$x^k = x^{p^{\theta}(p-1)} \equiv 1 \pmod{p^{\theta+1}}$$

by Theorem 72. Hence, if $p^{\theta+1}n$, where $p \nmid n$, is the sum of $p^{\theta+1} - 1$ or fewer kth powers, each of these powers must be divisible by $p^{\theta+1}$ and so by p^k. Hence $p^k \,|\, p^{\theta+1}n$, which is impossible; and therefore $G(k) \geqslant p^{\theta+1}$.

THEOREM 398. *If $p > 2$ and $\theta \geqslant 0$, then $G\{\frac{1}{2}p^{\theta}(p-1)\} \geqslant \frac{1}{2}(p^{\theta+1} - 1)$.*

For example, $G(10) \geqslant 12$.

It is plain that

$$k = \tfrac{1}{2}p^{\theta}(p-1) \geqslant p^{\theta} > \theta + 1,$$

except in the trivial case $p = 3,\ \theta = 0,\ k = 1$. Hence

$$x^k \equiv 0 \pmod{p^{\theta+1}}$$

if $p \,|\, x$. On the other hand, if $p \nmid x$, then

$$x^{2k} = x^{p^{\theta}(p-1)} \equiv 1 \pmod{p^{\theta+1}}$$

by Theorem 72. Hence $p^{\theta+1} | (x^{2k} - 1)$, i.e.

$$p^{\theta+1} | (x^k - 1)(x^k + 1).$$

Since $p > 2$, p cannot divide both $x^k - 1$ and $x^k + 1$, and so one of $x^k - 1$, and $x^k + 1$ is divisible by $p^{\theta+1}$. It follows that

$$x^k \equiv 0, 1, \text{ or } -1 \pmod{p^{\theta+1}}$$

for every x; and therefore that numbers of the form

$$p^{\theta+1}m \pm \tfrac{1}{2}(p^{\theta+1} - 1)$$

require at least $\frac{1}{2}\left(p^{\theta+1} - 1\right)$ kth powers.

THEOREM 399. *If* $\theta \geqslant 2$,† *then* $G(3.2^{\theta}) \geqslant 2^{\theta+2}$.

This is a trivial corollary of Theorem 396, since $G(3.2^{\theta}) \geqslant G(2^{\theta}) \geqslant 2^{\theta+2}$. We may sum up the results of this section in the following theorem.

THEOREM 400. $G(k)$ *has the lower bounds*

(i) $2^{\theta+2}$ *if* k *is* 2^{θ} *or* 3.2^{θ} *and* $\theta \geqslant 2$;
(ii) $p^{\theta+1}$ *if* $p > 2$ *and* $k = p^{\theta}(p-1)$;
(iii) $\frac{1}{2}(p^{\theta+1} - 1)$ *if* $p > 2$ *and* $k = \frac{1}{2}p^{\theta}(p-1)$;
(iv) $k+1$ *in any case.*

These are the best known lower bounds for $G(k)$. It is easily verified that none of them exceeds $4k$, so that the lower bounds for $G(k)$ are much smaller, for large k, than the lower bound for $g(k)$ assigned by Theorem 393. The value of $g(k)$ is, as we remarked in § 20.1, inflated by the difficulty of representing certain comparatively small numbers.

It is to be observed that k may be of several of the special forms mentioned in Theorem 400. Thus

$$6 = 3(3-1) = 7-1 = \tfrac{1}{2}(13-1),$$

so that 6 is expressible in two ways in the form (ii) and in one in the form (iii). The lower bounds assigned by the theorem are

$$3^2 = 9, \qquad 7^1 = 7, \qquad \tfrac{1}{2}(13-1) = 6, \qquad 6+1 = 7;$$

and the first gives the strongest result.

† The theorem is true for $\theta = 0$ and $\theta = 1$, but is then included in Theorems 394 and 397.

21.7. Sums affected with signs: the number $v(k)$. It is also natural to consider the representation of an integer n as the sum of s members of the set

$$0,\ 1^k,\ 2^k,\ \ldots,\ -1^k,\ -2^k,\ -3^k, \ldots, \tag{21.7.1}$$

or in the form

$$n = \pm x_1^k \pm x_2^k \pm \cdots \pm x_s^k. \tag{21.7.2}$$

We use $v(k)$ to denote the least value of s for which every n is representable in this manner.

The problem is in most ways more tractable than Waring's problem, but the solution is in one way still more incomplete. The value of $g(k)$ is known for many k, while that of $v(k)$ has not been found for any k but 2. The main difficulty here lies in the determination of a lower bound for $v(k)$; there is no theorem corresponding effectively to Theorem 393 or even to Theorem 394.

THEOREM 401: *$v(k)$ exists for every k.*

It is obvious that, if $g(k)$ exists, then $v(k)$ exists and does not exceed $g(k)$. But the direct proof of the existence of $v(k)$ is very much easier than that of the existence of $g(k)$.

We require a lemma.

THEOREM 402:

$$\sum_{r=0}^{k-1} (-1)^{k-1-r} \binom{k-1}{r} (x+r)^k = k!x + d,$$

where d is an integer independent of x.

The reader familiar with the elements of the calculus of finite differences will at once recognize this as a well-known property of the $(k-1)$th difference of x^k. It is plain that, if

$$Q_k(x) = A_k x^k + \cdots$$

is a polynomial of degree k, then

$$\Delta Q_k(x) = Q_k(x+1) - Q_k(x) = kA_k x^{k-1} + \cdots,$$
$$\Delta^2 Q_k(x) = k(k-1)A_k x^{k-2} + \cdots,$$
$$. \quad . \quad . \quad . \quad . \quad . \quad . \quad . \quad .$$
$$\Delta^{k-1} Q_k(x) = k!\, A_k x + d,$$

where d is independent of x. The lemma is the case $Q_k(x) = x^k$. In fact $d = \frac{1}{2}(k-1)(k!)$, but we make no use of this.

It follows at once from the lemma that any number of the form $k!x + d$ is expressible as the sum of

$$\sum_{r=0}^{k-1} \binom{k-1}{r} = 2^{k-1}$$

numbers of the set (21.7.1); and

$$n - d = k!\, x + l, \qquad -\tfrac{1}{2}(k!) < l \leqslant \tfrac{1}{2}(k!)$$

for any n and appropriate l and x. Thus

$$n = (k!\, x + d) + l,$$

and n is the sum of

$$2^{k-1} + l \leqslant 2^{k-1} + \tfrac{1}{2}(k!)$$

numbers of the set (21.7.1).

We have thus proved more than Theorem 401, viz.

THEOREM 403:

$$v(k) \leqslant 2^{k-1} + \tfrac{1}{2}(k!).$$

21.8. Upper bounds for $v(k)$. The upper bound in Theorem 403 is generally much too large.

It is plain, as we observed in § 21.7, that $v(k) \leqslant g(k)$. We can also find an upper bound for $v(k)$ if we have one for $G(k)$. For any number from a certain $N(k)$ onwards is the sum of $G(k)$ positive kth powers, and

$$n + y^k > N(k)$$

for some y, so that

$$n = \sum_{1}^{G(k)} x_i^k - y^k$$

and

$$v(k) \leqslant G(k) + 1. \tag{21.8.1}$$

For all but a few small k, this is a much better bound than $g(k)$.

The bound of Theorem 403 can also be improved substantially by more elementary methods. Here we consider only special values of k for which such elementary arguments give bounds better than (21.8.1).

(1) *Squares.* Theorem 403 gives $v(2) \leqslant 3$, which also follows from the identities

$$2x + 1 = (x + 1)^2 - x^2$$

and

$$2x = x^2 - (x - 1)^2 + 1^2.$$

On the other hand, 6 cannot be expressed by two squares, since it is not the sum of two, and $x^2 - y^2 = (x - y)(x + y)$ is either odd or a multiple of 4.

THEOREM 404:

$$v(2) = 3.$$

(2) *Cubes.* Since

$$n^3 - n = (n - 1)n(n + 1) \equiv 0 \pmod{6}$$

for any n, we have

$$n = n^3 - 6x = n^3 - (x+1)^3 - (x-1)^3 - 2x^3$$

for any n and some integral x. Hence $v(3) \leqslant 5$.

On the other hand,

$$y^3 \equiv 0, 1, \text{or } -1 \pmod{9};$$

and so numbers $9m \pm 4$ require at least 4 cubes. Hence $v(3) \geqslant 4$.

THEOREM 405: *$v(3)$ is 4 or 5.*

It is not known whether 4 or 5 is the correct value of $v(3)$. The identity

$$6x = (x+1)^3 + (x-1)^3 - 2x^3$$

shows that every multiple of 6 is representable by 4 cubes. Richmond and Mordell have given many similar identities applying to other arithmetical progressions. Thus the identity

$$6x + 3 = x^3 - (x-4)^3 + (2x-5)^3 - (2x-4)^3$$

shows that any odd multiple of 3 is representable by 4 cubes.

(3) *Biquadrates.* By Theorem 402, we have

$$(x+3)^4 - 3(x+2)^4 + 3(x+1)^4 - x^4 = 24x + d \tag{21.8.2}$$

(where $d = 36$). The residues of 0^4, 1^4, 3^4, 2^4 (mod 24) are 0, 1, 9, 16 respectively, and we can easily verify that every residue (mod 24) is the sum of 4 at most of 0, ± 1, ± 9, ± 16. We express this by saying that 0, 1, 9, 16 are fourth power residues (mod 24), and that any residue (mod 24) is representable by 4 of these fourth power residues. Now we can express any n in the form $n = 24x + d + r$, where $0 \leqslant r < 24$; and (21.8.2) then shows that any n is representable by $8 + 4 = 12$ numbers $\pm y^4$. Hence $v(4) \leqslant 12$. On the other hand the only fourth power residues (mod 16) are 0 and 1, and so a number $16m+8$ cannot be represented by 8 numbers $\pm y^4$ unless they are all odd and of the same sign. Since there are numbers of this form, e.g. 24, which are not sums of 8 biquadrates, it follows that $v(4) \geqslant 9$.

THEOREM 406:

$$9 \leqslant v(4) \leqslant 12.$$

(4) *Fifth powers.* In this case Theorem 402 does not lead to the best result; we use instead the identity

$$(x+3)^5 - 2(x+2)^5 + x^5 + (x-1)^5 - 2(x-3)^5 + (x-4)^5 = 720x - 360. \tag{21.8.3}$$

A little calculation shows that every residue (mod 720) can be represented by two fifth power residues. Hence $v(5) \leqslant 8 + 2 = 10$.

The only fifth power residues (mod 11) are 0, 1, and -1, and so numbers of the form $11m \pm 5$ require at least 5 fifth powers.

THEOREM 407:

$$5 \leqslant v(5) \leqslant 10.$$

21.9. The problem of Prouhet and Tarry: the number *P*(*k*,*j*). There is another curious problem which has some connexion with that of § 21.8 (though we do not develop this connexion here).

Suppose that the a and b are integers and that

$$S_h = S_h(a) = a_1^h + a_2^h + \cdots + a_s^h = \sum a_i^h;$$

and consider the system of k equations

$$S_h(a) = S_h(b) \qquad (1 \leqslant h \leqslant k). \tag{21.9.1}$$

It is plain that these equations are satisfied when the b are a permutation of the a; such a solution we call a trivial solution.

It is easy to prove that there are no other solutions when $s \leqslant k$. It is sufficient to consider the case $s = k$. Then

$$b_1 + b_2 + \cdots + b_k, \quad b_1^2 + \cdots + b_k^2, \quad \ldots, \quad b_1^k + \cdots + b_k^k$$

have the same values as the same functions of the a, and therefore† the elementary symmetric functions

$$\sum b_i, \ \sum b_i b_j, \ \ldots, \ b_1 b_2 \ldots b_k$$

† By Newton's relations between the coefficients of an equation and the sums of the powers of its roots.

have the same values as the same functions of the a. Hence the a and the b are the roots of the same algebraic equation, and the b are a permutation of the a.

When $s > k$ there may be non-trivial solutions, and we denote by $P(k, 2)$ the least value of s for which this is true. It is plain first (since there are no non-trivial solutions when $s \leqslant k$) that

(21.9.2) $$P(k, 2) \geqslant k + 1.$$

We may generalize our problem a little. Let us take $j \geqslant 2$, write

$$S_{hu} = a_{1u}^h + a_{2u}^h + \cdots + a_{su}^h$$

and consider the set of $k(j-1)$ equations

(21.9.3) $$S_{h1} = S_{h2} = \ldots = S_{hj} \ (1 \leqslant h \leqslant k).$$

A non-trivial solution of (21.9.3) is one in which no two sets $a_{iu}(1 \leqslant i \leqslant s)$ and $a_{iv}(1 \leqslant i \leqslant s)$ with $u \neq v$ are permutations of one another. We write $P(k,j)$ for the least value of s for which there is a non-trivial solution. Clearly a non-trivial solution of (21.9.3) for $j \geqslant 2$ includes a non-trivial solution of (21.9.1) for the same s. Hence, by (21.9.2),

THEOREM 408:

$$P(k,j) \geqslant P(k, 2) \geqslant k + 1.$$

In the other direction, we prove that

THEOREM 409:

$$P(k,j) \leqslant \tfrac{1}{2}k(k + 1) + 1.$$

Write $s = \frac{1}{2}k(k+1)+1$ and suppose that $n > s!s^k j$. Consider all the sets of integers

(21.9.4) $$a_1, a_2, \ldots, a_s$$

for which

$$1 \leqslant a_r \leqslant n \quad (1 \leqslant r \leqslant s).$$

There are n^s such sets.

Since $1 \leqslant a_r \leqslant n$ we have

$$s \leqslant S_h(a) \leqslant sn^h.$$

Hence there are at most

$$\prod_{h=1}^{k}(sn^h - s + 1) < s^k n^{\frac{1}{2}k(k+1)} = s^k n^{s-1}$$

different sets

(21.9.5) $$S_1(a),\ S_2(a), \ldots, S_k(a).$$

Now

$$s!\, j.\, s^k n^{s-1} \leqslant n^s,$$

and so at least $s!j$ of the sets (21.9.4) have the same set (21.9.5). But the number of permutations of s things, like or unlike, is at most $s!$, and so there are at least j sets (21.9.4), no two of which are permutations of one another and which have the same set (21.9.5). These provide a non-trivial solution of the equations (21.9.3) with

$$s = \tfrac{1}{2}k(k+1) + 1.$$

21.10. Evaluation of $P(k,j)$ for particular k and j. We prove

THEOREM 410. *$P(k,j) = k + 1$ for $k = 2$, 3, and 5 and all j.*

By Theorem 408, we have only to prove that $P(k,j) \leqslant k + 1$ and for this it is sufficient to construct actual solutions of (21.9.3) for any given j.

By Theorem 337, for any fixed j, there is an n such that

$$n = c_1^2 + d_1^2 = c_2^2 + d_2^2 = \ldots = c_j^2 + d_j^2,$$

where all the numbers $c_1, c_2, \ldots, c_j, d_1, \ldots, d_j$ are positive and no two are equal. If we put

$$a_{1u} = c_u, \quad a_{2u} = d_u, \quad a_{3u} = -c_u, \quad a_{4u} = -d_u,$$

it follows that

$$S_{1u} = 0, \quad S_{2u} = 2n, \quad S_{3u} = 0 \quad (1 \leqslant u \leqslant j),$$

and so we have a non-trivial solution of (21.9.3) for $k = 3$, $s = 4$. Hence $P(3,j) \leqslant 4$ and so $P(3,j) = 4$.

For $k = 2$ and $k = 5$, we use the properties of the quadratic field $k(\rho)$ found in Chapters XIII and XV. By Theorem 255, $\pi = 3+\rho$ and $\bar{\pi} = 3+\rho^2$ are conjugate primes with $\pi\bar{\pi} = 7$. They are not associates, since

$$\frac{\pi}{\bar{\pi}} = \frac{\pi^2}{\pi\bar{\pi}} = \frac{9+6\rho+\rho^2}{7} = \frac{8}{7} + \frac{5}{7}\rho,$$

which is not an integer and so, *a fortiori*, not a unity. Now let $u > 0$ and let $\pi^{2u} = A_u - B_{u\rho}$ where A_u, B_u are rational integers. If $7|A_u$, we have

$$\pi\bar{\pi}|A_u, \quad \pi|A_u, \quad \pi|B_u\rho$$

in $k(\rho)$, and $N\pi|B_u^2, 7|B_u^2, 7|B_u$ in $k(1)$. Finally $7|\pi^{2u}$, $\pi\bar{\pi}|\bar{\pi}^{2u}$, $\bar{\pi}|\pi 2^{u-1}$, $\bar{\pi}|\pi$ in $k(\rho)$, which is false. Hence $7 \nmid A_u$ and, similarly, $7 \nmid B_u$.

If we write $c_u = 7^{j-u}A_u$, $d_u = 7^{j-u}B_u$, we have

$$c_u^2 + c_u d_u + d_u^2 = N(c_u - d_u\rho) = 7^{2j-2u}N\pi^{2u} = 7^{2j}.$$

Hence, if we put $a_{1u} = c_u$, $a_{2u} = d_u$, $a_{3u} = -(c_u + d_u)$, we have $S_{1u} = 0$ and

$$S_{2u} = c_u^2 + d_u^2 + (c_u + d_u)^2 = 2(c_u^2 + c_u d_u + d_u^2) = 2 \,.\, 7^{2j}.$$

Since at least two of (a_{1u}, a_{2u}, a_{3u}) are divisible by 7^{j-u} but not by 7^{j-u+1}, no set is a permutation of any other set and we have a non-trivial solution of (21.9.3) with $k = 2$ and $s = 3$. Thus $P(2,j) = 3$.

Incidentally, we have also

$$S_{4u} = c_u^4 + d_u^4 + (c_u + d_u)^4 = 2(c_u^2 + c_u d_u + d_u^2)^2 = 2 \,.\, 7^{4j}$$

and so, for any j, we have a non-trivial solution of the equations

(21.10.1) $$x_1^2 + y_1^2 + z_1^2 = x_2^2 + y_2^2 + z_2^2 = \ldots = x_j^2 + y_j^2 + z_j^2$$

and

(21.10.2) $$x_1^4 + y_1^4 + z_1^4 = x_2^4 + y_2^4 + z_2^4 = \ldots = x_j^4 + y_j^4 + z_j^4.$$

For $k = 5$, we write

$$a_{1u} = c_u, \quad a_{2u} = d_u, \quad a_{3u} = -c_u - d_u, \quad a_{4u} = -a_{1u},$$
$$a_{5u} = -a_{2u}, \quad a_{6u} = -a_{3u}$$

and have $S_{1u} = S_{3u} = S_{5u} = 0, \quad S_{2u} = 4 \,.\, 7^{2j}, \quad S_{4u} = 4 \,.\, 7^{4j}$.
As before, we have no trivial solutions and so $P(5,j) = 6$.

The fact that, in the last solution for example, $S_{1u} = S_{3u} = S_{5u} = 0$ does not make the solution so special as appears at first sight. For, if

$$a_{ru} = A_{ru} \quad (1 \leqslant r \leqslant s,\ 1 \leqslant u \leqslant j)$$

is one solution of (21.9.3), it can easily be verified that, for any d,

$$a_{ru} = A_{ru} + d$$

is another such solution. Thus we can readily obtain solutions in which none of the S is zero.

The case $j = 2$ can be handled successfully by methods of little use for larger j. If $a_1, a_2, \dots, a_s, b_1, \dots, b_s$, is a solution of (21.9.1), then

(21.10.3)
$$\sum_{i=1}^{s} \left\{(a_i + d)^h + b_i^h\right\} = \sum_{i=1}^{s} \left\{a_i^h + (b_i + d)^h\right\} \ (1 \leqslant h \leqslant k+1)$$

for every d. For we may reduce these to

$$\sum_{l=1}^{h-1} \binom{h}{l} S_{h-l}(a) d^l = \sum_{l=1}^{h-1} \binom{h}{l} S_{h-l}(b) d^l \quad (2 \leqslant h \leqslant k+1)$$

and these follow at once from (21.9.1).

We choose d to be the number which occurs most frequently as a difference between two a or two b. We are then able to remove a good many terms which occur on both sides of the identity (21.10.3).

We write

$$[a_1, \dots, a_s]_k = [b_1, \dots, b_s]_k$$

to denote that $S_h(a) = S_h(b)$ for $1 \leqslant h \leqslant k$.
Then

$$[0, 3]_1 = [1, 2]_1.$$

Using (21.10.3), with $d = 3$, we get

$$[1, 2, 6]_2 = [0, 4, 5]_2.$$

Starting from the last equation and taking $d = 5$ in (21.10.3), we obtain

$$[0, 4, 7, 11]_3 = [1, 2, 9, 10]_3.$$

From this we deduce in succession

$$[1, 2, 10, 14, 18]_4 = [0, 4, 8, 16, 17]_4 \quad (d = 7),$$
$$[0, 4, 9, 17, 22, 26]_5 = [1, 2, 12, 14, 24, 25]_5 \quad (d = 8),$$
$$[1, 2, 12, 13, 24, 30, 35, 39]_6 = [0, 4, 9, 15, 26, 27, 37, 38]_6 \quad (d = 13),$$
$$[0, 4, 9, 23, 27, 41, 46, 50]_7 = [1, 2, 11, 20, 30, 39, 48, 49]_7 \quad (d = 11).$$

The example†

$$[0, 18, 27, 58, 64, 89, 101]_6 = [1, 13, 38, 44, 75, 84, 102]_6,$$

shows that $P(k, 2) \leqslant k + 1$ for $k = 6$; and these results, with Theorem 408, give

THEOREM 411. *If* $k \leqslant 7$, $P(k, 2) = k + 1$.

21.11. Further problems of Diophantine analysis. We end this chapter by a few unsystematic remarks about a number of Diophantine equations which are suggested by Fermat's problem of Ch. XIII.

(1) *A conjecture of Euler.* Can a kth power be the sum of s positive kth powers? Is

$$x_1^k + x_2^k + \cdots + x_s^k = y^k \tag{21.11.1}$$

soluble in positive integers? 'Fermat's last theorem' asserts the impossibility of the equation when $s = 2$ and $k > 2$, and Euler extended the conjecture to the values $3, 4, \ldots, k - 1$ of s. For $k = 5$, $s = 4$, however, the conjecture is false, since

$$27^5 + 84^5 + 110^5 + 133^5 = 144^5.$$

† This may be proved by starting with

$$[1, 8, 12, 15, 20, 23, 27, 34]_1 = [0, 7, 11, 17, 18, 24, 28, 35]_1$$

and taking $d = 7, 11, 13, 17, 19$ in succession.

The equation

$$x_1^k + x_2^k + \cdots + x_k^k = y^k \tag{21.11.2}$$

has also attracted much attention. The case $k = 2$ is familiar.[†] When $k = 3$, we can derive solutions from the analysis of § 13.7. If we put $\lambda = 1$ and $a = -3b$ in (13.7.8), and then write $-\frac{1}{2}q$ for b, we obtain

$$x = 1 - 9q^3, \quad y = -1, \quad u = -9q^4, \quad v = 9q^4 - 3q; \tag{21.11.3}$$

and so, by (13.7.2),

$$(9q^4)^3 + (3q - 9q^4)^3 + (1 - 9q^3)^3 = 1.$$

If we now replace q by ξ/η and multiply by η^{12}, we obtain the identity

$$(9\xi^4)^3 + (3\xi\eta^3 - 9\xi^4)^3 + (\eta^4 - 9\xi^3\eta)^3 = (\eta^4)^3. \tag{21.11.4}$$

All the cubes are positive if

$$0 < \xi < 9^{-\frac{1}{3}}\eta,$$

so that any twelfth power η^{12} can be expressed as a sum of three positive cubes in at least $\left[9^{-\frac{1}{3}}\eta\right]$ ways.

When $k > 3$, little is known. A few particular solutions of (21.11.2) are known for $k = 4$, the smallest of which is

$$30^4 + 120^4 + 272^4 + 315^4 = 353^4.\text{‡} \tag{21.11.5}$$

† See § 13.2.

‡ The identity

$$(4x^4 - y^4)^4 + 2(4x^3y)^4 + 2(2xy^3)^4 = (4x^4 + y^4)^4$$

gives an infinity of biquadrates expressible as sums of 5 biquadrates (with two equal pairs); and the identity

$$(x^2 - y^2)^4 + (2xy + y^2)^4 + (2xy + x^2)^4 = 2(x^3 + xy - y^2)^4$$

gives an infinity of solutions of

$$x_1^4 + x_2^4 + x_3^4 = y_1^4 + y_2^4$$

(all with $y_1 = y_2$).

For $k = 5$, there are an infinity included in the identity

(21.11.6) $$(75y^5 - x^5)^5 + (x^5 + 25y^5)^5 + (x^5 - 25y^5)^5 + (10x^3y^2)^5 + (50xy^4)^5 = (x^5 + 75y^5)^5.$$

All the powers are positive if $0 < 25y^5 < x^5 < 75y^5$. No solution is known with $k \geqslant 6$.

(2) *Equal sums of two kth powers.* Is

(21.11.7) $$x_1^k + y_1^k = x_2^k + y_2^k$$

soluble in positive integers? More generally, is

(21.11.8) $$x_1^k + y_1^k = x_2^k + y_2^k = \ldots = x_r^k + y_r^k$$

soluble for given k and r?

The answers are affirmative when $k = 2$, since, by Theorem 337, we can choose n so as to make $r(n)$ as large as we please. We shall now prove that they are also affirmative when $k = 3$.

THEOREM 412. *Whatever r, there are numbers which are representable as sums of two positive cubes in at least r different ways.*

We use two identities, viz.

(21.11.9) $$X^3 - Y^3 = x_1^3 + y_1^3$$

if

(21.11.10) $$X = \frac{x_1(x_1^3 + 2y_1^3)}{x_1^3 - y_1^3}, \quad Y = \frac{y_1(2x_1^3 + y_1^3)}{x_1^3 - y_1^3},$$

and

(21.11.11) $$x_2^3 + y_2^3 = X^3 - Y^3$$

if

(21.11.12) $$x_2 = \frac{X(X^3 - 2Y^3)}{X^3 + Y^3}, \quad y^2 = \frac{Y(2X^3 - Y^3)}{X^3 + Y^3}.$$

Each identity is an obvious corollary of the other, and either may be deduced from the formulae of § 13.7.† From (21.11.9) and (21.11.11) it follows that

(21.11.13) $$x_1^3 + y_1^3 = x_2^3 + y_2^3.$$

Here x_2, y_2 are rational if x_1, y_1 are rational.

Suppose now that r is given, that x_1 and y_1 are rational and positive and that

$$\frac{x_1}{4^{r-1}y_1}$$

is large. Then X, Y are positive, and X/Y is nearly $x_1/2y_1$; and x_2, y_2 are positive and x_2/y_2 is nearly $X/2Y$ or $x_1/4y_1$.

Starting now with x_2, y_2 in place of x_1, y_1, and repeating the argument, we obtain a third pair of rationals x_3, y_3 such that

$$x_1^3 + y_1^3 = x_2^3 + y_2^3 = x_3^3 + y_3^3$$

and x_3/y_3 is nearly $x_1/4^2y_1$. After r applications of the argument we obtain

(21.11.14) $$x_1^3 + y_1^3 = x_2^3 + y_2^3 = \ldots = x_r^3 + y_r^3,$$

all the numbers involved being positive rationals, and

$$\frac{x_1}{y_1},\ 4\frac{x_2}{y_2},\ 4^2\frac{x_3}{y_3},\ \ldots,\ 4^{r-1}\frac{x_r}{y_r}$$

all being nearly equal, so that the ratios $x_s/y_s (s = 1, 2, \ldots, r)$ are certainly unequal. If we multiply (21.11.14) by l^3, where l is the least common multiple of the denominators of $x_1, y_1, \ldots, x_r, y_r$, we obtain an integral solution of the system (21.11.14).

Solutions of

$$x_1^4 + y_1^4 = x_2^4 + y_2^4$$

† If we put $a = b$ and $\lambda = 1$ in (13.7.8), we obtain

$$x = 8a^3 + 1, \qquad y = 16a^3 - 1, \qquad u = 4a - 16a^4, \qquad v = 2a + 16a^4;$$

and if we replace u by $\frac{1}{2}q$, and use (13.7.2), we obtain

$$(q^4 - 2q)^3 + (2q^3 - 1)^3 = (q^4 + q)^3 - (q^3 + 1)^3,$$

an identity equivalent to (21.11.11).

can be deduced from the formulae (13.7.11); but no solution of

$$x_1^4 + y_1^4 = x_2^4 + y_2^4 = x_3^4 + y_3^4$$

is known. And no solution of (21.11.7) is known for $k \geqslant 5$.

We showed how to construct a solution of (21.10.2) for any j. Swinnerton-Dyer has found a parametric solution of

$$x_1^5 + x_2^5 + x_3^5 = y_1^5 + y_2^5 + y_3^5 \tag{21.11.15}$$

which yields solutions in positive integers. A numerical solution is

$$49^5 + 75^5 + 107^5 = 39^5 + 92^5 + 100^5. \tag{21.11.16}$$

The smallest result of this kind for sixth powers is

$$3^6 + 19^6 + 22^6 = 10^6 + 15^6 + 23^6. \tag{21.11.17}$$

NOTES

A great deal of work has been done on Waring's problem during the last hundred years, and it may be worth while to give a short summary of the results. We have already referred to Waring's original statement, to Hilbert's proof of the existence of $g(k)$, and to the proof that $g(3) = 9$ (Wieferich, *Math. Annalen,* 66 (1909), 99–101, corrected by Kempner, ibid. 72 (1912), 387–97 and simplified by Scholz, *Jber. Deutsch. Math. Ver.* 58 (1955), Abt. 1, 45–48).

Landau [*ibid.* 66 (1909), 102–5] proved that $G(3) \leqslant 8$ and it was not until 1942 that Linnik [*Comptes Rendus* (*Doklady*) *Acad. Sci. USSR,* 35 (1942), 162] announced a proof that $G(3) \leqslant 7$. Dickson [*Bull. Amer. Math. Soc.* 45 (1939) 588–91] showed that 8 cubes suffice for all but 23 and 239. See G. L. Watson, *Math. Gazette,* 37 (1953), 209–11, for a simple proof that $G(3) \leqslant 8$ and *Journ. London Math. Soc.* 26 (1951), 153–6 for one that $G(3) \leqslant 7$ and for further references. After Theorem 394, $G(3) \geqslant 4$, so that $G(3)$ is 4, 5, 6, or 7; it is still uncertain which, though the evidence of tables points very strongly to 4 or 5. See Western, *ibid.* 1 (1926), 244–50. Deshouillers, Hennecart, and Landreau (*Math. Comp.* 69 (2000), 421–39) have offered evidence to the effect that 7 373 170 279 850 is the largest integer that cannot be represented as the sum of four positive integral cubes.

Hardy and Littlewood, in a series of papers under the general title 'Some problems of partitio numerorum', published between 1920 and 1928, developed a new analytic method for the study of Waring's problem. They found upper bounds for $G(k)$ for any k, the first being

$$(k-2)2^{k-1} + 5,$$

and the second a more complicated function of k which is asymptotic to $k2^{k-2}$ for large k. In particular they proved that

$$(a) \qquad G(4) \leqslant 19, \quad G(5) \leqslant 41, \quad G(6) \leqslant 87, \quad G(7) \leqslant 193, \quad G(8) \leqslant 425.$$

Their method did not lead to any new result for $G(3)$; but they proved that 'almost all' numbers are sums of 5 cubes.

Davenport, *Acta Math.* 71 (1939), 123–43, has proved that almost all are sums of 4. Since numbers $9m\pm4$ require at least 4 cubes, this is the final result.

Hardy and Littlewood also found an asymptotic formula for the number of representations for n by s kth powers, by means of the so-called 'singular series'. Thus $r_{4,21}(n)$, the number of representations of n by 21 biquadrates, is approximately

$$\frac{\left\{2\Gamma\left(\frac{5}{4}\right)\right\}^{21}}{\Gamma\left(\frac{21}{4}\right)} n^{\frac{17}{4}} \left\{1 + 1{\cdot}331 \cos\left(\frac{1}{8}n\pi + \frac{11}{16}\pi\right) + 0{\cdot}379 \cos\left(\frac{1}{4}n\pi - \frac{5}{8}\pi\right) + \ldots\right\}$$

(the later terms of the series being smaller). There is a detailed account of all this work (except on its 'numerical' side) in Landau, *Vorlesungen,* i. 235–339.

As regards $g(k)$, the best results known, up to 1933, for small k, were

$$g(4) \leqslant 37, \quad g(5) \leqslant 58, \quad g(6) \leqslant 478, \quad g(7) \leqslant 3806, \quad g(8) \leqslant 31353$$

(due to Wieferich, Baer, Baer, Wieferich, and Kempner respectively). All these had been found by elementary methods similar to those used in §§ 21.1–4. The results of Hardy and Littlewood made it theoretically possible to find an upper bound for $g(k)$ for any k, though the calculations required for comparatively large k would have been impracticable. James, however, in a paper published in *Trans. Amer. Math. Soc.* 36 (1934), 395–444, succeeded in proving that

$$(b) \qquad g(6) \leqslant 183, \quad g(7) \leqslant 322, \quad g(8) \leqslant 595.$$

He also found bounds for $g(9)$ and $g(10)$.

The later work of Vinogradov made it possible to obtain much more satisfactory results. Vinogradov's earlier researches on Waring's problem had been published in 1924, and there is an account of his method in Landau, *Vorlesungen,* i. 340–58. The method then used by Vinogradov resembled that of Hardy and Littlewood in principle, but led more rapidly to some of their results and in particular to a comparatively simple proof of Hilbert's theorem. It could also be used to find an upper bound for $g(k)$. In his later work Vinogradov made very important improvements, based primarily on a new and powerful method for the estimation of certain trigonometrical sums, and obtained results which were, for large k, far better than any known before. Thus he proved that

$$G(k) \leqslant 6k \log k + (4 + \log 216)k;$$

so that $G(k)$ is at most of order $k\log k$. Vinogradov's proof was afterwards simplified considerably by Heilbronn, who proved that

$$(c) \qquad G(k) \leqslant 6k \log k + \left\{4 + 3\log\left(3 + \frac{2}{k}\right)\right\} k + 3.$$

The resulting upper bound for $G(k)$ is better than that of (a) for $k > 6$ (and naturally far better for large values of k). Vinogradov (1947) improved his result to $G(k) \leqslant k(3\log k + 11)$,

Tong (1957) and Chen (1958) replaced the number 11 in this by 9 and 5.2 respectively, while Vinogradov (*Izv. Akad. Nauk SSSR Ser. Mat.* 23 (1959), 637–42) proved that

$$\text{(d)} \qquad G(k) \leqslant k(2\log k + 4\log\log k + 2\log\log\log k + 13)$$

for all k in excess of 170,000.

More has been proved since concerning smaller k : in particular, the value of $G(4)$ is now known. Davenport [*Annals of Math.* 40 (1939), 731–47] proved that $G(4) \leqslant 16$, so that, after Theorem 395, $G(4) = 16$; and that any number not congruent to 14 or 15 (mod 16) is a sum of 14 biquadrates. He also proved [*Amer. Journal of Math.* 64 (1942), 199–207] that $G(5) \leqslant 23$ and $G(6) \leqslant 36$. It has been proved by Davenport's method that $G(7) \leqslant 53$ (Rao, *J. Indian Math. Soc.* 5 (1941), 117–21 and Vaughan, *Proc. London Math. Soc.* 28 (1974), 387). Narasimkamurti (*J. Indian Math. Soc.* 5 (1941), 11–12) proved that $G(8) \leqslant 73$ and found upper bounds for $k = 9$ and 10, subsequently improved by Cook and Vaughan (*Acta Arith.* 33 (1977), 231–53). The last-named proved that

$$G(9) \leqslant 91, \qquad G(10) \leqslant 107, \qquad G(11) \leqslant 122, \qquad G(12) \leqslant 137.$$

Vaughan's method leads to $G(k) \leqslant k(3\log k + 4.2)$ $(k \geqslant 9)$, which is better than (*d*) for $k \leqslant 2.131 \times 10^{10}$ (approx.) and otherwise worse.

Vinogradov's work also led to very remarkable results concerning $g(k)$. If we know that $G(k)$ does not exceed some upper bound $\bar{G}(k)$, so that numbers greater than $C(k)$ are representable by $\bar{G}(k)$ or fewer kth powers, then the way is open to the determination of an upper bound for $g(k)$. For we have only to study the representation of numbers up to $C(k)$, and this is logically, for a given k, a question of computation. It was thus that James determined the bounds set out in (*b*); but the results of such work, before Vinogradov's, were inevitably unsatisfactory, since the bounds (*a*) for $G(k)$ found by Hardy and Littlewood are (except for quite small values of k) much too large, and in particular larger than the lower bounds for $g(k)$ given by Theorem 393.

If

$$\underline{g}(k) = 2^k + \left[\left(\tfrac{3}{2}\right)^k\right] - 2$$

is the lower bound for $g(k)$ assigned by Theorem 393, and if, for the moment, we take $\bar{G}(k)$ to be the upper bound for $\bar{G}(k)$ assigned by (*d*), then $\bar{g}(k)$ is of much higher order of magnitude than $\bar{G}(k)$. In fact $\underline{g}(k) > \bar{G}(k)$ for $k \geqslant 7$. Thus if $k \geqslant 7$, if all numbers from $C(k)$ on are representable by $\bar{G}(k)$ powers, and all numbers below $C(k)$ by $\underline{g}(k)$ powers, then

$$g(k) = \underline{g}(k).$$

And it is not necessary to determine the $C(k)$ corresponding to this particular $\bar{G}(k)$; it is sufficient to know the $C(k)$ corresponding to any $\bar{G}(k) \leqslant \bar{g}(k)$, and in particular to $\bar{G}(k) = \underline{g}(k)$.

This type of argument led to an 'almost complete' solution of the original form of Waring's problem. The first, and deepest, part of the solution rests on an adaptation of Vinogradov's method. The second depends on an ingenious use of a 'method of ascent', a simple case of which appears in the proof, in § 21.3, of Theorem 390.

Let us write

$$A = \left[\left(\tfrac{3}{2}\right)^k\right], \quad B = 3^k - 2^k A, \quad D = \left[\left(\tfrac{4}{3}\right)^k\right].$$

The final result is that

$$(e) \qquad g(k) = 2^k + A - 2$$

for all $k \geqslant 2$ for which

$$(f) \qquad B \leqslant 2^k - A - 2.$$

In this case the value of $g(k)$ is fixed by the number

$$n = 2^k A - 1 = (A-1)2^k + (2^k - 1).1^k$$

used in the proof of Theorem 393, a comparatively small number representable only by powers of 1 and 2. The condition (f) is satisfied for $4 \leqslant k \leqslant 471\,600\,000$ (Kubina and Wunderlich, *Math. Comp.* 55 (1990), 815–20) and may well be true for all $k > 3$. It can only be false for at most a finite number of k (Mahler, *Mathematika* 4 (1957), 122–4).

It is known that $B \neq 2^k - A - 1$ and that $B \neq 2^k - A$ (except for $k = 1$). If $B \geqslant 2^k - A + 1$, the formula for $g(k)$ is different. In this case,

$$g(k) = 2^k + A + D - 3 \ \text{ if } \ 2^k < AD + A + D$$

and

$$g(k) = 2^k + A + D - 2 \ \text{ if } \ 2^k = AD + A + D.$$

It is readily shown that $2^k \leqslant AD + A + D$.

Most of these results were found independently by Dickson [*Amer. Journal of Math.* 58 (1936), 521–9, 530–5] and Pillai [*Journal Indian Math. Soc.* (2) 2 (1936), 16–44, and *Proc. Indian Acad. Sci.* (A), 4 (1936), 261]. They were completed by Pillai [*ibid.* 12 (1940), 30–40] who proved that $g(6) = 73$; by Rubugunday [*Journal Indian Math. Soc.* (2) 6 (1942), 192–8] who proved that $B \neq 2^k - A$; by Niven [*Amer. Journal of Math.* 66 (1944), 137–43] who proved (e) when $B = 2^k - A - 2$, a case previously unsolved; by Jing-run Chen (*Chinese Math. Acta* 6 (1965), 105–27) who proved that $g(5) = 37$, and by Balasubramanian, Deshouillers, and Dress, who have shown that $g(4) = 19$ (*C. R. Acad. Sci. Paris. Sér. I Math.* 303 (1986), 85–88 and 161–3).

It will be observed that there is much more uncertainty about the value of $G(k)$ than about that of $g(k)$; the most striking case is $k = 3$. This is natural, since the value of $G(k)$ depends on the deeper properties of the whole sequence of integers, and that of $g(k)$ on the more trivial properties of special numbers near the beginning.

Vaughan, *The Hardy–Littlewood Method,* gives an excellent account of the topic and a full bibliography.

Much progress has been accomplished on topics associated with Waring's problem over the past three decades. A fairly comprehensive survey may be found in the paper of Vaughan and Wooley in *Surveys in Number Theory, Papers from the Millenial Conference in Number*

Theory, (A. K. Peters, Ltd., MA, 2003). In brief, there have been two phases of activity. In the first phase, pursued more or less independently by Thanigasalam and Vaughan throughout the early 1980's, the methods originally developed by Davenport (as cited earlier) were refined to perfection. The papers of Vaughan (*Proc. London Math. Soc.* (3) 52 (1986), 45–63 and *J. London Math. Soc.* (2) 33 (1986), 227–36) represent the culmination of this activity, in which it is shown that $G(5) \leqslant 21$, $G(6) \leqslant 31$, $G(7) \leqslant 45$, $G(8) \leqslant 62$ and $G(9) \leqslant 82$. Vaughan also proved that 'almost all' positive integers are sums of 32 eighth powers, a conclusion that is best possible.

The landscape was then transformed at the end of the 1980's with the introduction by Vaughan of smooth numbers (that is, integers all of whose prime divisors are 'small') into the Hardy–Littlewood method (see *Acta Math.* 162 (1989), 1–71). This led *inter alia* to the bounds $G(5) \leqslant 19$, $G(6) \leqslant 29$, $G(7) \leqslant 41$, $G(8) \leqslant 57$, $G(9) \leqslant 75, \ldots,$ $G(20) \leqslant 248$. Subsequently, a new iterative element ('repeated efficient differencing') was found by Wooley (*Ann. of Math.* (2) 135 (1992), 131–64) that delivered the sharper bounds $G(6) \leqslant 27$, $G(7) \leqslant 36$, $G(8) \leqslant 47$, $G(9) \leqslant 55, \ldots, G(20) \leqslant 146$, and for larger exponents k, the upper bound $G(k) \leqslant k(\log k + \log\log k + O(1))$. The latter provided the first sizeable progress on Vinogradov's estimate (d), from 1959. Wooley also showed that 'almost all' positive integers are the sum of 64 16th powers, and also the sum of 128 32nd powers, each of which are best possible conclusions. The sharpest bounds currently (2007) available from this circle of ideas are

$$G(5) \leqslant 17, \quad G(6) \leqslant 24, \quad G(7) \leqslant 33, \quad G(8) \leqslant 42, \quad G(9) \leqslant 50, \ldots, \quad G(20) \leqslant 142$$

(see work of Vaughan and Wooley spanning the 1990's summarised in *Acta Arith.* (2000), 203–285), and

$$G(k) \leqslant k(\log k + \log\log k + 2 + O(\log\log k/\log k))$$

(see Wooley, *J. London Math. Soc.* (2) 51 (1995), 1–13).

Further progress has been made on the topic of sums of fourth powers beyond the conclusions of Davenport (1939) summarised above. Thus, Vaughan (*Acta Math.* 162 (1989), 1–71) has shown that whenever n is a large enough integer congruent to some number r modulo 16, with $1 \leqslant r \leqslant 12$, then n is the sum of 12 fourth powers. Kawada and Wooley (*J. Reine Angew. Math.* 512 (1999), 173–223) obtained a similar conclusion for sums of 11 fourth powers whenever n is congruent to some integer r modulo 16 with $1 \leqslant r \leqslant 10$.

§ 21.1. Liouville proved, in 1859, that $g(4) \leqslant 53$. This upper bound was improved gradually until Wieferich (1909) proved that $g(4) \leqslant 37$ by elementary methods. Dickson (1933) improved this to 35 by the methods described above and Dress (*Comptes Rendus* 272A (1971), 457–9) reduced it further to 30 by an adaptation of Hilbert's method of proof that $g(k)$ exists. We have already referred to the proof by Balasubramanian, Deshouillers, and Dress that $g(4) = 19$.

Complementing work of Davenport (*Ann. of Math.* (2) 40 (1939), 731–47) showing that $G(4) = 16$, Deshouillers, Hennecart, Kawada, Landreau, and Wooley (*J. Théor. Nombres Bordeaux* 12 (2000), 411–22 and *Mém. Soc. Fr.* (*N.S.*) No. 100 (2005), vi+120pp.) have recently established that the largest integer that is not the sum of 16 fourth powers is 13792. Amongst other devices, the proof makes use of the identity $x^4 + y^4 + (x+y)^4 = 2(x^2 + xy + y^2)^2$, which also appears in the display preceding equation (21.10.1) above.

References to the older literature relevant to this and the next few sections will be found in Bachmann, *Niedere Zahlentheorie,* ii. 328–48, or Dickson, *History,* ii, ch. xxv.

§§ 21.2–3. See the note on § 20.1 and the historical note above.

§ 21.4. The proof for $g(6)$ is due to Fleck. Maillet proved the existence of $g(8)$ by a more complicated identity than (21.4.2); the latter is due to Hurwitz. Schur found a similar proof for $g(10)$.

§ 21.5. The special numbers n considered here were observed by Euler (and probably by Waring).

§ 21.6. Theorem 394 is due to Maillet and Hurwitz, and Theorems 395 and 396 to Kempner. The other lower bounds for $G(k)$ were investigated systematically by Hardy and Littlewood, *Proc. London Math. Soc.* (2) 28 (1928), 518–42.

§§ 21.7–8. For the results of these sections see Wright, *Journal London Math. Soc.* 9 (1934), 267–72, where further references are given; Mordell, *ibid.* 11 (1936), 208–18; and Richmond, *ibid.* 12 (1937), 206.

Hunter, *Journal London Math. Soc.* 16: (1941), 177–9 proved that $9 \leqslant v(4) \leqslant 10$; we have incorporated in the text his simple proof that $v(4) \geqslant 9$. For inequalities satisfied by $v(k)$ for $6 \leqslant k \leqslant 20$, see Fuchs and Wright, *Quart. J. Math.* (*Oxford*), 10 (1939), 190–209 and Wright, *J. für Math.* 311/312 (1979), 170–3.

Vaserstein has shown that $v(8) \leqslant 28$ (*J. Number Theory* 28 (1988), 66–68), and A. Choudhry has proved that $v(7) \leqslant 12$ (*J. Number Theory* 81 (2000), 266–9). Both conclusions depend on the existence of remarkable polynomial identities too lengthy to record here.

§§ 21.9–10. Prouhet [*Comptes Rendus Paris,* 33 (1851), 225] found the first non-trivial result in this problem. He gave a rule to separate the first j^{k+1} positive integers into j sets of j^k members, which provide a solution of (21.9.3) with $s = j^k$. For a simple proof of Prouhet's rule, see Wright, *Proc. Edinburgh Math, Soc.* (2) 8 (1949), 138–42. See Dickson, *History,* ii, ch. xxiv, and Gloden and Palamà, *Bibliographie des Multigrades* (Luxembourg, 1948), for general references. Theorem 408 is due to Bastien [*Sphinx-Oedipe* 8 (1913), 171–2] and Theorem 409 to Wright [*Bull. American Math. Soc.* 54 (1948), 755–7].

§ 21.10. Theorem 410 is due to Gloden [*Mehrgradige Gleichungen*, Groningen, 1944, 71–90]. For Theorem 411, see Tarry, *L'intermédiaire des mathématiciens,* 20 (1913), 68–70, and Escott, *Quarterly Journal of Math.* 41 (1910), 152.

A. Létac found the examples

$$[1, 25, 31, 84, 87, 134, 158, 182, 198]_8 = [2, 18, 42, 66, 113, 116, 169, 175, 199]_8$$

and

$$\begin{aligned}&[\pm 12, \pm 11881, \pm 20231, \pm 20885, \pm 23738]_9\\&\quad = [\pm 436, \pm 11857, \pm 20449, \pm 20667, \pm 23750]_9,\end{aligned}$$

which show that $P(k, 2) = k + 1$ for $k = 8$ and $k = 9$. See A. Létac, *Gazeta Matematica* 48 (1942), 68–69, and A. Gloden, *loc. cit.*

P. Borwein, Lisoněk and Percival (*Math. Comp.* 72 (2003), 2063–70) found the example

$$[\pm 99, \pm 100, \pm 188, \pm 301, \pm 313]_9 = [\pm 71, \pm 131, \pm 180, \pm 307, \pm 308]_9,$$

which provides a smaller solution than that available earlier, again confirming that $P(k, 2) = k + 1$ for $k = 9$. As the result of what is probably best described as independently joint work of Shuwen Chen, Kuosa, and Meyrignac (see http://euler.free.fr/eslp/eslp.htm for more details), in 1999 an example equivalent to

$$[\pm 22, \pm 61, \pm 86, \pm 127, \pm 140, \pm 151]_{11} = [\pm 35, \pm 47, \pm 94, \pm 121, \pm 146, \pm 148]_{11}$$

was discovered that confirms that $P(k, 2) = k + 1$ for $k = 11$.

§ 21.11. The most important result in this section is Theorem 412. The relations (21.11.9)–(21.11.12) are due to Vieta; they were used by Fermat to find solutions of (21.11.14) for any r (see Dickson, *History*, ii. 550–1). Fermat assumed without proof that all the pairs x_s, y_s, $(s = 1, 2, \ldots, r)$ would be different. The first complete proof was found by Mordell, but not published.

Of the other identities and equations which we quote, (21.11.4) is due to Gérardin [*L'intermédiaire des math.* 19 (1912), 7] and the corollary to Mahler [*Journal London Math. Soc.* 11 (1936), 136–8], (21.11.6) to Sastry [*ibid.* 9 (1934), 242–6], the parametric solution of (21.11.15) to Swinnerton-Dyer [*Proc. Cambridge Phil. Soc.* 48 (1952), 516–8], (21.11.16) to Moessner [*Proc. Ind. Math. Soc.* A 10 (1939), 296–306], (21.11.17) to Subba Rao [*Journal London Math. Soc.* 9 (1934), 172–3], and (21.11.5) to Norrie. Patterson found a further solution and Leech 6 further solutions of (21.11.2) for $k = 4$ [*Bull. Amer. Math. Soc.* 48 (1942), 736 and *Proc. Cambridge Phil. Soc.* 54 (1958), 554–5]. The identities quoted in the footnote to p. 441 were found by Fauquembergue and Gérardin respectively. For detailed references to the work of Norrie and the last two authors and to much similar work, see Dickson, *History*, ii. 650–4. Lander and Parkin [*Math. Computation* 21 (1967), 101–3] found the result which disproves Euler's conjecture for $k = 5$, $s = 4$. Elkies (*Math. Comp.* 51 (1988), 825–35) has found solutions of (21.11.1) which disprove it for $k = 4$, $s = 3$. The smallest counter example, computed by Frye, is $95800^4 + 217519^4 + 414560^4 = 422481^4$. Brudno (*Math. Comp.* 30 (1976), 646–8) gives a two-parameter solution of the equation $x_1^6 + x_2^6 + x_3^6 = y_1^6 + y_2^6 + y_3^6$, of which (21.11.17) is a particular solution.

For a survey of the subject of equal sums of like powers see Lander, *American Math. Monthly* 75 (1968), 1061–73.

XXII

THE SERIES OF PRIMES (3)

22.1. The functions $\vartheta(x)$ and $\psi(x)$. In this chapter we return to the problems concerning the distribution of primes of which we gave a preliminary account in the first two chapters. There we proved nothing except Euclid's Theorem 4 and the slight extensions contained in §§ 2.1–6. Here we develop the theory much further and, in particular, prove Theorem 6 (the Prime Number Theorem). We begin, however, by proving the much simpler Theorem 7.

Our proof of Theorems 6 and 7 depends upon the properties of a function $\psi(x)$ and (to a lesser extent) of a function $\vartheta(x)$. We write†

$$\vartheta(x) = \sum_{p\leqslant x} \log p = \log \prod_{p\leqslant x} p \tag{22.1.1}$$

and

$$\psi(x) = \sum_{p^m\leqslant x} \log p = \sum_{n\leqslant x} \Lambda(n) \tag{22.1.2}$$

(in the notation of § 17.7). Thus

$$\psi(10) = 3\log 2 + 2\log 3 + \log 5 + \log 7,$$

there being a contribution $\log 2$ from 2, 4, and 8, and a contribution $\log 3$ from 3 and 9. If p^m is the highest power of p not exceeding x, $\log p$ occurs m times in $\psi(x)$. Also p^m is the highest power of p which divides any number up to x, so that

$$\psi(x) = \log U(x), \tag{22.1.3}$$

where $U(x)$ is the least common multiple of all numbers up to x. We can also express $\psi(x)$ in the form

$$\psi(x) = \sum_{p\leqslant x} \left[\frac{\log x}{\log p}\right] \log p. \tag{22.1.4}$$

† Throughout this chapter x (and y and t) are not necessarily integral. On the other hand, m, n, h, k, etc., are positive integers and p, as usual, is a prime. We suppose always that $x \geqslant 1$.

The definitions of $\vartheta(x)$ and $\psi(x)$ are more complicated than that of $\pi(x)$, but they are in reality more 'natural' functions. Thus $\psi(x)$ is, after (22.1.2), the 'sum function' of $\Lambda(n)$, and $\Lambda(n)$ has (as we saw in § 17.7) a simple generating function. The generating functions of $\vartheta(x)$, and still more of $\pi(x)$, are much more complicated. And even the arithmetical definition of $\psi(x)$, when written in the form (22.1.3), is very elementary and natural.

Since $p^2 \leqslant x, p^3 \leqslant x, \ldots$ are equivalent to $p \leqslant x^{\frac{1}{2}}, p \leqslant x^{\frac{1}{3}}, \ldots$, we have

$$\psi(x) = \vartheta(x) + \vartheta\left(x^{\frac{1}{2}}\right) + \vartheta\left(x^{\frac{1}{3}}\right) + \cdots = \sum \vartheta(x^{1/m}). \tag{22.1.5}$$

The series breaks off when $x^{1/m} < 2$, i.e. when

$$m > \frac{\log x}{\log 2}.$$

It is obvious from the definition that $\vartheta(x) < x \log x$ for $x \geqslant 2$. *A fortiori*

$$\vartheta\left(x^{1/m}\right) < x^{1/m} \log x \leqslant x^{\frac{1}{2}} \log x$$

if $m \geqslant 2$; and

$$\sum_{m \geqslant 2} \vartheta\left(x^{1/m}\right) = O\left\{x^{\frac{1}{2}} (\log x)^2\right\},$$

since there are only $O(\log x)$ terms in the series. Hence

THEOREM 413:

$$\psi(x) = \vartheta(x) + O\left\{x^{\frac{1}{2}} (\log x)^2\right\}.$$

We are interested in the order of magnitude of the functions. Since

$$\pi(x) = \sum_{p \leqslant x} 1, \quad \vartheta(x) = \sum_{p \leqslant x} \log p,$$

it is natural to expect $\vartheta(x)$ to be 'about $\log x$ times' $\pi(x)$. We shall see later that this is so. We prove next that $\vartheta(x)$ is of order x, so that Theorem 413 tells us that $\psi(x)$ is 'about the same as' $\vartheta(x)$ when x is large.

22.2. Proof that $\vartheta(x)$ and $\psi(x)$ are of order x. We now prove

THEOREM 414. *The functions $\vartheta(x)$ and $\psi(x)$ are of order x:*

$$Ax < \vartheta(x) < Ax, \quad Ax < \psi(x) < Ax \qquad (x \geqslant 2). \tag{22.2.1}$$

It is enough, after Theorem 413, to prove that

$$\vartheta(x) < Ax \tag{22.2.2}$$

and

$$\psi(x) > Ax \qquad (x \geqslant 2). \tag{22.2.3}$$

In fact, we prove a result a little more precise than (22.2.2), viz.

THEOREM 415:

$$\vartheta(n) < 2n \log 2 \quad \textit{for all } n \geqslant 1.$$

By Theorem 73,

$$M = \frac{(2m+1)!}{m!\,(m+1)!} = \frac{(2m+1)\,(2m)\ldots(m+2)}{m!}$$

is an integer. It occurs twice in the binomial expansion of $(1+1)^{2m+1}$ and so $2M < 2^{2m+1}$ and $M < 2^{2m}$.

If $m+1 < p \leqslant 2m+1$, p divides the numerator but not the denominator of M. Hence

$$\Bigl(\prod_{m+1<p\leqslant 2m+1} p\Bigr)\Big| M$$

and

$$\vartheta(2m+1) - \vartheta(m+1) = \sum_{m+1<p\leqslant 2m+1} \log p \leqslant \log M < 2m \log 2.$$

Theorem 415 is trivial for $n = 1$ and for $n = 2$. Let us suppose it true for all $n \leqslant n_0 - 1$. If n_0 is even, we have

$$\vartheta(n_0) = \vartheta(n_0 - 1) < 2\,(n_0 - 1)\log 2 < 2n_0 \log 2.$$

If n_0 is odd, say $n_0 = 2m + 1$, we have

$$\begin{aligned}\vartheta(n_0) &= \vartheta(2m+1) = \vartheta(2m+1) - \vartheta(m+1) + \vartheta(m+1)\\ &< 2m\log 2 + 2\,(m+1)\log 2\\ &= 2\,(2m+1)\log 2 = 2n_0\log 2,\end{aligned}$$

since $m + 1 < n_0$. Hence Theorem 415 is true for $n = n_0$ and so, by induction, for all n. The inequality (22.2.2) follows at once.

We now prove (22.2.3). The numbers $1, 2, \ldots, n$ include just $[n/p]$ multiples of p, just $[n/p^2]$ multiples of p^2, and so on. Hence

THEOREM 416:

$$n! = \prod_p p^{j(n,p)},$$

where

$$j\,(n,p) = \sum_{m\geqslant 1}\left[\frac{n}{p^m}\right].$$

We write

$$N = \frac{(2n)!}{(n!)^2} = \prod_{p\leqslant 2n} p^{k_p},$$

so that, by Theorem 416,

$$k_p = \sum_{m=1}^{\infty}\left(\left[\frac{2n}{p^m}\right] - 2\left[\frac{n}{p^m}\right]\right). \tag{22.2.4}$$

Each term in round brackets is 1 or 0, according as $[2n/p^m]$ is odd or even. In particular, the term is 0 if $p^m > 2n$. Hence

$$k_p \leqslant \left[\frac{\log 2n}{\log p}\right] \tag{22.2.5}$$

and

$$\log N = \sum_{p\leqslant 2n} k_p \log p \leqslant \sum_{p\leqslant 2n}\left[\frac{\log 2n}{\log p}\right]\log p = \psi(2n)$$

by (22.1.4). But

$$N = \frac{(2n)!}{(n!)^2} = \frac{n+1}{1}.\frac{n+2}{2}\cdots\frac{2n}{n} \geqslant 2^n \tag{22.2.6}$$

and so

$$\psi(2n) \geqslant n\log 2.$$

For $x \geqslant 2$, we put $n = \left[\frac{1}{2}x\right] \geqslant 1$ and have

$$\psi(x) \geqslant \psi(2n) \geqslant n\log 2 \geqslant \tfrac{1}{4}x\log 2,$$

which is (22.2.3).

22.3. Bertrand's postulate and a 'formula' for primes. From Theorem 414, we can deduce

THEOREM 417. *There is a number B such that, for every $x > 1$, there is a prime p satisfying*

$$x < p \leqslant Bx.$$

For, by Theorem 414,

$$C_1 x < \vartheta(x) < C_2 x \qquad (x \geqslant 2)$$

for some fixed C_1, C_2. Hence

$$\vartheta(C_2x/C_1) > C_1\,(C_2x/C_1) = C_2x > \vartheta(x)$$

and so there is a prime between x and C_2x/C_1. If we put $B = \max(C_2/C_1, 2)$, Theorem 417 is immediate.

We can, however, refine our argument a little to prove a more precise result.

THEOREM 418 (*Bertrand's Postulate*). *If $n \geqslant 1$, there is at least one prime p such that*

$$n < p \leqslant 2n; \tag{22.3.1}$$

that is, if p_r is the r-th prime,

$$p_{r+1} < 2p_r \tag{22.3.2}$$

for every r.

The two parts of the theorem are clearly equivalent. Let us suppose that, for some $n > 2^9 = 512$, there is no prime satisfying (22.3.1). With the notation of § 22.2, let p be a prime factor of N, so that $k_p \geqslant 1$. By our hypothesis, $p \leqslant n$. If $\frac{2}{3}n < p \leqslant n$, we have

$$2p \leqslant 2n < 3p, \quad p^2 > \tfrac{4}{9}n^2 > 2n$$

and (22.2.4) becomes

$$k_p = \left[\frac{2n}{p}\right] - 2\left[\frac{n}{p}\right] = 2 - 2 = 0.$$

Hence $p \leqslant \frac{2}{3}n$ for every prime factor p of N and so

$$\sum_{p|N} \log p \leqslant \sum_{p \leqslant \frac{2}{3}n} \log p = \vartheta\left(\tfrac{2}{3}n\right) \leqslant \tfrac{4}{3}n \log 2 \tag{22.3.3}$$

by Theorem 415.

Next, if $k_p \geqslant 2$, we have, by (22.2.5)

$$2 \log p \leqslant k_p \log p \leqslant \log(2n), \quad p \leqslant \surd(2n)$$

and so there are at most $\surd(2n)$ such values of p. Hence

$$\sum_{k_p \geqslant 2} k_p \log p \leqslant \surd(2n) \log(2n),$$

and so

$$\begin{aligned} \log N &\leqslant \sum_{k_p=1} \log p + \sum_{k_p \geqslant 2} k_p \log p \leqslant \sum_{p|N} \log p + \surd(2n) \log(2n) \\ &\leqslant \tfrac{4}{3}n \log 2 + \surd(2n) \log(2n) \end{aligned} \tag{22.3.4}$$

by (22.3.3)

On the other hand, N is the largest term in the expansion of $2^{2n} = (1+1)^{2n}$, so that

$$2^{2n} = 2 + \binom{2n}{1} + \binom{2n}{2} + \cdots + \binom{2n}{2n-1} \leqslant 2nN.$$

Hence, by (22.3.4),

$$2n \log 2 \leqslant \log(2n) + \log N \leqslant \tfrac{4}{3}n \log 2 + \left\{1 + \surd(2n)\right\} \log(2n),$$

which reduces to

$$2n \log 2 \leqslant 3\left\{1 + \surd(2n)\right\} \log(2n). \tag{22.3.5}$$

We now write

$$\zeta = \frac{\log(n/512)}{10 \log 2} > 0,$$

so that $2n = 2^{10(1+\zeta)}$. Since $n > 512$, we have $\zeta > 0$. (22.3.5) becomes

$$2^{10(1+\zeta)} \leqslant 30\left(2^{5+5\zeta} + 1\right)(1+\zeta),$$

whence

$$2^{5\zeta} \leqslant 30 \,.\, 2^{-5}\left(1+2^{-5-5\zeta}\right)(1+\zeta) < \left(1-2^{-5}\right)\left(1+2^{-5}\right)(1+\zeta) < 1+\zeta.$$

But

$$2^{5\zeta} = \exp(5\zeta \log 2) > 1 + 5\zeta \log 2 > 1 + \zeta,$$

a contradiction. Hence, if $n > 512$, there must be a prime satisfying (22.3.1).

Each of the primes

$$2, 3, 5, 7, 13, 23, 43, 83, 163, 317, 631$$

is less than twice its predecessor in the list. Hence one of them, at least, satisfies (22.3.1) for any $n \leqslant 630$. This completes the proof of Theorem 418.

We prove next

THEOREM 419. *If*

$$\alpha = \sum_{m=1}^{\infty} p_m 10^{-2^m} = \cdot 02030005000000070\ldots,$$

we have

$$p_n = \left[10^{2^n}\alpha\right] - 10^{2^{n-1}}\left[10^{2^{n-1}}\alpha\right]. \tag{22.3.6}$$

By (2.2.2),

$$p_m < 2^{2^m} = 4^{2^{m-1}}$$

and so the series for α is convergent. Again

$$0 < 10^{2^n} \sum_{m=n+1}^{\infty} p_m 10^{-2^m} < \sum_{m=n+1}^{\infty} 4^{2^{m-1}} 10^{-2^{m-1}}$$

$$= \sum_{m=n+1}^{\infty} \left(\tfrac{2}{5}\right)^{2^{m-1}} < \left(\tfrac{2}{5}\right)^{2^n} \frac{1}{\left(1-\frac{2}{5}\right)} < \tfrac{4}{15} < 1.$$

Hence

$$\left[10^{2^n}\alpha\right] = 10^{2^n} \sum_{m=1}^{n} p_m 10^{-2^m}$$

and, similarly,

$$\left[10^{2^{n-1}}\alpha\right] = 10^{2^{n-1}} \sum_{m=1}^{n-1} p_m 10^{-2^m}.$$

It follows that

$$\left[10^{2^n}\alpha\right] - 10^{2^{n-1}}\left[10^{2^n-1}\alpha\right] = 10^{2^n}\left(\sum_{m=1}^{n} p_m 10^{-2^m} - \sum_{m=1}^{n-1} p_m 10^{-2^m}\right) = p_n.$$

Although (22.3.6) gives a 'formula' for the nth prime p_n, it is not a very useful one. To calculate p_n from this formula, it is necessary to know the value of α correct to 2^n decimal places; and to do this, it is necessary to know the values of $p_1, p_2, \ldots, p_n$.

There are a number of similar formulae which suffer from the same defect. Thus, let us suppose that r is an integer greater than one. We have then

$$p_n \leqslant r^n$$

by (22.3.2). Indeed, for $r \geqslant 4$, this follows from Theorem 20. Hence we may write

$$\alpha_r = \sum_{m=1}^{\infty} p_m r^{-m^2}$$

and we can deduce that

$$p_n = \left[r^{n^2}\alpha_r\right] - r^{2n-1}\left[r^{(n-1)^2}\alpha_r\right]$$

by arguments similar to those used above.

Any one of these formulae (or any similar one) would attain a different status if the exact value of the number α or α_r which occurs in it could be expressed independently of the primes. There seems no likelihood of this, but it cannot be ruled out as entirely impossible.

For another formula for p_n, see § 1 of the Appendix.

22.4. Proof of Theorems 7 and 9. It is easy to deduce Theorem 7 from Theorem 414. In the first place

$$\vartheta(x) = \sum_{p \leqslant x} \log p \leqslant \log x \sum_{p \leqslant x} 1 = \pi(x) \log x$$

and so

$$\pi(x) \geqslant \frac{\vartheta(x)}{\log x} > \frac{Ax}{\log x}. \tag{22.4.1}$$

On the other hand, if $0 < \delta < 1$,

$$\vartheta(x) \geqslant \sum_{x^{1-\delta} < p \leqslant x} \log p \geqslant (1-\delta)\log x \sum_{x^{1-\delta} < p \leqslant x} 1$$

$$= (1-\delta)\log x\{\pi(x) - \pi(x^{1-\delta})\} \geqslant (1-\delta)\log x\{\pi(x) - x^{1-\delta}\}$$

and so

$$\pi(x) \leqslant x^{1-\delta} + \frac{\vartheta(x)}{(1-\delta)\log x} < \frac{Ax}{\log x}. \tag{22.4.2}$$

We can now prove

THEOREM 420:

$$\pi(x) \sim \frac{\vartheta(x)}{\log x} \sim \frac{\psi(x)}{\log x}.$$

After Theorems 413 and 414 we need only consider the first assertion. It follows from (22.4.1) and (22.4.2) that

$$1 \leqslant \frac{\pi(x)\log x}{\vartheta(x)} \leqslant \frac{x^{1-\delta}\log x}{\vartheta(x)} + \frac{1}{1-\delta}.$$

For any $\epsilon > 0$, we can choose $\delta = \delta(\epsilon)$ so that

$$\frac{1}{1-\delta} < 1 + \tfrac{1}{2}\epsilon$$

and then choose $x_0 = x_0(\delta, \epsilon) = x_0(\epsilon)$ so that

$$\frac{x^{1-\delta}\log x}{\vartheta(x)} < \frac{A\log x}{x^{\delta}} < \tfrac{1}{2}\epsilon$$

for all $x > x_0$. Hence

$$1 \leqslant \frac{\pi(x)\log x}{\vartheta(x)} < 1 + \epsilon$$

for all $x > x_0$. Since ϵ is arbitrary, the first part of Theorem 420 follows at once.

Theorem 9 is (as stated in § 1.8) a corollary of Theorem 7. For, in the first place,

$$n = \pi(p_n) < \frac{Ap_n}{\log p_n}, \quad p_n > An\log p_n > An\log n.$$

Secondly,

$$n = \pi(p_n) > \frac{Ap_n}{\log p_n},$$

so that

$$\sqrt{p_n} < \frac{Ap_n}{\log p_n} < An, \quad p_n < An^2,$$

and

$$p_n < An\log p_n < An\log n.$$

22.5. Two formal transformations. We introduce here two elementary formal transformations which will be useful throughout this chapter.

THEOREM 421. *Suppose that* $c_1, c_2, \ldots$ *is a sequence of numbers, that*

$$C(t) = \sum_{n \leqslant t} c_n,$$

and that $f(t)$ *is any function of* t. *Then*

$$\sum_{n \leqslant x} c_n f(n) = \sum_{n \leqslant x-1} C(n)\{f(n) - f(n+1)\} + C(x)f([x]). \tag{22.5.1}$$

If, in addition, $c_j = 0$ *for* $j < n_1$† *and* $f(t)$ *has a continuous derivative for* $t \geqslant n_1$, *then*

$$\sum_{n \leqslant x} c_n f(n) = C(x)f(x) - \int_{n_1}^{x} C(t)f'(t)\,dt. \tag{22.5.2}$$

If we write $N = [x]$, the sum on the left of (22.5.1) is

$$\begin{aligned} &C(1)f(1) + \{C(2) - C(1)\}f(2) + \cdots + \{C(N) - C(N-1)\}f(N) \\ &\quad = C(1)\{f(1) - f(2)\} + \cdots + C(N-1)\{f(N-1) - f(N)\} \\ &\qquad + C(N)f(N). \end{aligned}$$

Since $C(N) = C(x)$, this proves (22.5.1). To deduce (22.5.2) we observe that $C(t) = C(n)$ when $n \leqslant t < n+1$ and so

$$C(n)\{f(n) - f(n+1)\} = -\int_{n}^{n+1} C(t)f'(t)dt.$$

Also $C(t) = 0$ when $t < n_1$.

† In our applications, $n_1 = 1$ or 2. If $n_1 = 1$, there is, of course, no restriction on the c_n. If $n_1 = 2$, we have $c_1 = 0$.

If we put $c_n = 1$ and $f(t) = 1/t$, we have $C(x) = [x]$ and (22.5.2) becomes

$$\sum_{n\leqslant x} \frac{1}{n} = \frac{[x]}{x} + \int_1^x \frac{[t]}{t^2} dt$$
$$= \log x + \gamma + E,$$

where

$$\gamma = 1 - \int_1^\infty \frac{(t-[t])}{t^2} dt$$

is independent of x and

$$E = \int_x^\infty \frac{(t-[t])}{t^2} dt - \frac{x-[x]}{x} = \int_x^\infty \frac{O(1)}{t^2} dt + O\left(\frac{1}{x}\right) = O\left(\frac{1}{x}\right).$$

Thus we have

THEOREM 422:

$$\sum_{n\leqslant x} \frac{1}{n} = \log x + \gamma + O\left(\frac{1}{x}\right),$$

where γ is a constant (known as Euler's constant).

22.6. An important sum. We prove first the lemma

THEOREM 423:

$$\sum_{n\leqslant x} \log^h \left(\frac{x}{n}\right) = O(x) \qquad (h > 0).$$

Since $\log t$ increases with t, we have, for $n \geqslant 2$,

$$\log^h \left(\frac{x}{n}\right) \leqslant \int_{n-1}^n \log^h \left(\frac{x}{t}\right) dt.$$

Hence

$$\sum_{n=2}^{[x]} \log^h\left(\frac{x}{n}\right) \leqslant \int_1^x \log^h\left(\frac{x}{t}\right) dt = x\int_1^x \frac{\log^h u}{u^2} du$$

$$< x\int_1^\infty \frac{\log^h u}{u^2} du = Ax,$$

since the infinite integral is convergent. Theorem 423 follows at once.

If we put $h = 1$, we have

$$\sum_{n\leqslant x} \log n = [x]\log x + O(x) = x\log x + O(x).$$

But, by Theorem 416,

$$\sum_{n\leqslant x} \log n = \sum_{p\leqslant x} j([x], p)\log p = \sum_{p^m\leqslant x} \left[\frac{x}{p^m}\right]\log p = \sum_{n\leqslant x}\left[\frac{x}{n}\right]\Lambda(n)$$

in the notation of § 17.7. If we remove the square brackets in the last sum, we introduce an error less than

$$\sum_{n\leqslant x} \Lambda(n) = \psi(x) = O(x)$$

and so

$$\sum_{n\leqslant x} \frac{x}{n}\Lambda(n) = \sum_{n\leqslant x} \log n + O(x) = x\log x + O(x).$$

If we remove a factor x, we have

THEOREM 424:

$$\sum_{n\leqslant x} \frac{\Lambda(n)}{n} = \log x + O(1).$$

From this we can deduce

THEOREM 425:

$$\sum_{p\leqslant x} \frac{\log p}{p} = \log x + O(1).$$

For

$$\sum_{n\leqslant x}\frac{\Lambda(n)}{n}-\sum_{p\leqslant x}\frac{\log p}{p}=\sum_{m\geqslant 2}\sum_{p^m\leqslant x}\frac{\log p}{p^m}$$
$$<\sum_{p}\left(\frac{1}{p^2}+\frac{1}{p^3}+\cdots\right)\log p=\sum_{p}\frac{\log p}{p(p-1)}$$
$$<\sum_{n=2}^{\infty}\frac{\log n}{n(n-1)}=A.$$

If, in (22.5.2), we put $f(t)=1/t$ and $c_n=\Lambda(n)$, so that $C(x)=\psi(x)$, we have

$$\sum_{n\leqslant x}\frac{\Lambda(n)}{n}=\frac{\psi(x)}{x}+\int_2^x\frac{\psi(t)}{t^2}dt$$

and so, by Theorems 414 and 424, we have

$$\int_2^x\frac{\psi(t)}{t^2}dt=\log x+O(1). \tag{22.6.1}$$

From (22.6.1) we can deduce

$$\underline{\lim}\,\{\psi(x)/x\}\leqslant 1,\quad \overline{\lim}\,\{\psi(x)/x\}\geqslant 1. \tag{22.6.2}$$

For, if $\underline{\lim}\,\{\psi(x)/x\}=1+\delta$, where $\delta>0$, we have $\psi(x)>\left(1+\frac{1}{2}\delta\right)x$ for all x greater than some x_0. Hence

$$\int_2^x\frac{\psi(t)}{t^2}dt>\int_2^{x_0}\frac{\psi(t)}{t^2}dt+\int_{x_0}^x\frac{\left(1+\frac{1}{2}\delta\right)}{t}dt>\left(1+\tfrac{1}{2}\delta\right)\log x-A,$$

in contradiction to (22.6.1). If we suppose that $\overline{\lim}\{\psi(x)/x\}=1-\delta$, we get a similar contradiction.

By Theorem 420, we can deduce from (22.6.2)

THEOREM 426:

$$\underline{\lim}\left\{\pi(x)\Big/\frac{x}{\log x}\right\}\leqslant 1,\quad \overline{\lim}\left\{\pi(x)\Big/\frac{x}{\log x}\right\}\geqslant 1.$$

If $\pi(x)\big/\frac{x}{\log x}$ tends to a limit as $x\to\infty$, the limit is 1.

Theorem 6 would follow at once if we could prove that $\pi(x)\big/\frac{x}{\log x}$ tends to a limit. Unfortunately this is the real difficulty in the proof of Theorem 6,

22.7. The sum Σp^{-1} and the product $\Pi(1-p^{-1})$. Since

$$0 < \log\left(\frac{1}{1-p^{-1}}\right) - \frac{1}{p} = \frac{1}{2p^2} + \frac{1}{3p^3} + \cdots \tag{22.7.1}$$

$$< \frac{1}{2p^2} + \frac{1}{2p^3} + \cdots = \frac{1}{2p(p-1)}$$

and

$$\sum \frac{1}{p(p-1)}$$

is convergent, the series

$$\sum \left\{\log\left(\frac{1}{1-p^{-1}}\right) - \frac{1}{p}\right\}$$

must be convergent. By Theorem 19, Σp^{-1} is divergent and so the product

$$\prod (1-p^{-1}) \tag{22.7.2}$$

must diverge also (to zero).

From the divergence of the product (22.7.2) we can deduce that

$$\pi(x) = o(x),$$

i.e. almost all numbers are composite, without using any of the results of §§ 22.1–6. Of course, this result is weaker than Theorem 7, but the very simple proof is of some interest.

We choose r so that

$$M = p_1 p_2 \dots p_r \leqslant x < p_1 \dots p_r p_{r+1}$$

and k the positive integer such that $kM \leqslant x < (k+1)M$. Let H be the number of positive integers which (i) do not exceed $(k+1)M$ and (ii) are not divisible by any of the primes $p_1, \dots, p_r$, i.e. are prime to M. These numbers clearly include all the primes $p_{r+1}, \dots, p_{\pi(x)}$. Hence

$$\pi(x) \leqslant r + H.$$

By definition $\phi(M)$ is the number of integers prime to M and less than or equal to M, so that $H = (k+1)\phi(M)$. But $x \geqslant kM$ and so, by (16.1.3),

$$\frac{H}{x} \leqslant \frac{(k+1)\phi(M)}{kM} \leqslant \frac{2\phi(M)}{M} = 2\prod_{i=1}^{r}\left(1 - \frac{1}{p_i}\right) \to 0$$

as $r \to \infty$, since the product (22.7.2) diverges. Also

$$\frac{r}{x} \leqslant \frac{r}{p_{r-1}p_r} \leqslant \frac{1}{p_{r-1}} \to 0.$$

As $x \to \infty$, so does r and we have

$$\frac{\pi(x)}{x} \leqslant \frac{r}{x} + \frac{H}{x} \to 0$$

that is, $\pi(x) = o(x)$.

We can prove the divergence of $\Pi(1 - p^{-1})$ independently of that of $\sum p^{-1}$ as follows. It is plain that

$$\prod_{p \leqslant N} \left(\frac{1}{1 - p^{-1}}\right) = \prod_{p \leqslant N} \left(1 + \frac{1}{p} + \frac{1}{p^2} + \cdots\right) = \sum_{(N)} \frac{1}{n},$$

the last sum being extended over all n composed of prime factors $p \leqslant N$. Since all $n \leqslant N$ satisfy this condition,

$$\prod_{p \leqslant N} \left(\frac{1}{1 - p^{-1}}\right) \geqslant \sum_{n=1}^{N} \frac{1}{n} > \log N - A$$

by Theorem 422. Hence the product (22.7.2) is divergent.

If we use the results of the last two sections, we can obtain much more exact information about $\sum p^{-1}$. In Theorem 421, let us put $c_p = \log p/p$, and $c_n = 0$ if n is not a prime, so that

$$C(x) = \sum_{p \leqslant x} \frac{\log p}{p} = \log x + \tau(x),$$

where $\tau(x) = O(1)$ by Theorem 425. With $f(t) = 1/\log t$, (22.5.2) becomes

$$\text{(22.7.3)} \qquad \sum_{p \leqslant x} \frac{1}{p} = \frac{C(x)}{\log x} + \int_2^x \frac{C(t)}{t \log^2 t} dt$$

$$= 1 + \frac{\tau(x)}{\log x} + \int_2^x \frac{dt}{t \log t} + \int_2^x \frac{\tau(t)dt}{t \log^2 t}$$

$$= \log\log x + B_1 + E(x),$$

where

$$B_1 = 1 - \log\log 2 + \int_2^\infty \frac{\tau(t)dt}{t\log^2 t}$$

and

(22.7.4)

$$E(x) = \frac{\tau(x)}{\log x} - \int_x^\infty \frac{\tau(t)dt}{t\log^2 t} = O\left(\frac{1}{\log x}\right) + O\left(\int_x^\infty \frac{dt}{t\log^2 t}\right) = O\left(\frac{1}{\log x}\right).$$

Hence we have

THEOREM 427:

$$\sum_{p\leqslant x} \frac{1}{p} = \log\log x + B_1 + o(1),$$

where B_1 *is a constant.*

22.8. Mertens's theorem. It is interesting to push our study of the series and product of the last section a little further.

THEOREM 428. *In Theorem* 427,

(22.8.1) $$B_1 = \gamma + \sum \left\{\log\left(1 - \frac{1}{p}\right) + \frac{1}{p}\right\},$$

where γ *is Euler's constant.*

THEOREM 429 (MERTENS'S THEOREM):

$$\prod_{p\leqslant x} \left(1 - \frac{1}{p}\right) \sim \frac{e^{-\gamma}}{\log x}.$$

As we saw in § 22.7, the series in (22.8.1) converges. Since

$$\sum_{p\leqslant x} \frac{1}{p} + \sum_{p\leqslant x} \log\left(1 - \frac{1}{p}\right) = \sum_{p\leqslant x} \left\{\log\left(1 - \frac{1}{p}\right) + \frac{1}{p}\right\},$$

Theorem 429 follows from Theorems 427 and 428. Hence it is enough to prove Theorem 428. We shall assume that†

$$\gamma = -\Gamma'(1) = -\int_0^\infty e^{-x}\log x\,dx. \tag{22.8.2}$$

If $\delta \geqslant 0$, we have

$$0 < -\log\left(1 - \frac{1}{p^{1+\delta}}\right) - \frac{1}{p^{1+\delta}} < \frac{1}{2p^{1+\delta}(p^{1+\delta}-1)} \leqslant \frac{1}{2p(p-1)}$$

by calculations similar to those of (22.7.1). Hence the series

$$F(\delta) = \sum_p \left\{\log\left(1 - \frac{1}{p^{1+\delta}}\right) + \frac{1}{p^{1+\delta}}\right\}$$

is uniformly convergent for all $\delta \geqslant 0$ and so

$$F(\delta) \to F(0)$$

as $\delta \to 0$ through positive values.

We now suppose $\delta > 0$. By Theorem 280,

$$F(\delta) = g(\delta) - \log\zeta(1+\delta),$$

where

$$g(\delta) = \sum_p p^{-1-\delta}.$$

If, in Theorem 421, we put $c_p = 1/p$ and $c_n = 0$ when n is not prime, we have

$$C(x) = \sum_{p \leqslant x} \frac{1}{p} = \log\log x + B_1 + E(x)$$

by (22.7.3). Hence, if $f(t) = t^{-\delta}$, (22.5.2) becomes

$$\sum_{p \leqslant x} p^{-1-\delta} = x^{-\delta}C(x) + \delta\int_2^x t^{-1-\delta}C(t)\,dt.$$

† See, for example, Whittaker and Watson, *Modern analysis,* ch. xii.

Letting $x \to \infty$, we have

$$g(\delta) = \delta \int_2^\infty t^{-1-\delta} C(t) dt$$

$$= \delta \int_2^\infty t^{-1-\delta} (\log\log t + B_1) dt + \delta \int_2^\infty t^{-1-\delta} E(t)\, dt.$$

Now, if we put $t = e^{u/\delta}$,

$$= \delta \int_1^\infty t^{-1-\delta} \log\log t\, dt = \int_2^\infty e^{-u} \log\left(\frac{u}{\delta}\right) du = -\gamma - \log\delta$$

by (22.8.2), and

$$\delta \int_1^\infty t^{-1-\delta} dt = 1.$$

Hence

$$g(\delta) + \log\delta - B_1 + \gamma = \delta \int_2^\infty t^{-1-\delta} E(t)\, dt - \delta \int_1^2 t^{-1-\delta} (\log\log t + B_1) dt.$$

Now, by (22.7.4), if $T = \exp(1/\sqrt{\delta})$,

$$\left| \delta \int_2^\infty \frac{E(t)}{t^{1+\delta}} dt \right| < A\delta \int_2^T \frac{dt}{t} + \frac{A\delta}{\log T} \int_T^\infty \frac{dt}{t^{1+\delta}}$$

$$< A\delta \log T + \frac{A}{\log T} < A\sqrt{\delta} \to 0,$$

as $\delta \to 0$. Also

$$\left| \int_1^2 t^{-1-\delta} (\log\log t + B_1)\, dt \right| < \int_1^2 t^{-1} (|\log\log t| + |B_1|) dt = A,$$

since the integral converges at $t = 1$. Hence

$$g(\delta) + \log \delta \to B_1 - \gamma$$

as $\delta \to 0$.

But, by Theorem 282,

$$\log \zeta(1 + \delta) + \log \delta \to 0$$

as $\delta \to 0$ and so

$$F(\delta) \to B_1 - \gamma.$$

Hence

$$B_1 = \gamma + F(0),$$

which is (22.8.1).

22.9. Proof of Theorems 323 and 328. We are now able to prove Theorems 323 and 328. If we write

$$f_1(n) = \frac{\phi(n)e^{\gamma} \log\log n}{n}, \quad f_2(n) = \frac{\sigma(n)}{ne^{\gamma} \log\log n},$$

we have to show that

$$\underline{\lim}\, f_1(n) = 1, \quad \overline{\lim}\, f_2(n) = 1.$$

It will be enough to find two functions $F_1(t)$, $F_2(t)$, each tending to 1 as $t \to \infty$ and such that

$$f_1(n) \geqslant F_1(\log n), \quad f_2(n) \leqslant \frac{1}{F_1(\log n)} \tag{22.9.1}$$

for all $n \geqslant 3$ and

$$f_2(n_j) \geqslant F_2(j), \quad f_1(n_j) \leqslant \frac{1}{F_2(j)} \tag{22.9.2}$$

for an infinite increasing sequence $n_2, n_3, n_4, \ldots$.

By Theorem 329, $f_1(n)f_2(n) < 1$ and so the second inequality in (22.9.1) follows from the first; similarly for (22.9.2).

Let $p_1, p_2, \ldots, p_{r-\rho}$ be the primes which divide n and which do not exceed $\log n$ and let $p_{r-\rho+1}, \ldots, p_r$ be those which divide n and are greater than $\log n$. We have

$$(\log n)^\rho < p_{r-\rho+1} \ldots p_r \leqslant n, \quad \rho < \frac{\log n}{\log\log n}$$

and so

$$\frac{\phi(n)}{n} = \prod_{i=1}^{r}\left(1 - \frac{1}{p_i}\right) \geqslant \left(1 - \frac{1}{\log n}\right)^\rho \prod_{i=1}^{r-\rho}\left(1 - \frac{1}{p_i}\right)$$
$$> \left(1 - \frac{1}{\log n}\right)^{\log n/\log\log n} \prod_{p \leqslant \log n}\left(1 - \frac{1}{p}\right).$$

Hence the first part of (22.9.1) is true with

$$F_1(t) = e^\gamma \log t \left(1 - \frac{1}{t}\right)^{t/\log t} \prod_{p \leqslant t}\left(1 - \frac{1}{p}\right).$$

But, by Theorem 429, as $t \to \infty$,

$$F_1(t) \sim \left(1 - \frac{1}{t}\right)^{t/\log t} = 1 + O\left(\frac{1}{\log t}\right) \to 1.$$

To prove the first part of (22.9.2), we write

$$n_j = \prod_{p \leqslant e^j} p^j \qquad (j \geqslant 2),$$

so that

$$\log n_j = j\vartheta(e^j) \leqslant Aje^j$$

by Theorem 414. Hence

$$\log\log n_j \leqslant A_0 + j + \log j.$$

Again

$$\prod_{p \leqslant e^j} (1 - p^{-j-1}) > \prod (1 - p^{-j-1}) = \frac{1}{\zeta(j+1)}$$

by Theorem 280. Hence

$$f_2(n_j) = \frac{\sigma(n_j)}{n_j e^{\gamma} \log\log n_j} = \frac{e^{-\gamma}}{\log\log n_j} \prod_{p \leqslant e^j} \left(\frac{1 - p^{-j-1}}{1 - p^{-1}} \right)$$

$$\geqslant \frac{e^{-\gamma}}{\zeta(j+1)(A_0 + j + \log j)} \prod_{p \leqslant e^j} \left(\frac{1}{1 - p^{-1}} \right) = F_2(j)$$

(say). This is the first part of (22.9.2). Again, as $j \to \infty$, $\zeta(j+1) \to 1$ and, by Theorem 429,

$$F_2(j) \sim \frac{j}{\zeta(j+1)(A_1 + j + \log j)} \to 1.$$

22.10. The number of prime factors of *n*. We define $\omega(n)$ as the number of different prime factors of n, and $\Omega(n)$ as its total number of prime factors; thus

$$\omega(n) = r, \quad \Omega(n) = a_1 + a_2 + \cdots + a_r,$$

when $n = p_1^{a_1} \ldots p_r^{a_r}$.

Both $\omega(n)$ and $\Omega(n)$ behave irregularly for large n. Thus both functions are 1 when n is prime, while

$$\Omega(n) = \frac{\log n}{\log 2}$$

when n is a power of 2. If

$$n = p_1 p_2 \ldots p_r$$

is the product of the first r primes, then

$$\omega(n) = r = \pi(p_r), \ \log n = \vartheta(p_r)$$

and so, by Theorems 420 and 414,

$$\omega(n) \sim \frac{\vartheta(p_r)}{\log p_r} \sim \frac{\log n}{\log\log n}$$

(when $n \to \infty$ through this particular sequence of values).

Theorem 430. *The average order of both* $\omega(n)$ *and* $\Omega(n)$ *is* $\log\log n$. *More precisely*

$$\sum_{n\leqslant x}\omega(n) = x\log\log x + B_1x + o(x), \tag{22.10.1}$$

$$\sum_{n\leqslant x}\Omega(n) = x\log\log x + B_2x + o(x), \tag{22.10.2}$$

where B_1 *is the number in Theorems* 427 *and* 428 *and*

$$B_2 = B_1 + \sum_p \frac{1}{p(p-1)}.$$

We write

$$S_1 = \sum_{n\leqslant x}\omega(n) = \sum_{n\leqslant x}\sum_{p|n} 1 = \sum_{p\leqslant x}\left[\frac{x}{p}\right],$$

since there are just $[x/p]$ values of $n \leqslant x$ which are multiples of p. Removing the square brackets, we have

$$S_1 = \sum_{p\leqslant x}\frac{x}{p} + O\{\pi(x)\} = x\log\log x + B_1x + o(x) \tag{22.10.3}$$

by Theorems 7 and 427.

Similarly

$$S_2 = \sum_{n\leqslant x}\Omega(n) = \sum_{n\leqslant x}\ \sum_{p^m|n} 1 = \sum_{p^m\leqslant x}\left[\frac{x}{p^m}\right], \tag{22.10.4}$$

so that

$$S_2 - S_1 = \sum{}'[x/p^m],$$

where $\sum'$ denotes summation over all $p^m \leqslant x$ for which $m \geqslant 2$. If we remove the square brackets in the last sum the error introduced is less than

$$\sum{}' 1 \leqslant \sum{}' \frac{\log p}{\log 2} = \frac{\psi(x) - \vartheta(x)}{\log 2} = o(x)$$

by Theorem 413. Hence

$$S_2 - S_1 = x \sum{}' p^{-m} + o(x).$$

The series

$$\sum_{m=2}^{\infty} \sum_{p} \frac{1}{p^m} = \sum_{p} \left(\frac{1}{p^2} + \frac{1}{p^3} + \cdots \right) = \sum \frac{1}{p(p-1)} = B_2 - B_1$$

is convergent and so

$$\sum{}' p^{-m} = B_2 - B_1 + o(1)$$

as $x \to \infty$. Hence

$$S_2 - S_1 = (B_2 - B_1)x + o(x)$$

and (22.10.2) follows from (22.10.3).

22.11. The normal order of $\omega(n)$ and $\Omega(n)$. The functions $\omega(n)$ and $\Omega(n)$ are irregular, but have a definite 'average order' loglog n. There is another interesting sense in which they may be said to have 'on the whole' a definite order. We shall say, roughly, that $f(n)$ has the *normal order* $F(n)$ if $f(n)$ is approximately $F(n)$ for almost all values of n. More precisely, suppose that

$$(1-\epsilon)F(n) < f(n) < (1+\epsilon)F(n) \tag{22.11.1}$$

for every positive ϵ and almost all values of n. Then we say that *the normal order of* $f(n)$ is $F(n)$. Here 'almost all' is used in the sense of §§ 1.6 and 9.9. There may be an exceptional 'infinitesimal' set of n for which (22.11.1) is false, and this exceptional set will naturally depend upon ϵ.

A function may possess an average order, but no normal order, or conversely. Thus the function

$$f(n) = 0 \ (n \text{ even}), \qquad f(n) = 2 \ (n \text{ odd})$$

has the average order 1, but no normal order. The function

$$f(n) = 2^m \quad (n = 2^m), \quad f(n) = 1 \quad (n \neq 2^m)$$

has the normal order 1, but no average order.

THEOREM 431. *The normal order of $\omega(n)$ and $\Omega(n)$ is* loglog n. *More precisely, the number of n, not exceeding x, for which*

$$|f(n) - \log\log n| > (\log\log n)^{\frac{1}{2}+\delta}, \tag{22.11.2}$$

where $f(n)$ is $\omega(n)$ or $\Omega(n)$, is $o(x)$ for every positive δ.

It is sufficient to prove that the number of n for which

$$|f(n) - \log\log x| > (\log\log x)^{\frac{1}{2}+\delta} \tag{22.11.3}$$

is $o(x)$; the distinction between loglog n and loglog x has no importance. For

$$\log\log x - 1 \leqslant \log\log n \leqslant \log\log x$$

when $x^{1/e} \leqslant n \leqslant x$, so that loglog n is practically loglog x for all such values of n; and the number of other values of n in question is

$$O(x^{1/e}) = o(x).$$

Next, we need only consider the case $f(n) = \omega(n)$. For $\Omega(n) \geqslant \omega(n)$ and, by (22.10.1) and (22.10.2),

$$\sum_{n \leqslant x} \{\Omega(n) - \omega(n)\} = O(x).$$

Hence the number of $n \leqslant x$ for which

$$\Omega(n) - \omega(n) > (\log\log x)^{\frac{1}{2}}$$

is

$$O\left(\frac{x}{(\log\log x)^{\frac{1}{2}}}\right) = o(x);$$

so that one case of Theorem 431 follows from the other.

Let us consider the number of pairs of different prime factors p, q of n (i.e. $p \neq q$), counting the pair q, p distinct from p, q. There are $\omega(n)$

possible values of p and, with each of these, just $\omega(n) - 1$ possible values of q. Hence

$$\omega(n)\{\omega(n) - 1\} = \sum_{\substack{pq|n \\ p \neq q}} 1 = \sum_{pq|n} 1 - \sum_{p^2|n} 1.$$

Summing over all $n \leqslant x$, we have

$$\begin{aligned}\sum_{n \leqslant x} \{\omega(n)\}^2 - \sum_{n \leqslant x} \omega(n) &= \sum_{n \leqslant x} \left(\sum_{pq|n} 1 - \sum_{p^2|n} 1 \right) \\ &= \sum_{pq \leqslant x} \left[\frac{x}{pq}\right] - \sum_{p^2 \leqslant x} \left[\frac{x}{p^2}\right].\end{aligned}$$

First

$$\sum_{p^2 \leqslant x} \left[\frac{x}{p^2}\right] \leqslant \sum_{p^2 \leqslant x} \frac{x}{p^2} \leqslant x \sum_{p} \frac{1}{p^2} = O(x),$$

since the series is convergent. Next

$$\sum_{pq \leqslant x} \left[\frac{x}{pq}\right] = x \sum_{pq \leqslant x} \frac{1}{pq} + O(x).$$

Hence, using (22.10.1), we have

$$\sum_{n \leqslant x} \{\omega(n)\}^2 = x \sum_{pq \leqslant x} \frac{1}{pq} + O(x \operatorname{loglog} x). \tag{22.11.4}$$

Now

$$\left(\sum_{p \leqslant \sqrt{x}} \frac{1}{p} \right)^2 \leqslant \sum_{pq \leqslant x} \frac{1}{pq} \leqslant \left(\sum_{p \leqslant x} \frac{1}{p} \right)^2, \tag{22.11.5}$$

since, if $pq \leqslant x$ then $p < x$ and $q < x$, while, if $p \leqslant \sqrt{x}$ and $q \leqslant \sqrt{x}$, then $pq \leqslant x$. The outside terms in (22.11.5) are each

$$\{\log\log x + O(1)\}^2 = (\log\log x)^2 + O(\log\log x)$$

and therefore

$$\sum_{n \leqslant x} \{\omega(n)\}^2 = x(\log\log x)^2 + O(x \log\log x). \tag{22.11.6}$$

It follows that

(22.11.7)

$$\begin{aligned}
&\sum_{n \leqslant x} \{\omega(n) - \log\log x\}^2 \\
&\quad = \sum_{n \leqslant x} \{\omega(n)\}^2 - 2\log\log x \sum_{n \leqslant x} \omega(n) + [x](\log\log x)^2 \\
&\quad = x(\log\log x)^2 + O(x\log\log x) \\
&\qquad - 2\log\log x\,\{x\log\log x + O(x)\} + \{x + O(1)\}\,(\log\log x)^2 \\
&\quad = x(\log\log x)^2 - 2x(\log\log x)^2 + x(\log\log x)^2 + O(x\log\log x) \\
&\quad = O(x\log\log x),
\end{aligned}$$

by (22.10.1) and (22.11.6).

If there are more than ηx numbers, not exceeding x, which satisfy (22.11.3) with $f(n) = \omega(n)$, then

$$\sum_{n \leqslant x} \{\omega(n) - \log\log x\}^2 \geqslant \eta x(\log\log x)^{1+2\delta},$$

which contradicts (22.11.7) for sufficiently large x; and this is true for every positive η. Hence the number of n which satisfy (22.11.3) is $o(x)$; and this proves the theorem.

22.12. A note on round numbers. A number is usually called 'round' if it is the product of a considerable number of comparatively small factors. Thus $1200 = 2^4 \,.\, 3 \,.\, 5^2$ would certainly be called round. The roundness of a number like $2187 = 3^7$ is obscured by the decimal notation.

It is a matter of common observation that *round numbers are very rare*; the fact may be verified by any one who will make a habit of factorizing numbers which, like numbers of taxi-cabs or railway carriages, are presented to his attention in a random manner. Theorem 431 contains the mathematical explanation of this phenomenon.

Either of the functions $\omega(n)$ or $\Omega(n)$ gives a natural measure of the 'roundness' of n, and each of them is usually about loglog n, a function of n which increases very slowly. Thus loglog 10^7 is a little less than 3, and loglog 10^{80} is a little larger than 5. A number near 10^7 (the limit of the factor tables) will usually have about 3 prime factors; and a number near 10^{80} (the number, approximately, of protons in the universe) about 5 or 6. A number like

$$6092087 = 37 \,.\, 229 \,.\, 719$$

is in a sense a 'typical' number.

These facts seem at first very surprising, but the real paradox lies a little deeper. What is really surprising is that most numbers should have *so many* factors and not that they should have so few. Theorem 431 contains two assertions, that $\omega(n)$ is usually not much larger than loglog n and that it is usually not much smaller; and it is the second assertion which lies deeper and is more difficult to prove. That $\omega(n)$ is usually not much larger than loglog n can be deduced from Theorem 430 without the aid of (22.11.6).†

22.13. The normal order of $d(n)$. If $n = p_1^{a_1} p_2^{a_2} \ldots p_r^{a_r}$, then

$$\omega(n) = r, \quad \Omega(n) = a_1 + a_2 + \cdots + a_r,$$
$$d(n) = (1 + a_1)(1 + a_2)\ldots(1 + a_r).$$

Also

$$2 \leqslant 1 + a \leqslant 2^a$$

and

$$2^{\omega(n)} \leqslant d(n) \leqslant 2^{\Omega(n)}.$$

Hence, after Theorem 431, the normal order of $\log d(n)$ is

$$\log 2 \log\log n.$$

† Roughly, if $\chi(x)$ were of higher order than loglog x, and $\omega(n)$ were larger than $\chi(n)$ for a fixed proportion of numbers less than x, then

$$\sum_{n \leqslant x} \omega(n)$$

would be larger than a fixed multiple of $x\chi(x)$, in contradiction to Theorem 430.

THEOREM 432. *If ϵ is positive, then*

$$2^{(1-\epsilon)\log\log n} < d(n) < 2^{(1+\epsilon)\log\log n} \tag{22.13.1}$$

for almost all numbers n.

Thus $d(n)$ is 'usually' about

$$2^{\log\log n} = (\log n)^{\log 2} = (\log n)^{\cdot 69\ldots}.$$

We cannot quite say that 'the normal order of $d(n)$ is $2^{\log\log n}$' since the inequalities (22.13.1) are of a less precise type than (22.11.1); but one may say, more roughly, that 'the normal order of $d(n)$ is about $2^{\log\log n}$'.

It should be observed that this normal order is notably less than the average order $\log n$. The average

$$\frac{1}{n}\{d(1) + d(2) + \cdots + d(n)\}$$

is dominated, not by the 'normal' n for which $d(n)$ has its most common magnitude, but by the small minority of n for which $d(n)$ is very much larger than $\log n$.[†] The irregularities of $\omega(n)$ and $\Omega(n)$ are not sufficiently violent to produce a similar effect.

22.14. Selberg's theorem. We devote the next three sections to the proof of Theorem 6. Of the earlier results of this chapter we use only Theorems 420–4 and the fact that

$$\psi(x) = O(x), \tag{22.14.1}$$

which is part of Theorem 414. We prove first

THEOREM 430 (SELBERG'S THEOREM):

$$\psi(x)\log x + \sum_{n\leqslant x}\Lambda(n)\psi\left(\frac{x}{n}\right) = 2x\log x + O(x) \tag{22.14.2}$$

and

$$\sum_{n\leqslant x}\Lambda(n)\log n + \sum_{mn\leqslant x}\Lambda(m)\Lambda(n) = 2x\log x + O(x). \tag{22.14.3}$$

[†] See the remarks at the ends of §§ 18.1 and 18.2.

It is easy to see that (22.14.2) and (22.14.3) are equivalent. For

$$\sum_{n\leqslant x}\Lambda(n)\psi\left(\frac{x}{n}\right)=\sum_{n\leqslant x}\Lambda(n)\sum_{m\leqslant x/n}\Lambda(m)=\sum_{mn\leqslant x}\Lambda(m)\Lambda(n)$$

and, if we put $c_n=\Lambda(n)$ and $f(t)=\log t$ in (22.5.2),

(22.14.4)

$$\sum_{n\leqslant x}\Lambda(n)\log n=\psi(x)\log x-\int_2^x\frac{\psi(t)}{t}dt=\psi(x)\log x+O(x)$$

by (22.14.1).

In our proof of (22.14.3) we use the Möbius function $\mu(n)$ defined in § 16.3. We recall Theorems 263, 296, and 298 by which

$$\sum_{d|n}\mu(d)=1\quad(n=1),\quad\sum_{d|n}\mu(d)=0\quad(n>1),\tag{22.14.5}$$

$$\Lambda(n)=-\sum_{d|n}\mu(d)\log d,\quad\log n=\sum_{d|n}\Lambda(d).\tag{22.14.6}$$

Hence

$$\begin{aligned}\sum_{h|n}\Lambda(h)\Lambda\left(\frac{n}{h}\right)&=-\sum_{h|n}\Lambda(h)\sum_{d|\frac{n}{h}}\mu(d)\log d\\&=-\sum_{d|n}\mu(d)\log d\sum_{h|\frac{n}{d}}\Lambda(h)=-\sum_{d|n}\mu(d)\log d\log\left(\frac{n}{d}\right)\\&=\Lambda(n)\log n+\sum_{d|n}\mu(d)\log^2 d.\end{aligned}\tag{22.14.7}$$

Again, by (22.14.5),

$$\sum_{d|1}\mu(d)\log^2\left(\frac{x}{d}\right)=\log^2 x,$$

but, for $n > 1$,

$$\sum_{d|n} \mu(d) \log^2 \left(\frac{x}{d}\right) = \sum_{d|n} \mu(d)(\log^2 d - 2 \log x \log d)$$

$$= 2\Lambda(n) \log x - \Lambda(n) \log n + \sum_{hk=n} \Lambda(h)\Lambda(k)$$

by (22.14.6) and (22.14.7). Hence, if we write

$$S(x) = \sum_{n \leqslant x} \sum_{d|n} \mu(d) \log^2 \left(\frac{x}{d}\right),$$

we have

$$S(x) = \log^2 x + 2\psi(x) \log x - \sum_{n \leqslant x} \Lambda(n) \log n + \sum_{hk \leqslant x} \Lambda(h)\Lambda(k)$$

$$= \sum_{n \leqslant x} \Lambda(n) \log n + \sum_{mn \leqslant x} \Lambda(m)\Lambda(n) + O(x)$$

by (22.14.4). To complete the proof of (22.14.3), we have only to show that

(22.14.8) $$S(x) = 2x \log x + O(x).$$

By (22.14.5),

$$S(x) - \gamma^2 = \sum_{n \leqslant x} \sum_{d|n} \mu(d) \left\{\log^2 \left(\frac{x}{d}\right) - \gamma^2\right\}$$

$$= \sum_{d \leqslant x} \mu(d) \left[\frac{x}{d}\right] \left\{\log^2 \left(\frac{x}{d}\right) - \gamma^2\right\},$$

since the number of $n \leqslant x$, for which $d|n$, is $[x/d]$. If we remove the square brackets, the error introduced is less than

$$\sum_{d \leqslant x} \left\{\log^2 \left(\frac{x}{d}\right) + \gamma^2\right\} = O(x)$$

by Theorem 423. Hence

(22.14.9) $$S(x) = x \sum_{d \leqslant x} \frac{\mu(d)}{d} \left\{\log^2 \left(\frac{x}{d}\right) - \gamma^2\right\} + O(x).$$

Now, by Theorem 422,

(22.14.10)
$$\sum_{d\leqslant x}\frac{\mu(d)}{d}\left\{\log^2\left(\frac{x}{d}\right)-\gamma^2\right\}$$
$$=\sum_{d\leqslant x}\frac{\mu(d)}{d}\left\{\log\left(\frac{x}{d}\right)-\gamma\right\}\left\{\sum_{k\leqslant x/d}\frac{1}{k}+O\left(\frac{d}{x}\right)\right\}.$$

The sum of the various error terms is at most

(22.14.11)
$$\sum_{d\leqslant x}\frac{1}{d}\left\{\log\left(\frac{x}{d}\right)+\gamma\right\}O\left(\frac{d}{x}\right)=O\left(\frac{1}{x}\right)\sum_{d\leqslant x}\log\left(\frac{x}{d}\right)+O(1)$$
$$=O(1)$$

by Theorem 423. Also

(22.14.12)
$$\sum_{d\leqslant x}\frac{\mu(d)}{d}\left\{\log\left(\frac{x}{d}\right)-\gamma\right\}\sum_{k\leqslant x|d}\frac{1}{k}$$
$$=\sum_{dk\leqslant x}\frac{\mu(d)}{dk}\left\{\log\left(\frac{x}{d}\right)-\gamma\right\}=\sum_{n\leqslant x}\frac{1}{n}\sum_{d|n}\mu(d)\left\{\log\left(\frac{x}{d}\right)-\gamma\right\}$$
$$=\log x-\gamma+\sum_{2\leqslant n\leqslant x}\frac{\Lambda(n)}{n}=2\log x+O(1)$$

by (22.14.5), (22.14.6), and Theorem 424. (22.14.8) follows when we combine (22.14.9)–(22.14.12).

22.15. The functions $R(x)$ and $V(\xi)$. After Theorem 420 the Prime Number Theorem (Theorem 6) is equivalent to

THEOREM 434:
$$\psi(x)\sim x,$$

and it is this last theorem that we shall prove. If we put
$$\psi(x)=x+R(x)$$

in (22.14.2) and use Theorem 424, we have

$$R(x)\log x + \sum_{n\leqslant x} \Lambda(n) R\left(\frac{x}{n}\right) = O(x). \tag{22.15.1}$$

Our object is to prove that $R(x) = o(x)$.†

If we replace n by m and x by x/n in (22.15.1), we have

$$R\left(\frac{x}{n}\right)\log\left(\frac{x}{n}\right) + \sum_{m\leqslant x/n} \Lambda(m) R\left(\frac{x}{mn}\right) = O\left(\frac{x}{n}\right).$$

Hence

$$\log x\left\{R(x)\log x + \sum_{n\leqslant x}\Lambda(n)R\left(\frac{x}{n}\right)\right\}$$

$$-\sum_{n\leqslant x}\Lambda(n)\left\{R\left(\frac{x}{n}\right)\log\left(\frac{x}{n}\right) + \sum_{m\leqslant x/n}\Lambda(m)R\left(\frac{x}{mn}\right)\right\}$$

$$= O(x\log x) + O\left(x\sum_{n\leqslant x}\frac{\Lambda(n)}{n}\right) = O(x\log x),$$

that is

$$R(x)\log^2 x = -\sum_{n\leqslant x}\Lambda(n)R\left(\frac{x}{n}\right)\log n$$

$$+\sum_{mn\leqslant x}\Lambda(m)\Lambda(n)R\left(\frac{x}{mn}\right) + O(x\log x),$$

whence

$$|R(x)|\log^2 x \leqslant \sum_{n\leqslant x} a_n\left|R\left(\frac{x}{n}\right)\right| + O(x\log x), \tag{22.15.2}$$

where

$$a_n = \Lambda(n)\log n + \sum_{hk=n}\Lambda(h)\Lambda(k)$$

† Of course, this would be a trivial deduction if $R(x) \geqslant 0$ for all x (or if $R(x) \leqslant 0$ for all x). Indeed, more would follow, viz. $R(x) = O(x/\log x)$. But it is possible, so far as we know at this stage of our argument, that $R(x)$ is usually of order x, but that its positive and negative values are so distributed that the sum over n on the left-hand side of (22.15.1) is of opposite sign to the first term and largely offsets it.

and

$$\sum_{n \leqslant x} a_n = 2x \log x + O(x)$$

by (22.14.3).

We now replace the sum on the right-hand side of (22.15.2) by an integral. To do so, we shall prove that

$$\sum_{n \leqslant x} a_n \left| R\left(\frac{x}{n}\right) \right| = 2 \int_1^x \left| R\left(\frac{x}{t}\right) \right| \log t \, dt + O(x \log x). \tag{22.15.3}$$

We remark that, if $t > t' \geqslant 0$,

$$\begin{aligned} ||R(t)| - |R(t')|| &\leqslant |R(t) - R(t')| = |\psi(t) - \psi(t') - t + t'| \\ &\leqslant \psi(t) - \psi(t') + t - t' = F(t) - F(t'), \end{aligned}$$

where

$$F(t) = \psi(t) + t = O(t)$$

and $F(t)$ is a steadily increasing function of t. Also

$$\begin{aligned} \sum_{n \leqslant x-1} n \left\{ F\left(\frac{x}{n}\right) - F\left(\frac{x}{n+1}\right) \right\} &= \sum_{n \leqslant x} F\left(\frac{x}{n}\right) - [x] F\left(\frac{x}{[x]}\right) \\ &= O\left(x \sum_{n \leqslant x} \frac{1}{n} \right) = O(x \log x). \end{aligned} \tag{22.15.4}$$

We prove (22.15.3) in two stages. First, if we put

$$c_1 = 0, \qquad c_n = a_n - 2 \int_{n-1}^{n} \log t \, dt, \qquad f(n) = \left| R\left(\frac{x}{n}\right) \right|$$

in (22.5.1), we have

$$C(x) = \sum_{n \leqslant x} a_n - 2 \int_1^{[x]} \log t \, dt = O(x)$$

and

(22.15.5)

$$\sum_{n\leqslant x} a_n \left|R\left(\frac{x}{n}\right)\right| - 2\sum_{2\leqslant n\leqslant x}\left|R\left(\frac{x}{n}\right)\right|\int_{n-1}^{n}\log t\,dt$$

$$= \sum_{n\leqslant x-1} C(n)\left\{\left|R\left(\frac{x}{n}\right)\right| - \left|R\left(\frac{x}{n+1}\right)\right|\right\} + C(x)R\left(\frac{x}{[x]}\right)$$

$$= O\left(\sum_{n\leqslant x-1} n\left\{F\left(\frac{x}{n}\right) - F\left(\frac{x}{n+1}\right)\right\}\right) + O(x) = O(x\log x)$$

by (22.15.4).

Next

$$\left|\left|R\left(\frac{x}{n}\right)\right|\int_{n-1}^{n}\log t\,dt - \int_{n-1}^{n}\left|R\left(\frac{x}{t}\right)\right|\log t\,dt\right|$$

$$\leqslant \int_{n-1}^{n}\left|\left|R\left(\frac{x}{n}\right)\right| - \left|R\left(\frac{x}{t}\right)\right|\right|\log t\,dt$$

$$\leqslant \int_{n-1}^{n}\left\{F\left(\frac{x}{t}\right) - F\left(\frac{x}{n}\right)\right\}\log t\,dt \leqslant (n-1)\left\{F\left(\frac{x}{n-1}\right) - F\left(\frac{x}{n}\right)\right\}.$$

Hence

(22.15.6)

$$\sum_{2\leqslant n\leqslant x}\left|R\left(\frac{x}{n}\right)\right|\int_{n-1}^{n}\log t\,dt - \int_{1}^{x}\left|R\left(\frac{x}{t}\right)\right|\log t\,dt$$

$$= O\left(\sum_{n\leqslant x-1} n\left\{F\left(\frac{x}{n}\right) - F\left(\frac{x}{n+1}\right)\right\}\right) + O(x\log x) = O(x\log x).$$

Combining (22.15.5) and (22.15.6) we have (22.15.3).

Using (22.15.3) in (22.15.2) we have

$$|R(x)|\log^2 x \leqslant 2\int_1^x \left|R\left(\frac{x}{t}\right)\right| \log t\, dt + O(x\log x). \tag{22.15.7}$$

We can make the significance of this inequality a little clearer if we introduce a new function, viz.

$$\begin{aligned} V(\xi) &= e^{-\xi}R(e^{\xi}) = e^{-\xi}\psi(e^{\xi}) - 1 \\ &= e^{-\xi}\left\{\sum_{n\leqslant e^{\xi}} \Lambda(n)\right\} - 1. \end{aligned} \tag{22.15.8}$$

If we write $x = e^{\xi}$ and $t = xe^{-\eta}$, we have

$$\begin{aligned} \int_1^x \left|R\left(\frac{x}{t}\right)\right| \log t\, dt &= x\int_0^{\xi} |V(\eta)|(\xi-\eta)d\eta = x\int_0^{\xi} |V(\eta)| \int_{\eta}^{\xi} d\zeta d\eta \\ &= x\int_0^{\xi}\int_0^{\zeta} |V(\eta)|\, d\eta d\zeta, \end{aligned}$$

on changing the order of integration. (22.15.7) becomes

$$\xi^2\,|V(\xi)| \leqslant 2\int_0^{\xi}\int_0^{\zeta} |V(\eta)|\, d\eta d\zeta + O(\xi). \tag{22.15.9}$$

Since $\psi(x) = O(x)$, it follows from (22.15.8) that $V(\xi)$ is bounded as $\xi \to \infty$. Hence we may write

$$\alpha = \overline{\lim_{\xi\to\infty}}\, |V(\xi)|, \qquad \beta = \overline{\lim}\, \frac{1}{\xi}\int_0^{\xi} |V(\eta)|\, d\eta,$$

since both these upper limits exist. Clearly

$$|V(\xi)| \leqslant \alpha + o(1) \tag{22.15.10}$$

and

$$\int_0^\xi |V(\eta)| d\eta \leqslant \beta\xi + o(\xi).$$

Using this in (22.15.9), we have

$$\xi^2 |V(\xi)| \leqslant 2 \int_0^\xi \{\beta\xi + o(\zeta)\} d\zeta + O(\xi) = \beta\xi^2 + o(\xi^2)$$

and so

$$|V(\xi)| \leqslant \beta + o(1).$$

Hence

(22.15.11) $$\alpha \leqslant \beta.$$

22.16. Completion of the proof of Theorems 434, 6, and 8. By (22.15.8), Theorem 434 is equivalent to the statement that $V(\xi) \to 0$ as $\xi \to \infty$, that is, that $\alpha = 0$. We now suppose that $\alpha > 0$ and prove that, in that case, $\beta < \alpha$ in contradiction to (22.15.11). We require two further lemmas.

THEOREM 435. *There is a fixed positive number A_1, such that, for every positive ξ_1, ξ_2, we have*

$$\left| \int_{\xi_1}^{\xi_2} V(\eta) d\eta \right| < A_1.$$

If we put $x = e^\xi$, $t = e^\eta$, we have

$$\int_0^\xi V(\eta) d\eta = \int_1^x \left\{ \frac{\psi(t)}{t^2} - \frac{1}{t} \right\} dt = O(1)$$

by (22.6.1). Hence

$$\int_{\xi_1}^{\xi_2} V(\eta)d\eta = \int_0^{\xi_2} V(\eta)d\eta - \int_0^{\xi_1} V(\eta)d\eta = O(1)$$

and this is Theorem 435.

THEOREM 436. *If* $\eta_0 > 0$ *and* $V(\eta_0) = 0$, *then*

$$\int_0^{\alpha} |V(\eta_0 + \tau)|\, d\tau \leqslant \tfrac{1}{2}\alpha^2 + O(\eta_0^{-1}).$$

We may write (22.14.2) in the form

$$\psi(x)\log x + \sum_{mn \leqslant x} \Lambda(m)\Lambda(n) = 2x\log x + O(x).$$

If $x > x_0 \geqslant 1$, the same result is true with x_0 substituted for x. Subtracting, we have

$$\psi(x)\log x - \psi(x_0)\log x_0 + \sum_{x_0 < mn \leqslant x} \Lambda(m)\Lambda(n)$$
$$= 2(x\log x - x_0\log x_0) + O(x).$$

Since $\Lambda(n) \geqslant 0$,

$$0 \leqslant \psi(x)\log x - \psi(x_0)\log x_0 \leqslant 2(x\log x - x_0\log x_0) + O(x),$$

whence

$$|R(x)\log x - R(x_0)\log x_0| \leqslant x\log x - x_0\log x_0 + O(x).$$

We put $x = e^{\eta_0+\tau}$, $x_0 = e^{\eta_0}$, so that $R(x_0) = 0$. We have, since $0 \leqslant \tau \leqslant \alpha$,

$$|V(\eta_0 + \tau)| \leqslant 1 - \left(\frac{\eta_0}{\eta_0 + \tau}\right)e^{-\tau} + O\left(\frac{1}{\eta_0}\right)$$
$$= 1 - e^{-\tau} + O(1/\eta_0) \leqslant \tau + O(1/\eta_0)$$

and so

$$\int_0^\alpha |V(\eta_0+\tau)|\,d\tau \leqslant \int_0^\alpha \tau\,d\tau + O\left(\frac{1}{\eta_0}\right) = \tfrac{1}{2}\alpha^2 + O\left(\frac{1}{\eta_0}\right).$$

We now write

$$\delta = \frac{3\alpha^2+4A_1}{2\alpha} > \alpha,$$

take ζ to be any positive number and consider the behaviour of $V(\eta)$ in the interval $\zeta \leqslant \eta \leqslant \zeta+\delta-\alpha$. By (22.15.8), $V(\eta)$ decreases steadily as η increases, except at its discontinuities, where $V(\eta)$ increases. Hence, in our interval, either $V(\eta_0) = 0$ for some η_0 or $V(\eta)$ changes sign at most once. In the first case, we use (22.15.10) and Theorem 436 and have

$$\begin{aligned}\int_\zeta^{\zeta+\delta} |V(\eta)|\,d\eta &= \int_\zeta^{\eta_0} + \int_{\eta_0}^{\eta_0+\alpha} + \int_{\eta_0+\alpha}^{\zeta+\delta} |V(\eta)|d\eta \\ &\leqslant \alpha(\eta_0-\zeta) + \tfrac{1}{2}\alpha^2 + \alpha(\zeta+\delta-\eta_0-\alpha) + o(1) \\ &= \alpha\left(\delta - \tfrac{1}{2}\alpha\right) + o(1) = \alpha'\delta + o(1)\end{aligned}$$

for large ζ, where

$$\alpha' = \alpha\left(1 - \frac{\alpha}{2\delta}\right) < \alpha.$$

In the second case, if $V(\eta)$ changes sign just once at $\eta = \eta_1$ in the interval $\zeta \leqslant \eta \leqslant \zeta+\delta-\alpha$, we have

$$\int_\zeta^{\zeta+\delta-\alpha} |V(\eta)|\,d\eta = \left|\int_\zeta^{\eta_1} V(\eta)d\eta\right| + \left|\int_{\eta_1}^{\zeta+\delta-\alpha} V(\eta)d\eta\right| < 2A_1,$$

while, if $V(\eta)$ does not change sign at all in the interval, we have

$$\int_\zeta^{\zeta+\delta-\alpha} |V(\eta)|\,d\eta = \left|\int_\zeta^{\zeta+\delta-\alpha} V(\eta)d\eta\right| < A_1$$

by Theorem 435. Hence

$$\int_{\zeta}^{\zeta+\delta} |V(\eta)|\,d\eta = \int_{\zeta}^{\zeta+\delta-\alpha} + \int_{\zeta+\delta-\alpha}^{\zeta+\delta} |V(\eta)|\,d\eta$$
$$< 2A_1 + \alpha^2 + o(1) = \alpha''\delta + o(1),$$

where

$$\alpha'' = \frac{2A_1+\alpha^2}{\delta} = \alpha\left(\frac{4A_1+2\alpha^2}{4A_1+3\alpha^2}\right) = \alpha\left(1-\frac{\alpha}{2\delta}\right) = \alpha'.$$

Hence we have always

$$\int_{\zeta}^{\zeta+\delta} |V(\eta)|\,d\eta \leqslant \alpha'\delta + o(1),$$

where $o(1) \to 0$ as $\zeta \to \infty$. If $M = [\xi/\delta]$,

$$\int_0^{\xi} |V(\eta)|\,d\eta = \sum_{m=0}^{M-1} \int_{m\delta}^{(m+1)\delta} |V(\eta)|\,d\eta + \int_{M\delta}^{\xi} |V(\eta)|\,d\eta$$
$$\leqslant \alpha' M\delta + o(M) + O(1) = \alpha'\xi + o(\xi).$$

Hence

$$\beta = \overline{\lim}\frac{1}{\xi}\int_0^{\xi} |V(\eta)|d\eta \leqslant \alpha' < \alpha,$$

in contradiction to (22.15.11). It follows that $\alpha = 0$, whence we have Theorem 434 and Theorem 6. As we saw on p. 10, Theorem 8 is a trivial deduction from Theorem 6.

22.17. Proof of Theorem 335. Theorem 335 is a simple consequence of Theorem 434. We have

$$\sum_{n\leqslant x} \mu(n)\log\left(\frac{x}{n}\right) = O(x)$$

by Theorem 423 and so

$$M(x)\log x = \sum_{n \leqslant x} \mu(n) \log n + O(x).$$

By Theorem 297, with the notation of § 22.15,

$$\begin{aligned}
-\sum_{n \leqslant x} \mu(n) \log n &= \sum_{n \leqslant x} \sum_{d|n} \mu\left(\frac{n}{d}\right) \Lambda(d) = \sum_{dk \leqslant x} \mu(k)\Lambda(d) \\
&= \sum_{k \leqslant x} \mu(k)\psi\left(\frac{x}{k}\right) = \sum_{k \leqslant x} \mu(k)\psi\left(\left[\frac{x}{k}\right]\right) \\
&= \sum_{k \leqslant x} \mu(k)\left[\frac{x}{k}\right] + \sum_{k \leqslant x} \mu(k) R\left(\left[\frac{x}{k}\right]\right) = S_3 + S_4
\end{aligned}$$

(say). Now, by (22.14.5),

$$S_3 = \sum_{k \leqslant x} \mu(k) \left[\frac{x}{k}\right] = \sum_{n \leqslant x} \sum_{k|n} \mu(k) = 1.$$

By Theorem 434, $R(x) = o(x)$; that is, for any $\epsilon > 0$, there is an integer $N = N(\epsilon)$ such that $|R(x)| < \epsilon x$ for all $x \geqslant N$. Again, by Theorem 414, $|R(x)| < Ax$ for all $x \geqslant 1$. Hence

$$\begin{aligned}
|S_4| &\leqslant \sum_{k \leqslant x} \left| R\left(\left[\frac{x}{k}\right]\right)\right| \leqslant \sum_{k \leqslant x/N} \epsilon \left[\frac{x}{k}\right] + \sum_{x/N < k \leqslant x} A \left[\frac{x}{k}\right] \\
&\leqslant \epsilon x \log(x/N) + Ax\left\{\log x - \log(x/N)\right\} + O(x) \\
&= \epsilon x \log x + O(x).
\end{aligned}$$

Since ϵ is arbitrary, it follows that $S_4 = o(x \log x)$ and so

$$-M(x)\log x = S_3 + S_4 + O(x) = o(x \log x),$$

whence Theorem 335.

22.18. Products of *k* prime factors. Let $k \geqslant 1$ and consider a positive integer n which is the product of just k prime factors, i.e.

$$n = p_1 p_2 \ldots p_k. \tag{22.18.1}$$

In the notation of § 22.10, $\Omega(n) = k$. We write $\tau_k(x)$ for the number of such $n \leqslant x$. If we impose the additional restriction that all the p in (22.18.1) shall be different, n is squarefree and $\omega(n) = \Omega(n) = k$. We write $\pi_k(x)$ for the number of these (squarefree) $n \leqslant x$. We shall prove

THEOREM 437:

$$\pi_k(x) \sim \tau_k(x) \sim \frac{x(\log\log x)^{k-1}}{(k-1)!\log x} \quad (k \geqslant 2).$$

For $k = 1$, this result would reduce to Theorem 6, if, as usual, we take $0! = 1$.

To prove Theorem 437, we introduce three auxiliary functions, viz.

$$L_k(x) = \sum \frac{1}{p_1p_2\dots p_k}, \Pi_k(x) = \sum 1, \vartheta_k(x) = \sum \log(p_1p_2\dots p_k),$$

where the summation in each case extends over all sets of primes $p_1, p_2, \dots, p_k$ such that $p_1 \dots p_k \leqslant x$, two sets being considered different even if they differ only in the order of the p. If we write c_n for the number of ways in which n can be represented in the form (22.18.1), we have

$$\Pi_k(x) = \sum_{n \leqslant x} c_n, \qquad \vartheta_k(x) = \sum_{n \leqslant x} c_n \log n.$$

If all the p in (22.18.1) are different, $c_n = k!$, while in any case $c_n \leqslant k!$. If n is not of the form (22.18.1), $c_n = 0$. Hence

$$k!\pi_k(x) \leqslant \Pi_k(x) \leqslant k!\tau_k(x) \quad (k \geqslant 1). \tag{22.18.2}$$

Again, for $k \geqslant 2$, consider those n which are of the form (22.18.1) with at least two of the p equal. The number of these $n \leqslant x$ is $\tau_k(x) - \pi_k(x)$. Every such n can be expressed in the form (22.18.1) with $p_{k-1} = p_k$ and so

(22.18.3)

$$\tau_k(x) - \pi_k(x) \leqslant \sum_{p_1p_2\dots p_{k-1}^2 \leqslant x} 1 \leqslant \sum_{p_1p_2\dots p_{k-1} \leqslant x} 1 = \Pi_{k-1}(x) \quad (k \geqslant 2).$$

We shall prove below that

$$\vartheta_k(x) \sim kx(\log\log x)^{k-1} \quad (k \geqslant 2). \tag{22.18.4}$$

By (22.5.2) with $f(t) = \log t$, we have

$$\vartheta_k(x) = \Pi_k(x)\log x - \int_2^x \frac{\Pi_k(t)}{t}\,dt.$$

Now $\tau_k(x) \leqslant x$ and so, by (22.18.2), $\Pi_k(t) = O(t)$ and

$$\int_2^x \frac{\Pi_k(t)}{t}\,dt = O(x).$$

Hence, for $k \geqslant 2$,

$$\Pi_k(x) = \frac{\vartheta_k(x)}{\log x} + O\left(\frac{x}{\log x}\right) \sim \frac{kx(\log\log x)^{k-1}}{\log x} \tag{22.18.5}$$

by (22.18.4). But this is also true for $k = 1$ by Theorem 6, since $\Pi_1(x) = \pi(x)$. When we use (22.18.5) in (22.18.2) and (22.18.3), Theorem 437 follows at once.

We have now to prove (22.18.4). For all $k \geqslant 1$,

$$\begin{aligned} k\vartheta_{k+1}(x) &= \sum_{p_1\ldots p_{k+1}\leqslant x} \{\log(p_2p_3\ldots p_{k+1}) + \log(p_1p_3p_4\ldots p_{k+1}) \\ &\qquad + \cdots + \log(p_1p_2\ldots p_k)\} \\ &= (k+1)\sum_{p_1\ldots p_{k+1}\leqslant x} \log(p_2p_3\ldots p_{k+1}) = (k+1)\sum_{p_1\leqslant x}\vartheta_k\left(\frac{x}{p_1}\right) \end{aligned}$$

and, if we put $L_0(x) = 1$,

$$L_k(x) = \sum_{p_1\ldots p_k\leqslant x}\frac{1}{p_1\ldots p_k} = \sum_{p_1\leqslant x}\frac{1}{p_1}L_{k-1}\left(\frac{x}{p_1}\right).$$

Hence, if we write

$$f_k(x) = \vartheta_k(x) - kxL_{k-1}(x),$$

we have

$$kf_{k+1}(x) = (k+1)\sum_{p \leqslant x} f_k\left(\frac{x}{p}\right). \tag{22.18.6}$$

We use this to prove by induction that

$$f_k(x) = o\left\{x(\log\log x)^{k-1}\right\} \quad (k \geqslant 1). \tag{22.18.7}$$

First

$$f_1(x) = \vartheta_1(x) - x = \vartheta(x) - x = o(x)$$

by Theorems 6 and 420, so that (22.18.7) is true for $k = 1$. Let us suppose (22.18.7) true for $k = K \geqslant 1$ so that, for any $\epsilon > 0$, there is an $x_0 = x_0(K, \epsilon)$ such that

$$|f_K(x)| < \epsilon x(\log\log x)^{K-1}$$

for all $x \geqslant x_0$. From the definition of $f_K(x)$, we see that

$$|f_K(x)| < D$$

for $1 \leqslant x < x_0$, where D depends only on K and ϵ. Hence

$$\sum_{p \leqslant x/x_0} \left|f_K\left(\frac{x}{p}\right)\right| < \epsilon(\log\log x)^{K-1} \sum_{p \leqslant x/x_0} \frac{x}{p}$$
$$< 2\epsilon x(\log\log x)^K$$

for large enough x, by Theorem 427. Again

$$\sum_{x/x_0 < p \leqslant x} \left|f_K\left(\frac{x}{p}\right)\right| < D\pi(x) < Dx.$$

Hence, by (22.18.6), since $K + 1 \leqslant 2K$,

$$|f_{K+1}(x)| < 2x\left\{2\epsilon(\log\log x)^K + D\right\} < 5\epsilon x(\log\log x)^K$$

for $x > x_1 = x_1(\epsilon, D, K) = x_1(\epsilon, K)$. Since ϵ is arbitrary, this implies (22.18.7) for $k = K + 1$ and it follows for all $k \geqslant 1$ by induction.

After (22.18.7), we can complete the proof of (22.18.4) by showing that

(22.18.8) $$L_k(x) \sim (\log\log x)^k \quad (k \geqslant 1).$$

In (22.18.1), if every $p_i \leqslant x^{1/k}$, then $n \leqslant x$; conversely, if $n \leqslant x$, then $p_i \leqslant x$ for every i. Hence

$$\left(\sum_{p \leqslant x^{1/k}} \frac{1}{p}\right)^k \leqslant L_k(x) \leqslant \left(\sum_{p \leqslant x} \frac{1}{p}\right)^k.$$

But, by Theorem 427,

$$\sum_{p \leqslant x} \frac{1}{p} \sim \log\log x, \qquad \sum_{p \leqslant x^{1/k}} \frac{1}{p} \sim \log\left(\frac{\log x}{k}\right) \sim \log\log x$$

and (22.18.8) follows at once.

22.19. Primes in an interval. Suppose that $\epsilon > 0$, so that

(22.19.1)
$$\pi(x + \epsilon x) - \pi(x) = \frac{x + \epsilon x}{\log x + \log(1 + \epsilon)} - \frac{x}{\log x} + o\left(\frac{x}{\log x}\right)$$
$$= \frac{\epsilon x}{\log x} + o\left(\frac{x}{\log x}\right).$$

The last expression is positive provided that $x > x_0(\epsilon)$. Hence there is always a prime p satisfying

(22.19.2) $$x < p < (1 + \epsilon)x$$

when $x > x_0(\epsilon)$. This result may be compared with Theorem 418. The latter corresponds to the case $\epsilon = 1$ of (22.19.2), but holds *for all* $x \geqslant 1$.

If we put $\epsilon = 1$ in (22.19.1), we have

(22.19.3) $$\pi(2x) - \pi(x) = \frac{x}{\log x} + o\left(\frac{x}{\log x}\right) \sim \pi(x).$$

Thus, to a first approximation, the number of primes between x and $2x$ is the same as the number less than x. At first sight this is surprising, since we know that the primes near x 'thin out' (in some vague sense) as x increases.

In fact, $\pi(2x) - 2\pi(x) \to \infty$ as $x \to \infty$ (though we cannot prove this here), but this is not inconsistent with (22.19.3), which is equivalent to

$$\pi(2x) - 2\pi(x) = O\{\pi(x)\}.$$

22.20. A conjecture about the distribution of prime pairs $p, p+2$. Although, as we remarked in § 1.4, it is not known whether there is an infinity of prime-pairs $p, p+2$, there is an argument which makes it plausible that

$$P_2(x) \sim \frac{2C_2 x}{(\log x)^2}, \tag{22.20.1}$$

where $P_2(x)$ is the number of these pairs with $p \leqslant x$ and

$$C_2 = \prod_{p\geqslant 3}\left\{\frac{p(p-2)}{(p-1)^2}\right\} = \prod_{p\geqslant 3}\left\{1 - \frac{1}{(p-1)^2}\right\}. \tag{22.20.2}$$

We take x any large positive number and write

$$N = \prod_{p\leqslant\sqrt{x}} p.$$

We shall call any integer n which is prime to N, i.e. any n not divisible by any prime p not exceeding $\sqrt{x}$, a *special* integer and denote by $S(X)$ the number of special integers which are less than or equal to X. By Theorem 62,

$$S(N) = \phi(N) = N\prod_{p\leqslant\sqrt{x}}\left(1 - \frac{1}{p}\right) = NB(x)$$

(say). Hence the proportion of special integers in the interval $(1, N)$ is $B(x)$. It is easily seen that the proportion is the same in any complete set of residues (mod N) and so in any set of rN consecutive integers for any positive integral r.

If the proportion were the same in the interval $(1, x)$, we should have

$$S(x) = xB(x) \sim \frac{2e^{-\gamma}x}{\log x}$$

by Theorem 429. But this is false. For every composite n not exceeding x has a prime factor not exceeding $\sqrt{x}$ and so the special n not exceeding x are just the primes between $\sqrt{x}$ (exclusive) and x (inclusive). We have then

$$S(x) = \pi(x) - \pi(\sqrt{x}) \sim \frac{x}{\log x}$$

by Theorem 6. Hence the proportion of special integers in the interval $(1, x)$ is about $\frac{1}{2}e^{\gamma}$ times the proportion in the interval $(1, N)$.

There is nothing surprising in this, for, in the notation of § 22.1,

$$\log N = \vartheta(\sqrt{x}) \sim \sqrt{x}$$

by Theorems 413 and 434, and so N is much greater than x. The proportion of special integers in every interval of length N need not be the same as that in a particular interval of (much shorter) length x.† Indeed, $S(\sqrt{x}) = 0$, and so in the particular interval $(1, \sqrt{x})$ the proportion is 0. We observe that the proportion in the interval $(N - x, N)$ is again about $1/\log x$, and that in the interval $(N - \sqrt{x}, N)$ is again 0.

Next we evaluate the number of pairs $n, n+2$ of special integers for which $n \leqslant N$. If n and $n+2$ are both special, we must have

$$n \equiv 1 \pmod 2), \qquad n \equiv 2 \pmod 3)$$

and

$$n \equiv 1, 2, 3, \ldots, p-3, \text{ or } p-1 \pmod p) \quad (3 < p \leqslant \sqrt{x})$$

The number of different possible residues for n (mod N) is therefore

$$\prod_{3 \leqslant p \leqslant \sqrt{x}} (p-2) = \tfrac{1}{2}N \prod_{3 \leqslant p \leqslant \sqrt{x}} \left(1 - \frac{2}{p}\right) = N B_1(x)$$

(say) and this is the number of special pairs $n, n+2$ with $n \leqslant N$.

Thus the proportion of special pairs in the interval $(1, N)$ is $B_1(x)$ and the same is clearly true in any interval of rN consecutive integers. In the smaller interval $(1, x)$, however, the proportion of special integers is about $\frac{1}{2}e^{\gamma}$ times the proportion in the longer intervals. We may therefore expect (and it is here only that we 'expect' and cannot prove) that the proportion

† Considerations of this kind explain why the usual 'probability' arguments lead to the wrong asymptotic value for $\pi(x)$.

of special *pairs* $n, n+2$ in the interval $(1, x)$ is about $\left(\frac{1}{2}e^{\gamma}\right)^2$ times the proportion in the longer intervals. But the special pairs in the interval $(1, x)$ are the prime pairs $p, p+2$ in the interval $(\surd x, x)$. Hence we should expect that

$$P_2(x) - P_2(\surd x) \sim \tfrac{1}{4}e^{2\gamma} x B_1(x).$$

By Theorem 429,

$$B(x) \sim \frac{2e^{-\gamma}}{\log x}$$

and so

$$\tfrac{1}{4}e^{2\gamma} B_1(x) \sim \frac{1}{(\log x)^2} \frac{B_1(x)}{\{B(x)\}^2}.$$

But

$$\frac{B_1(x)}{\{B(x)\}^2} = 2 \prod_{3 \leqslant p \leqslant \surd x} \frac{(1-2/p)}{(1-1/p)^2} = 2 \prod_{3 \leqslant p \leqslant \surd x} \frac{p(p-2)}{(p-1)^2} \to 2C_2$$

as $x \to \infty$. Since $P_2(\surd x) = O(\surd x)$, we have finally the result (22.20.1).

NOTES

§§ 22.1, 2, and 4. The theorems of these sections are essentially Tchebychef's. Theorem 416 was found independently by de Polignac. Theorem 415 is an improvement of a result of Tchebychef's; the proof we give here is due to Erdős and Kalmar.

There is full information about the history of the theory of primes in Dickson's *History* (i, ch. xviii), in Ingham's tract (introduction and ch. i), and in Landau's *Handbuch* (3–102 and 883–5); and we do not give detailed references.

There is also an elaborate account of the early history of the theory in Torelli, *Sulla totalità dei numeri primi, Atti della R. Acad. di Napoli* (2) 11 (1902), 1–222; and shorter ones in the introductions to Glaisher's *Factor table for the sixth million* (London, 1883) and Lehmer's table referred to in the note on § 1.4.

§22.2 Various authors have given versions of Theorem 414 with explicit numerical constants. Thus Tchebychef (*Mem. Acad. Sc. St. Petersburg* 7, (1850–1854), 15–33) showed that

$$(0.921\ldots)x \leqslant \theta(x) \leqslant (1.105\ldots)x$$

for large enough x, and used this in his proof of Bertrand's postulate. Diamond and Erdős (*Enseign. Math.* (2) 26 (1980), 313–21) have shown that elementary methods of the kind used by Tchebychef allow one to get upper and lower bound constants as close to 1 as desired. Unfortunately, since their paper actually uses the Prime Number Theorem in the course of the argument, their result does not produce an independent proof of the theorem.

§ 22.3. 'Bertrand's postulate' is that, for every $n > 3$, there is a prime p satisfying $n < p < 2n - 2$. Bertrand verified this for $n < 3,000,000$ and Tchebychef proved it for all $n > 3$ in 1850. Our Theorem 418 states a little less but the proof could be modified to prove the better result. Our proof is due to Erdős, *Acta Litt. Ac. Sci. (Szeged)*, 5 (1932), 194–8.

For Theorem 419, see L. Moser, *Math. Mag.* 23 (1950), 163–4. See also Mills, *Bull. American Math. Soc.* 53 (1947), 604; Bang, *Norsk. Mat. Tidsskr.* 34 (1952), 117–18; and Wright, *American Math. Monthly,* 58 (1951), 616–18 and 59 (1952), 99 and *Journal London Math. Soc.* 29 (1954), 63–71.

§ 22.7. Euler proved in 1737 that $\sum p^{-1}$ and $\prod(1 - p^{-1})$ are divergent.

§ 22.8. For Theorem 429 see Mertens, *Journal für Math.* 78 (1874), 46–62. For another proof (given in the first two editions of this book) see Hardy, *Journal London Math. Soc.* 10 (1935), 91–94.

§ 22.10. Theorem 430 is stated, in a rather more precise form, by Hardy and Ramanujan, *Quarterly Journal of Math.* 48 (1917), 76–92 (no. 35 of Ramanujan's *Collected papers*). It may be older, but we cannot give any reference.

§§ 22.11–13. These theorems were first proved by Hardy and Ramanujan in the paper referred to in the preceding note. The proof given here is due to Turán, *Journal London Math. Soc.* 9 (1934), 274–6, except for a simplification suggested to us by Mr. Marshall Hall. Turán [*ibid.* 11 (1936), 125–33] has generalized the theorems in two directions.

In fact the function $(\omega(n) - \log\log n)/\sqrt{\log\log n}$ is normally distributed, in the sense that, for any fixed real z, one has

$$x^{-1}\#\left\{n \leqslant x : \frac{\omega(n) - \log\log n}{\sqrt{\log\log n}} \leqslant z\right\} \to \frac{1}{\sqrt{2\pi}}\int_{-\infty}^{z} \exp\left\{-w^2/2\right\}dw$$

as $x \to \infty$. The same is true if $\omega(n)$ is replaced by $\Omega(n)$. These results are due to Erdős and Kac (*Amer. J. Math.* 62, (1940) 738–42).

There is a massive literature on the distribution of values of additive functions. See, for example, Kubilius, *Probabilistic methods in the theory of numbers* (Providence, R.I., A.M.S., 1964) and Kac, *Statistical independence in probability, analysis and number theory* (Washington, D.C., Math. Assoc. America, 1959).

§§ 22.14–16. A. Selberg gives his theorem in the forms

$$\vartheta(x)\log x + \sum_{p \leqslant x} \vartheta\left(\frac{x}{p}\right)\log p = 2x\log x + O(x)$$

and

$$\sum_{p \leqslant x} \log^2 p + \sum_{pp' \leqslant x} \log p \log p' = 2x\log x + O(x).$$

These may be deduced without difficulty from Theorem 433. There are two essentially different methods by which the Prime Number Theorem may be deduced from Selberg's theorem. For the first, due to Erdős and Selberg jointly, see *Proc. Nat. Acad. Sci.* 35 (1949), 374–84 and for the second, due to Selberg alone, see *Annals of Math.* 50 (1949), 305–13. Both methods are more 'elementary' (in the logical sense) than the one we give, since they avoid the use of the integral calculus at the cost of a little complication of detail. The method which we use in §§ 22.15 and 16 is based essentially on Selberg's own method. For the use of $\psi(x)$ instead of $\vartheta(x)$, the introduction of the integral calculus and other minor changes, see Wright, *Proc. Roy. Soc. Edinburgh*, 63 (1951), 257–67.

For an alternative exposition of the elementary proof of Theorem 6, see van der Corput, *Colloques sur la théorie des nombres* (Liège 1956). See Errera (*ibid.* 111–18) for a short (non-elementary) proof. The same volume (pp. 9–66) contains a reprint of the original paper in which de la Vallée Poussin (contemporaneously with Hadamard, but independently) gave the first proof (1896).

Later work by de la Vallée Poussin showed that

$$\pi(x) = \int_2^x \frac{dt}{\log t} + O\left(x \exp\left\{-c\sqrt{\log c}\right\}\right)$$

$$\psi(x) = x + O\left(x \exp\left\{-c\sqrt{\log c}\right\}\right)$$

for a certain positive constant c. These have been improved by subsequent authors, the best known error term now being $O\left(x \exp\left\{-c\,(\log x)^{3/5}\,(\log\log x)^{-1/5}\right\}\right)$, due independently to Korobov (*Uspehi Mat. Nauk* 13 (1958). no. 4 (82), 185–92) and Vinogradov (*Izv. Akad. Nauk SSSR. Ser. Mat.* 22 (1958), 161–64).

For an alternative to the work of § 22.15, see V. Nevanlinna, *Soc. Sci. Fennica: Comm. Phys. Math.* 27/3 (1962), 1–7. The same author (*Ann. Acad. Sci. Fennicae A* I343 (1964), 1–52) gives a comparative account of the various elementary proofs.

Two other, quite different, elementary proofs of the prime number theorem have also been given. These are by Daboussi (*C. R. Acad. Sci. Paris Sér. I Math.* 298 (1984), 161–64) and Hildebrand (*Mathematika* 33 (1986), 23–30) respectively.

Various authors have shown that the elementary proof based on Selberg's formulae can be adapted to prove an explicit error term in the Prime Number Theorem. In particular Diamond and Steinig (*Invent. Math.* 11 (1970), 199–258) showed in this way that

$$\pi(x) = \int_2^x \frac{dt}{\log t} + O\left(x \exp\left(-\log^{\theta} x\right)\right)$$

and

$$\psi(x) = x + O(x \exp(-\log^{\theta} x))$$

for any fixed $\theta < \frac{1}{7}$. See also Lavrik and Sobirov (*Dokl. Akad. Nauk SSSR*, 211 (1973), 534–6), Srinivasan and Sampath (*J. Indian Math. Soc.* (*N.S.*), 53 (1988), 1–50), and Lu (*Rocky Mountain J. Math.* 29 (1999), 979–1053).

§ 22.18. Landau proved Theorem 437 in 1900 and found more detailed asymptotic expansions for $\pi_k(x)$ and $\tau_k(x)$ in 1911. Subsequently Shah (1933) and S. Selberg (1940) obtained results of the latter type by more elementary means. For our proof and references to the literature, see Wright, *Proc. Edinburgh Math. Soc.* 9 (1954), 87–90.

§ 22.20. This type of argument can be applied to obtain similar conjectural asymptotic formulae for the number of prime-triplets and of longer blocks of primes. See Cherwell and Wright, *Quart. J. Math.* 11 (1960), 60–63 amd Pólya *American Math. Monthly* 66 (1959), 375–84. Hardy and Littlewood [*Acta Math.* 44 (1923), 1–70 (43)] found these formulae by a different (analytic) method (also subject to an unproved hypothesis). They give references to work by Staeckel and others. See also Cherwell, *Quarterly Journal of Math.* (Oxford), 17 (1946), 46–62, for another simple heuristic method.

The formulae agree very well with the results of counts. D. H. and E. Lehmer have carried these out for various prime pairs, triplets, and quadruplets up to 40 million and Golubew has counted quintuplets,…, 9-plets up to 20 million (*Osterreich Akad. Wiss. Math.-Naturwiss. Kl.* 1971, no. 1, 19–22). See also Leech (*Math. Comp.* 13 (1959), 56) and Bohman (*BIT, Nordisk Tidskr. Inform. behandl.* 13 (1973), 242–4).

XXIII
KRONECKER'S THEOREM

23.1. Kronecker's theorem in one dimension. Dirichlet's Theorem 201 asserts that, given any set of real numbers $\vartheta_1, \vartheta_2, \ldots, \vartheta_k$, we can make $n\vartheta_1, n\vartheta_2, \ldots, n\vartheta_k$ all differ from integers by as little as we please. This chapter is occupied by the study of a famous theorem of Kronecker which has the same general character as this theorem of Dirichlet but lies considerably deeper. The theorem is stated, in its general form, in § 23.4, and proved, by three different methods, in §§ 23.7–9. For the moment we consider only the simplest case, in which we are concerned with a single ϑ.

Suppose that we are given two numbers ϑ and α. *Can we find an integer n for which*

$$n\vartheta - \alpha$$

is nearly an integer? The problem reduces to the simplest case of Dirichlet's problem when $\alpha = 0$.

It is obvious at once that the answer is no longer unrestrictedly affirmative. If ϑ is a rational number a/b, in its lowest terms, then $(n\vartheta) = n\vartheta - [n\vartheta]$ has always one of the values

$$0,\ \frac{1}{b},\ \frac{2}{b},\ \ldots,\ \frac{b-1}{b}. \tag{23.1.1}$$

If $0 < \alpha < 1$, and α is not one of (23.1.1), then

$$\left|\frac{r}{b} - \alpha\right| \quad (r = 0,\ 1, \ldots, b)$$

has a positive minimum μ, and $n\vartheta - \alpha$ cannot differ from an integer by less than μ.

Plainly $\mu \leqslant 1/2b$, and $\mu \to 0$ when $b \to \infty$; and this suggests the truth of the theorem which follows.

THEOREM 438. *If ϑ is irrational, α is arbitrary, and N and ϵ are positive, then there are integers n and p such that $n > N$ and*

$$|n\vartheta - p - \alpha| < \epsilon. \tag{23.1.2}$$

We can state the substance of the theorem more picturesquely by using the language of § 9.10. It asserts that there are n for which $(n\vartheta)$ is as near as we please to any number in (0, 1), or, in other words,

THEOREM 439. *If ϑ is irrational, then the set of points $(n\vartheta)$ is dense in the interval* (0, 1).†

Either of Theorems 438 and 439 may be called 'Kronecker's theorem in one dimension'.

23.2. Proofs of the one-dimensional theorem. Theorems 438 and 439 are easy, but we give several proofs, to illustrate different ideas important in this field of arithmetic. Some of our arguments are, and some are not, extensible to space of more dimensions.

(i) By Theorem 201, with $k = 1$, there are integers n_1 and p such that $|n_1\vartheta - p| < \epsilon$. The point $(n_1\vartheta)$ is therefore within a distance ϵ of either 0 or 1. The series of points

$$(n_1\vartheta),\ (2n_1\vartheta),\ (3n_1\vartheta), \ldots,$$

continued so long as may be necessary, mark a chain (in one direction or the other) across the interval (0, 1) whose mesh‡ is less than ϵ. There is therefore a point $(kn_1\vartheta)$ or $(n\vartheta)$ within a distance ϵ of any α of (0, 1).

(ii) We can restate (i) so as to avoid an appeal to Theorem 201, and we do this explicitly because the proof resulting will be the model of our first proof in space of several dimensions.

We have to prove the set S of points P_n or $(n\vartheta)$ with $n = 1, 2, 3, \ldots$, dense in (0, 1). Since ϑ is irrational, no point falls at 0, and no two points coincide. The set has therefore a limit point, and there are pairs (P_n, P_{n+r}), with $r > 0$, and indeed with arbitrarily large r, as near to one another as we please.

We call the directed stretch $P_n\,P_{n+r}$ a *vector*. If we mark off a stretch $P_m\,Q$, equal to $P_n\,P_{n+r}$ and in the same direction, from any P_m, then Q is another point of S, and in fact P_{m+r}. It is to be understood, when we make this construction, that if the stretch $P_m\,Q$ would extend beyond 0 or 1, then the part of it so extending is to be replaced by a congruent part measured from the other end 1 or 0 of the interval (0, 1).

There are vectors of length less than ϵ, and such vectors, with $r > N$, extending from any point of S and in particular from P_1. If we measure off

† We may seem to have lost something when we state the theorem thus (viz. the inequality $n > N$). But it is plain that, if there are points of the set as near as we please to every α of (0, 1), then among these points there are points for which n is as large as we please.

‡ The distance between consecutive points of the chain.

such a vector repeatedly, starting from P_1 we obtain a chain of points with the same properties as the chain of (i), and can complete the proof in the same way.

(iii) There is another interesting 'geometrical' proof which cannot be extended, easily at any rate, to space of many dimensions.

We represent the real numbers, as in § 3.8, on a circle of unit circumference instead of on a straight line. This representation automatically rejects integers; 0 and 1 are represented by the same point of the circle and so, generally, are $(n\vartheta)$ and $n\vartheta$.

To say that S is dense on the circle is to say that every α belongs to the derived set S'. If α belongs to S but not to S', there is an interval round α free from points of S, except for α itself, and therefore there are points near α belonging neither to S nor to S'. It is therefore sufficient to prove that every α belongs either to S or to S'.

If α belongs neither to S nor to S', there is an interval $(\alpha - \delta, \alpha + \delta')$, with positive δ and δ', which contains no point of S inside it; and among all such intervals there is a *greatest.*† We call this maximum interval $I(\alpha)$ *the excluded interval of* α.

It is plain that, if α is surrounded by an excluded interval $I(\alpha)$, then $\alpha - \vartheta$ is surrounded by a congruent excluded interval $I(\alpha - \vartheta)$. We thus define an infinite series of intervals

$$I(\alpha),\ I(\alpha - \vartheta),\ I(\alpha - 2\vartheta),\ \ldots$$

similarly disposed about the points $\alpha,\ \alpha - \vartheta,\ \alpha - 2\vartheta, \ldots$. No two of these intervals can coincide, since ϑ is irrational; and no two can overlap, since two overlapping intervals would constitute together a larger interval, free from points of S, about one of the points. This is a contradiction, since the circumference cannot contain an infinity of non-overlapping intervals of equal length. The contradiction shows that there can be no interval $I(\alpha)$, and so proves the theorem.

(iv) Kronecker's own proof is rather more sophisticated, but proves a good deal more. It proves

THEOREM 440. *If* ϑ *is irrational,* α *is arbitrary, and* N *positive, then there is an* $n > N$ *and a* p *for which*

$$|n\vartheta - p - \alpha| < \frac{3}{n}.$$

† We leave the formal proof, which depends upon the construction of 'Dedekind sections' of the possible values of δ and δ', and is of a type familiar in elementary analysis, to the reader.

It will be observed that this theorem, unlike Theorem 438, gives a definite bound for the 'error' in terms of n, of the same kind (though not so precise) as those given by Theorems 183 and 193 when $\alpha = 0$.

By Theorem 193 there are coprime integers $q > 2N$ and r such that

$$|q\vartheta - r| < \frac{1}{q}. \tag{23.2.1}$$

Suppose that Q is the integer, or one of the two integers, such that

$$|q\alpha - Q| \leqslant \tfrac{1}{2}. \tag{23.2.2}$$

We can express Q in the form

$$Q = vr - uq, \tag{23.2.3}$$

where u and v are integers and

$$|v| \leqslant \tfrac{1}{2}q. \tag{23.2.4}$$

Then

$$q(v\vartheta - u - \alpha) = v(q\vartheta - r) - (q\alpha - Q),$$

and therefore

$$|q(v\vartheta - u - \alpha)| < \tfrac{1}{2}q \cdot \frac{1}{q} + \frac{1}{2} = 1, \tag{23.2.5}$$

by (23.2.1), (23.2.2), and (23.2.4). If now we write

$$n = q + v, \qquad p = r + u,$$

then

$$N < \tfrac{1}{2}q \leqslant n \leqslant \tfrac{3}{2}q \tag{23.2.6}$$

and

$$|n\vartheta - p - \alpha| \leqslant |v\vartheta - u - \alpha| + |q\vartheta - r| < \frac{1}{q} + \frac{1}{q} = \frac{2}{q} \leqslant \frac{3}{n},$$

by (23.2.1), (23.2.5), and (23.2.6).

It is possible to refine upon the 3 of the theorem, but not, by this method, in a very interesting way. We return to this question in Ch. XXIV.

23.3. The problem of the reflected ray. Before we pass to the general proof of Kronecker's theorem, we shall apply the special case already proved to a simple but entertaining problem of plane geometry solved by König and Szücs.

The sides of a square are reflecting mirrors. A ray of light leaves a point inside the square and is reflected repeatedly in the mirrors. What is the nature of its path?†

THEOREM 441. *Either the path is closed and periodic or it is dense in the square, passing arbitrarily near to every point of the square. A necessary and sufficient condition for periodicity is that the angle between a side of the square and the initial direction of the ray should have a rational tangent.*

In Fig. 9 the parallels to the axes are the lines

$$x = l + \tfrac{1}{2}, \qquad y = m + \tfrac{1}{2},$$

where l and m are integers. The thick square, of side 1, round the origin is the square of the problem and P, or (a, b), is the starting-point. We construct all images of P in the mirrors, for direct or repeated reflection. A moment's thought will show that they are of four types, the coordinates of the images of the different types being

$$\begin{array}{ll} \text{(A)}\ a+2l, b+2m; & \text{(B)}\ a+2l, -b+2m+1; \\ \text{(C)}\ -a+2l+1,\ b+2m; & \text{(D)}\ -a+2l+1, -b+2m+1; \end{array}$$

where l and m are arbitrary integers.‡ Further, if the velocity at P has direction cosines λ, μ, then the corresponding images of the velocity have direction cosines

$$\text{(A)}\ \lambda, \mu; \quad \text{(B)}\ \lambda, -\mu; \quad \text{(C)}\ -\lambda, \mu; \quad \text{(D)}\ -\lambda, -\mu.$$

We may suppose, on grounds of symmetry, that μ is positive.

† It may happen exceptionally that the ray passes through a *corner* of the square. In this case we assume that it returns along its former path. This is the convention suggested by considerations of continuity.

‡ The x-coordinate takes all values derived from a by the repeated use of the substitutions $x' = 1 - x$ and $x' = -1 - x$. The figure shows the images corresponding to non-negative l and m.

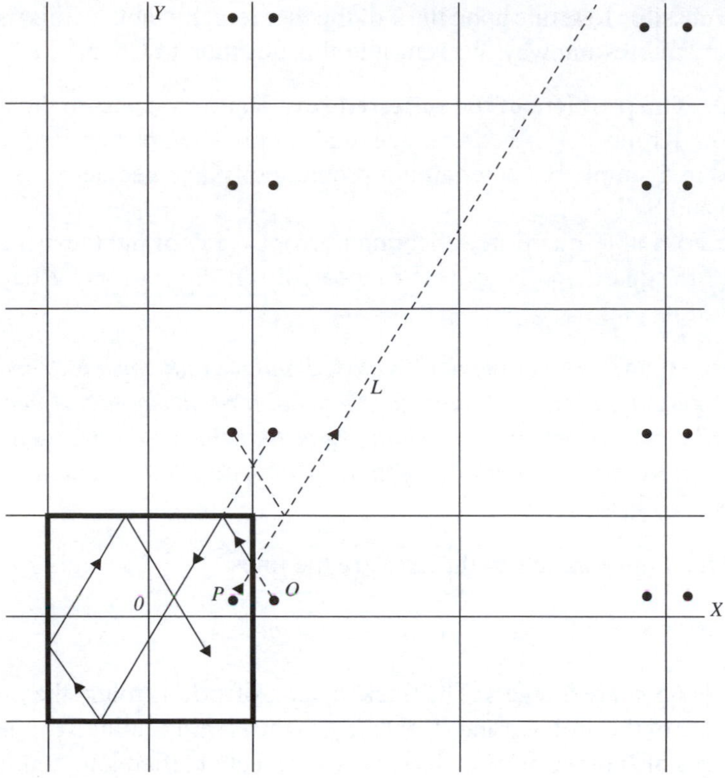

FIG. 9.

If we think of the plane as divided into squares of unit side, the interior of a typical square being

$$l - \tfrac{1}{2} < x < l + \tfrac{1}{2}, \qquad m - \tfrac{1}{2} < y < m + \tfrac{1}{2}, \tag{23.3.1}$$

then each square contains just one image of every point in the original square

$$-\tfrac{1}{2} < x < \tfrac{1}{2}, \qquad -\tfrac{1}{2} < y < \tfrac{1}{2};$$

and, if the image in (23.3.1) of any point in the original square is of type A, B, C, or D, then the image in (23.3.1) of any other point in the original square is of the same type.

We now imagine P moving with the ray. When P meets a mirror at Q, it coincides with an image; and the image of P which momentarily coincides with P continues the motion of P, in its original direction, in one of the squares adjacent to the fundamental square. We follow the motion of the image, in this square, until it in its turn meets a side of the square. It is plain that the original path of P will be continued indefinitely in the same line L, by a series of different images.

The segment of L in any square (23.3.1) is the image of a straight portion of the path of P in the original square. There is a one-to-one correspondence between the segments of L, in different squares (23.3.1), and the portions of the path of P between successive reflections, each segment of L being an image of the corresponding portion of the path of P.

The path of P in the original square will be periodic if P returns to its original position moving in the same direction; and this will happen if and only if L passes through an image of type A of the original P. The coordinates of an arbitrary point of L are

$$x = a + \lambda t, \qquad y = b + \mu t.$$

Hence the path will be periodic if and only if

$$\lambda t = 2l, \qquad \mu t = 2m$$

for some t and integral l, m; i.e. if λ/μ *is rational.*

It remains to show that, when λ/μ is irrational, the path of P approaches arbitrarily near to every point (ξ, η) of the square. It is necessary and sufficient for this that L should pass arbitrarily near to some image of (ξ, η) and sufficient that it should pass near some image of (ξ, η) of type A, and this will be so if

$$|a + \lambda t - \xi - 2l| < \epsilon, \qquad |b + \mu t - \eta - 2m| < \epsilon \tag{23.3.2}$$

for every ξ and η, any positive ϵ, some positive t, and appropriate integral l and m.

We take

$$t = \frac{\eta + 2m - b}{\mu},$$

when the second of (23.3.2) is satisfied automatically. The first inequality then becomes

$$|m\vartheta - \omega - l| < \tfrac{1}{2}\epsilon, \tag{23.3.3}$$

where

$$\vartheta = \frac{\lambda}{\mu}, \qquad \omega = (b - \eta)\frac{\lambda}{2\mu} - \tfrac{1}{2}(a - \xi).$$

Theorem 438 shows that, when ϑ is irrational, there are l and m, large enough to make t positive, which satisfy (23.3.3).

23.4. Statement of the general theorem. We pass to the general problem in space of k dimensions. The numbers $\vartheta_1, \vartheta_2, \ldots, \vartheta_k$ are given, and we wish to approximate to an arbitrary set of numbers $\alpha_1, \alpha_2, \ldots, \alpha_k$, integers apart, by equal multiples of $\vartheta_1, \vartheta_2, \ldots, \vartheta_k$. It is plain, after § 23.1, that the ϑ must be irrational, but this condition is not a sufficient condition for the possibility of the approximation.

Suppose for example, to fix our ideas, that $k = 2$, that ϑ, ϕ, α, β are positive and less than 1, and that ϑ and ϕ (whether rational or irrational) satisfy a relation

$$a\vartheta + b\phi + c = 0$$

with integral a, b, c. Then

$$a.n\vartheta + b.n\phi$$

and

$$a(n\vartheta) + b(n\phi)$$

are integers, and the point whose coordinates are $(n\vartheta)$ and $(n\phi)$ lies on one or other of a finite number of straight lines. Thus Fig. 10 shows the case $a = 2, b = 3$, when the point lies on one or other of the lines $2x + 3y = \nu\ (\nu = 1, 2, 3, 4)$. It is plain that, if (α, β) does not lie on one of these lines, it is impossible to approximate to it with more than a certain accuracy.

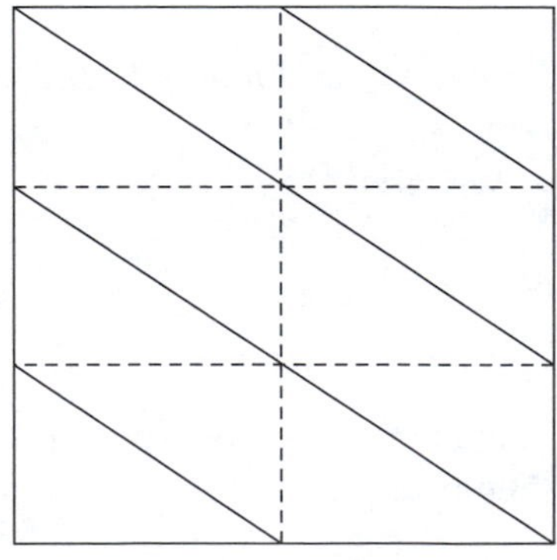

Fig. 10

We shall say that a set of numbers

$$\xi_1, \xi_2, \ldots, \xi_r$$

is *linearly independent* if no linear relation

$$a_1\xi_1 + a_2\xi_2 + \cdots + a_r\xi_r = 0,$$

with integral coefficients, not all zero, holds between them. Thus, if $p_1, p_2, \ldots, p_r$ are different primes, then

$$\log p_1, \log p_2, \ldots, \log p_r$$

are linearly independent; for

$$a_1 \log p_1 + a_2 \log p_2 + \cdots + a_r \log p_r = 0$$

is

$$p_1^{a_1} p_2^{a_2} \ldots p_r^{a_r} = 1,$$

which contradicts the fundamental theorem of arithmetic.

We now state Kronecker's theorem in its general form.

THEOREM 442. *If*

$$\vartheta_1, \vartheta_2, \ldots, \vartheta_k, 1$$

are linearly independent, $\alpha_1, \alpha_2, \ldots, \alpha_k$ *are arbitrary, and* N *and* ϵ *are positive, then there are integers*

$$n > N, \quad p_1, p_2, \ldots, p_k$$

such that

$$|n\vartheta_m - p_m - \alpha_m| < \epsilon \quad (m = 1, 2, \ldots, k).$$

We can also state the theorem in a form corresponding to Theorem 439, but for this we must extend the definitions of § 9.10 to k-dimensional space.

If the coordinates of a point P of k-dimensional space are $x_1, x_2, \ldots, x_k$, and δ is positive, then the set of points $x_1', x_2', \ldots, x_k'$ for which

$$|x_m' - x_m| \leqslant \delta \quad (m = 1, 2, \ldots, k)$$

is called a *neighbourhood* of P. The phrases *limit point, derivative, closed, dense in itself,* and *perfect* are then defined exactly as in § 9.10. Finally, if we describe the set defined by

$$0 \leqslant x_m \leqslant 1 \quad (m = 1, 2, \ldots, k)$$

as the 'unit cube', then a set of points S is *dense in the unit cube* if every point of the cube is a point of the derived set S'.

THEOREM 443. *If* $\vartheta_1, \vartheta_2, \ldots, \vartheta_k, 1$ *are linearly independent, then the set of points*

$$(n\vartheta_1), (n\vartheta_2), \ldots, (n\vartheta_k)$$

is dense in the unit cube.

23.5. The two forms of the theorem. There is an alternative form of Kronecker's theorem in which both hypothesis and conclusion assert a little less.

THEOREM 444. *If* $\vartheta_1, \vartheta_2, \ldots, \vartheta_k$ *are linearly independent,* $\alpha_1, \alpha_2, \ldots, \alpha_k$ *are arbitrary, and* T *and* ϵ *are positive, then there is a real number* t*, and integers* $p_1, p_2, \ldots, p_k$*, such that*

$$t > T$$

and

$$|t\vartheta_m - p_m - \alpha_m| < \epsilon \quad (m = 1, 2, \ldots, k).$$

The fundamental hypothesis in Theorem 444 is weaker than in Theorem 442, since it only concerns linear relations homogeneous in the ϑ. Thus $\vartheta_1 = \sqrt{2}, \vartheta_2 = 1$ satisfy the condition of Theorem 444 but not that of Theorem 442; and, in Theorem 444, just one of the ϑ may be rational. The conclusion is also weaker, because t is not necessarily integral.

It is easy to prove that the two theorems are equivalent. It is useful to have both forms, since some proofs lead most naturally to one form and some to the other.

(1) *Theorem* 444 *implies Theorem* 442. We suppose, as we may, that every ϑ lies in (0, 1) and that $\epsilon < 1$. We apply Theorem 444, with $k + 1$ for $k, N + 1$ for T, and $\frac{1}{2}\epsilon$ for ϵ, to the systems

$$\vartheta_1, \vartheta_2, \ldots, \vartheta_k, 1; \quad \alpha_1, \alpha_2, \ldots, \alpha_k, 0.$$

The hypothesis of linear independence is then that of Theorem 442; and the conclusion is expressed by

$$t > N + 1, \tag{23.5.1}$$

(23.5.2) $$|t\vartheta_m - p_m - \alpha_m| < \tfrac{1}{2}\epsilon \quad (m = 1, 2, \ldots, k),$$

(23.5.3) $$|t - p_{k+1}| < \tfrac{1}{2}\epsilon.$$

From (23.5.1) and (23.5.3) it follows that $p_{k+1} > N$, and from (23.5.2) and (23.5.3) that

$$|p_{k+1}\vartheta_m - p_m - \alpha_m| \leqslant |t\vartheta_m - p_m - \alpha_m| + |t - p_{k+1}| < \epsilon.$$

These are the conclusions of Theorem 442, with $n = p_{k+1}$.

(2) *Theorem* 442 *implies Theorem* 444. We now deduce Theorem 444 from Theorem 442. We observe first that Kronecker's theorem (in either form) is 'additive in the α'; if the result is true for a set of ϑ and for $\alpha_1, \ldots, \alpha_k$, and also for the same set of ϑ and for $\beta_1, \ldots, \beta_k$, then it is true for the same ϑ and for $\alpha_1 + \beta_1, \ldots, \alpha_k + \beta_k$. For if the differences of $p\vartheta$ from α, and of $q\vartheta$ from β, are nearly integers, then the difference of $(p + q)\vartheta$ from $\alpha + \beta$ is nearly an integer.

If $\vartheta_1, \vartheta_2, \ldots, \vartheta_{k+1}$ are linearly independent, then so are

$$\frac{\vartheta_1}{\vartheta_{k+1}}, \ldots, \frac{\vartheta_k}{\vartheta_{k+1}}, 1.$$

We apply Theorem 442, with $N = T$, to the system

$$\frac{\vartheta_1}{\vartheta_{k+1}}, \ldots, \frac{\vartheta_k}{\vartheta_{k+1}}; \quad \alpha_1, \ldots, \alpha_k.$$

There are integers $n > N, p_1, \ldots, p_k$ such that

(23.5.4) $$\left|\frac{n\vartheta_m}{\vartheta_{k+1}} - p_m - \alpha_m\right| < \epsilon \quad (m = 1, 2, \ldots, k).$$

If we take $t = n/\vartheta_{k+1}$, then the inequalities (23.5.4) are k of those required, and

$$|t\vartheta_{k+1} - n| = 0 < \epsilon.$$

Also $t \geqslant n > N = T$. We thus obtain Theorem 444, for

$$\vartheta_1, \ldots, \vartheta_k, \vartheta_{k+1}; \quad \alpha_1, \ldots, \alpha_k, 0.$$

We can prove it similarly for

$$\vartheta_1, \ldots, \vartheta_k, \vartheta_{k+1}; \quad 0, \ldots, 0, \alpha_{k+1},$$

and the full theorem then follows from the remark at the beginning of (2).

23.6. An illustration. Kronecker's theorem is one of those mathematical theorems which assert, roughly, that 'what is not impossible will happen some times however improbable it may be'. We can illustrate this 'astronomically'.

Suppose that k spherical planets revolve round a point O in concentric coplanar circles, their angular velocities being $2\pi\omega_1$, $2\pi\omega_2,\ldots$, $2\pi\omega_k$, that there is an observer at O, and that the apparent diameter of the inmost planet P, observed from O, is greater than that of any outer planet.

If the planets are all in conjunction at time $t = 0$ (so that P occults all the other planets), then their angular coordinates at time t are $2\pi t\omega_1,\ldots$. Theorem 201 shows that we can choose a t, as large as we please, for which all these angles are as near as we please to integral multiples of 2π. Hence occultation of the whole system by P will recur continually. This conclusion holds for *all* angular velocities.

If the angular coordinates are initially $\alpha_1, \alpha_2,\ldots, \alpha_k$, then such an occultation may never occur. For example, two of the planets might be originally in opposition and have equal angular velocities. Suppose, however, that *the angular velocities are linearly independent.* Then Theorem 444 shows that, for appropriate t, as large as we please, all of

$$2\pi t\omega_1 + \alpha_1, \ldots, 2\pi t\omega_k + \alpha_k$$

will be as near as we please to multiples of 2π; and then occultations will recur whatever the initial positions.

23.7. Lettenmeyer's proof of the theorem. We now suppose that $k = 2$, and prove Kronecker's theorem in this case by a 'geometrical' method due to Lettenmeyer. When $k = 1$, Lettenmeyer's argument reduces to that used in § 23.2 (ii).

We take the first form of the theorem, and write ϑ, ϕ for ϑ_1, ϑ_2. We may suppose

$$0 < \vartheta < 1, \quad 0 < \phi < 1;$$

and we have to show that if ϑ, ϕ, 1 are linearly independent then the points P_n whose coordinates are

$$(n\vartheta), \quad (n\phi) \quad (n = 1, 2, \ldots)$$

are dense in the unit square. No two P_n coincide, and no P_n lies on a side of the square.

We call the directed stretch

$$P_n P_{n+r} \quad (n > 0, r > 0)$$

a *vector.* If we take any point P_m, and draw a vector P_mQ equal and parallel to the vector P_nP_{n+r}, then the other end Q of this vector is a point of the set (and in fact P_{m+r}). Here naturally we adopt the convention corresponding to that of § 23.2 (ii), viz. that, if P_mQ meets a side of the square, then it is continued in the same direction from the corresponding point on the opposite side of the square.

Since no two points P_n coincide, the set (P_n) has a limit point; there are therefore vectors whose length is less than any positive ϵ, and vectors of this kind for which r is as large as we please. We call these vectors *ϵ-vectors.* There are ϵ-vectors, and ϵ-vectors with arbitrarily large r, issuing from every P_n, and in particular from P_1. If

$$\epsilon < \min(\vartheta, \phi, 1-\vartheta, 1-\phi),$$

then all ϵ-vectors issuing from P_1 are unbroken, i.e. do not meet a side of the square.

Two cases are possible *a priori.*

(1) *There are two ϵ-vectors which are not parallel.*† In this case we mark them off from P_1 and construct the lattice based upon P_1 and the two other ends of the vectors. Every point of the square is then within a distance ϵ of some lattice point, and the theorem follows.

(2) *All ϵ-vectors are parallel.* In this case all ϵ-vectors issuing from P_1 lie along the same straight line, and there are points P_r, P_s on this line with arbitrarily large suffixes r, s. Since P_1, P_r, P_s are collinear,

$$0 = \begin{vmatrix} \vartheta & \phi & 1 \\ (r\vartheta) & (r\phi) & 1 \\ (s\vartheta) & (s\phi) & 1 \end{vmatrix} = \begin{vmatrix} \vartheta & \phi & 1 \\ r\vartheta - [r\vartheta] & r\phi - [r\phi] & 1 \\ s\vartheta - [s\vartheta] & s\phi - [s\phi] & 1 \end{vmatrix},$$

and so

$$\begin{vmatrix} \vartheta & \phi & 1 \\ [r\vartheta] & [r\phi] & r-1 \\ [s\vartheta] & [s\phi] & s-1 \end{vmatrix} = 0,$$

† In the sense of elementary geometry, where we do not distinguish two directions on one straight line.

or

$$a\vartheta + b\phi + c = 0,$$

where a, b, c are integers. But ϑ, ϕ, 1 are linearly independent, and therefore a, b, c are all zero. Hence, in particular,

$$\begin{vmatrix} [r\phi] & r-1 \\ [s\phi] & s-1 \end{vmatrix} = 0,$$

or

$$\frac{[s\phi]}{s-1} = \frac{[r\phi]}{r-1}.$$

We can make $s \to \infty$, since there are P_s with arbitrarily large s; and we then obtain

$$\phi = \lim \frac{[s\phi]}{s-1} = \frac{[r\phi]}{r-1},$$

which is impossible because ϕ is irrational.

It follows that case (2) is impossible, so that the theorem is proved.

23.8. Estermann's proof of the theorem. Lettenmeyer's argument may be extended to space of k dimensions, and leads to a general proof of Kronecker's theorem; but the ideas which underlie it are illustrated adequately in the two-dimensional case. In this and the next section we prove the general theorem by two other quite different methods.

Estermann's proof is inductive. His argument shows that the theorem is true in space of k dimensions if it is true in space of $k-1$. It also shows incidentally that the theorem is true in one-dimensional space, so that the proof is self-contained; but this we have proved already, and the reader may, if he pleases, take it for granted.

The theorem in its first form states that, if $\vartheta_1, \vartheta_2, \ldots, \vartheta_k$, 1 are linearly independent, $\alpha_1, \alpha_2, \ldots, \alpha_k$ are arbitrary, and ϵ and ω are positive, then there are integers $n, p_1, p_2, \ldots, p_k$ such that

$$n > \omega \tag{23.8.1}$$

and

$$|n\vartheta_m - p_m - \alpha_m| < \epsilon \quad (m = 1, 2, \ldots, k). \tag{23.8.2}$$

Here the emphasis is on large positive values of n. It is convenient now to modify the enunciation a little, and consider both positive and negative values of n. We therefore assert a little more, viz. that, given a positive ϵ and ω, and a λ of either sign, then we can choose n and the p to satisfy (23.8.2) and

$$|n| > \omega, \qquad \operatorname{sign} n = \operatorname{sign} \lambda, \tag{23.8.3}$$

the second equation meaning that n has the same sign as λ. We have to show (*a*) that this is true for k if it is true for $k-1$, and (*b*) that it is true when $k = 1$.

There are, by Theorem 201, integers

$$s > 0, \qquad b_1, b_2, \ldots, b_k$$

such that

$$|s\vartheta_m - b_m| < \tfrac{1}{2}\epsilon \quad (m = 1, 2, \ldots, k). \tag{23.8.4}$$

Since ϑ_k is irrational, $s\vartheta_k - b_k \neq 0$; and the k numbers

$$\phi_m = \frac{s\vartheta_m - b_m}{s\vartheta_k - b_k}$$

(of which the last is 1) are linearly independent, since a linear relation between them would involve one between $\vartheta_1, \ldots, \vartheta_k, 1$.

Suppose first that $k > 1$, and assume the truth of the theorem for $k-1$. We apply the theorem, with $k-1$ for k, to the system

$$\phi_1, \phi_2, \ldots, \phi_{k-1} \quad (\text{for } \vartheta_1, \vartheta_2, \ldots, \vartheta_{k-1}),$$

$$\beta_1 = \alpha_1 - \alpha_k\phi_1, \quad \beta_2 = \alpha_2 - \alpha_k\phi_2, \quad \ldots, \quad \beta_{k-1} = \alpha_{k-1} - \alpha_k\phi_{k-1}$$
$$(\text{for } \alpha_1, \alpha_2, \ldots, \alpha_{k-1}),$$

$$\tfrac{1}{2}\epsilon \quad (\text{for } \epsilon), \qquad \lambda(s\vartheta_k - b_k) \quad (\text{for } \lambda),$$

$$\Omega = (\omega + 1)|s\vartheta_k - b_k| + |\alpha_k| \quad (\text{for } \omega). \tag{23.8.5}$$

There are integers $c_k, c_1, c_2, \ldots, c_{k-1}$ such that

$$|c_k| > \Omega, \qquad \operatorname{sign} c_k = \operatorname{sign}\{\lambda(s\vartheta_k - b_k)\}, \tag{23.8.6}$$

and

$$(23.8.7) \qquad |c_k\phi_m - c_m - \beta_m| < \tfrac{1}{2}\epsilon \quad (m = 1, 2, \ldots, k-1).$$

The inequality (23.8.7), when expressed in terms of the ϑ is

$$(23.8.8) \qquad \left|\frac{c_k + \alpha_k}{s\vartheta_k - b_k}(s\vartheta_m - b_m) - c_m - \alpha_m\right| < \tfrac{1}{2}\epsilon \quad (m = 1, 2, \ldots, k).$$

Here we have included the value k of m, as we may do because the left-hand side of (23.8.8) vanishes when $m = k$.

We have supposed $k > 1$. When $k = 1$, (23.8.8) is trivial, and we have only to choose c_k to satisfy (23.8.6), as plainly we may.

We now choose an integer N so that

$$(23.8.9) \qquad \left|N - \frac{c_k + \alpha_k}{s\vartheta_k - b_k}\right| < 1,$$

and take

$$n = Ns, \quad p_m = Nb_m + c_m.$$

Then

$$\begin{aligned} |n\vartheta_m - p_m - \alpha_m| &= |N(s\vartheta_m - b_m) - c_m - \alpha_m| \\ &\leqslant \left|\frac{c_k + \alpha_k}{s\vartheta_k - b_k}(s\vartheta_m - b_m) - c_m - \alpha_m\right| + |s\vartheta_m - b_m| \\ &< \tfrac{1}{2}\epsilon + \tfrac{1}{2}\epsilon = \epsilon \quad (m = 1, 2, \ldots, k), \end{aligned}$$

by (23.8.4), (23.8.8), and (23.8.9). This is (23.8.2). Next

$$(23.8.10) \qquad \left|\frac{c_k + \alpha_k}{s\vartheta_k - b_k}\right| \geqslant \frac{|c_k| - |\alpha_k|}{|s\vartheta_k - b_k|} > \omega + 1,$$

by (23.8.5) and (23.8.6); so that $|N| > \omega$ and

$$|n| = |N|s \geqslant |N| > \omega.$$

Finally, n has the sign of N, and so, after (23.8.9) and (23.8.10), the sign of

$$\frac{c_k}{s\vartheta_k - b_k}.$$

This, by (23.8.6), is the sign of λ.

Hence n and the p satisfy all our demands, and the induction from $k-1$ to k is established.

23.9. Bohr's proof of the theorem. There are also a number of 'analytical' proofs of Kronecker's theorem, of which perhaps the simplest is one due to Bohr. All such proofs depend on the facts that

$$e(x) = e^{2\pi ix}$$

has the period 1 and is equal to 1 if and only if x is an integer.

We observe first that

$$\lim_{T\to\infty} \frac{1}{T}\int_0^T e^{cit}\,dt = \lim_{T\to\infty} \frac{e^{ciT}-1}{ciT} = 0$$

if c is real and not zero, and is 1 if $c = 0$. It follows that, if

$$\chi(t) = \sum_{\nu=1}^{r} b_\nu e^{c_\nu it}, \tag{23.9.1}$$

where no two c_ν are equal, then

$$b_\nu = \lim_{T\to\infty} \frac{1}{T}\int_0^T \chi(t)e^{-c_\nu it}\,dt. \tag{23.9.2}$$

We take the second form of Kronecker's theorem (Theorem 444), and consider the function

$$\phi(t) = |F(t)|, \tag{23.9.3}$$

where

$$F(t) = 1 + \sum_{m=1}^{k} e(\vartheta_m t - \alpha_m), \tag{23.9.4}$$

of the real variable t. Obviously

$$\phi(t) \leqslant k+1.$$

If Kronecker's theorem is true, we can find a large t for which every term in the sum is nearly 1 and $\phi(t)$ is nearly $k+1$. Conversely, if $\phi(t)$ is nearly $k+1$ for some large t, then (since no term can exceed 1 in absolute value) every term must be nearly 1 and Kronecker's theorem must be true. We shall therefore have proved Kronecker's theorem if we can prove that

$$\overline{\lim_{t\to\infty}}\ \phi(t) = k+1. \tag{23.9.5}$$

The proof is based on certain formal relations between $F(t)$ and the function

$$\psi(x_1, x_2, \ldots, x_k) = 1 + x_1 + x_2 + \cdots + x_k \tag{23.9.6}$$

of the k variables x. If we raise ψ to the pth power by the multinomial theorem, we obtain

$$\psi^p = \sum a_{n_1,n_2,\ldots,n_k} x_1^{n_1} x_2^{n_2} \ldots x_k^{n_k}. \tag{23.9.7}$$

Here the coefficients a are positive; their individual values are irrelevant, but their *sum* is

$$\sum a = \psi^p(1,1,\ldots,1) = (k+1)^p. \tag{23.9.8}$$

We also require an upper bound for their *number.* There are $p+1$ of them when $k=1$; and

$$(1+x_1+\cdots+x_k)^p$$
$$= (1+x_1+\cdots+x_{k-1})^p + \binom{p}{1}(1+x_1+\cdots+x_{k-1})^{p-1}x_k + \cdots + x_k^p,$$

so that the number is multiplied at most by $p+1$ when we pass from $k-1$ to k. Hence the number of the a does not exceed $(p+1)^k$.†

We now form the corresponding power

$$F^p = \{1 + e(\vartheta_1 t - \alpha_1) + \cdots + e(\vartheta_k t - \alpha_k)\}^p$$

of F. This is a sum of the form (23.9.1), obtained by replacing x_r in (23.9.7) by $e(\vartheta_r t - \alpha_r)$. When we do this, *every product* $x_1^{n_1} \ldots x_k^{n_k}$ *in* (23.9.7) *will give rise to a different* c_ν, since the equality of two c_ν would imply a linear

† The actual number is $\binom{p+k}{k}$.

relation between the ϑ.[†] It follows that every coefficient b_ν has an absolute value equal to the corresponding coefficient a, and that

$$\sum |b_\nu| = \sum a = (k+1)^p.$$

Suppose now that, in contradiction to (23.9.5),

$$\overline{\lim}\, \phi(t) < k+1. \tag{23.9.9}$$

Then there is a λ and a t_0 such that, for $t > t_0$,

$$|F(t)| \leqslant \lambda < k+1,$$

and

$$\overline{\lim} \frac{1}{T} \int_0^T |F(t)|^p dt \leqslant \lim \frac{1}{T} \int_0^T \lambda^p dt = \lambda^p.$$

Hence

$$|b_\nu| = \left| \lim \frac{1}{T} \int_0^T \{F(t)\}^p e^{-c_\nu i t} dt \right| \leqslant \overline{\lim} \frac{1}{T} \int_0^T |F(t)|^p dt \leqslant \lambda^p;$$

and therefore $a \leqslant \lambda^p$ for every a. Hence, since there are at most $(p+1)^k$ of the a, we deduce

$$(k+1)^p = \sum a \leqslant (p+1)^k \lambda^p,$$

$$\left(\frac{k+1}{\lambda}\right)^p \leqslant (p+1)^k. \tag{23.9.10}$$

But $\lambda < k+1$, and so

$$\left(\frac{k+1}{\lambda}\right)^p = e^{\delta p},$$

[†] It is here only that we use the linear independence of the ϑ, and this is naturally the kernel of the proof.

where $\delta > 0$. Thus

$$e^{\delta p} \leqslant (p+1)^k,$$

which is impossible for large p because

$$e^{-\delta p}(p+1)^k \to 0$$

when $p \to \infty$. Hence (23.9.9) involves a contradiction for large p, and this proves the theorem.

23.10. Uniform distribution. Kronecker's theorem, important as it is, does not tell the full truth about the sets of points $(n\vartheta)$ or $(n\vartheta_1), (n\vartheta_2), \ldots$ with which it is concerned. These sets are not merely *dense* in the unit interval, or cube, but 'uniformly distributed'.

Returning for the moment to one dimension, we say that a set of points P_n in (0,1) is *uniformly distributed* if, roughly, every sub-interval of (0,1) contains its proper quota of points. To put the definition precisely, we suppose that I is a sub-interval of (0, 1), and use I both for the interval and for its length. If n_I is the number of the points $P_1, P_2, \ldots, P_n$ which fall in I, and

$$\frac{n_I}{n} \to I, \tag{23.10.1}$$

whatever I, when $n \to \infty$, then the set is uniformly distributed. We can also write (23.10.1) in either of the forms

$$n_I \sim nI, \qquad n_I = nI + o(n). \tag{23.10.2}$$

THEOREM 445. *If ϑ is irrational then the points $(n\vartheta)$ are uniformly distributed in* (0, 1).

Let $0 < \epsilon < \frac{1}{10}$. By Theorem 439, we can choose j so that $0 < (j\vartheta) = \delta < \epsilon$. We write $K = [1/\delta]$. If $0 \leqslant h < K$, the interval I_h is that in which

$$(hj\vartheta) < x \leqslant (\{h+1\}j\vartheta).$$

Here I_K extends beyond the point 1 and we are using the circular representation of § 23.2 (iii). We denote by $\eta_h(n)$ the number of $(\vartheta), (2\vartheta), \ldots, (n\vartheta)$, which lie in I_h. If $(t\vartheta)$ lies in I_0, where t is a positive integer, then $(\{t+hj\}\vartheta)$ lies in I_h and conversely. Hence, if $n > hj$,

$$\eta_h(n) - \eta_h(hj) = \eta_0(n - hj).$$

But $\eta_h(hj) \leqslant hj$ and $\eta_0(n - hj) \geqslant \eta_0(n) - hj$. Hence

$$\eta_0(n) - hj \leqslant \eta_h(n) \leqslant \eta_0(n) + hj$$

and so

$$\lim_{n\to\infty} \frac{\eta_h(n)}{\eta_0(n)} = 1 \qquad (0 \leqslant h \leqslant K). \tag{23.10.3}$$

Now

$$\sum_{h=0}^{K-1} \eta_h(n) \leqslant n \leqslant \sum_{h=0}^{K} \eta_h(n)$$

and we deduce from (23.10.3) that

$$\frac{1}{K+1} \leqslant \varliminf_{n\to\infty} \frac{\eta_0(n)}{n} \leqslant \varlimsup_{n\to\infty} \frac{\eta_0(n)}{n} \leqslant \frac{1}{K}. \tag{23.10.4}$$

If I is the interval (α, β) and $\beta - \alpha \geqslant \epsilon$, there are integers u, k such that

$$0 \leqslant (uj\vartheta) \leqslant \alpha \leqslant (\{u+1\}j\vartheta) \leqslant (\{u+k\}j\vartheta) \leqslant \beta < (\{u+k+1\}j\vartheta),$$

so that

$$\sum_{h=u+1}^{u+k-1} \eta_h(n) \leqslant n_I \leqslant \sum_{h=u}^{u+k} \eta_h(n).$$

Hence, by (23.10.3), we have

$$k - 1 \leqslant \varliminf_{n\to\infty} \frac{n_I}{\eta_0(n)} \leqslant \varlimsup_{n\to\infty} \frac{n_I}{\eta_0(n)} \leqslant k+1$$

and so, using (23.10.4),

$$\frac{k-1}{K+1} \leqslant \varliminf \frac{n_I}{n} \leqslant \varlimsup \frac{n_I}{n} \leqslant \frac{k+1}{K}.$$

But

$$K\delta \leqslant 1 \leqslant (K+1)\delta, \qquad (k-1)\delta < I < (k+1)\delta.$$

Hence

$$\frac{I - 2\delta}{1+\delta} \leqslant \varliminf \frac{n_I}{n} \leqslant \varlimsup \frac{n_I}{n} \leqslant \frac{I + 2\delta}{1 - \delta}.$$

Since we can choose ϵ (and so δ) as small as we please, (23.10.1) follows.

The definition of uniform distribution may be extended at once to space of k dimensions, and Kronecker's general theorem may be sharpened in the same way. But the proof is more complicated.

It is natural to inquire what happens in the exceptional cases when the ϑ are connected by one or more linear relations. Suppose, to fix our ideas, that $k = 3$. If there is *one* relation, the points P_n are limited to certain planes, as they were limited to certain lines in § 23.4; if there are *two,* they are limited to lines. Analogy suggests that the distribution on these planes or lines should be dense, and indeed uniform; and it can be proved that this is so, and that the corresponding theorems in space of k dimensions are also true.

NOTES

§ 23.1. Kronecker first stated and proved his theorem in the *Berliner Sitzungs berichte,* 1884 [*Werke,* iii (i), 47–110]. For a fuller account and a bibliography of later work inspired by the theorem, see Cassels, *Diophantine approximation.* The one-dimensional theorem seems to be due to Tchebychef: see Koksma, 76.

§ 23.2. For proof (iii) see Hardy and Littlewood, *Acta Math.* 37 (1914), 155–91, especially 161–2.

§ 23.3. König and Szőucs, *Rendiconti del circolo matematico di Palermo,* 36 (1913), 79–90.

§ 23.7. Lettenmeyer, *Proc. London Math. Soc.* (2), 21 (1923), 306–14.

§ 23.8. Estermann, *Journal London Math. Soc.* 8 (1933), 18–20.

§ 23.9. H. Bohr, *Journal London Math. Soc.* 9 (1934), 5–6; for a variation see *Proc. London Math. Soc.* (2) 21 (1923), 315–16. There is another simple proof by Bohr and Jessen in *Journal London Math. Soc.* 7 (1932), 274–5.

§ 23.10. Theorem 445 seems to have been found independently, at about the same time, by Bohl, Sierpiński, and Weyl. See Koksma, 92. The particular form of the proof given was suggested by Dr. Miclavc (*Proc. American Math. Soc.* 39 (1973), 279–80).

The best proof of the theorem is no doubt that given by Weyl in a very important paper in *Math. Annalen,* 77 (1916), 313–52. Weyl proves that a necessary and sufficient condition for the uniform distribution of the numbers

$$(f(1)),\quad (f(2)),\quad (f(3)),\quad \ldots$$

in (0, 1) is that

$$\sum_{\nu=1}^{n} e\{hf(\nu)\} = o(n)$$

for every integral h. This principle has many important applications, particularly to the problems mentioned at the end of the chapter.

For a detailed account of the subject of uniform distribution, see Kuipers and Niederreiter.

XXIV

GEOMETRY OF NUMBERS

24.1. Introduction and restatement of the fundamental theorem. This chapter is an introduction to the 'geometry of numbers', the subject created by Minkowski on the basis of his fundamental Theorem 37 and its generalization in space of n dimensions.

We shall need the n-dimensional generalizations of the notions which we used in §§ 3.9–11; but these, as we said in § 3.11, are straightforward. We define a lattice, and equivalence of lattices, as in § 3.5, parallelograms being replaced by n-dimensional parallelepipeds; and a convex region as in the first definition of § 3.9.† Minkowski's theorem is then

THEOREM 446. *Any convex region in n-dimensional space, symmetrical about the origin and of volume greater than* 2^n, *contains a point with integral coordinates, not all zero.*

Any of the proofs of Theorem 37 in Ch. III may be adapted to prove Theorem 446: we take, for example, Mordell's. The planes

$$x_r = 2p_r/t \quad (r = 1, 2, \ldots, n)$$

divide space into cubes of volume $(2/t)^n$. If $N(t)$ is the number of corners of these cubes in the region R under consideration, and V the volume of R, then

$$(2/t)^n N(t) \to V$$

when $t \to \infty$; and $N(t) > t^n$ if $V > 2^n$ and t is sufficiently large. The proof may then be completed as before.

If $\xi_1, \xi_2, \ldots, \xi_n$ are linear forms in $x_1, x_2, \ldots, x_n$, say

$$\xi_r = \alpha_{r,1}x_1 + \alpha_{r,2}x_2 + \cdots + \alpha_{r,n}x_n \quad (r = 1, 2, \ldots, n), \tag{24.1.1}$$

with real coefficients and determinant

$$\Delta = \begin{vmatrix} \alpha_{1,1} & \alpha_{1,2} & . & . & . & \alpha_{1,n} \\ . & . & . & . & . & . \\ \alpha_{n,1} & \alpha_{n,2} & . & . & . & \alpha_{n,n} \end{vmatrix} \neq 0, \tag{24.1.2}$$

† The second definition can also be adapted to n dimensions, the line l becoming an $(n-1)$-dimensional 'plane' (whereas the line of the first definition remains a 'line'). We shall use three-dimensional language: thus we shall call the region $|x_1| < 1, |x_2| < 1, \ldots, |x_n| < 1$ the 'unit cube'.

then the points in ξ-space corresponding to integral $x_1, x_2, \ldots, x_n$ form a lattice Λ[†]: we call Δ the determinant of the lattice. A region R of x-space is transformed into a region P of ξ-space, and a convex R into a convex P.[‡] Also

$$\int\int \ldots \int d\xi_1 d\xi_2 \ldots d\xi_n = |\Delta| \int\int \ldots \int dx_1 dx_2 \ldots dx_n,$$

so that the volume of P is $|\Delta|$ times that of R. We can therefore restate Theorem 446 in the form

THEOREM 447. *If Λ is a lattice of determinant Δ, and* P *is a convex region symmetrical about O and of volume greater than $2^n|\Delta|$, then* P *contains a point of Λ other than O.*

We assume throughout the chapter that $\Delta \neq 0$.

24.2. Simple applications. The theorems which follow will all have the same character. We shall be given a system of forms ξ_r, usually linear and homogeneous, but sometimes (as in Theorem 455) non-homogeneous, and we shall prove that there are integral values of the x_r (usually not all 0) for which the ξ_r satisfy certain inequalities. We can obtain such theorems at once by applying Theorem 447 to various simple regions P.

(1) Suppose first that P is the region defined by

$$|\xi_1| < \lambda_1, \; |\xi_2| < \lambda_2, \ldots, \; |\xi_n| < \lambda_n.$$

This is convex and symmetrical about O, and its volume is $2^n \lambda_1 \lambda_2 \ldots \lambda_n$. If $\lambda_1 \lambda_2 \ldots \lambda_n > |\Delta|$, P contains a lattice point other than O; if $\lambda_1 \lambda_2 \ldots \lambda_n \geqslant |\Delta|$, there is a lattice point, other than O, inside P or on its boundary.[||] We thus obtain

THEOREM 448. *If $\xi_1, \xi_2, \ldots, \xi_n$ are homogeneous linear forms in $x_1, x_2, \ldots, x_n$, with real coefficients and determinant Δ, and $\lambda_1, \lambda_2, \ldots, \lambda_n$*

[†] In § 3.5 we used L for a lattice of lines, Λ for the corresponding point-lattice. It is more convenient now to reserve Greek letters for configurations in 'ξ-space'.

[‡] The invariance of convexity depends on two properties of linear transformations viz. (1) that lines and planes are transformed into lines and planes, and (2) that the order of points on a line is unaltered.

[||] We pass here, by an appeal to continuity, from a result concerning an open region to one concerning the corresponding closed region. We might, of course, make a similar change in the general theorems 446 and 447: thus any closed convex region, symmetrical about O, and of volume not less than 2^n, has a lattice point, other than O, inside it or on its boundary. We shall not again refer explicitly to such trivial appeals to continuity.

are positive, and

$$\lambda_1\lambda_2\ldots\lambda_n \geqslant |\Delta|, \tag{24.2.1}$$

then there are integers $x_1, x_2, \ldots, x_n$, *not all* 0, *for which*

$$|\xi_1| \leqslant \lambda_1,\ |\xi_2| \leqslant \lambda_2, \ldots,\ |\xi_n| \leqslant \lambda_n. \tag{24.2.2}$$

In particular we can make $|\xi_1| \leqslant \sqrt[n]{|\Delta|}$ *for each r.*

(2) Secondly, suppose that P is defined by

$$|\xi_1| + |\xi_2| + \cdots + |\xi_n| < \lambda. \tag{24.2.3}$$

If $n = 2$, P is a square; if $n = 3$, an octahedron. In the general case it consists of 2^n congruent parts, one in each 'octant'. It is obviously symmetrical about O, and it is convex because

$$|\mu\xi + \mu'\xi'| \leqslant \mu|\xi| + \mu'|\xi'|$$

for positive μ and μ'. The volume in the positive octant $\xi_r > 0$ is

$$\lambda^n \int_0^1 d\xi_1 \int_0^{1-\xi_1} d\xi_2 \cdots \int_0^{1-\xi_1-\cdots-\xi_{n-1}} d\xi_n = \frac{\lambda_n}{n!}.$$

If $\lambda^n > n!|\Delta|$ then the volume of P exceeds $2^n|\Delta|$, and there is a lattice point, besides O, in P. Hence we obtain

THEOREM 449. *There are integers* $x_1, x_2, \ldots, x_n$, *not all* 0, *for which*

$$|\xi_1| + |\xi_2| + \cdots + |\xi_n| \leqslant (n!|\Delta|)^{1/n}. \tag{24.2.4}$$

Since, by the theorem of the arithmetic and geometric means,

$$n|\xi_1\xi_2\ldots\xi_n|^{1/n} \leqslant |\xi_1| + |\xi_2| + \cdots + |\xi_n|,$$

we have also

THEOREM 450. *There are integers* $x_1, x_2, \ldots, x_n$, *not all* 0, *for which*

$$|\xi_1\xi_2\ldots\xi_n| \leqslant n^{-n}n!|\Delta|. \tag{24.2.5}$$

(3) As a third application, we define P by

$$\xi_1^2 + \xi_2^2 + \cdots + \xi_n^2 < \lambda^2:$$

this region is convex because

$$(\mu\xi + \mu'\xi')^2 \leqslant (\mu + \mu')(\mu\xi^2 + \mu'\xi'^2)$$

for positive μ and μ'. The volume of P is $\lambda^n J_n$, where†

$$J_n = \underset{\xi_1^2+\xi_2^2+\cdots+\xi_n^2\leqslant 1}{\int\int\cdots\int} d\xi_1 d\xi_2 \ldots d\xi_n = \frac{\pi^{\frac{1}{2}n}}{\Gamma\left(\frac{1}{2}n+1\right)}.$$

Hence we obtain

THEOREM 451. *There are integers $x_1, x_2, \ldots, x_n$, not all 0, for which*

$$\xi_1^2 + \xi_2^2 + \cdots + \xi_n^2 \leqslant 4\left(\frac{|\Delta|}{J_n}\right)^{2/n} \tag{24.2.6}$$

Theorem 451 may be expressed in a different way. *A quadratic form Q* in $x_1, x_2, \ldots, x_n$ is a function

$$Q(x_1, x_2, \ldots, x_n) = \sum_{r=1}^{n}\sum_{s=1}^{n} a_{r,s}x_r x_s$$

with $a_{s,r} = a_{r,s}$. The *determinant D* of Q is the determinant of its coefficients. If $Q > 0$ for all $x_1, x_2, \ldots, x_n$, not all 0, then Q is said to be *positive definite*. It is familiar‡ that Q can then be expressed in the form

$$Q = \xi_1^2 + \xi_2^2 + \cdots + \xi_n^2,$$

where $\xi_1, \xi_2, \ldots, \xi_n$ are linear forms with real coefficients and determinant $\surd D$. Hence Theorem 451 may be restated as

THEOREM 452. *If Q is a positive definite quadratic form in $x_1, x_2, \ldots, x_n$, with determinant D, then there are integral values of $x_1, x_2, \ldots, x_n$, not all 0, for which*

$$Q \leqslant 4D^{1/n}J_n^{-2/n}. \tag{24.2.7}$$

† See, for example, Whittaker and Watson, *Modern analysis*, ed. 3 (1920), 258. For $n = 2$ and $n = 3$ we get the values $\pi\lambda^2$ and $\frac{4}{3}\pi\lambda^3$ for the volumes of a circle or a sphere.

‡ See, for example, Bôcher, *Introduction to higher algebra*, ch. 10, or Ferrar, *Algebra*, ch. 11.

24.3. Arithmetical proof of Theorem 448. There are various proofs of Theorem 448 which do not depend on Theorem 446, and the great importance of the theorem makes it desirable to give one here. We confine ourselves for simplicity to the case $n = 2$. Thus we are given linear forms

$$\xi = \alpha x + \beta y, \qquad \eta = \gamma x + \delta y, \tag{24.3.1}$$

with real coefficients and determinant $\Delta = \alpha\delta - \beta\gamma \neq 0$, and positive numbers λ, μ for which $\lambda\mu \geqslant |\Delta|$; and we have to prove that

$$|\xi| \leqslant \lambda, \qquad |\eta| \leqslant \mu, \tag{24.3.2}$$

for some integral x and y not both 0. We may plainly suppose $\Delta > 0$.

We prove the theorem in three stages: (1) when the coefficients are integral and each of the pairs α, β and γ, δ is coprime; (2) when the coefficients are rational; and (3) in the general case.

(1) We suppose first that α, β, γ, and δ are integers and that

$$(\alpha, \beta) = (\gamma, \delta) = 1.$$

Since $(\alpha, \beta) = 1$, there are integers p and q for which $\alpha q - \beta p = 1$. The linear transformation

$$\alpha x + \beta y = X, \quad px + qy = Y$$

establishes a (1, 1) correlation between integral pairs x, y and X, Y; and

$$\xi = X, \quad \eta = rX + \Delta Y,$$

where $r = \gamma q - \delta p$ is an integer. It is sufficient to prove that $|\xi| \leqslant \lambda$ and $|\eta| \leqslant \mu$ for some integral X and Y not both 0.

If $\lambda \leqslant 1$ then $\mu \geqslant \Delta$, and $X = 0, Y = 1$ gives $\xi = 0$, $|\eta| = \Delta \leqslant \mu$.

If $\lambda > 1$, we take

$$n = [\lambda], \quad \xi = -\tfrac{r}{\Delta}, \quad h = Y, \quad k = X,^{\dagger}$$

in Theorem 36. Then

$$0 < x \leqslant [\lambda] \leqslant \lambda$$

† The ξ here is naturally not the ξ of this section.

and

$$|rX + \Delta Y| = \Delta X\left|-\frac{r}{\Delta} - \frac{Y}{X}\right| \leqslant \frac{\Delta}{n+1} = \frac{\Delta}{[\lambda]+1} < \frac{\Delta}{\lambda} \leqslant \mu,$$

so that $X = k$ and $Y = h$ satisfy our requirements.

(2) We suppose next that α, β, γ, and δ are any rational numbers. Then we can choose ρ and σ so that

$$\xi' = \rho\xi = \alpha' x + \beta' y, \quad \eta' = \sigma\eta = \gamma' x + \delta' y,$$

where α', β', γ', and δ' are integers, $(\alpha', \beta') = 1, (\gamma', \delta') = 1$, and $\Delta' = \alpha'\delta' - \beta'\gamma' = \rho\sigma\Delta$. Also $\rho\lambda \,.\, \sigma\mu \geqslant \Delta'$, and therefore, after (1), there are integers x, y, not both 0, for which

$$|\xi'| \leqslant \rho\lambda, \quad |\eta'| \leqslant \sigma\mu.$$

These inequalities are equivalent to (24.3.2), so that the theorem is proved in case (2).

(3) Finally, we suppose α, β, γ, and δ unrestricted. If we put $\alpha = \alpha'\surd\Delta, \ldots, \xi = \xi'\surd\Delta, \ldots$, then $\Delta' = \alpha'\delta' - \beta'\gamma' = 1$. If the theorem has been proved when $\Delta = 1$, and $\lambda'\mu' \geqslant 1$, then there are integral x, y, not both 0, for which

$$|\xi'| \leqslant \lambda', \quad |\eta'| \leqslant \mu';$$

and these inequalities are equivalent to (24.3.2), with $\lambda = \lambda'\surd\Delta, \mu = \mu'\surd\Delta$, $\lambda\mu \geqslant \Delta$. We may therefore suppose without loss of generality that $\Delta = 1$.†

We can choose a sequence of rational sets $\alpha_n, \beta_n, \gamma_n, \delta_n$ such that

$$\alpha_n\delta_n - \beta_n\gamma_n = 1$$

and $\alpha_n \to \alpha, \beta_n \to \beta, \ldots$, when $n \to \infty$. It follows from (2) that there are integers x_n and y_n, not both 0, for which

$$|\alpha_n x_n + \beta_n y_n| \leqslant \lambda, \quad |\gamma_n x_n + \delta_n y_n| \leqslant \mu. \tag{24.3.3}$$

Also

$$|x_n| = |\delta_n(\alpha_n x_n + \beta_n y_n) - \beta_n(\gamma_n x_n + \delta_n y_n)| \leqslant \lambda|\delta_n| + \mu|\beta_n|,$$

† A similar appeal to homogeneity would enable us to reduce the proof of any of the theorems of this chapter to its proof in the case in which Δ has any assigned value.

so that x_n is bounded; and similarly y_n is bounded. It follows, since x_n and y_n are integral, that some pair of integers x, y must occur infinitely often among the pairs x_n, y_n. Taking $x_n = x, y_n = y$ in (24.3.3), and making $n \to \infty$, through the appropriate values, we obtain (24.3.2).

It is important to observe that this method of proof, by reduction to the case of rational or integral coefficients, cannot be used for such a theorem as Theorem 450. This (when $n = 2$) asserts that $|\xi\eta| \leqslant \frac{1}{2}|\Delta|$ for appropriate x, y. If we try to use the argument of (3) above, it fails because x_n and y_n are not necessarily bounded. The failure is natural, since the theorem is trivial when the coefficients are rational: we can obviously choose x and y so that $\xi = 0$, $|\xi\eta| = 0 < \frac{1}{2}|\Delta|$.

24.4. Best possible inequalities. It is easy to see that Theorem 448 is the best possible theorem of its kind, in the sense that it becomes false if (24.2.1) is replaced by

$$\lambda_1\lambda_2 \ldots \lambda_n \geqslant k|\Delta| \tag{24.4.1}$$

with any $k < 1$. Thus if $\xi_r = x_r$, for each r, so that $\Delta = 1$, and $\lambda_r = \sqrt[n]{k}$, then (24.4.1) is satisfied; but $|\xi_r| \leqslant \lambda_r < 1$ implies $x_r = 0$, and there is no solution of (24.2.2) except $x_1 = x_2 = \ldots = 0$.

It is natural to ask whether Theorems 449–51 are similarly 'best possible'. Except in one special case, the answer is negative; the numerical constants on the right of (24.2.4), (24.2.5), and (24.2.6) can be replaced by smaller numbers.

The special case referred to is the case $n = 2$ of Theorem 449. This asserts that we can make

$$|\xi| + |\eta| \leqslant \surd(2|\Delta|), \tag{24.4.2}$$

and it is easy to see that this is the best possible result. If $\xi = x+y, \eta = x-y$, then $\Delta = -2$, and (24.4.2) is $|\xi| + |\eta| \leqslant 2$. But

$$|\xi| + |\eta| = \max(|\xi + \eta|, |\xi - \eta|) = \max(|2x|, |2y|),$$

and this cannot be less than 2 unless $x = y = 0$.[†]

Theorem 450 is not a best possible theorem even when $n = 2$. It then asserts that

$$|\xi\eta| \leqslant \tfrac{1}{2}|\Delta|, \tag{24.4.3}$$

[†] Actually the case $n = 2$ of Theorem 449 is equivalent to the corresponding case of Theorem 448.

and we shall show in § 24.6 that the $\frac{1}{2}$ here may be replaced by the smaller constant $5^{-\frac{1}{2}}$. We shall also make a corresponding improvement in Theorem 451. This asserts (when $n = 2$) that

$$\xi^2 + \eta^2 \leqslant 4\pi^{-1}|\Delta|,$$

and we shall show that $4\pi^{-1} = 1{\cdot}27\ldots$ may be replaced by $\left(\frac{4}{3}\right)^{\frac{1}{2}} = 1{\cdot}15\ldots$.

We shall also show that $5^{-\frac{1}{2}}$ and $\left(\frac{4}{3}\right)^{\frac{1}{2}}$ are the best possible constants. When $n > 2$, the determination of the best possible constants is difficult.

24.5. The best possible inequality for $\xi^2 + \eta^2$. If

$$Q(x,y) = ax^2 + 2bxy + cy^2$$

is a quadratic form in x and y (with real, but not necessarily integral, coefficients);

$$x = px' + qy', \quad y = rx' + sy' \quad (ps - qr = \pm 1)$$

is a unimodular substitution in the sense of § 3.6; and

$$Q(x, y) = a'x'^2 + 2b'x'y' + c'y'^2 = Q'(x', y'),$$

then we say that Q is equivalent to Q', and write $Q \sim Q'$. It is easily verified that $a'c' - b'^2 = ac - b^2$, so that equivalent forms have the same determinant. It is plain that the assertions that $|Q| \leqslant k$ for appropriate integral x, y, and that $|Q'| \leqslant k$ for appropriate integral x', y', are equivalent to one another.

Now let x_0, y_0 be coprime integers such that $M = Q(x_0, y_0) \neq 0$. We can choose x_1, y_1 so that $x_0y_1 - x_1y_0 = 1$. The transformation

$$x = x_0x' + x_1y', \quad y = y_0x' + y_1y' \tag{24.5.1}$$

is unimodular and transforms $Q(x,y)$ into $Q'(x',y')$ with

$$a' = ax_0^2 + 2bx_0y_0 + cy_0^2 = Q(x_0, y_0) = M.$$

If we make the further unimodular transformation

$$x' = x'' + ny'', \quad y' = y'', \tag{24.5.2}$$

where n is an integer, $a' = M$ is unchanged and b' becomes

$$b'' = b' + na' = b' + nM.$$

Since $M \neq 0$, we can choose n so that $-|M| < 2b'' \leqslant |M|$. Thus we transform $Q(x,y)$ by unimodular substitutions into

$$Q''(x'',y'') = Mx''^2 + 2b''x''y'' + c''y''^2$$

with $-|M| < 2b'' \leqslant |M|$.†

We can now improve the results of Theorems 450 and 451, for $n = 2$. We take the latter theorem first.

THEOREM 453. *There are integers x, y, not both 0, for which*

$$\xi^2 + \eta^2 \leqslant \left(\tfrac{4}{3}\right)^{\frac{1}{2}} |\Delta|; \tag{24.5.3}$$

and this is true with inequality unless

$$\xi^2 + \eta^2 \sim \left(\tfrac{4}{3}\right)^{\frac{1}{2}} |\Delta|(x^2 + xy + y^2). \tag{24.5.4}$$

We have

$$\xi^2 + \eta^2 = ax^2 + 2bxy + cy^2 = Q(x,y), \tag{24.5.5}$$

where

$$\begin{cases} a = \alpha^2 + \gamma^2, \quad b = \alpha\beta + \gamma\delta, \quad c = \beta^2 + \delta^2, \\ \qquad ac - b^2 = (\alpha\delta - \beta\gamma)^2 = \Delta^2 > 0. \end{cases} \tag{24.5.6}$$

Then $Q > 0$ except when $x = y = 0$, and there are at most a finite number of integral pairs x, y for which Q is less than any given k. It follows that, among such integral pairs, not both 0, there is one, say (x_0, y_0), for which Q assumes a positive minimum value m. Clearly x_0 and y_0 are coprime and so, by what we have just said, Q is equivalent to a form Q'', with $a'' = m$ and $-m < 2b'' \leqslant m$. Thus (dropping the dashes) we may suppose that the form is

$$mx^2 + 2bxy + cy^2,$$

† A reader familiar with the elements of the theory of quadratic forms will recognize Gauss's method for transforming Q into a 'reduced' form.

where $-m < 2b \leqslant m$. Then $c \geqslant m$, since otherwise $x = 0, y = 1$ would give a value less than m; and

$$\Delta^2 = mc - b^2 \geqslant m^2 - \tfrac{1}{4}m^2 = \tfrac{3}{4}m^2, \tag{24.5.7}$$

so that $m \leqslant \left(\tfrac{4}{3}\right)^{\frac{1}{2}} |\Delta|$.

This proves (24.5.3). There can be equality throughout (24.5.7) only if $c = m$ and $b = \tfrac{1}{2}m$, in which case $Q \sim m(x^2 + xy + y^2)$. For this form the minimum is plainly $\left(\tfrac{4}{3}\right)^{\frac{1}{2}} |\Delta|$.

24.6. The best possible inequality for $|\xi\eta|$. Passing to the product $|\xi\eta|$, we prove

THEOREM 454. *There are integers x, y, not both* 0, *for which*

$$|\xi\eta| \leqslant 5^{-\frac{1}{2}} |\Delta|; \tag{24.6.1}$$

and this is true with inequality unless

$$\xi\eta \sim 5^{-\frac{1}{2}} |\Delta| (x^2 + xy - y^2). \tag{24.6.2}$$

The proof is a little less straightforward than that of Theorem 453 because we are concerned with an 'indefinite form'. We write

$$\xi\eta = ax^2 + 2bxy + cy^2 = Q(x, y), \tag{24.6.3}$$

where

$$\begin{cases} a = \alpha\gamma, \quad 2b = \alpha\delta + \beta\gamma, \quad c = \beta\delta, \\ \qquad 4(b^2 - ac) = \Delta^2 > 0. \end{cases} \tag{24.6.4}$$

We write m for the lower bound of $|Q(x, y)|$, for x and y not both zero; we may plainly suppose that $m > 0$ since there is nothing to prove if $m = 0$. There may now be no pair x, y such that $|Q(x, y)| = m$, but there must be pairs for which $|Q(x, y)|$ is as near to m as we please. Hence we can find a coprime pair x_0 and y_0 so that $m \leqslant |M| < 2m$, where $M = Q(x_0, y_0)$. Without loss of generality we may take $M > 0$. If we transform as in § 24.5, and drop the dashes, our new quadratic form is

$$Q(x, y) = Mx^2 + 2bxy + cy^2,$$

where

$$(24.6.5) \qquad m \leqslant M < 2m, \qquad -M < 2b \leqslant M$$

and

$$(24.6.6) \qquad 4(b^2 - Mc) = \Delta^2 > 0.$$

By the definition of m, $|Q(x,y)| \geqslant m$ for all integral pairs x, y other than $0, 0$. Hence if, for a particular pair, $Q(x,y) < m$, it follows that $Q(x,y) \leqslant -m$. Now, by (24.6.5) and (24.6.6),

$$Q(0,1) = c < \frac{b^2}{M} \leqslant \tfrac{1}{4}M < m.$$

Hence $c \leqslant -m$ and we write $C = -c \geqslant m > 0$. Again

$$Q\left(1, \frac{-b}{|b|}\right) = M - |2b| - C \leqslant M - C \leqslant M - m < m$$

and so $M - |2b| - C \leqslant -m$, that is

$$(24.6.7) \qquad |2b| \geqslant M + m - C.$$

If $M + m - C < 0$, we have $C > M + m \geqslant 2m$ and

$$\Delta^2 = 4(b^2 + MC) \geqslant 4MC \geqslant 8m^2 > 5m^2.$$

If $M + m - C \geqslant 0$, we have from (24.6.7)

$$\begin{aligned} \Delta^2 = 4b^2 + 4MC &\geqslant (M + m - C)^2 + 4MC \\ &= (M - m + C)^2 + 4Mm \geqslant 5m^2. \end{aligned}$$

Equality can occur only if $M - m + C = m$ and $M = m$, so that $M = C = m$ and $|b| = m$. This corresponds to one or other of the two (equivalent) forms $m(x^2 + xy - y^2)$ and $m(x^2 - xy - y^2)$. For these, $|Q(1,0)| = m = 5^{-\frac{1}{2}}\Delta$. For all other forms, $5m^2 < \Delta^2$ and so we may choose x_0, y_0 so that

$$5m^2 \leqslant 5M^2 < \Delta^2.$$

This is Theorem 454.

24.7. A theorem concerning non-homogeneous forms. We prove next an important theorem of Minkowski concerning non-homogeneous forms

$$\xi - \rho = \alpha x + \beta y - \rho, \quad \eta - \sigma = \gamma x + \delta y - \sigma. \tag{24.7.1}$$

THEOREM 455. *If ξ and η are homogeneous linear forms in x, y, with determinant $\Delta \neq 0$, and ρ and σ are real, then there are integral x, y for which*

$$|(\xi - \rho)(\eta - \sigma)| \leqslant \tfrac{1}{4}|\Delta|; \tag{24.7.2}$$

and this is true with inequality unless

(24.7.3)
$$\xi = \theta u, \quad \eta = \phi v, \quad \theta\phi = \Delta, \quad \rho = \theta\left(f + \tfrac{1}{2}\right), \quad \sigma = \phi\left(g + \tfrac{1}{2}\right),$$

where u and v are forms with integral coefficients (*and determinant* 1), *and f and g are integers.*

It will be observed that this theorem differs from all which precede in that we do not exclude the values $x = y = 0$. It would be false if we did not allow this possibility, for example if ξ and η are the special forms of Theorem 454 and $\rho = \sigma = 0$.

It will be convenient to restate the theorem in a different form. The points in the plane ξ, η corresponding to integral x, y form a lattice Λ of determinant Δ. Two points P, Q are equivalent with respect to Λ if the vector PQ is equal to the vector from the origin to a point of Λ;† and $(\xi - \rho, \eta - \sigma)$, with integral x, y, is equivalent to $(-\rho, -\sigma)$. Hence the theorem may be restated as

THEOREM 456. *If Λ is a lattice of determinant Δ in the plane of (ξ, η), and Q is any given point of the plane, then there is a point equivalent to Q for which*

$$|\xi\eta| \leqslant \tfrac{1}{4}|\Delta|, \tag{24.7.4}$$

with inequality except in the special case (24.7.3).

† See p. 42. It is the same thing to say that the corresponding points in the (x, y) plane are equivalent with respect to the fundamental lattice.

In what follows we shall be concerned with three sets of variables, (x, y), (ξ, η), and (ξ', η') We call the planes of the last two sets of variables π and π'.

We may suppose $\Delta = 1$.[†] By Theorem 450 (and *a fortiori* by Theorem 454), there is a point P_0 of Λ, other than the origin, and corresponding to x_0, y_0, for which

$$|\xi_0\eta_0| \leqslant \tfrac{1}{2}. \tag{24.7.5}$$

We may suppose x_0 and y_0 coprime (so that P_0 is 'visible' in the sense of § 3.6). Since ξ_0 and η_0 satisfy (24.7.5), and are not both 0, there is a real positive λ for which

$$(\lambda\xi_0)^2 + \left(\lambda^{-1}\eta_0\right)^2 = 1. \tag{24.7.6}$$

We put

$$\xi' = \lambda\xi, \quad \eta' = \lambda^{-1}\eta. \tag{24.7.7}$$

Then the lattice Λ in π corresponds to a lattice Λ' in π', also of determinant 1. If O' and P'_0 correspond to O and P_0, then P'_0, like P_0, is visible; and $O'P'_0 = 1$, by (24.7.6). Thus the points of Λ' on $O'P'$, are spaced out at unit distances, and, since the area of the basic parallelogram of Λ' is 1, the other points of Λ' lie on lines parallel to $O'P'_0$ which are at unit distances from one another.

We denote by S' the square whose centre is O' and one of whose sides bisects $O'P'_0$ perpendicularly.[‡] Each side of S' is 1; S' lies in the circle

$$\xi'^2 + \eta'^2 = 2\left(\tfrac{1}{2}\right)^2 = \tfrac{1}{2},$$

and

$$\left|\xi'\eta'\right| \leqslant \tfrac{1}{2}\left(\xi'^2 + \eta'^2\right) \leqslant \tfrac{1}{4} \tag{24.7.8}$$

at all points of S'.

If A' and B' are two points inside S', then each component of the vector $A'B'$ (measured parallel to the sides of the square) is less than 1, so that A' and B' cannot be equivalent with respect to Λ'. It follows from Theorem

[†] See the footnote to p. 528.

[‡] The reader should draw a figure.

42 that there is a point of S' equivalent to Q' (the point of π' corresponding to Q). The corresponding point of π is equivalent to Q, and satisfies

$$|\xi\eta| = |\xi'\eta'| \leqslant \tfrac{1}{4}. \tag{24.7.9}$$

This proves the main clause of Theorem 456 (or 455).

If there is equality in (24.7.9), there must be equality in (24.7.8), so that $|\xi'| = |\eta'| = \frac{1}{2}$. This is only possible if S' has its sides parallel to the coordinate axes and the point of S' in question is at a corner. In this case P'_0 must be one of the four points $(\pm 1, 0)$, $(0, \pm 1)$: let us suppose, for example, that it is $(1, 0)$.

The lattice Λ' can be based on $O'P'_0$ and $O'P'_1$, where P'_1 is on $\eta' = 1$. We may suppose, selecting P'_1 appropriately, that it is $(c, 1)$, where $0 \leqslant c < 1$. If the point of S' equivalent to Q' is, say, $\left(\frac{1}{2}, \frac{1}{2}\right)$, then $\left(\frac{1}{2} - c, \frac{1}{2} - 1\right)$, i.e. $\left(\frac{1}{2} - c, -\frac{1}{2}\right)$, is another point equivalent to Q' and this can only be at a corner of S', as it must be, if $c = 0$. Hence P'_1 is (0,1), Λ' is the fundamental lattice in π', and Q', being equivalent to $\left(\frac{1}{2}, \frac{1}{2}\right)$, has coordinates

$$\xi' = f + \tfrac{1}{2}, \quad \eta' = g + \tfrac{1}{2},$$

where f and g are integers. We are thus led to the exceptional case (24.7.3), and it is plain that in this case the sign of equality is necessary.

24.8. Arithmetical proof of Theorem 455. We also give an arithmetical proof of the main clause of Theorem 455. We transform it as in Theorem 456, and we have to show that, given μ and ν, we can satisfy (24.7.4) with an x and a y congruent to μ and ν to modulus 1.

We again suppose $\Delta = 1$. As in § 24.7, there are integers x_0, y_0, which we may suppose coprime, for which

$$|(\alpha x_0 + \beta y_0)(\gamma x_0 + \delta y_0)| \leqslant \tfrac{1}{2}.$$

We choose x_1 and y_1 so that $x_0 y_1 - x_1 y_0 = 1$. The transformation

$$x = x_0 x' + x_1 y', \quad y = y_0 x' + y_1 y'$$

changes ξ and η into forms $\xi' = \alpha' x' + \beta' y', \eta' = \gamma' x' + \delta' y'$ for which

$$|\alpha'\gamma'| = |(\alpha x_0 + \beta y_0)(\gamma x_0 + \delta y_0)| \leqslant \tfrac{1}{2}.$$

Hence, reverting to our original notation, we may suppose without loss of generality that

(24.8.1) $$|\alpha\gamma| \leqslant \tfrac{1}{2}.$$

It follows from (24.8.1) that there is a real λ for which

$$\lambda^2\alpha^2 + \lambda^{-2}\gamma^2 = 1;$$

and

$$\begin{aligned} 2\,|(\alpha x + \beta y)(\gamma x + \delta y)| &\leqslant \lambda^2(\alpha x + \beta y)^2 + \lambda^{-2}(\gamma x + \delta y)^2 \\ &= x^2 + 2bxy + cy^2 = (x + by)^2 + py^2, \end{aligned}$$

for some b, c, p. The determinant of this quadratic form is, on the one hand, the square of that of $\lambda(\alpha x + \beta y)$ and $\lambda^{-1}(\gamma x + \delta y)$,[†] that is to say 1, and on the other the square of that of $x + by$ and $p^{\frac{1}{2}}y$, that is to say p; and therefore $p = 1$. Thus

$$2\,|(\alpha x + \beta y)(\gamma x + \delta y)| \leqslant (x + by)^2 + y^2.$$

We can choose $y \equiv \nu \pmod 1$ so that $|y| \leqslant \frac{1}{2}$, and then $x \equiv \mu \pmod 1$ so that $|x + by| \leqslant \frac{1}{2}$; and then

$$|\xi\eta| \leqslant \tfrac{1}{2}\left\{\left(\tfrac{1}{2}\right)^2 + \left(\tfrac{1}{2}\right)^2\right\} = \tfrac{1}{4}.$$

We leave it to the reader to discriminate the cases of equality in this alternative proof.

24.9. Tchebotaref's theorem. It has been conjectured that Theorem 455 could be extended to n dimensions, with 2^{-n} in place of $\frac{1}{4}$; but this has been proved only for $n = 3$ and $n = 4$. There is, however, a theorem of Tchebotaref which goes some way in this direction.

THEOREM 457. *If $\xi_1, \xi_2, \ldots, \xi_n$ are homogeneous linear forms in $x_1, x_2, \ldots, x_n$, with real coefficients and determinant Δ; $\rho_1, \rho_2, \ldots, \rho_n$ are real; and m is the lower bound of*

$$|(\xi_1 - \rho_1)(\xi_2 - \rho_2)\ldots(\xi_n - \rho_n)|,$$

† See (24.5.5) and (24.5.6).

then

(24.9.1) $$m \leqslant 2^{-\frac{1}{2}n}\,|\Delta|\,.$$

We may suppose $\Delta = 1$ and $m > 0$. Then, given any positive ϵ, there are integers $x_1^*, x_2^*, \ldots, x_n^*$ for which

(24.9.2)
$$\prod |\xi_i^* - \rho_i| = |(\xi_1^* - \rho_1)(\xi_2^* - \rho_2)\ldots(\xi_n^* - \rho_n)| = \frac{m}{1-\theta}, \quad 0 \leqslant \theta < \epsilon.$$

We put

$$\xi_i' = \frac{\xi_i - \xi_1^*}{\xi_i^* - \rho_i} \quad (i = 1, 2, \ldots, n).$$

Then $\xi_1', \ldots, \xi_n'$ are linear forms in $x_1 - x_1^*, \ldots, x_n - x_n^*$, with a determinant D whose absolute value is

$$|D| = \left(\prod |\xi_i^* - \rho_i|\right)^{-1} = \frac{1-\theta}{m};$$

and the points in ξ'-space corresponding to integral x form a lattice Λ' whose determinant is of absolute value $(1-\theta)/m$. Since

$$\prod |\xi_i - \rho_i| \geqslant m,$$

every point of Λ' satisfies

$$\prod |\xi_i' + 1| = \prod \left|\frac{\xi_i - \rho_i}{\xi_i^* - \rho_i}\right| \geqslant 1 - \theta.$$

The same inequality is satisfied by the point symmetrical about the origin, so that $\prod |\xi_i' - 1| \geqslant 1 - \theta$ and

(24.9.3) $$\prod |\xi_i'^2 - 1| = |(\xi_1'^2 - 1)(\xi_2'^2 - 1)\ldots(\xi_n'^2 - 1)| \geqslant (1-\theta)^2.$$

We now prove that *when ϵ and θ are small, there is no point of Λ', other than the origin, in the cube C' defined by*

(24.9.4) $$|\xi_i'| < \surd\{1 + (1-\theta)^2\}.$$

If there is such a point, it satisfies

$$-1 \leqslant \xi_i'^2 - 1 < (1-\theta)^2 \leqslant 1 \quad (i = 1, 2, \ldots, n). \tag{24.9.5}$$

If

$$\xi_i'^2 - 1 > -(1-\theta)^2 \tag{24.9.6}$$

for some i, then $|\xi_i'^2 - 1| < (1-\theta)^2$ for that i, and $|\xi_i'^2 - 1| \leqslant 1$ for every i, so that

$$\prod |\xi_i'^2 - 1| < (1-\theta)^2,$$

in contradiction to (24.9.3). Hence (24.9.6) is impossible, and therefore

$$-1 \leqslant \xi_i'^2 - 1 \leqslant -(1-\theta)^2 \quad (i = 1, 2, \ldots, n);$$

and hence

$$|\xi_i'| \leqslant \surd\{1 - (1-\theta)^2\} \leqslant \surd(2\theta) \quad (i = 1, 2, \ldots, n). \tag{24.9.7}$$

Thus every point of Λ' in C' is very near to the origin when ϵ and θ are small.

But this leads at once to a contradiction. For if $(\xi_1', \ldots, \xi_n')$ is a point of Λ', then so is $(N\xi_1', \ldots, N\xi_n')$ for every integral N. If θ is small, every coordinate of a lattice point in C' satisfies (24.9.7), and at least one of them is not 0, then plainly we can choose N so that $(N\xi_1', \ldots, N\xi_n')$, while still in C', is at a distance at least $\frac{1}{2}$ from the origin, and therefore cannot satisfy (24.9.7). The contradiction shows that, as we stated, there is no point of Λ', except the origin, in C'.

It is now easy to complete the proof of Theorem 457. Since there is no point of Λ', except the origin, in C', it follows from Theorem 447 that the volume of C' does not exceed

$$2^n\,|D| = 2^n(1-\theta)/m;$$

and therefore that

$$2^n m\left\{1 + (1-\theta)^2\right\}^{\frac{1}{2}n} \leqslant 2^n(1-\theta).$$

Dividing by 2^n, and making $\theta \to 0$, we obtain

$$m \leqslant 2^{-\frac{1}{2}n},$$

the result of the theorem.

24.10. A converse of Minkowski's Theorem 446. There is a partial converse of Theorem 446, which we shall prove for the case $n = 2$. The result is not confined to convex regions and we therefore first redefine the area of a bounded region P, since the definition of §3.9 may no longer be applicable.

For every $\rho > 0$, we denote by $\Lambda(\rho)$ the lattice of points $(\rho x, \rho y)$, where x, y take all integral values, and write $g(\rho)$ for the number of points of $\Lambda(\rho)$ (apart from the origin O) which belong to the bounded region P. We call

$$V = \lim_{\rho \to 0} \rho^2 g(\rho) \tag{24.10.1}$$

the area of P, if the limit exists. This definition embodies the only property of area which we require in what follows. It is clearly equivalent to any natural definition of area for elementary regions such as polygons, ellipses, etc.

We prove first

THEOREM 458. *If* P *is a bounded plane region with an area V which is less than* 1, *there is a lattice of determinant* 1 *which has no point* (*except perhaps O*) *belonging to* P.

Since P is bounded, there is a number N such that

$$-N \leqslant \xi \leqslant N, \quad -N \leqslant \eta \leqslant N \tag{24.10.2}$$

for every point (ξ, η) of P. Let p be any prime such that

$$p > N^2. \tag{24.10.3}$$

Let u be any integer and Λ_u the lattice of points (ξ, η) where

$$\xi = \frac{X}{\sqrt{p}}, \qquad \eta = \frac{uX + pY}{\sqrt{p}}$$

and X, Y take all integral values. The determinant of Λ_u is 1. If Theorem 458 is false, there is a point T_u belonging to both Λ_u and P and not coinciding with O. Let the coordinates of T_u be

$$\xi_u = \frac{X_u}{\sqrt{p}}, \qquad \eta_u = \frac{uX_u + pY_u}{\sqrt{p}}.$$

If $X_u = 0$, we have

$$\sqrt{p}\,|Y_u| = |\eta_u| \leqslant N < \sqrt{p}$$

by (24.10.2) and (24.10.3). It follows that $Y_u = 0$ and T_u is O, contrary to our hypothesis. Hence $X_u \neq 0$ and

$$0 < |X_u| = \sqrt{p}\,|\xi_u| \leqslant N\sqrt{p} < p.$$

Thus

(24.10.4) $$X_u \not\equiv 0 \pmod{p}.$$

If T_u and T_v coincide, we have

$$X_u = X_v, \quad uX_u + pY_u = vX_v + pY_v$$

and so

$$X_u(u - v) \equiv 0, \quad u \equiv v \pmod{p}$$

by (24.10.4). Hence the p points

(24.10.5) $$T_0, T_1, T_2, \ldots, T_{p-1}$$

are all different. Since they all belong to P and to $\Lambda\left(p^{-\frac{1}{2}}\right)$, it follows that

$$g\left(p^{-\frac{1}{2}}\right) \geqslant p.$$

But this is false for large enough p, since

$$p^{-1}g\left(p^{-\frac{1}{2}}\right) \to V < 1$$

by (24.10.1). Hence Theorem 458 is true.

For our next result we require the idea of *visible* points of a lattice introduced in Ch. III. A point T of $\Lambda(\rho)$ is *visible* (i.e. visible from the origin) if T is not O and if there is no point of $\Lambda(\rho)$ on OT between O and T. We write $f(\rho)$ for the number of visible points of $\Lambda(\rho)$ belonging to P and prove the following lemma.

THEOREM 459:

$$\rho^2 f(\rho) \to \frac{V}{\zeta(2)} \quad \textit{as} \quad \rho \to 0.$$

The number of points of $\Lambda(\rho)$ other than O, whose coordinates satisfy (24.10.2) is

$$(2\,[N/\rho]+1)^2-1.$$

Hence

$$f(\rho)=g(\rho)=0 \quad (\rho>N) \tag{24.10.6}$$

and

$$f(\rho)\leqslant g(\rho)<9N^2/\rho^2 \tag{24.10.7}$$

for all ρ.

Clearly $(\rho x, \rho y)$ is a visible point of $\Lambda(\rho)$ if, and only if, x, y are coprime. More generally, if m is the highest common factor of x and y, the point $(\rho x, \rho y)$ is a visible point of $\Lambda(m\rho)$ but not of $\Lambda(k\rho)$ for any integral $k\neq m$. Hence

$$g(\rho)=\sum_{m=1}^{\infty} f(m\rho).$$

By Theorem 270, it follows that

$$f(\rho)=\sum_{m=1}^{\infty}\mu(m)g(m\rho)\,.$$

The convergence condition of that theorem is satisfied trivially since, by (24.10.6), $f(m\rho)=g(m\rho)=0$ for $m\rho>N$. Again, by Theorem 287,

$$\frac{1}{\zeta(2)}=\sum_{m=1}^{\infty}\frac{\mu(m)}{m^2}$$

and so

$$\rho^2 f(\rho)-\frac{V}{\zeta(2)}=\sum_{m=1}^{\infty}\frac{\mu(m)}{m^2}\left\{m^2\rho^2 g(m\rho)-V\right\}. \tag{24.10.8}$$

Now let $\epsilon>0$. By (24.10.1), there is a number $\rho_1=\rho_1(\epsilon)$ such that

$$|m^2\rho^2 g(m\rho)-V|<\epsilon$$

whenever $m\rho < \rho_1$. Again, by (24.10.7),

$$|m^2\rho^2 g(m\rho) - V| < 9N^2 + V$$

for all m. If we write $M = [\rho_1/\rho]$, we have, by (24.10.8),

$$\left|\rho^2 f(\rho) - \frac{V}{\zeta(2)}\right| < \epsilon \sum_{m=1}^{M} \frac{1}{m^2} + \left(9N^2 + V\right) \sum_{m=M+1}^{\infty} \frac{1}{m^2}$$
$$< \frac{\epsilon\pi^2}{6} + \frac{9N^2 + V}{M+1} < 3\epsilon,$$

if ρ is small enough to make

$$M = [\rho_1/\rho] > (9N^2 + V)/\epsilon.$$

Since ϵ is arbitrary, Theorem 459 follows at once.

We can now show that the condition $V < 1$ of Theorem 458 can be relaxed if we confine our result to regions of a certain special form. We say that the bounded region P is a *star region* provided that (i) O belongs to P, (ii) P has an area V defined by (24.10.1), and (iii) if T is any point of P, then so is every point of OT between O and T. Every convex region containing O is a star region; but there are star regions which are not convex. We can now prove

THEOREM 1. *If* P *is a star region, symmetrical about* O *and of area* $V < 2\zeta(2) = \frac{1}{3}\pi^2$ *there is a lattice of determinant* 1 *which has no point* (*except* O) *in* P.

We use the same notation and argument as in the proof of Theorem 458. If Theorem 460 is false, there is a T_u, different from O, belonging to Λ_u and to P.

If T_u is not a visible point of $\Lambda(p^{-\frac{1}{2}})$, we have $m > 1$, where m is the highest common factor of X_u and $uX_u + pY_u$. By (24.10.4), $p \nmid X_u$ and so $p \nmid m$. Hence $m | Y_u$. If we write $X_u = mX_u'$, $Y_u = mY_u'$, the numbers X_u' and $uX_u' + pY_u'$ are coprime. Thus the point T_u', whose coordinates are

$$\frac{X_u'}{\sqrt{p}}, \quad \frac{uX_u' + pY_u'}{\sqrt{p}},$$

belongs to Λ_u and is a visible point of $\Lambda(p^{-\frac{1}{2}})$. But T_u' lies on OT_u and so belongs to the star region P. Hence, if T_u is not visible, we may replace it by a visible point.

Now P contains the p points

(24.10.9) $$T_0,\ T_1, \ldots,\ T_{p-1},$$

all visible points of $\Lambda(p^{-\frac{1}{2}})$, all different (as before) and none coinciding with O. Since P is symmetrical about O, P also contains the p points

(24.10.10) $$\overline{T}_0, \overline{T}_1, \ldots,\ \overline{T}_{p-1},$$

where $\bar{T}_u$ is the point $(-\xi_u, -\eta_u)$. All these p points are visible points of $\Lambda(p^{-\frac{1}{2}})$, all are different and none is O. Now T_u and $\overline{T}_u$ cannot coincide (for then each would be O). Again, if $u \neq v$ and T_u and $\overline{T}_v$ coincide, we have

$$X_u = -X_v, \quad uX_u + pY_u = -vX_v - pY_v,$$
$$(u-v)X_u \equiv 0, \quad X_u \equiv 0 \quad \text{or} \quad u \equiv v \pmod{p},$$

both impossible. Hence the $2p$ points listed in (24.10.9) and (24.10.10) are all different, all visible points of $\Lambda(p^{-\frac{1}{2}})$ and all belong to P so that

(24.10.11) $$f\left(p^{-\frac{1}{2}}\right) \geqslant 2p.$$

But, by Theorem 459, as $p \to \infty$,

$$p^{-1}f\left(p^{-\frac{1}{2}}\right) \to 6V/\pi^2 < 2$$

by hypothesis, and so (24.10.11) is false for large enough p. Theorem 460 follows.

The above proofs of Theorems 458 and 460 extend at once to n dimensions. In Theorem 460, $\zeta(2)$ is replaced by $\zeta(n)$.

NOTES

§ 24.1. Minkowski's writings on the geometry of numbers are contained in his books *Geometrie der Zahlen* and *Diophantische Approximationen,* already referred to in the note on § 3.10, and in a number of papers reprinted in his *Gesammelte Abhandlungen* (Leipzig, 1911). The fundamental theorem was first stated and proved in a paper of 1891 (*Gesammelte Abhandlungen,* i. 265). There is a very full account of the history and bibliography of the subject, up to 1936, in Koksma, chs. 2 and 3, and a survey of later progress by Davenport in *Proc. International Congress Math.* (Cambridge, Mass., 1950), 1 (1952), 166–74. More recent accounts of the whole subject are given by Cassels, *Geometry of numbers*; Gruber and Lekkerkerker, *Geometry of Numbers* (North Holland, Amsterdam, 1987); and Erdős, Gruber, and Hammer, *Lattice points* (Longman Scientific, Harlow, 1989).

Siegel [*Acta Math.* 65 (1935), 307–23] has shown that if V is the volume of a convex and symmetrical region R containing no lattice point but O, then

$$2^n = V + V^{-1} \sum |I|^2,$$

where each I is a multiple integral over R. This formula makes Minkowski's theorem evident.

Minkowski (*Geometrie der Zahlen,* 211–19) proved a further theorem which includes and goes beyond the fundamental theorem. We suppose R convex and symmetrical, and write λR for R magnified linearly about O by a factor λ. We define $\lambda_1, \lambda_2, \ldots, \lambda_n$ as follows: λ_1 is the least λ for which λR has a lattice point P_1 on its boundary; λ_2 the least for which λR has a lattice point P_2, not collinear with O and P_1, on its boundary; λ_2 the least for which λR has a lattice point P_3, not coplanar with O, P_1, and P_2, on its boundary; and so on. Then

$$0 < \lambda_1 \leqslant \lambda_2 \leqslant \ldots \leqslant \lambda_n$$

(λ_2, for example, being equal to λ_1 if $\lambda_1 R$ has a second lattice point, not collinear with O and P_1, on its boundary); and

$$\lambda_1 \lambda_2 \ldots \lambda_n V \leqslant 2^n.$$

The fundamental theorem is equivalent to $\lambda_1^n V \leqslant 2^n$. Davenport [*Quarterly Journal of Math.* (Oxford), 10 (1939), 117–21] has given a short proof of the more general theorem. See also Bambah, Woods, and Zassenhaus (*J. Australian Math. Soc.* 5 (1965), 453–62) and Henk (*Rend. Circ. Mat. Palermo* (II) Vol 1, Suppl.70 (2002) 377–84).

§ 24.2. All these applications of the fundamental theorem were made by Minkowski.

Siegel, *Math. Annalen,* 87 (1922), 36–8, gave an analytic proof of Theorem 448: see also Mordell, *ibid.* 103 (1930), 38–47.

Hajós, *Math. Zeitschrift,* 47 (1941), 427–67, has proved an interesting conjecture of Minkowski concerning the 'boundary case' of Theorem 448. Suppose that $\Delta = 1$, so that there are integral $x_1, x_2, \ldots, x_n$ such that $|\xi_r| \leqslant 1$ for $r = 1, 2, \ldots, n$. Can the x_r be chosen so that $|\xi_r| < 1$ for every r? Minkowski's conjecture, now established by Hajós, was that this is true except when the ξ_r can be reduced, by a change of order and a unimodular substitution, to the forms

$$\xi_1 = x_1, \quad \xi_2 = \alpha_{2,1} x_1 + x_2, \quad \ldots, \quad \xi_n = \alpha_{n,1} x_1 + \alpha_{n,2} x_2 + \cdots + x_n.$$

The conjecture had been proved before only for $n \leqslant 7$.

The first general results concerning the minima of definite quadratic forms were found by Hermite in 1847 (*Œuvres,* i, 100 *et seq.*): these are not quite so sharp as Minkowski's.

§ 24.3. The first proof of this character was found by Hurwitz, *Göttinger Nachrichten* (1897), 139–45, and is reproduced in Landau, *Algebraische Zahlen,* 34–40. The proof was afterwards simplified by Weber and Wellstein, *Math. Annalen,* 73 (1912), 275–85, Mordell, *Journal London Math. Soc.* 8 (1933), 179–82, and Rado, *ibid.* 9 (1934), 164–5 and 10 (1933), 115. The proof given here is substantially Rado's (reduced to two dimensions).

§ 24.5. Theorem 453 is in Gauss, *D.A.*, § 171. The corresponding results for forms in n variables are known only for $n \leqslant 8$: see Koksma, 24, and Mordell, *Journal London Math. Soc.* 19 (1944), 3–6.

§ 24.6. Theorem 454 was first proved by Korkine and Zolotareff, *Math. Annalen* 6 (1873), 366–89 (369). Our proof is due to Professor Davenport. See Macbeath, *Journal*

London Math. Soc. 22 (1947), 261–2, for another simple proof. There is a close connexion between Theorems 193 and 454.

Theorem 454 is the first of a series of theorems, due mainly to Markoff, of which there is a systematic account in Dickson, *Studies,* ch. 7. If $\xi\eta$ is not equivalent either to the form in (24.6.2) or to

$$(a) \qquad 8^{-\frac{1}{2}}\,|\Delta|\left(x^2+2xy-y^2\right),$$

then

$$|\xi\eta| < 8^{-\frac{1}{2}}\,|\Delta|$$

for appropriate x, y; if it is not equivalent either to the form in (24.6.2), to (*a*), or to

$$(b) \qquad (221)^{-\frac{1}{2}}\,|\Delta|\left(5x^2+11xy-5y^2\right),$$

then

$$|\xi\eta| < 5\,(221)^{-\frac{1}{2}}\,|\Delta|;$$

and so on. The numbers on the right of these inequalities are

$$(c) \qquad m\left(9m^2-4\right)^{-\frac{1}{2}},$$

where m is one of the 'Markoff numbers' 1, 2, 5, 13, 29,...; and the numbers (*c*) have the limit $\frac{1}{3}$. See also Cassels, *Diophantine approximation,* ch. 2 for an alternative proof of these theorems.

There is a similar set of theorems associated with rational approximations to an irrational ξ, of which the simplest is Theorem 193: see §§ 11.8–10, and Koksma, 31–33.

Davenport [*Proc. London Math. Soc.* (2) 44 (1938), 412–31, and *Journal London Math. Soc.* 16 (1941), 98–101] has solved the corresponding problem for $n = 3$. We can make

$$|\xi_1\xi_2\xi_3| < \tfrac{1}{7}\,|\Delta|$$

unless

$$\xi_1\xi_2\xi_3 \sim \tfrac{1}{7}\prod\left(x_1+\theta x_2+\theta^2 x_3\right),$$

where the product extends over the roots θ of $\theta^3+\theta^2-2\theta-1=0$. Mordell, in *Journal London Math. Soc.* 17 (1942), 107–15, and a series of subsequent papers in the *Journal* and *Proceedings,* has obtained the best possible inequality for the minimum of a general binary cubic form with given determinant, and has shown how Davenport's result can be deduced from it; and this has been the starting-point for a considerable body of work, by Mordell, Mahler, and Davenport, on lattice points in non-convex regions.

The corresponding problem for $n > 3$ has not yet been solved.

Minkowski [*Göttinger Nachrichten* (1904), 311–35; *Gesammelte Abhandlungen,* ii. 3–42] found the best possible result for $|\xi_1|+|\xi_2|+|\xi_3|$, viz.

$$|\xi_1|+|\xi_2|+|\xi_3| \leqslant \left(\tfrac{108}{19}\,|\Delta|\right)^{\frac{1}{2}}.$$

No simple proof of this result is known, nor any corresponding result with $n > 3$.

An alternative formulation of Theorem 454 states that if $Q(x,y)$ is an indefinite quadratic form of determinant D, then there are integer values x_0, y_0, not both zero, for which $|Q(x_0,y_0)| \leqslant 2\sqrt{|D|/5}$. It is natural to ask what happens for quadratic forms in more than 2 variables. It was conjectured by Oppenheim in 1929 that if Q is an indefinite form in $n \geqslant 3$ variables, and not proportional to an integral form, then $Q(x_1,\ldots,x_n)$ attains arbitrarily small values at integral arguments $x_1,\ldots,x_n$ not all zero. This was proved by Margulis, (*Dynamical systems and ergodic theory* (Warsaw, 1986), 399–409).

§§ 24.7–8. Minkowski proved Theorem 455 in *Math. Annalen,* 54 (1901), 91–124 (*Gesammelte Abhandlungen,* i. 320–56, and *Diophantische Approximationen,* 42–7). The proof in § 24.7 is due to Heilbronn and that in § 24.8 to Landau, *Journal für Math.* 165 (1931), 1–3: the two proofs, though very different in form, are based on the same idea. Davenport [*Acta Math.* 80 (1948), 65–95] solved the corresponding problem for indefinite ternary quadratic forms.

§ 24.9. The conjecture mentioned at the beginning of this section is usually attributed to Minkowski, but Dyson [*Annals of Math.* 49 (1948), 82–109] remarks that he can find no reference to it in Minkowski's published work. The statement is easy to prove when the coefficients of the forms are rational. Remak [*Math. Zeitschrift,* 17 (1923), 1–34 and 18 (1923), 173–200] proved the truth of the conjecture for $n = 3$, Dyson [*loc. cit.*] for $n = 4$. Davenport [*Journal London Math. Soc.* 14 (1939), 47–51] gave a much shorter proof for $n = 3$.

The Remak–Davenport–Dyson approach depends on the observation that Minkowski's conjecture follows from the following two conjectures.

Conjecture I : *For each lattice L in n-dimensional Euclidean space, there is an ellipsoid of the form*

$$a_1x_1^2 + \cdots + a_nx_n^2 \leqslant 1$$

which contains n linearly independent points of L on its boundary and has no point of L in its interior other than O.

Conjecture II: *Let L be a lattice of determinant* 1 *in n-dimensional Euclidean space and let S be a sphere centred at O which contains n linearly independent points of L on its boundary but no point of L in its interior other than O. Then the family* $\{(\sqrt{n}/2)S + A : A \in L\}$ *covers the whole space.*

Woods in a series of three papers (*Mathematika* 12 (1965), 138–42, 143–50 and *J. Number Theory* 4 (1972), 157–80) gave a simple proof of Conjecture II for $n = 4$ and proved it for $n = 5, 6$. For Conjecture I, Bambah and Woods (*J. Number Theory* 12 (1980), 27–48) gave a simple proof for $n = 4$. Around the same time, Skubenko (*Zap. Naučn. Sem. Leningrad. Otdel. Mat. Inst. Steklov.* (*LOMI*) 33 (1973), 6–36 and *Trudy Mat. Inst. Steklov* 142 (1976), 240–53) outlined a proof for $n \leqslant 5$. A complete proof for $n = 5$, on the lines suggested by Skubenko, was given by Bambah and Woods (*J. Number Theory* 12 (1980), 27–48). McMullen (*J. Amer. Math. Soc.* 18 (2005), 711–34) later proved Conjecture I for all n. This, together with the results on Conjecture II mentioned above, implies that Minkowski's conjecture is proved for all $n \leqslant 6$. Another proof for $n = 3$ was given by Birch and Swinnerton-Dyer (*Mathematica* 3 (1956), 25–39) and still another approach via factorization of matrices was explored by Macbeath (*Proc. Glasgow Math. Assoc.* 5 (1961), 86–89) and later by Narzullaev in a series of papers. Gruber (1976) and Ahmedov (1977) showed however that this approach will not be successful for large n.

Tchebotaref's theorem appeared in *Bulletin Univ. Kasan* (2) 94 (1934), Heft 7, 3–16; the proof is reproduced in *Zentralblatt für Math.* 18 (1938), 110–11. Mordell [*Vierteljahrsschrift d. Naturforschenden Ges. in Zürich,* 85 (1940), 47–50] has shown that the result may be sharpened a little. See also Davenport, *Journal London Math. Soc.* 21 (1946), 28–34.

For more details, including asymptotic results and references, the reader is referred to Gruber and Lekkerkerker, *Geometry of Numbers*; and Bambah, Dumir, and Hans-Gill, (*Number Theory*, 15–41, Birkhauser, Basel 2000).

Minkowski's conjecture for $n = 2$ (i.e. Theorem 455) can be interpreted as a problem on non-homogeneous binary indefinite quadratic forms. Its generalization to indefinite quadratic forms in n variables has aroused the interest of various writers including Bambah, Birch, Blaney, Davenport, Dumir, Foster, Hans-Gill, Madhu Raka, Watson, and Woods. In particular, Watson (*Proc. London Math. Soc.* (3) 12 (1962), 564–76) found the optimal result for $n \geqslant 21$ and made a corresponding conjecture for $4 \leqslant n \leqslant 21$. This conjecture was later proved by Dumir, Hans-Gill, and Woods (*J. Number Theory* 4 (1994), 190–197). Positive values of quadratic forms and asymmetric inequalities have also been studied and analogous results obtained. For references and related results see Bambah, Dumir, and Hans-Gill *loc. cit.*

§ 24.10. Minkowski [*Gesammelte Abhandlungen* (Leipzig, 1911), i. 265, 270, 277] first conjectured the n-dimensional generalizations of Theorems 458 and 460 and proved the latter for the n-dimensional sphere [*loc. cit.* ii. 95]. The first proof of the general theorems was given by Hlawka [*Math. Zeitschrift,* 49 (1944), 285–312]. Our proof is due to Rogers [*Annals of Math.* 48 (1947), 994–1002 and *Nature* 159 (1947), 104–5]. See also Rogers, *Packing and Covering* for an account of the Minkowski–Hlawka theorems and subsequent improvements.

XXV

ELLIPTIC CURVES

25.1. The congruent number problem. A *congruent number* is a rational number q that is the area of a right triangle, all of whose sides have rational length. We observe that if the triangle has sides a, b, and c, and if s is a rational number, then s^2q is also a congruent number whose associated triangle has sides sa, sb, and sc. So it is enough to ask which squarefree integers n are congruent numbers.

If we take c to be the length of the hypotenuse, then we are looking for squarefree integers n such that there are rational numbers a, b, c satisfying

$$a^2 + b^2 = c^2 \quad \text{and} \quad \frac{1}{2}ab = n. \tag{25.1.1}$$

A simple algebraic calculation shows that the positive solutions to the simultaneous equations (25.1.1) are in one-to-one correspondence with the positive solutions to the equation

$$y^2 = x^3 - n^2x \tag{25.1.2}$$

via the transformations

$$x = \frac{n\,(a+c)}{b}, \quad y = \frac{2n^2\,(a+c)}{b^2}, \quad a = \frac{y}{x}, \quad b = \frac{2nx}{y}, \quad c = \frac{x^2+n^2}{y}.$$

Thus n is a congruent number if and only if (25.1.2) has a solution in positive rational numbers x and y.

Equation (25.1.2) is an example of a Diophantine equation, similar to those discussed in Chapter XIII. Equations of this shape are called *elliptic curves*, although we must note that the name is somewhat unfortunate, since elliptic curves and ellipses have very little to do with one another. More generally, an *elliptic curve* is given by an equation of the form

$$E\colon y^2 = x^3 + Ax + B, \tag{25.1.3}$$

with the one further requirement that the *discriminant*

$$\Delta = 4A^3 + 27B^2 \tag{25.1.4}$$

should not vanish. The discriminant condition ensures that the cubic polynomial has distinct (complex) roots and that the locus of E in the real plane is nonsingular. For convenience, we shall generally assume that the coefficients A and B are integers. It is also convenient to write $E(\mathbb{R})$ for the solutions to (25.1.3) in real numbers, $E(\mathbb{Q})$ for the solutions in rational numbers, and so on.

Elliptic curves form a family of Diophantine equations. They have many fascinating properties, some of which we shall touch upon in this chapter. Elliptic curves have provided the testing ground for numerous theorems and conjectures in number theory, and there are many number theoretic problems, such as the congruent number problem, whose solution leads naturally to one or more elliptic curves. Most notable among the recent applications of elliptic curves is Wiles' proof of Fermat's Last Theorem. Wiles makes extensive use of elliptic curves, despite the fact that when $n \geqslant 4$, the Fermat equation $x^n + y^n = z^n$ is itself most defintely not an elliptic curve.

25.2. The addition law on an elliptic curve. In studying the solutions of equation (25.1.3), each nonzero number u gives an equivalent equation

$$Y^2 = X^3 + u^4 AX + u^6 B \tag{25.2.1}$$

via the identification $(x, y) = (u^{-2}X, u^{-3}Y)$. We say that (25.1.3) and (25.2.1) define *isomorphic* elliptic curves. If A, B, and u are all in a given field k, we say that the curves are *isomorphic over k,* in which case there is a natural bijection between the solutions of (25.1.3) and (25.2.1) with coordinates in k.

The *j-invariant of E* is the quantity

$$j(E) = \frac{4A^3}{4A^3 + 27B^2} = \frac{4A^3}{\Delta}.$$

If E and E' are isomorphic, then $j(E) = j(E')$, and over an algebraically closed field such as $\mathbb{C}$, the converse is true. Over other fields, such as $\mathbb{Q}$, the situation is slightly more complicated, since the value of u is restricted. There are three cases, depending on whether one of A or B vanishes.

THEOREM 461. *Let E and E′ be elliptic curves given by equations*

$$E\colon y^2 = x^3 + Ax + B \quad \textit{and} \quad E'\colon y^2 = x^3 + A'x + B'$$

having coefficients in some field k. Then E and E′ are isomorphic over k if and only if $j(E) = j(E')$ and one of the following conditions holds:

(*a*) $A = A' = 0$ *and* B/B' *is a* 6*th power in k;*

(*b*) $B = B' = 0$ *and* A/A' *is a* 4*th power in k*;

(*c*) $ABA'B' \neq 0$ *and* $AB'/A'B$ *is a square in k.*

Suppose first that $AB \neq 0$, so $j(E) \neq 0$ and $j(E) \neq 1$. If E and E' are isomorphic over k, then the relations $A' = u^4 A$ and $B' = u^4 B$ immediately imply that $j(E') = j(E)$, so $A'B' \neq 0$, and also

$$\frac{AB'}{A'B} = \frac{Au^6 B}{u^4 AB} = u^2$$

is a square in k.

Conversely, suppose that $j(E) = j(E')$ and $AB'/A'B = u^2$ for some $u \in K$. The j-invariant assumption implies that

$$\frac{A^3}{B^2} = \frac{27j(E)}{4 - 4j(E)} = \frac{27j(E')}{4 - 4j(E')} = \frac{A'^3}{B'^2}.$$

Hence

$$A' = \frac{A^3 B'^2}{A'^2 B^2} = \left(\frac{AB'}{A'B}\right)^2 A = u^4 A \quad \text{and} \quad B' = \frac{A^3 B'^3}{A'^3 B^2} = \left(\frac{AB'}{A'B}\right)^3 B = u^6 B,$$

so E and E' are isomorphic over k. The cases $A = 0$ and $B = 0$ are handled similarly.

One of the properties that makes an elliptic curve E such a fascinating object is the existence of a composition law that allows us to 'add' points to one another. In order to do this, we visualize the real solutions (x, y) of (25.1.3) as points in the Cartesian plane. The geometric description of the addition law on E is then quite simple. Let P and Q be distinct points on E and let L be the line through P and Q. Then the fact that E is given by an equation (25.1.3) of degree 3 means that L intersects E in three points.† Two of these points are P and Q. If we let R denote the third point in $L \cap E$, then the sum of P and Q is defined by

$$P + Q = (\text{the reflection of } R \text{ across the } x\text{-axis}).$$

In order to add P to itself, we let Q approach P, so L becomes the tangent line to E at P. The addition law on E is illustrated in Figure 11.

† The intersection points must be counted with appropriate multiplicity, and there are some special cases that we shall deal with presently.

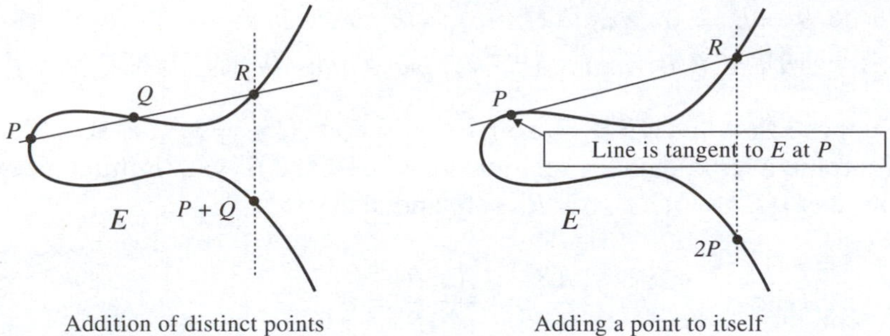

FIG. 11. The addition law on an elliptic curve

The one situation in which addition fails is when the line L is vertical. For later convenience, we define the negation of a point $P = (x, y)$ to be its reflection across the x-axis,

$$-P = (x, -y).$$

The line L through P and $-P$ intersects E in only these two points, so there is no third point R to use in the addition law. To remedy this situation, we adjoin an idealized point $\mathcal{O}$ to the plane. This point $\mathcal{O}$, which we call the *point at infinity*, has the property that it lies on every vertical line and on no other lines.† Further, the tangent line to E at $\mathcal{O}$ is defined to have a triple order contact with E at $\mathcal{O}$. Then the geometric addition law on E is defined for all pairs of points. In particular, the special rules relating to the point $\mathcal{O}$ are

$$\text{(25.2.2)} \qquad P + (-P) = \mathcal{O} \quad \text{and} \quad P + \mathcal{O} = P \quad \text{for all points } P \text{ on } E.$$

We now use a small amount of analytic geometry and calculus to derive formulae for the addition law. Let $P = (x_P, y_P)$ and $Q = (x_Q, y_Q)$ be two points on the curve E. If $P = -Q$, then $P + Q = \mathcal{O}$, so we assume that $P \neq -Q$. We denote by

$$L\colon y = \lambda x + \nu$$

† Those who are familiar with the projective plane $\mathbb{P}^2$ will recognize that $\mathcal{O}$ is one of the points on the line at infinity. The projective plane may be constructed by adjoining to the affine plane $\mathbb{A}^2$ one additional point for each direction, i.e. for each line through $(0, 0)$.

the line through P and Q if they are distinct, or the tangent line to E at P if they coincide. Explicitly,

$$\lambda = \frac{y_Q - y_P}{x_Q - x_P} \quad \text{and} \quad \nu = \frac{y_P x_Q - y_Q x_P}{x_Q - x_P}, \text{ if } P \neq Q, \tag{25.2.3}$$

$$\lambda = \frac{3x_P^2 + A}{2y_P} \quad \text{and} \quad \nu = \frac{-x_P^3 + Ax_P - 2B}{2y_P}, \text{ if } P = Q. \tag{25.2.4}$$

We compute the intersection of E and L by solving the equation

$$(\lambda x + \nu)^2 = x^3 + Ax + B. \tag{25.2.5}$$

The intersection of E and L includes the points P and Q, so two of the roots of the cubic equation (25.2.5) are x_P and x_Q. (If $P = Q$, then x_P will appear as a double root, since L is tangent to E at P). Letting $R = (x_R, y_R)$ denote the third intersection point of E and L, equation (25.2.5) factors as

$$\begin{aligned} x^3 - \lambda^2 x^2 + (A - 2\lambda\nu)\, x + \left(B - \beta^2\right) \\ = (x - x_P)\,(x - x_Q)\,(x - x_R)\,. \end{aligned} \tag{25.2.6}$$

Comparing the quadratic terms of (25.2.6) gives the formula

$$x_R = \lambda^2 - x_P - x_Q, \tag{25.2.7}$$

and then the formula for L gives the corresponding $y_R = \lambda x_R + \nu$. Finally, the sum of P and Q is computed by reflecting across the y-axis,

$$P + Q = (x_R, -y_R)\,. \tag{25.2.8}$$

For later use, we compute explicitly the *duplication formula*

$$x_{2P} = \left(\frac{3x_P^2 + A}{2y_P}\right)^2 - 2x_P = \frac{x_P^4 - 2Ax_P^2 - 8Bx_P + A^2}{4x_P^3 + 4Ax_P + 4B}. \tag{25.2.9}$$

THEOREM 462. *Let E be an elliptic curve. The addition law described above has the following properties:*

(*a*) [Identity] $P + \mathcal{O} = \mathcal{O} + P = P$ *for all* $P \in E$.

(*b*) [Inverse] $P + (-P) = \mathcal{O}$ *for all* $P \in E$.

(*c*)* [Associativity] $(P + Q) + R = P + (Q + R)$ *for all* $P, Q, R \in E$.

(*d*) [Commutativity] $P + Q = Q + P$ *for all* $P, Q \in E$.

The identity and inverse formulae are true by construction, since we have placed $\mathcal{O}$ to lie on every vertical line and to have a tangent line with a triple order contact. Commutativity is also clear, since $P+Q$ is computed using the line through P and Q, while $Q+P$ is computed using the line through Q and P, which is the same line. The associative law is more difficult. It may be proven by a long and tedious algebraic calculation using the addition formulae and considering many special cases, or it may be proven using more advanced techniques from algebraic geometry or complex analysis.

The content of Theorem 462 is that the set of points of E forms a commutative group with identity element $\mathcal{O}$. Repeated addition and negation allows us to 'multiply' points of E by an arbitrary integer m. This function from E to itself is called the *multiplication-by-m* map,

$$(25.2.10)\quad \phi_m : E \to E, \quad \phi_m(P) = mP = \operatorname{sign}(m)\,\overbrace{(P+P+\cdots+P)}^{|m| \text{ terms}}.$$

(By convention, we also define $\phi_0(P) = \mathcal{O}$).

Theorem 462 says that the set of points of E forms a commutative group. The next result says that the same is true if we take points whose coordinates lie in any field.

THEOREM 463. *Let E be an elliptic curve given by an equation* (25.1.3) *whose coefficients A and B are in a field k and let*

$$E(k) = \left\{(x,y) \in k^2 : y^2 = x^3 + Ax + B\right\} \cup \{\mathcal{O}\}.$$

Then the sum and difference of two points in $E(k)$ is again in $E(k)$, so $E(k)$ is a commutative group.

The proof is immediate, since a brief examination of the formulae for addition on E show that if A and B are in k and if the coordinates of P and Q are in k, then the coordinates of $P \pm Q$ are also in k. The crucial feature of the addition formulae is that they are all given by rational functions; at no stage are we required to take roots. Thus $E(k)$ is closed under addition and subtraction, and Theorem 462 says that the addition law has the requisite properties to make $E(k)$ into a commutative group.

If k is a field of arithmetic interest, for example $\mathbb{Q}$ or $k(i)$ or a finite field $\mathbb{F}_p$, then a description of the solutions to the Diophantine equation

$$y^2 = x^3 + Ax + B \quad \text{with } x, y \in k$$

may be accomplished by describing the group $E(k)$. To illustrate, we describe (without proof) the group of points with rational coordinates on the four curves

$$E_1: y^2 = x^3 + 7, \quad E_2: y^2 = x^3 - 43x + 166,$$
$$E_3: y^2 = x^3 - 2, \quad E_4: y^2 = x^3 + 17.$$

The curve E_1 has no nontrivial rational points, so $E_1(\mathbb{Q}) = \{\mathcal{O}\}$ .The curve E_2 has finitely many rational points. More precisely, $E_2(\mathbb{Q})$ is a cyclic group with 7 elements,

$$E_2(\mathbb{Q}) = \{(3, \pm 8), (-5, \pm 16), (11, \pm 32), \mathcal{O}\}.$$

The curves E_3 and E_4, by way of contrast, have infinitely many rational points. The group $E_3(\mathbb{Q})$ is freely generated by the single point $P = (3, 5)$, in the sense that every point in $E_3(\mathbb{Q})$ has the form nP for a unique $n \in \mathbb{Z}$. Similarly, the points $P = (-2, 3)$ and $Q = (2, 5)$ freely generate $E_4(\mathbb{Q})$ in the sense that every point in $E_4(\mathbb{Q})$ has the form $mP + nQ$ for a unique pair of integers $m, n \in \mathbb{Z}$. We note that none of these assertions concerning E_1, E_2, E_3, E_4 is obvious.

It is quite easy to characterize the points of order 2 on an elliptic curve.

THEOREM 464. *A point $P = (x, y) \neq \mathcal{O}$ on an elliptic curve E is a point of order 2, i.e. satisfies $2P = \mathcal{O}$, if and only if $y = 0$.*

According to the geometric description of the addition law, a point P has order 2 if and only if the tangent line to E at P is vertical. The slope of the tangent line L at $P = (x, y)$ satisfies

$$2y\frac{dy}{dx} = 3x^2 + A,$$

hence L is vertical if and only if $y = 0$. (Note that it is not possible to have both $y = 0$ and $3x^2 + A = 0$, since $y = 0$ implies that $x^3 + Ax + B = 0$, and the condition $\Delta \neq 0$ ensures that $x^3 + Ax + B = 0$ and its derivative do not have a common root.)

The multiplication-by-m map (25.2.10) is defined by rational functions in the sense that x_{mP} and y_{mP} can be expressed as elements of $\mathbb{Q}(A, B, x_P, y_P)$. For example, the duplication formula (25.2.9) gives such an expression for x_{2P}. Maps $E \to E$ defined by rational functions and sending $\mathcal{O}$ to $\mathcal{O}$ are called *endomorphisms of E.* Endomorphisms can be added and multiplied (composed) according to the rules

$$(\phi + \psi)(P) = \phi(P) + \psi(P) \quad \text{and} \quad (\phi\psi)(P) = \phi(\psi(P)),$$

and one can show that with these operations, the set of endomorphisms $\mathrm{End}(E)$ becomes a ring.†

For most elliptic curves (over fields of characteristic 0), the only endomorphisms are the multiplication-by-m maps, so for these curves $\mathrm{End}(E) = \mathbb{Z}$. Curves that admit additional endomorphisms are said to have *complex multiplication* (or CM, for short). Examples of such curves include

$$E_5 : y^2 = x^3 + Ax, \quad \text{which has the endomorphism } \phi_i(x, y) = (-x, iy),$$

and

$$E_6 : y^2 = x^3 + B, \quad \text{which has the endomorphism } \phi_\rho(x, y) = (\rho x, y).$$

(Here $i = \sqrt{-1}$ and $\rho = e^{\frac{2}{3}\pi i}$ are as in Chapter XII.) These endomorphisms satisfy

$$\phi_i^2(P) = -P \quad \text{and} \quad \phi_\rho^2(P) + \phi_\rho(P) + P = 0.$$

One can show that $\mathrm{End}(E_5)$ is isomorphic to the ring of Gaussian integers and that $\mathrm{End}(E_6)$ is the ring of integers in $k(\rho)$. This is typical in the sense that the endomorphism ring of a CM elliptic curve over a field of characteristic 0 is always a subring of a quadratic imaginary field. In particular, the composition of endomorphisms is commutative, i.e. $\phi(\psi(P)) = \psi(\phi(P))$ for all $P \in E$.‡

25.3. Other equations that define elliptic curves. A homogeneous polynomial equation

$$F(X, Y, Z) = \sum_{i+j+k=d} A_{ijk} X^i Y^j Z^k = 0 \tag{25.3.1}$$

† The hardest part of the proof is the distributive law, i.e. to show that the mere fact that ϕ is defined by rational functions implies that ϕ satisfies $\phi(P + Q) = \phi(P) + \phi(Q)$.

‡ However, it should be noted that there are elliptic curves defined over finite fields whose endomorphism rings are noncommutative.

is *nonsingular* if the simultaneous equations

$$F(X,Y,Z) = \frac{\partial}{\partial X}F(X,Y,Z) = \frac{\partial}{\partial Y}F(X,Y,Z) = \frac{\partial}{\partial Z}F(X,Y,Z) = 0$$

have no (complex) solutions other than $X = Y = Z = 0$. One can show that any nonsingular equation (25.3.1) of degree 3 with a specified nontrivial solution $P_0 = (x_0 : y_0 : z_0)$ is an elliptic curve in the sense that it may be transformed by rational functions into an equation of the form

$$y^2 + a_1xy + a_3y = x^3 + a_2x^2 + a_4x + a_6, \tag{25.3.2}$$

with the point P_0 being sent to the point $\mathcal{O}$ sitting at infinity. Further, if k is a field containing all of the A_{ijk} and containing the coordinates x_0, y_0, z_0 of P_0, then k also contains the new coefficients $a_1, \ldots, a_6$. An equation of the form (25.3.2) is called a *generalized Weierstrass equation.*

The following example illustrates this general principle and is useful for applications.

Theorem 465. *The nonzero solutions to the equation*

$$X^3 + Y^3 = A \tag{25.3.3}$$

are mapped bijectively, via the function

$$(X, Y) \longmapsto \left(\frac{12A}{X+Y}, 36A\frac{X-Y}{X+Y}\right), \tag{25.3.4}$$

to the solutions (with $x \neq 0$) of the equation

$$y^2 = x^3 - 432A^2. \tag{25.3.5}$$

The inverse map is given by

$$(x, y) \longmapsto \left(\frac{36A+y}{6x}, \frac{36A-y}{6x}\right). \tag{25.3.6}$$

It is an elementary calculation to verify that the maps (25.3.4) and (25.3.6) take the curves (25.3.3) and (25.3.5) to one another and that the composition of the maps is the identity. The curve (25.3.3) has three points at infinity, corresponding to setting $Z = 0$ in the homogeneous form $X^3 + Y^3 = AZ^3$. The transformation (25.3.4) identifies the point $(1 : -1 : 0)$ on (25.3.3) with the unique point at infinity on (25.3.5).

The *discriminant* of a generalized Weierstrass equation (25.3.2) is given by the rather complicated expression[†]

$$(25.3.7)\quad \begin{aligned}\Delta = {} & -a_1^6a_6 + a_1^5a_3a_4 + a_1^4a_2a_3^2 - 12a_1^4a_2a_6 + a_1^4a_4^2 \\ & + 8a_1^3a_3a_2a_4 + a_1^3a_3^3 + 36a_1^3a_3a_6 - 8a_1^2a_2^2a_3^2 \\ & - 48a_1^2a_2^2a_6 + 8a_1^2a_2a_4^2 - 30a_1^2a_3^2a_4 + 72a_1^2a_4a_6 \\ & + 16a_1a_2^2a_3a_4 + 36a_1a_2a_3^3 + 144a_1a_2a_3a_6 - 96a_1a_3a_4^2 \\ & - 16a_2^3a_3^2 - 64a_2^3a_6 + 16a_2^2a_4^2 + 72a_2a_3^2a_4 + 288a_2a_4a_6 \\ & - 27a_3^4 - 216a_3^2a_6 - 432a_6^2 - 64a_4^3.\end{aligned}$$

One can check at some length that the curve is nonsingular if and only if $\Delta \neq 0$.

The most general transformation preserving the Weierstrass equation form (25.3.2) is

$$(25.3.8)\quad x = u^2x' + r \quad \text{and} \quad y = u^3y' + u^2sx' + t \quad \text{with} \qquad u \neq 0.$$

The effect of the transformation (25.3.8) on the discriminant is $\Delta' = u^{-12}\Delta$.

When investigating integral or rational points on an elliptic curve (25.3.2), it is often advantageous to impose a minimality condition on the equation that is analogous to writing a fraction in lowest terms. An equation (25.3.2) is called a (*global*) *minimal Weierstrass equation* if for all transformations (25.3.8) with $r, s, t \in \mathbb{Q}$ and $u \in \mathbb{Q}^*$, the discriminant

$$|\Delta| \text{ is minimized subject to the condition } a_1, \ldots, a_6 \in \mathbb{Z}.$$

If the characteristic of k is not equal to 2 or 3, then the substitution

$$x = x' - \frac{1}{12}a_1^2 - \frac{1}{3}a_2, \quad y = y' - \frac{1}{2}a_1x' - \frac{1}{24}a_1^3 - \frac{1}{6}a_1a_2 - \frac{1}{2}a_3,$$

[†] The astute reader will have noted that this new discriminant (25.3.7) is 16 times our old discriminant (25.1.4). The extra factor is of importance only when working with the prime $p = 2$, in which case the new version is the more appropriate.

transforms (25.3.2) into the shorter Weierstrass form (25.1.3) with

$$A = \frac{1}{48}a_1^4 + \frac{1}{6}a_1^2 a_2 - \frac{1}{2}a_1 a_3 + \frac{1}{3}a_2^2 - a_4,$$

$$B = -\frac{1}{864}a_1^6 - \frac{1}{72}a_1^4 a_2 + \frac{1}{24}a_1^3 a_3 - \frac{1}{18}a_1^2 a_2^2 + \frac{1}{12}a_1^2 a_4 + \frac{1}{6}a_1 a_2 a_3$$
$$- \frac{1}{4}a_3^2 - \frac{2}{27}a_2^3 + \frac{1}{3}a_2 a_4 - a_6.$$

25.4. Points of finite order. A point $P \in E$ has *finite order* if some positive multiple mP of P is equal to $\mathcal{O}$. The order of P is the smallest such value of m. For example, Theorem 464 says that P has order 2 if and only if $y_P = 0$. Using the theory of elliptic functions, one can show that the points of order m in $E(\mathbb{C})$ form a product of two cyclic groups of order m. In this section, we prove an elegant theorem of Nagell and Lutz that characterizes the points of finite order in $E(\mathbb{Q})$. In particular, there are only finitely many such points, and the theorem gives an effective method for finding all of them.

THEOREM 466. *Let E be an elliptic curve given by an equation* (25.1.3) *having integer coefficients and let* $P = (x, y) \in E(\mathbb{Q})$ *be a point of finite order. Then the coordinates of P are integers, and either* $y = 0$ *or else* $y^2 | \Delta$.

It is often convenient to move the 'point at infinity' on the equation (25.1.3) to the point $(0, 0)$ by introducing the change of coordinates

$$z = \frac{x}{y}, \qquad w = \frac{1}{y}. \tag{25.4.1}$$

The new equation for the elliptic curve is

$$E\colon w = z^3 + Azw^2 + Bw^3, \tag{25.4.2}$$

and the point $\mathcal{O}$ is now the point $(z, w) = (0, 0)$. (The three points on the curve with $y = 0$, i.e. the points of order 2, have been moved 'to infinity'.) We observe that the transformation (25.4.1) sends lines to lines; for example, the line $y = \lambda x + \nu$ in the (x, y)-plane becomes the line $1 = \lambda z + \nu w$ in the (z, w)-plane. This means that we can add points on E in the (z, w)-plane using the same procedure that we used in the (x, y)-plane. We now derive explicit formulae for the (z, w) addition law.

THEOREM 467. *Let E be an elliptic curve given by* (25.4.2) *and let* $P = (z_P, w_P)$ *and* $Q = (z_Q, w_Q)$ *be points on E. Set*

$$\alpha = \frac{z_Q^2 + z_P z_Q + z_P^2 + A w_P^2}{1 - A z_Q \left(w_Q + w_P\right) - B\left(w_Q^2 + w_P w_Q + w_P^2\right)}, \tag{25.4.3}$$

$$\beta = w_P - \alpha z_P.$$

Then the z-coordinate of $P + Q$ *is given by the formula*

$$z_{P+Q} = \frac{2A\alpha\beta + 3B\alpha^2\beta}{1 + A\alpha^2 + B\alpha^3} + z_P + z_Q. \tag{25.4.4}$$

(*If* $z_P = z_Q$ *and* $w_P \neq w_Q$, *then* α *is formally equal to* ∞, *so* (25.4.4) *must be interpreted as* $\alpha \to \infty$ *and* $\beta/\alpha \to -z_P$, *which yields* $z_{P+Q} = -z_P$ *in this case.*†)

The proof of Theorem 467 is not difficult, but it requires a certain amount of algebraic manipulation of formulae. Suppose first that $z_P \neq z_Q$, so the line $w = \alpha z + \beta$ through P and Q has slope

$$\alpha = \frac{w_Q - w_P}{z_Q - z_P}.$$

The points P and Q both satisfy (25.4.2). Subtracting gives

$$\begin{aligned} w_Q - w_P &= \left(z_Q^3 - z_P^3\right) + A\left(z_Q w_Q^2 - z_P w_P^2\right) + B\left(w_Q^3 - w_P^3\right) \\ &= \left(z_Q^3 - z_P^3\right) + A z_Q\left(w_Q^2 - w_P^2\right) \\ &\quad + A\left(z_Q - z_P\right) w_P^2 + B\left(w_Q^3 - w_P^3\right). \end{aligned} \tag{25.4.5}$$

Every term in (25.4.5) is divisible by either $w_Q - w_P$ or $z_Q - z_P$, so a small amount of algebra yields

(25.4.6)

$$\alpha = \frac{w_Q - w_P}{z_Q - z_P} = \frac{z_Q^2 + z_P z_Q + z_P^2 + A w_P^2}{1 - A z_Q\left(w_Q + w_P\right) - B\left(w_Q^2 + w_P w_Q + w_P^2\right)}.$$

† If also $B = 0$, then the formulae need a small further modification that we leave to the reader.

Similarly, if $P = Q$, then the slope of the tangent line is

$$\alpha = \frac{dw}{dz}(P) = \frac{3z_P^2 + Aw_P^2}{1 - 2Az_Pw_P - 3Bw_P^2}. \tag{25.4.7}$$

We observe that (25.4.6) becomes equal to (25.4.7) if we make the substitution $(z_Q, w_Q) = (z_P, w_P)$, so we may also use (25.4.6) in this case.

The line $L: w = \alpha z + \beta$ intersects the curve E at the points P and Q and a third point R. Substituting $w = \alpha z + \beta$ into (25.4.2) gives a cubic equation whose roots, with appropriate multiplicities, are z_P, z_Q, and z_R. Thus there is a constant C so that

$$\begin{aligned} z^3 + Az(\alpha z + \beta)^2 + B(\alpha z + \beta)^3 - (\alpha z + \beta) \\ = C(z - z_P)(z - z_Q)(z - z_R). \end{aligned}$$

Comparing the coefficients of z^2 and z^3 yields

$$-z_P - z_Q - z_R = \frac{2A\alpha\beta + 3B\alpha^2\beta}{1 + A\alpha^2 + B\alpha^3}.$$

The points P, Q, and R satisfy $P + Q + R = \mathcal{O}$, so $P + Q = -R$. Finally we note that the negative of a point on E in the (z, w) plane is given by $-(z, w) = (-z, -w)$, so the z-coordinate of $P + Q$ is $-z_R$.

It remains to deal with the case $z_P = z_Q$ and $w_P \neq w_Q$. Then the line L through P and Q is the line $z = z_P$, and, provided $B \neq 0$, the line L intersects E at 3 points in the zw-plane. The third point $R = (z_R, w_R)$ necessarily satisfies $z_R = z_P$, since it lies on L, and then $z_{P+Q} = z_{-R} = -z_R = -z_P$. This completes the proof of Theorem 467.

We shall prove that points of finite order have integral coordinates by demonstrating that there are no primes dividing their denominators. For this purpose we fix a prime p and let

$$R_p = \left\{ \frac{a}{b} \in \mathbb{Q} : p \nmid b \right\}.$$

It is easily verified that R_p is closed under addition, subtraction, and multiplication, so R_p is a subring of $\mathbb{Q}$. Further, divisibility may be defined in R_p just as it was for $\mathbb{Z}$. The unities in R_p, i.e. the elements with multiplicative inverses, are precisely those rational numbers whose numerators and denominators are both relatively prime to p. We may reduce elements of

R_p modulo p, and the theory of congruences described in §§ 5.2 and 5.3 remains valid.[†]

We define the *p-adic valuation* $v_p(a)$ of a nonzero integer a to be the exponent of the largest power of p that divides a, and we extend the definition to rational numbers by setting

$$v_p\left(\frac{a}{b}\right) = v_p(a) - v_p(b).$$

We also formally set $v_p(0) = \infty$ to be larger than every real number. Notice that R_p is characterized by

$$R_p = \{\alpha \in \mathbb{Q} : v_p(\alpha) \geqslant 0\}.$$

The following properties of v_p are easily verified:[‡]

(25.4.8) $$v_p(\alpha\beta) = v_p(\alpha) + v_p(\beta),$$

(25.4.9) $$v_p(\alpha + \beta) \geqslant \min\{v_p(\alpha), v_p(\beta)\}.$$

Further, in the case of unequal valuation we have equality in (25.4.9),

(25.4.10) $$v_p(\alpha) \neq v_p(\beta) \Longrightarrow v_p(\alpha + \beta) = \min\{v_p(\alpha), v_p(\beta)\}.$$

THEOREM 468. *Let E be an elliptic curve given by equations* (25.1.3) *and* (25.4.2) *having integer coefficients and let* $P = (x, y) = (z, w)$ *be a point on E having rational coordinates. Then*

$$v_p(x) < 0 \Longleftrightarrow v_p(y) < 0 \Longleftrightarrow v_p(z) > 0 \Longleftrightarrow v_p(w) > 0.$$

If any of these equivalent conditions is true, then

$$v_p(x) = -2v_p(z), \quad v_p(y) = -3v_p(z), \quad \textit{and} \quad v_p(w) = 3v_p(z).$$

All of the assertions of Theorem 468 are immediate consequences of the basic valuation rules (25.4.8), (25.4.9), and (25.4.10) applied to equations (25.1.3) and (25.4.2) defining E.

THEOREM 469. *Let E be an elliptic curve given by an equation* (25.4.2) *having integer coefficients. Let P and Q be points of E whose* (z, w)*-coordinates are in* R_p*, and suppose that these points satisfy*

(25.4.11) $$z_P \equiv z_Q \equiv 0 \pmod{p^k} \quad \textit{for some } k \geqslant 1.$$

[†] R_p is an example of a *local ring*, i.e. a ring with a single maximal ideal.

[‡] Properties (25.4.8) and (25.4.9) say that the function $v_p : \mathbb{Q}^* \to \mathbb{Z}$ is a *discrete valuation*.

Then the z-coordinate of their sum satisfies

$$z_{P+Q} \equiv z_P + z_Q \pmod{p^{5k}}. \tag{25.4.12}$$

In particular, (25.4.11) *implies that* $z_{P+Q} \equiv 0 \pmod{p^k}$.

Theorem 468 and (25.4.11) tell us that $w_P \equiv w_Q \equiv 0 \pmod{p^{3k}}$. We begin by ruling out the exceptional case in Theorem 467. Suppose that $z_P = z_Q$. Subtracting (25.4.2) evaluated at P from (25.4.2) evaluated at Q yields

$$(w_Q - w_P)\left(1 - Az_P\,(w_Q + w_P) - B\left(w_Q^2 + w_P w_Q + w_Q^2\right)\right) = 0.$$

The second factor is congruent to 1 modulo p, hence $w_Q = w_P$.

Having ruled out the case $z_P = z_Q$ and $w_P \neq w_Q$, we see that the quantities α and β defined by (25.4.3) of Theorem 467 satisfy

$$\alpha \equiv 0 \pmod{p^{2k}} \quad \text{and} \quad \beta \equiv 0 \pmod{p^{3k}}.$$

Then (25.4.4) in Theorem 467 gives

$$z_{P+Q} = \frac{2A\alpha\beta + 3B\alpha^2\beta}{1 + A\alpha^2 + B\alpha^3} + z_P + z_Q \equiv z_P + z_Q \pmod{p^{5k}}.$$

Theorem 469 provides the tools needed to prove the integrality statement in Theorem 466. Let $P = (x_P, y_P) \in E(\mathbb{Q})$ be a point of finite order. We are required to prove that x_P and y_P are integers. If $y_P = 0$, so $2P = \mathcal{O}$ from Theorem 464, then equation (25.1.3) of E shows that x_P is an integer and we are done. We assume henceforth that $y_P \neq 0$.

Suppose to the contrary that there is some prime p dividing the denominator of x_P. Switching to (z, w) coordinates, Theorem 469 tells us that $p|z_P$. Let $k = v(z_P) > 0$, so $p^k|z_P$ and $p^{k+1} \nmid z_P$. Repeated application of (25.4.12) from Theorem 469 yields

$$z_{nP} \equiv nz_P \pmod{p^{5k}} \quad \text{for all } n \geqslant 1. \tag{25.4.13}$$

We now make use of the assumption that P has finite order, so $mP = \mathcal{O}$ for some $m \geqslant 1$. Setting $n = m$ in (25.4.13) and using the fact that $z_{\mathcal{O}} = 0$ gives

$$0 = z_{\mathcal{O}} = z_{mP} \equiv mz_P \pmod{p^{5k}}. \tag{25.4.14}$$

If $p \nmid m$, then (25.4.14) contradicts our assumption that $p^{k+1} \nmid z_P$, which proves that p does not divide the denominator of x_P and y_P.

It remains to deal with the case that p divides m. We write $m = pm'$, set $P' = m'P$, and let $k' = v(z_{P'})$. (Note that $k' \geqslant k \geqslant 1$ from (25.4.13) with $n = m'$.) Since P' has order p, the same argument yields

$$0 = z_{\mathcal{O}} = z_{pP'} \equiv pz_{P'} \pmod{p^{5k'}}.$$

Hence $p^{5k'-1}$ divides $z_{P'}$, which is again a contradiction. This completes the proof that the (x, y)-coordinates of points of finite order are integers.

Now that we know that points of finite order have integral coordinates, the second part of Theorem 466 is easy. First, Theorem 464 says that $2P = \mathcal{O}$ if and only if $y = 0$, so we may assume that $P = (x, y)$ has order $m \geqslant 3$. Then P and $2P$ are both points of finite order, so from our previous work we know that they both have integral coordinates. The duplication formula (25.2.9) says that

$$x_{2P} = \frac{x_P^4 - 2Ax_P^2 - 8Bx_P + A^2}{4x_P^3 + 4Ax_P + 4B}, \tag{25.4.15}$$

and a standard Euclidean algorithm or resultant calculation yields the identity

$$\begin{aligned}(3x^2 + 4A)\left(x^4 - 2Ax^2 - 8Bx + A^2\right)& \\ - \left(3x^3 - 5Ax - 27B\right)\left(x^3 + Ax + B\right) &= 4A^3 + 27B^2 = \Delta.\end{aligned} \tag{25.4.16}$$

Combining (25.4.15) and (25.4.16) with the basic relation $y^2 = x^3 + Ax + B$ gives

$$y_P^2\left(4\left(3x_P^2 + 4A\right)x_{2P} - \left(3x_P^2 - 5Ax_P - 27B\right)\right) = \Delta. \tag{25.4.17}$$

All of the quantities in (25.4.17) are integers, which proves that $y_P^2 | \Delta$.

25.5. The group of rational points. Points of finite order in $E(\mathbb{Q})$ are effectively determined by Theorem 466. Points of infinite order are far more difficult to characterize. A fundamental theorem, due to Mordell for $E(\mathbb{Q})$ and generalized by Weil, states that every point in $E(\mathbb{Q})$ can be written as a linear combination of points taken from a finite set of generators, where note that addition is always via the composition law on the elliptic curve E.

THEOREM 470. *Let E be an elliptic curve given by an equation* (25.1.3) *having rational coefficients. Then the group of rational points $E(\mathbb{Q})$ is finitely generated.*

A standard algebraic result says that every finitely generated abelian group is the direct sum of a finite group and a freely generated group. Thus Theorem 470 implies the following more precise statement.

THEOREM 471. *Let E be an elliptic curve given by an equation* (25.1.3) *having rational coefficients. There exists a finite set of points $P_1, \ldots, P_r$ in $E(\mathbb{Q})$ such that every point in $P \in E(\mathbb{Q})$ can be uniquely written in the form*

$$P = n_1P_1 + n_2P_2 + \cdots + n_rP_r + T,$$

with $n_1, \ldots, n_r \in \mathbb{Z}$ and T a point of finite order. The nonnegative integer r, which is uniquely determined by $E(\mathbb{Q})$, is called the rank of $E(\mathbb{Q})$.

We begin with an elementary lemma and some rank 0 cases of Theorem 470, after which we state a weak form of the theorem and use it to deduce the full theorem via a Fermat-style descent argument.

THEOREM 472. *Let E be an elliptic curve given by an equation* (25.1.3) *having rational coefficients and let $P = (x, y)$ be a point of E with rational coordinates. Then the coordinates of E may be written in the form*

$$P = \left(\frac{a}{d^2}, \frac{b}{d^3}\right) \quad \text{with } \gcd(a, d) = (b, d) = 1.$$

Theorem 472 is a consequence of Theorem 468, but we give a short direct proof. We write the coordinates of $P = (a/u, b/v)$ as fractions in lowest terms with positive denominators and substitute into (25.1.3) to obtain

$$\frac{(\text{a number prime to } v)}{v^2} = \frac{(\text{a number prime to } u)}{u^3}$$

Hence $v^2 = u^3$, and on comparing the prime factorizations of v and u, we see that there is an integer d such that $v = d^3$ and $u = d^2$.

Some of the Diophantine equations that we studied in Chapter XIII were elliptic curves. The next two theorems reformulate those results to prove a few rank 0 cases of Theorem 470.

THEOREM 473. *The elliptic curve $E\colon y^2 = x^3 + x$ has rank zero. Its group of rational points $E(\mathbb{Q}) = \{(0,0), \mathcal{O}\}$ is a cyclic group of order* 2.

Let $P = (a/d^2, b/d^3) \in E(\mathbb{Q})$. Then

$$b^2 = a^3 + ad^4 = a\left(a^2 + d^4\right), \tag{25.5.1}$$

and the fact that $\gcd(a, d) = 1$ implies that the factors in (25.5.1) are squares, say

$$a = u^2 \quad \text{and} \quad a^2 + d^4 = v^2.$$

Eliminating a yields $u^4 + d^4 = v^2$, and then Theorem 226 tells us that $udv = 0$. By assumption, $d \neq 0$, and $v = 0$ forces $u = d = 0$, so the only solution is $u = 0$. Hence $a = 0$ and $P = (0, 0)$.

THEOREM 474. *For each value of* $B \in \{16, -144, -432, 3888\}$, *the elliptic curve*

$$E_B : y^2 = x^3 + B$$

has rank 0, *that is,* $E_B(\mathbb{Q})$ *is finite.*

Theorem 465 gives a map from the curve

$$C_A : X^3 + Y^3 = A$$

to the curve E_{-432A^2}. This map, with at most a couple of exceptions, identifies the set of rational points $C_A(\mathbb{Q})$ with the set of rational points $E_{-432A^2}(\mathbb{Q})$.

An argument similar to that given in the proof of Theorem 472 shows that every rational point in $C_A(\mathbb{Q})$ has the form $(a/c, b/c)$, where the fractions are in lowest terms. Thus

$$a^3 + b^3 = Ac^3.$$

Theorem 228 for $A = 1$ and Theorem 232 for $A = 3$ tell us that

$$C_1(\mathbb{Q}) = \{(1, 0), (0, 1)\} \quad \text{and} \quad C_3(\mathbb{Q}) = \emptyset,$$

from which it follows that $E_{-432}(\mathbb{Q})$ and $E_{3888}(\mathbb{Q})$ are finite.

It is an algebraic exercise to verify that the following formula gives a well-defined map from E_B to E_{-27B} that is at most 3-to-1 on $E_B(\mathbb{Q})$,[†]

$$\begin{aligned} E_B\colon y^2 = x^3 + B &\longrightarrow E_{-27B}\colon y^2 = x^3 - 27B, \\ (x, y) \qquad\qquad &\longmapsto \left(\left(x^3 + 4B\right)/x^2, y\left(x^3 - 8B\right)/x^3\right). \end{aligned}$$

Taking $B = 16$ gives $E_{16}(\mathbb{Q}) \to E_{-432}(\mathbb{Q})$, so $E_{16}(\mathbb{Q})$ is finite, and similarly taking $B = -144$ shows that $E_{-144}(\mathbb{Q})$ is finite.

We now take up the proof of Theorem 470, which is traditionally divided into two parts. The first part we state without proof, since it requires tools beyond our disposal.[‡]

THEOREM 475. *Let E be an elliptic curve given by an equation* (25.1.3) *having rational coefficients. Then the quotient group $E(\mathbb{Q})/2E(\mathbb{Q})$ is finite, i.e. there is a finite set of points $Q_1, \ldots, Q_k \in E(\mathbb{Q})$ such that every Q in $E(\mathbb{Q})$ can be written in the form*

$$Q = Q_i + 2Q'$$

for some $1 \leqslant i \leqslant k$ and some $Q' \in E(\mathbb{Q})$.

The second part of the proof of Theorem 470 is a descent argument very much in the spirit of Fermat. Making a change of varibles of the form $x = u^2x'$ and $y = u^3y'$ for an appropriate rational number u, we may assume that the equation (25.1.3) defining E has integer coefficients.

For the descent, we shall use height functions to measure the arithmetic size of points in $E(\mathbb{Q})$. The *height* of a rational number $t \in \mathbb{Q}$ is the quantity

$$H(t)=H\left(\frac{a}{b}\right)=\max\{|a|, |b|\} \quad \text{for} \quad t=\frac{a}{b} \in \mathbb{Q} \text{ with } \gcd(a, b) = 1,$$

and the height of a point $P = (x_P, y_P) \in E(\mathbb{Q})$ is then defined by

$$H(P) = H(x_P) \text{ if } P \neq \mathcal{O}, \quad \text{and} \quad H(\mathcal{O}) = 1.$$

It is clear that there are only finitely many rational numbers of height less than any given bound, and similarly for points in $E(\mathbb{Q})$, since each rational x-coordinate gives at most two rational y-coordinates.

† The map is exactly 3-to-1 on complex points $E_B(\mathbb{C}) \to E_{-27B}(\mathbb{C})$. Maps between elliptic curves defined by rational functions are called *isogenies.*

‡ If the cubic equation $x^3 + Ax + B$ in (25.1.3) has a rational root, then Theorem 470 admits an elementary, albeit lengthy, proof, which may be found, for example, in Silverman–Tate, *Rational Points on Elliptic Curves,* Chapter III.

The key to performing the descent is to understand the effect of the group law on the heights of points.

THEOREM 476. *Let E be an elliptic curve given by an equation* (25.1.3) *having integer coefficients. There are constants c_1 and $c_2 > 0$ so that*

$$H(P+Q) \leqslant c_1 H(P)^2 H(Q)^2 \quad \text{for all } P, Q \in E(\mathbb{Q}), \tag{25.5.2}$$

$$H(2P) \geqslant c_2 H(P)^4 \quad \text{for all } P \in E(\mathbb{Q}). \tag{25.5.3}$$

The height function satisfies $H \geqslant 1$, so both (25.5.2) and (25.5.3) are true with $c_1 = c_2 = 1$ if either $P = \mathcal{O}$ or $Q = \mathcal{O}$. Similarly, if $P+Q = \mathcal{O}$, then (25.5.2) is true with $c_1 = 1$. We consider the remaining cases.

We use Theorem 472 to write

$$P = (x_P, y_P) = \left(\frac{a_P}{d_P^2}, \frac{b_P}{d_P^3}\right) \quad \text{and} \quad Q = (x_Q, y_Q) = \left(\frac{a_Q}{d_Q^2}, \frac{b_Q}{d_Q^3}\right).$$

Assuming that $P \neq Q$, the addition formulae (25.2.3), (25.2.7), (25.2.8) give

(25.5.4)

$$\begin{aligned} x_{P+Q} &= \left(\frac{y_Q - y_P}{x_Q - x_P}\right)^2 - x_P - x_Q \\ &= \frac{\left(x_P x P_Q + A\right)\left(x_P + x_Q\right) + 2B - 2y_P y_Q}{\left(x_P - x_Q\right)^2} \\ &= \frac{\left(a_P a_Q + A d_P^2 d_Q^2\right)\left(a_P d_Q^2 + a_Q d_P^2\right) + 2B d_P^4 d_Q^4 - 2b_P d_P b_Q d_Q}{\left(a_P d_Q^2 - a_Q d_P^2\right)^2}. \end{aligned}$$

The height of a rational number can only decrease if there is cancellation between numerator and denominator, so (25.5.4) and the triangle inequality yield

$$\begin{aligned} H\left(x_{P+Q}\right) \leqslant c_3 &\max\left\{|a_P|^2, |d_P|^4, |b_P d_P|\right\} \\ &\times \max\left\{|a_Q|^2, |d_Q|^4, |b_Q d_Q|\right\}. \end{aligned} \tag{25.5.5}$$

(Explicitly, we may take $c_3 = 4 + 2|A| + 2|B|$.) Next we observe that since P and Q are points on the curve, their coordinates satisfy

$$b_P^2 = a_P^3 + Aa_P d_P^4 + Bd_P^6 \quad \text{and} \quad b_Q^2 = a_Q^3 + Aa_Q d_Q^3 + Bd_Q^6.$$

Hence

$$\text{(25.5.6)} \qquad \begin{aligned} |b_P| &\leqslant c_4 \max\left\{|a_P|^{3/2}, |d_P|^3\right\} \quad \text{and} \\ |b_Q| &\leqslant c_4 \max\left\{|a_Q|^{3/2}, |d_Q|^3\right\}. \end{aligned}$$

(Explicitly $c_4 = 1 + |A| + |B|$.) Substituting (25.5.6) into (25.5.5) yields

$$\begin{aligned} H\left(x_{P+Q}\right) &\leqslant c_3 c_4^2 \max\left\{|a_P|^2, |d_P|^4\right\} \max\left\{|a_Q|^2, |d_Q|^4\right\} \\ &= c_1 H(P)^2 H(Q)^2, \end{aligned}$$

which completes the proof of (25.5.2) for $P \neq Q$. The proof for $P = Q$ is similar using the duplication formula (25.2.9) and may safely be left to the reader.

We turn now to the lower bound (25.5.3). If the polynomial $x^3 + Ax + B$ has any rational roots, then we first insist that the positive constant c_2 satisfies

$$\text{(25.5.7)} \qquad c_2 < \min\left\{H(\xi)^{-4} : \xi \in \mathbb{Q} \quad \text{and} \quad \xi^3 + A\xi + B = 0\right\}.$$

Theorem 464 then tells us that (25.5.3) is true if $2P = \mathcal{O}$, so we may assume that $2P \neq \mathcal{O}$.

To ease notation, we write

$$x_P = \frac{\alpha}{\delta}$$

as a fraction in lowest terms. We define polynomials

$$\begin{aligned} F(X, Z) &= X^4 - 2AX^2Z^2 - 8BXZ^3 + A^2Z^4, \\ G(X, Z) &= 4X^3Z + 4AXZ^3 + 4BZ^4, \end{aligned}$$

and we use them to homogenize the duplication formula (25.2.9). Thus the x-coordinate of $2P$ is given by

$$\text{(25.5.8)} \qquad x_{2P} = \frac{F(\alpha, \delta)}{G(\alpha, \delta)}.$$

The Euclidean algorithm or the theory of resultants tells us how to find relationships that eliminate either X or Z from F and G, cf. (25.4.16). Explicitly, if we define polynomials

$$f_1(X,Z) = 12X^2Z + 16AZ^3, \tag{25.5.9}$$

$$g_1(X,Z) = 3X^3 - 5AXZ^2 - 27BZ^3, \tag{25.5.10}$$

$$\begin{aligned} f_2(X,Z) = 4\left(4A^3 + 27B^2\right)X^3 - 4A^2BX^2Z \\ + 4A\left(3A^3 + 22B^2X\right)Z^2 + 12B\left(A^3 + 8B^2\right)Z^3, \end{aligned} \tag{25.5.11}$$

$$\begin{aligned} g_2(X,Z) = A^2BX^3 + A\left(5A^3 + 32B^2\right)X^2Z \\ + 2B\left(13A^3 + 96B^2\right)XZ^2 - 3A^2\left(A^3 + 8B^2\right)Z^3, \end{aligned} \tag{25.5.12}$$

then an elementary, but tedious, calculation verifies the two formal identities

$$f_1(X,Z)F(X,Z) + g_1(X,Z)G(X,Z) = 4\Delta Z^7, \tag{25.5.13}$$

$$f_2(X,Z)F(X,Z) + g_2(X,Z)G(X,Z) = 4\Delta X^7. \tag{25.5.14}$$

Here $\Delta = 4A^3 + 27B^2 \neq 0$ is the discriminant of E, as usual.

We substitute $X = \alpha$ and $Z = \delta$ into (25.5.13) and (25.5.14) to obtain

$$f_1(\alpha,\delta)F(\alpha,\delta) + g_1(\alpha,\delta)G(\alpha,\delta) = 4\Delta\delta^7. \tag{25.5.15}$$

$$f_2(\alpha,\delta)F(\alpha,\delta) + g_2(\alpha,\delta)G(\alpha,\delta) = 4\Delta\alpha^7. \tag{25.5.16}$$

From (25.5.15) and (25.5.16) and the fact that $\gcd(\alpha, \delta) = 1$, we see that

$$\gcd(F(\alpha,\delta), G(\alpha,\delta)) \mid 4\Delta.$$

Hence there is at most a factor of 4Δ cancellation between the numerator and the denominator of (25.5.8), so

$$H(x_{2P}) \geqslant \frac{\max\{F(\alpha,\delta), G(\alpha,\delta)\}}{|4\Delta|}. \tag{25.5.17}$$

The identities (25.5.15) and (25.5.16) also allow us to estimate

$$|4\Delta\delta^7| \leqslant 2\max\{|f_1(\alpha,\delta)|, |g_1(\alpha,\delta)|\} \times \max\{|F(\alpha,\delta)|, |G(\alpha,\delta)|\}, \tag{25.5.18}$$

$$|4\Delta\delta^7| \leqslant 2\max\{|f_2(\alpha,\delta)|, |g_2(\alpha,\delta)|\} \times \max\{|F(\alpha,\delta)|, |G(\alpha,\delta)|\}. \tag{25.5.19}$$

Looking at the explicit expressions (25.5.9)–(25.5.12) for f_1, g_1, f_2, and g_2, we see that

$$\max\{|f_1(\alpha,\delta)|, |g_1(\alpha,\delta)|, |f_2(\alpha,\delta)|, |g_2(\alpha,\delta)|\} \leqslant c_5 \max\left\{|\alpha|^3, |\delta|^3\right\}, \tag{25.5.20}$$

where c_5 depends only on A and B. Combining (25.5.18), (25.5.19), and (25.5.20) yields

$$4|\Delta|\max\{|\alpha|, |\delta|\}^7 \leqslant 2c_5 \max\{|\alpha|, |\delta|\}^3 \cdot \max\{|F(\alpha,\delta)|, |G(\alpha,\delta)|\}, \tag{25.5.21}$$

and then (25.5.17) and (25.5.21) imply that

$$H(x_{2P}) \geqslant (2c_5)^{-1}\max\{|\alpha|, |\delta|\}^4 \geqslant c_2 H(x_P)^4,$$

where we may take any positive $c_2 \leqslant (2c_5)^{-1}$ satisfying (25.5.7). This completes the proof of (25.5.3).

Theorem 476 is written in multiplicative form, in the sense that it relates sums of points on E to products of their heights. It is convenient to rewrite it using the *logarithmic height*

$$h(P) = \log H(P).$$

With this notation, the two inequalities of Theorem 476 become

$$h(P+Q) \leqslant 2h(P) + 2h(Q) + C_1 \quad \text{for all } P, Q \in E(\mathbb{Q}), \tag{25.5.22}$$

$$h(2P) \geqslant 4h(P) - C_2 \quad \text{for all } P \in E(\mathbb{Q}), \tag{25.5.23}$$

where C_1 and C_2 are nonnegative constants depending only on E.

We shall now prove that there is a set of points $\mathcal{S} \subset E(\mathbb{Q})$ of bounded height such that every point in $E(\mathbb{Q})$ is a linear combination of the points in $\mathcal{S}$. This implies finite generation of $E(\mathbb{Q})$ (Theorem 470), since sets of bounded height are finite.

Theorem 475 tells us that there is a finite set of points $Q_1, \ldots, Q_k \in E(\mathbb{Q})$ such that every point in $E(\mathbb{Q})$ differs from some Q_j by a point in $2E(\mathbb{Q})$. We set

$$C_3 = \frac{1}{2}\max\left\{h\left(Q_j\right) : 1 \leqslant j \leqslant k\right\} + \frac{C_1 + C_2}{4}, \tag{25.5.24}$$

where C_1 and C_2 are the constants appearing in (25.5.22) and (25.5.23), respectively, and we define our finite set of points $\mathcal{S} \subset E(\mathbb{Q})$ by

$$\mathcal{S} = \{R \in E(\mathbb{Q}) : h(R) \leqslant 2C_3 + 1\}. \tag{25.5.25}$$

Note in particular that $Q_1, \ldots, Q_k$ are in $\mathcal{S}$.

Let $P_0 \in E(\mathbb{Q})$ be an arbitrary nonzero point in $E(\mathbb{Q})$. We inductively define a sequence of indices $j_0, j_1, j_2, \ldots$ and points $P_0, P_1, P_2, \ldots$ in $E(\mathbb{Q})$ satisfying

$$P_0 = 2P_1 + Q_{j_1}, \quad P_1 = 2P_2 + Q_{j_2}, \quad P_2 = 2P_3 + Q_{j_3}, \ldots. \tag{25.5.26}$$

The choice of the successive P_i and j_i need not be unique, but Theorem 475 ensures that at each stage there is at least one choice. We apply first (25.5.23) and then (25.5.22) to show that the heights of the P_i are rapidly decreasing. Thus

$$\begin{aligned} h(P_i) &\leqslant \frac{1}{4}\left(h(2P_i) + C_2\right) = \frac{1}{4}\left(h(P_{i-1} - Q_{j_i}) + C_2\right) \\ &\leqslant \frac{1}{4}\left(2h(P_{i-1}) + 2h(Q_{j_i}) + C_1 + C_2\right) \\ &\leqslant \frac{1}{4}h(P_{i-1}) + C_3, \end{aligned} \tag{25.5.27}$$

where C_3 is defined by (25.5.24), and we have used the fact that $h(-Q) = h(Q)$, since $h(Q)$ depends only on x_Q.

We apply (25.5.27), starting at P_n and working backwards to P_0,

$$h(P_n) \leqslant \frac{1}{2^n}h(P_0) + \left(1 + \frac{1}{2} + \frac{1}{4} + \cdots + \frac{1}{2^{n-1}}\right)C_3 \leqslant \frac{1}{2^n}h(P_0) + 2C_3.$$

Hence if we choose n to satisfy $2^n \geqslant h(P_0)$, then the point P_n is in the set $\mathcal{S}$ defined by (25.5.25). Finally, using back-substitution on the sequence of equations (25.5.26) shows that

$$P_0 = 2^n P_n + \sum_{i=1}^{n} 2^{i-1} Q_{j_i},$$

so the original point P_0 is a linear combination of points in $\mathcal{S}$. This completes the proof that the finite set $\mathcal{S}$ is a generating set for the group $E(\mathbb{Q})$.

25.6. The group of points modulo p. It is instructive to investigate elliptic curves whose coefficients lie in other fields, for example the field of p elements, which we denote by $\mathbb{F}_p$.† The mod p points on the curve,

$$E(\mathbb{F}_p) = \left\{(x, y) \in \mathbb{F}_p^2 \colon y^2 \equiv x^3 + Ax + B \pmod{p}\right\} \cup \{\mathcal{O}\},$$

can be added to one another via the usual addition formulae (25.2.2)–(25.2.8), and they satisfy the usual properties as described in Theorem 462.

We can use the Legendre symbol (§ 6.5) to count the number of points in $E(\mathbb{F}_p)$ by applying the fact that the congruence $y^2 \equiv a \pmod{p}$ has $1 + \left(\frac{a}{p}\right)$ solutions. Thus

$$\#E(\mathbb{F}_p) = 1 + \sum_{x=0}^{p-1}\left(1 + \left(\frac{x^3 + Ax + B}{p}\right)\right) = p + 1 + \sum_{x=0}^{p-1}\left(\frac{x^3 + Ax + B}{p}\right).$$

We would expect the quantity $\left(\frac{x^3+Ax+B}{p}\right)$ to be $+1$ and -1 approximately equally often, so $\#E(\mathbb{F}_p)$ should be approximately $p + 1$. The validity of this heuristic argument is put into a precise form in a theorem due to Hasse.

THEOREM 477*. *Let p be a prime number and let E be an elliptic curve with coefficients in the finite field $\mathbb{F}_p$ of p elements. Then the number of points of E with coordinates in $\mathbb{F}_p$ satisfies the estimate*

$$\left|\#E\left(\mathbb{F}_p\right) - (p + 1)\right| < 2\sqrt{p}.$$

† For simplicity, we assume that p is an odd prime. In order to work with elliptic curves over $\mathbb{F}_2$ or over other fields of characteristic 2, it is necessary to use a generalized Weierstrass equation (25.3.2) with a correspondingly more complicated expression (25.3.7) for the discriminant as discussed in § 25.3.

25.7. Integer points on elliptic curves. Elliptic curves frequently have infinitely many points with rational coordinates, since the sum of two rational points is again a rational point. The situation for points with integer coordinates is much different, since a perusal of the rational functions used in the addition formulae (25.2.2)–(25.2.8) makes it clear that the sum of integer points need not be an integer point.

The principal theorem in this area, due to Siegel, says that an elliptic curve has only finitely many integer points. We start by proving three elementary cases of Siegel's theorem, continue with an example showing the close connection between integer points on (elliptic) curves and the theory of Diophantine approximation (Chapter XI), and conclude with the full statement of Siegel's result.

THEOREM 478*. *The equation*

$$y^2 = x^3 + 7 \tag{25.7.1}$$

has no solutions in integers.†

Suppose that (x, y) is an integer solution to (25.7.1). Note that x cannot be even, since a number of the form $8k + 7$ cannot be a square. We rewrite (25.7.1) as

$$y^2 + 1 = x^3 + 8 = (x + 2)\left(x^2 - 2x + 4\right). \tag{25.7.2}$$

Since x is odd, we have

$$x^2 - 2x + 4 = (x - 1)^2 + 3 \equiv 3 \pmod 4,$$

so there exists some prime $p \equiv 3 \pmod 4$ dividing $x^2 - 2x + 4$. Then (25.7.2) implies that

$$y^2 \equiv -1 \pmod p,$$

which is a contradiction of Theorem 82. Hence (25.7.1) has no integer solutions.

THEOREM 479*. *The only solutions in integers to the equation*

$$y^2 = x^3 - 2 \tag{25.7.3}$$

are $(x, y) = (3, \pm 5)$.

† In fact, equation (25.7.1) has no solutions in rational numbers, but the proof requires different methods and is significantly more difficult.

We work in the ring of integers in the quadratic field $k(\sqrt{-2})$, which according to Theorem 238 is the set of numbers of the form

$$a + b\sqrt{-2} \quad \text{with} \quad a, b \in \mathbb{Z}.$$

The field $k(\sqrt{-2})$ is a Euclidean field (Theorem 246), so its elements have unique factorization into primes, and its only unities are ± 1 (Theorem 240).

We now suppose that (x, y) is a solution in rational integers to (25.7.3). Our first observation is that x and y must be odd, since if $2 \mid x$, then

$$y^2 \equiv -2 \pmod 8,$$

which is not possible.

In the ring of integers of $k(\sqrt{-2})$ we have the factorization

$$x^3 = y^2 + 2 = (y + \sqrt{-2})(y - \sqrt{-2}). \tag{25.7.4}$$

Any common factor of $y + \sqrt{-2}$ and $y - \sqrt{-2}$ must divide their sum $2y$ and their difference $2\sqrt{-2}$. But neither factor in (25.7.4) is divisible by $\sqrt{-2}$, since y is odd, so they have no common prime factors. Hence (25.7.4) implies that each factor is a cube in the ring of integers of $k(\sqrt{-2})$, say

$$y + \sqrt{-2} = \xi^3 \quad \text{and} \quad y - \sqrt{-2} = \eta^3. \tag{25.7.5}$$

Subtracting the second equation in (25.7.5) from the first yields

$$2\sqrt{-2} = \xi^3 - \eta^3 = (\xi - \eta)\left(\xi^2 + \xi\eta + \eta^2\right). \tag{25.7.6}$$

The equations (25.7.5) are complex conjugates of one another, so if we write $\xi = a + b\sqrt{-2}$, then $\eta = a - b\sqrt{-2}$, and (25.7.6) becomes

$$2\sqrt{-2} = 2b\sqrt{-2}\left(3a^2 - 2b^2\right).$$

Hence $b = 1$ and $a = \pm 1$, which yields $y = \pm 5$ and $x = 3$.

THEOREM 480*. *Let A be a nonzero integer. Then every solution in integers to the equation*

$$x^3 + y^3 = A \quad \textit{satisfies} \quad x^2 + y^2 \leqslant 2|A|.$$

The elementary proof of Theorem 480 hinges on the fact that the cubic form $x^3 + y^3$ factors as

$$x^3 + y^3 = (x + y)(x^2 - xy + y^2) = A.$$

Since $x + y \neq 0$, we have $|x + y| \geqslant 1$, so

$$|A| \geqslant |x^2 - xy + y^2| \geqslant \frac{1}{2}(x^2 + y^2).$$

It is natural to attempt to repeat the proof of Theorem 480 for equations such as

$$x^3 + 2y^3 = A$$

by using the factorization

$$(x + \sqrt[3]{2}y)(x^2 - \sqrt[3]{2}xy + \sqrt[3]{4}y^2) = A.$$

It turns out that the integers in the field $k(\sqrt[3]{2})$ satisfy the fundamental theorem, but the existence of infinitely many unities prevents the elementary proof from succeeding. In general, the existence of integral points on elliptic curves is closely tied up with the theory of Diophantine approximation.

THEOREM 481*. *Let d be an integer that is not a perfect cube and let A be a nonzero integer. Then the equation*

$$x^3 + dy^3 = A \tag{25.7.7}$$

has only finitely many solutions in integers.

In order to prove Theorem 481, we require a result on Diophantine approximation that is stronger than Theorem 191. Such estimates were proven by Thue, Siegel, Gelfond, and Dyson before culminating in the following theorem of Roth (see the Notes to Chapter XI).

THEOREM 482*. *Let ξ be an algebraic number of degree at least 2 as defined in § 11.5. Then for every $\epsilon > 0$ there is a positive constant C, depending on ξ and ϵ, so that*

$$\left|\frac{a}{b} - \xi\right| \geqslant \frac{C}{b^{2+\epsilon}}$$

for all rational numbers a/b written in lowest terms with $b > 0$.

The proof of Theorem 482, or even a weaker version in which the exponent on b is any value strictly smaller than the degree of ξ, would take us too far afield. So we shall be content to use Theorem 482 in order to prove Theorem 481.

To ease notation, we let $\delta = \sqrt[3]{d}$,and we let $\rho = \frac{1}{2}\left(-1+\sqrt{-3}\right)$be a cube root of unity as in Chapter XII. We also replace y by $-y$, so equation (25.7.7) factors completely as

$$x^3 - dy^3 = (x-\delta y)(x-\rho\delta y)(x-\rho^2\delta y) = A.$$

We divide by y^3 to obtain

$$\left(\frac{x}{y}-\delta\right)\left(\frac{x}{y}-\rho\delta\right)\left(\frac{x}{y}-\rho^2\delta\right) = \frac{A}{y^3}. \tag{25.7.8}$$

The real number x/y cannot be close to either of the complex numbers $\rho\delta$ or $\rho^2\delta$. Indeed,

$$\left|\frac{x}{y}-\rho\delta\right| \geqslant \operatorname{Im}(\rho\delta) = \frac{\sqrt{3}\delta}{2},$$

and similarly for $|x/y - \rho^2\delta|$. Hence (25.7.8) leads to the estimate

$$\frac{|A|}{|y|^3} \geqslant \left|\frac{x}{y}-\sqrt[3]{d}\right|\left(\frac{\sqrt{3}\delta}{2}\right)^2.$$

Thus there is a constant C', which is independent of x and y, such that

$$\frac{C'}{|y|^3} \geqslant \left|\frac{x}{y}-\sqrt[3]{d}\right|. \tag{25.7.9}$$

We now apply Theorem 482 with $\epsilon = \frac{1}{2}$ to the algebraic number$\sqrt[3]{d}$,which gives a corresponding lower bound

$$\left|\frac{x}{y}-\sqrt[3]{d}\right| \geqslant \frac{C}{|y|^{5/2}}. \tag{25.7.10}$$

Combining (25.7.9) and (25.7.10) yields

$$(C'/C)^2 \geqslant |y|,$$

which shows that y takes on only finitely many values. Finally, the equation $x^3 + 2y^3 = A$ shows that each value of y leads to only finitely many values for x.

An argument similar to, but significantly more complicated than, the proof of Theorem 481 was used by Siegel to show that an analogous result is true for all elliptic curves.

THEOREM 483*. *Let E be an elliptic curve given by an equation having rational coefficients. Then E has only finitely many points with integer coordinates. In particular, the equation*

$$y^2 = x^3 + Ax + B \quad \text{with } A, B \in \mathbb{Z} \quad \text{and} \quad 4A^3 + 27B^2 \neq 0$$

has only finitely many solutions in integers.

Siegel's proof of Theorem 483 yields a stronger result saying, in effect, that the numerators and the denominators of the coordinates of rational points have approximately the same size.

THEOREM 484*. *Let E be an elliptic curve given by an equation having rational coefficients and let $P_1, P_2, P_3, \ldots \in E(\mathbb{Q})$ be a sequence of distinct rational points. Write the x-coordinate of P_i as a fraction $x_{P_i} = \alpha_i/\beta_i$. Then*

$$\lim_{i\to\infty} \frac{\log|\alpha_i|}{\log|\beta_i|} = 1.$$

25.8. The L-series of an elliptic curve. Let E be an elliptic curve given by a minimal Weierstrass equation[†] (25.3.2). For every prime p, we reduce the coefficients of (25.3.2) modulo p and, provided that $p \nmid \Delta$, we obtain an elliptic curve E_p defined over the finite field $\mathbb{F}_p$. Theorem 477 tells us that the quantity

$$a_p = p + 1 - \#E(\mathbb{F}_p) \quad \text{satisfies} \quad |a_p| < 2\sqrt{p}. \tag{25.8.1}$$

(If $p|\Delta$, we still define a_p using (25.8.1). One can show in this case that $a_p \in \{-1, 0, 1\}$.)

It is convenient to encapsulate all of this mod p information into a generating function. The *L-series of E* is the infinite product

$$L(E, s) = \prod_{p|\Delta} \frac{1}{1 - a_p p^{-s}} \times \prod_{p\nmid\Delta} \frac{1}{1 - a_p p^{-s} + p^{1-2s}}. \tag{25.8.2}$$

[†] If we ignore the primes $p = 2$ and $p = 3$, then it suffices to take an equation (25.1.3) with $A, B \in \mathbb{Z}$ and $\gcd(A^3, B^2)$ 12th power free.

The product (25.8.2) defining the *L-series* can be formally expanded into a Dirichlet series

$$L(E,s) = \sum_{n\geqslant 1} \frac{a_n}{n^s} \tag{25.8.3}$$

using the geometric series

$$\frac{1}{1-a_p p^{-s}} = \sum_{k\geqslant 0} \frac{a_p^k}{p^{ks}} \quad \text{and} \quad \frac{1}{1-a_p p^{-s}+p^{1-2s}} = \sum_{k\geqslant 0} \left(\frac{a_p}{p^s} + \frac{1}{p^{2s-1}}\right)^k.$$

THEOREM 485*. *The coefficients a_n of the L-series $L(E,s)$ have the following properties:*

(25.8.4) $a_{mn} = a_m a_n$ *for all relatively prime m and n.*

(25.8.5) $a_p a_{p^k} = a_{p^{k+1}} + p a_{p^{k-1}}$ *for all prime powers p^k with $k \geqslant 1$.*

(25.8.6) $|a_n| \leqslant d(n)\sqrt{n}$ *for all $n \geqslant 1$.*

(Here $d(n)$ is the number of divisors of n, see § 16.7.)

The proofs of (25.8.4) and (25.8.5) are formal computations. First, comparing (25.8.2) and (25.8.3), we see that

$$L(E,s) = \prod_p \sum_{k\geqslant 0} \frac{a_{p^k}}{p^{sk}}. \tag{25.8.7}$$

Hence if we factor n as $n = p_1^{k_1} p_2^{k_2} \cdots p_t^{k_t}$, then

$$a_n = a_{p_1^{k_1}} a_{p_2^{k_2}} \cdots a_{p_t^{k_t}}.$$

In particular, $a_{mn} = a_m a_n$ if $\gcd(m,n) = 1$.

Next, for each prime $p \nmid \Delta$, we factor

$$1 - a_p X + pX^2 = (1-\alpha_p X)(1-\beta_p X) \quad \text{with } \alpha_p, \beta_p \in \mathbb{C}. \tag{25.8.8}$$

For $p|\Delta$, we set $\alpha_p = a_p$ and $\beta_p = 0$, and then in all cases, the p-factor in (25.8.2) is equal to

$$\text{(25.8.9)} \qquad \frac{1}{1-\alpha_p p^{-s}} \cdot \frac{1}{1-\beta_p p^{-s}} = \left(\sum_{i=0}^{\infty} \frac{\alpha_p^i}{p^{si}}\right) \cdot \left(\sum_{j=0}^{\infty} \frac{\beta_p^j}{p^{sj}}\right) = \sum_{k=0}^{\infty} \frac{1}{p^{sk}} \sum_{i+j=k} \alpha_p^i \beta_p^j.$$

(For $p|\Delta$, we set $0^0 = 1$ by convention.)

Comparing (25.8.9) and (25.8.7) yields

$$\text{(25.8.10)} \qquad a_{p^k} = \sum_{i+j=k} \alpha_p^i \beta_p^j = \frac{\alpha_p^{k+1} - \beta_p^{k+1}}{\alpha_p - \beta_p}.$$

Using (25.8.10) and the relation $\alpha_p \beta_p = p$ from (25.8.8), we compute

$$a_p a_{p^k} = (\alpha_p + \beta_p)\left(\frac{\alpha_p^{k+1} - \beta_p^{k+1}}{\alpha_p - \beta_p}\right) = \frac{\alpha_p^{k+2} - \beta_p^{k+2} + \alpha_p\beta_p\left(\alpha_p^k - \beta_p^k\right)}{\alpha_p - \beta_p} = a_{p^{k+1}} + p a_{p^{k-1}}.$$

We verify (25.8.6) by applying Theorem 477, which tells us that $|a_p| < 2\sqrt{p}$. This implies that the roots of the quadratic polynomial (25.8.8) are complex conjugates, hence α_p and β_p are complex conjugates whose product is equal to p. They thus satisfy

$$\text{(25.8.11)} \qquad |\alpha_p| = |\beta_p| = \sqrt{p}.$$

Applying (25.8.11) to (25.8.10) gives

$$\left|a_{p^k}\right| \leqslant \sum_{i+j=k} \left|\alpha_p^i \beta_p^j\right| = \sum_{i+j=k} p^{k/2} = (k+1)\,p^{k/2} = d(p^k)p^{k/2}.$$

Then the multiplicativity (25.8.4) of the a_n and the multiplicativity of $d(n)$ from Theorem 273 imply that $|a_n| \leqslant d(n)\sqrt{n}$.

THEOREM 486*. *The L-series $L(E,s)$ defined by* (25.8.2) *and* (25.8.3), *considered as a function of the complex variable s, is absolutely convergent* *for all* $\mathrm{Re}(s) > \frac{3}{2}$*and defines a nonvanishing holomorphic function in that region.*

The estimate (25.8.6) in Theorem 485 says that the Dirichlet coefficients of $L(E,s)$ satisfy $|a_n| \leqslant d(n)\sqrt{n}$. Theorem 315 tells us that the sum of divisors function is quite small,

$$d(n) = O(n^{\delta}) \quad \text{for any } \delta > 0.$$

We write $\sigma = \mathrm{Re}(s)$ and estimate the Dirichlet series (25.8.3) by

$$\sum_{n\geqslant 1} \left|\frac{a_n}{n^s}\right| \leqslant \sum_{n\geqslant 1} \frac{d(n)n^{1/2}}{n^{\sigma}} = O\left(\sum_{n\geqslant 1} \frac{1}{n^{\sigma-\frac{1}{2}-\delta}}\right).$$

Hence the Dirichlet series is absolutely convergent for $\mathrm{Re}(s) > \frac{3}{2} + \delta$, and since δ is arbitrary, $L(E,s)$ defines a holomorphic function on $\mathrm{Re}(s) > \frac{3}{2}$. Finally, the nonvanishing of $L(E,s)$ on the region $\mathrm{Re}(s) > \frac{3}{2}$ follows from its product expansion (25.8.2).

Although the series (25.8.2) defining $L(E,s)$ only converges for $\mathrm{Re}(s) > \frac{3}{2}$, the function that it defines is similar to the Riemann ζ-function in the sense that it has an analytic continuation and satisfies a functional equation. The next theorem represents a pinnacle of modern number theory, but its proof is far beyond the scope of this book.

THEOREM 487*. *The L-series $L(E,s)$ has an analytic continuation to the entire complex plane. Further, there is an integer N_E, the* conductor *of E, that divides the discriminant Δ such that the function*

$$\xi(E,s) = N_E^{s/2}\,(2\pi)^{-2}\,\Gamma(s)L(E,s)$$

satisfies the functional equation

$$\xi(E,2-s) = \pm\xi(E,s) \quad \textit{for all } s \in \mathbb{C}.$$

The L-series of an elliptic curve is built up out of purely local (mod p) information. A conjecture of Birch and Swinnerton-Dyer predicts that $L(E,s)$ contains a significant amount of global information concerning the rational points on the curve. For example, they conjecture that the order of vanishing of $L(E,s)$ at $s = 1$ equals the rank of the group of rational points $E(\mathbb{Q})$. In particular, $L(E,\ 1)$ should vanish if and only if $E(\mathbb{Q})$ contains infinitely many points. The small amount of progress that has been made on the conjecture of Birch and Swinnerton-Dyer, as described in the next theorem, requires a vast panoply of mathematical tools for its proof.

Theorem 488*. *If $L(E, 1) \neq 0$, then $E(\mathbb{Q})$ has rank 0; and if $L(E, 1) = 0$ and $L'(E, 1) \neq 0$, then $E(\mathbb{Q})$ has rank 1.*

25.9. Points of finite order and modular curves. We have seen in § 25.4 that any particular elliptic curve has only finitely many points of finite order having rational coordinates. In this section, we change our perspective and attempt to classify all elliptic curves having a point of a given finite order. Thus, for a given integer $N \geqslant 1$, we aim to describe the set of ordered pairs

$$\left\{(E, P) : \begin{array}{l} E \text{ is an elliptic curve and } P \text{ is} \\ \text{a point of exact order } N \text{ on } E \end{array}\right\}, \tag{25.9.1}$$

up to the natural equivalence relation in which any two pairs (E_1, P_1) and (E_2, P_2) are considered to be identical if there is an isomorphism $\phi : E_1 \to E_2$ satisfying $\phi(P_1) = P_2$. This is an example of what is known as a moduli problem.

For example, if $N = 1$, then we simply want to classify elliptic curves up to isomorphism. We already know how to do this using the j-invariant, since two curves E_1 and E_2 are isomorphic if and only if their j-invariants $j(E_1)$ and $j(E_2)$ are equal, cf. Theorem 461.

Theorem 489. *Let E be an elliptic curve given by an equation* (25.1.3) *with coefficients in a field k, and let $P \in E(k)$ be a point with coordinates in k and satisfying $2P \neq \mathcal{O}$ and $3P \neq \mathcal{O}$. Then there is a change of coordinates* (25.3.8) *with u, r, s, $t \in k$ that transforms E into an equation of the form*

$$y^2 + (w+1)\,xy + vy = x^3 + vx^2 \quad \textit{with } P = (0, 0)\,. \tag{25.9.2}$$

The discriminant of the elliptic curve (25.9.2) *is*

$$\Delta = -v^3\left(w^4 + 3w^3 + 8vw^2 + 3w^2 - 20vw + w + 16v^2 - v\right). \tag{25.9.3}$$

The values of w and v are uniquely determined by E and P.

Proof. We begin with the transformation

$$x \longmapsto x + x_P \quad \text{and} \quad y \longmapsto y + y_P,$$

which has the effect of moving P to the point (0, 0) and puts E into the form

$$y^2 + A_1 y = x^3 + B_1 x^2 + C_1 x.$$

The assumption that $2P \neq \mathcal{O}$ tells us that $A_1 \neq 0$ (cf. Theorem 464), so the substitution

$$y \longmapsto y + (C_1/A_1)\,x$$

puts E into the form

$$y^2 + A_2 xy + B_2 y = x^3 + C_2 x^2. \tag{25.9.4}$$

We note that the nonvanishing of the discriminant of (25.9.4) implies that $B_2 \neq 0$. Further, since $2P = (-C_2, A_2C_2 - B_2)$, we see that

$$3P = \mathcal{O} \Longleftrightarrow 2P = -P \Longleftrightarrow x_{2P} = x_P \Longleftrightarrow C_2 = 0.$$

Thus our assumption that $3P \neq \mathcal{O}$ implies that $C_2 \neq 0$, so we may make the substitutions

$$x \longmapsto (B_2/C_2)^2\,x \quad \text{and} \quad y \longmapsto (B_2/C_2)^3\,y.$$

This puts E into the desired form (25.9.2) with $w = A_2C_2/B_2 - 1$ and $v = C_2^3/B_2^2$.

The formula for the discriminant of (25.9.2) follows directly from the general discriminant formula (25.3.7).

In order to see that w and v are uniquely determined, we look at which change of variables (25.3.8) preserves the form of the equation (25.9.2) while simultaneously fixing the point (0, 0). The assumption that (0, 0) is fixed means that $r = t = 0$ in (25.3.8), and then the substitutions $x \to u^2x$ and $y \to u^3y + u^2sx$ transform (25.9.2) into

$$\begin{aligned} y^2 + u^{-1}\,(w + 1 + 2s)\,xy + u^{-3}vy \\ = x^3 + u^{-2}\left(v + s^2 + (w+1)\,s\right)x^2 + u^{-4}vsx. \end{aligned} \tag{25.9.5}$$

Comparing the x terms of (25.9.2) and (25.9.5) shows that $s = 0$ (note that $v \neq 0$ since $\Delta \neq 0$), and then the y and x^2 terms show that $u^3 = u^2 = 1$, so $u = 1$. Hence only the identity transformation preserves both equation (25.9.2) and the point (0, 0), and thus w and v are uniquely determined by E and P. □

We now show that solving our moduli problem (25.9.1) is equivalent to describing the solutions to a certain polynomial equation. In other words, the set of pairs (E,P) consisting of an elliptic curve E and a point P of order N is naturally parametrized by the solutions of a polynomial equation $\Psi_N(W,V)=0$.

THEOREM 490. *For any given values of w and v such that the discriminant* (25.9.3) *does not vanish, let $E_{w,v}$ be the elliptic curve*

$$E_{w,v}\colon y^2+(w+1)\,xy+vy=x^3+vx^2 \tag{25.9.6}$$

and let $P_{w,v}=(0,0)\in E_{w,v}$. Let $N\geqslant 4$ be an integer.

(*a*) *There is a nonzero polynomial $\Psi_N(W,V)$ with integer coefficients having the property that $P_{w,v}$ is a point of order N if and only if $\Psi_N(w,v)=0$.*

(*b*) *Let E be any elliptic curve given by an equation with coefficients in a field k and let $Q\in E(k)$ be a point of exact order N. Then there is a change of variables* (25.3.8) *with $u,r,s,t\in k$ that puts E into the form* (25.9.6) *and sends Q to $P=(0,0)$. The curve E and point Q uniquely determine w and v.*

Proof. (*a*) We treat $E_{W,V}$ as an elliptic curve over the field $\mathbb{Q}(W,V)$ of rational functions in two variables. Then the coordinates of the multiples of

$$P_{W,V}=(0,0)\in E_{W,V}$$

are quotients of polynomials in $\mathbb{Q}[W,V]$. More precisely, since the ring $\mathbb{Q}[W,V]$ has unique factorization, an argument similar to that used in Theorem 472 shows that if $NP_{W,V}\neq\mathcal{O}$, then we can write $NP_{W,V}$ as

$$NP_{W,V}=\left(\frac{\Phi_N(W,V)}{\Psi_N(W,V)^2},\frac{\Omega_N(W,V)}{\Psi_N(W,V)^3}\right)\quad\text{with }\Psi_N,\Phi_N,\Omega_N\in\mathbb{Z}[W,Z].$$

The polynomial $\Psi_N(W,V)$ vanishes at $(W,V)=(w,\ v)$ if and only if $P_{w,v}\ \in\ E_{w,v}$ is a point of order N, so it remains to prove that $NP_{W,V}\neq\mathcal{O}$.

We first consider the multiple

$$4P_{W,V}=\left(\frac{V^2-VW}{W^2},\frac{-V^2W^2+V^2W-V^3}{W^3}\right).$$

From this formula for $4P_{W,V}$ we see that for most choices of integers w and v, the coordinates of the point $4P_{w,v}$ are fractions that are *not* integers. For example, this is the case if $|w| > 1$ and $\gcd(2, v) = 1$. It follows from Theorem 466 that for such integer values of w and v, the point $4P_{w,v}$ is not a point of finite order, and hence that $nP_{w,v} \neq \mathcal{O}$ for all $n \geqslant 1$. This implies that $nP_{W,V} \neq \mathcal{O}$ for all $n \geqslant 1$ when we treat W and V as indeterminates, since otherwise $P_{w,v} \in E_{w,v}$ would have finite order when we substitute particular values for W and V.
(*b*) This is the special case of Theorem 489 in which we start with a point of finite order $N \geqslant 4$. □

Here are the polynomials $\Psi_N(W, V)$ for some small values of N:

$$\begin{aligned}
\Psi_5(W,V) &= W - V,\\
\Psi_6(W,V) &= W^2 - W + V,\\
\Psi_7(W,V) &= W^3 - VW + V^2,\\
\Psi_8(W,V) &= VW^3 + W^3 - 3VW^2 + 2V^2W,\\
\Psi_9(W,V) &= W^5 - W^4 + VW^3 + W^3 - 3VW^2 + 3V^2W - V^3.
\end{aligned}$$

The polynomials Ψ_5 and Ψ_6 are linear in V, so we can eliminate V from the equation $\Psi_N(W, V) = 0$ and create a universal one-parameter family of elliptic curves with a point of order 5 or 6. For example, up to isomorphism, every elliptic curve with a point P of order 6 can be put into the form

$$y^2 + (w+1)xy + (w - w^2)y = x^3 + (w - w^2)x^2, \quad P = (0,0).$$

It is also possible to parametrize the solutions to $\Psi_N(W, V) = 0$ for $N = 7$, 8, and 9. For example, the curve $\Psi_7(W, V) = 0$ may be parametrized using the parameter $Z = V/W$. Then $W = Z - Z^2$ and $V = Z^2 - Z^3$, so every elliptic curve with a point P of order 7 can be put in the form

$$y^2 + (1 + z - z^2)xy + (z^2 - z^3)y = x^3 + (z^2 - z^3)x^2, \quad P = (0,0).$$

However, as the value of N increases, it is no longer possible to describe the solutions to $\Psi_N(W, V) = 0$ using a single parameter. The *modular curve*

$X_1(N)$ is defined to be the plane curve given by the equation[†]

$$X_1(N) = \{(w, v) \colon \Psi_N(w, v) = 0\}.$$

The increasing complexity of $X_1(N)$ as N increases may be measured by studying the points of $X_1(N)$ having complex coordinates, i.e. the complex solutions to the equation $\Psi_N = 0$. For $N \leqslant 10$ and $N = 12$, the complex points $X_1(N)(\mathbb{C})$ form a sphere (a 0-holed torus),[‡] and it is exactly in these cases that $X_1(N)$ is parametrizable by a single parameter. The curves $X_1(11)$ and $X_1(13)$ turn out themselves to be elliptic curves, so their complex points are 1-holed tori. As N increases, the complex points $X_1(N)(\mathbb{C})$ form a g_N-holed torus, where the genus g_N goes to infinity with N. For prime values of N, the genus g_N is approximately $N/12$.

Mazur used modular curves to prove the following strong uniformity bound for rational points of finite order on elliptic curves.

THEOREM 491*. *Let E be an elliptic curve given by an equation with rational coefficients and let $P \in E(\mathbb{Q})$ be a point of exact order N. Then either $N \leqslant 10$ or $N = 12$.*

In order to prove Theorem 491, one shows that if $N = 11$ or $N \geqslant 13$, then the only solutions to $\Psi_N(w, v) = 0$ in rational numbers w and v are solutions for which the discriminant (25.9.3) vanishes. Since such solutions (w, v) do not correspond to actual elliptic curves, Theorem 491 then follows from Theorem 490. The proof that $\Psi_N(w, v) = 0$ has no nontrivial rational solutions requires a detailed analysis of the curve $X_1(N)$ and deep tools from modern algebraic geometry.

25.10. Elliptic curves and Fermat's last theorem. Fermat's last theorem, already alluded to in Chapter XIII, was stated by Fermat in the 17th century and proven by Andrew Wiles in the 20th.

THEOREM 492*. *Let $n \geqslant 3$ be an integer. Then the equation*

$$a^n + b^n = c^n$$

has no solutions in nonzero integers a, b, c.

[†] This definition of $X_1(N)$ is not quite accurate, although it will suffice for our purposes. In general, the equation $\Psi_N = 0$ has singularities and is missing points 'at infinity.' The correct definition of $X_1(N)$ is that it is the desingularization of the compactification of the curve $\Psi_N = 0$.

[‡] For example, $X_1(5)(\mathbb{C})$ is the compactification of the set $\{(w, v) \in \mathbb{C}^2 : w - v = 0\}$. This set is a copy of the complex plane $\mathbb{C}$, and the (one point) compactification of $\mathbb{C}$ is a two-dimensional sphere.

It clearly suffices to prove Theorem 492 for $n = 4$ and $n = p$ an odd prime, and since Theorems 226 and 228 cover the cases $n = 4$ and $n = 3$, respectively, it suffices to prove that there are no solutions in nonzero integers to the equation

$$a^p + b^p = c^p, \qquad \text{where } p \geqslant 5 \text{ is prime.} \tag{25.10.1}$$

Dividing by any common factor, we may further assume that a, b, and c are pairwise relatively prime.

Setting $u = a/c$ and $v = b/c$, Fermat's last theorem reduces to the statement that the equation

$$u^p + v^p = 1 \tag{25.10.2}$$

has no solutions in nonzero rational numbers u and v. This equation defines a curve, but it is most definitely not an elliptic curve.† So instead of working directly with (25.10.2), we use a hypothetical solution to (25.10.1) to define an elliptic curve

$$E_{a,b,c}\colon Y^2 = X(X + a^p)(X - b^p).$$

Using the general discriminant formula (25.3.7) from § 25.3, we find that the discriminant of $E_{a,b,c}$ is‡

$$\Delta_{a,b,c} = 16a^{2p}b^{2b}\left(a^p + b^p\right)^2 = 16\,(abc)^{2p}. \tag{25.10.3}$$

An elliptic curve whose discriminant is (essentially) a perfect $2p$th power would be a strange animal, indeed! The proof of Fermat's last theorem lies in showing that such a curve cannot exist and comes down to proving the following two statements:

- The elliptic curve $E_{a,b,c}$ is not modular.
- The elliptic curve $E_{a,b,c}$ is modular.

There are a number of equivalent definitions of what it means for an elliptic curve to be modular, but unfortunately, as bare definitions, they are not very illuminating. In keeping with the scope of this book, we give a definition that is purely algebraic, but we note that the underlying motivation lies in the analytic theory of modular forms and L-series.

† The complex points of the compactified Fermat curve $u^n + v^n = 1$ form an $\frac{(n-1)(n-2)}{2}$-holed torus, so the Fermat curve is an elliptic curve only for $n = 3$.

‡ After a simple change of variables, the discriminant (25.3.7) becomes simply $(abc)^{2p}$.

For each $N \geqslant 1$ we defined in § 25.9 the modular curve $X_1(N)$ whose points classify pairs (C, P) consisting of an elliptic curve C and a point P of order N. (We call the elliptic curve C to distinguish it from E.) We now say that an elliptic curve E is *modular* if E can be covered by some modular curve, i.e. if there is a covering map

$$X_1(N) \to E \tag{25.10.4}$$

defined by rational functions. The smallest N for which there exists a covering map (25.10.4) is called the *conductor* of E.

After Frey suggested that the elliptic curves $E_{a,b,c}$ created from putative Fermat equation solutions should not be modular, Serre described a 'level-lowering' conjecture which implied that if $E_{a,b,c}$ were modular, then the special form (25.10.3) of its discriminant would force the conductor to divide 4. But the complex points of $X_1(N)$ for $N \leqslant 4$ are spheres (0-holed tori), and a sphere cannot be continuously mapped onto the complex points of an elliptic curve (a 1-holed torus). Ribet subsequently proved Serre's conjecture, which showed that Frey's intuition was correct: the elliptic curve $E_{a,b,c}$ is not modular.

It is not clear why this should be surprising. The points of $X_1(N)$ solve a classification problem related to elliptic curves, but there is no reason, *a priori,* to expect any particular elliptic curve to admit a covering map from some $X_1(N)$. However, earlier work of Eichler, Shimura, Taniyama, and Weil suggested that every elliptic curve given by an equation with rational coefficients should be modular.

Thus the final step in the proof of Fermat's last theorem was to show that all, or at least most, elliptic curves are modular. This was done by Wiles, who, with assistance by Taylor for one step of the proof, proved that every semistable elliptic curve is modular.† Since the $E_{a,b,c}$ curves, if they existed, would be semistable, this completed the proof of Fermat's last theorem. Building on Wiles' work, Breuil, Conrad, Diamond, and Taylor subsequently completed the proof of the full modularity conjecture, whose proof is far beyond the scope of this book.

THEOREM 493*. *Every elliptic curve given by an equation with rational coefficients is modular.*

† Aside from some special conditions at 2 and 3, an elliptic curve $Y^2 = X^3 + AX + B$ is semistable if $\gcd(A, B) = 1$.

NOTES

§ 25.1. Some cases of rational right triangles with rational area were studied in ancient Greece, but the systematic study of congruent numbers began with Arab scholars during the 10th century. Arab mathematicians tended to use the equivalent characterization, also known to the Greeks, that n is a congruent number if and only if there is a rational number x such that both $x^2 + n$ and $x^2 - n$ are squares of rational numbers. See Dickson *History,* ii, ch. xvi, for additional information on the mathematical history of congruent numbers.

There exists a vast literature on elliptic curves,† including many textbooks devoted to their number theoretic properties. The reader may consult the books of Cassels, Knapp, Koblitz, Lang, Silverman, and Silverman–Tate for proofs of the unproven theorems in this chapter (other than those in §§ 25.8–25.10) and for much additional basic material.

§ 25.2. The genesis of the name 'elliptic curve' is from the integrals that arise when computing the arc length of an ellipse. After an algebraic substitution, such integrals take the form $\int R(x)dx/\sqrt{x^3 + Ax + B}$ for some rational function $R(x)$. These *elliptic integrals* may be viewed as integrals $\int R(x)dx/y$ on the curve (Riemann surface) $y^2 = x^3 + Ax + B$, hence the name elliptic curve.

Special cases of the duplication and composition law on elliptic curves, described algebraically, date back to Diophantus, but it appears that the first geometric description via secant lines is due to Newton, *Mathematical Papers,* iv, 1674–1684, Camb. Univ. Press, 1971, 110–115. A nice historical survey of the composition law is given by Schappacher, *Sém. Théor. Nomb. Paris* 1988–1989, *Progr. Math.* 91 (1990), 159–84.

A proof that addition on an elliptic curve is associative (Theorem 462(c)) may be found in the standard texts listed earlier.

Theorem 463 was first observed by Poincaré, *Jour. Math. Pures Appl.* 7 (1901).

Elliptic curves with complex multiplication have many special properties not shared by general elliptic curves. In particular, if the endomorphism ring of such a curve E is a subring of the quadratic imaginary field k, then Abel, Jacobi, Kronecker,... proved that the coordinates of the points of finite order in E can be used to generate abelian extensions of k that are natural analogues of the cyclotomic extensions of $\mathbb{Q}$, i.e. the extensions of $\mathbb{Q}$ generated by roots of unity. In particular, $k(j(E))$ is the *Hilbert class field of k,* the maximal abelian unramified extension of k.

§ 25.3. It is easy to create a Weierstrass equation that is minimal except possibly for the primes 2 and 3. An algorithm of Tate (*Lecture Notes in Math.* (Springer), 476 (1975), 33–52) handles all primes.

§ 25.4. Theorem 466 was proven independently by Nagell (*Wid Akad. Skrifter Oslo I,* 1 (1935)) and Lutz (*J. Reine Angew. Math.* 177 (1937), 237–47). The proof that we give follows Tate's 1961 Haverford lectures as they appear in Silverman–Tate, *Rational points on elliptic curves.*

A modern formulation of Theorem 469 says that the group of p-adic points $E(\mathbb{Q}_p)$ has a filtration by subgroups $E_k\left(\mathbb{Q}_p\right) = \left\{(z,w) \in E\left(\mathbb{Q}_p\right) : v_p(z) \geqslant k\right\}$ for $k = 1, 2, \ldots$. Further, the map $P \mapsto z_P$ induces an isomorphism $E_k(\mathbb{Q}_p)/E_{k+1}(\mathbb{Q}_p) \to p^k\mathbb{Z}/p^{k+1}\mathbb{Z}$. The groups $E_1(\mathbb{Q}_p)$ and $p\mathbb{Z}_p$ are isomorphic as p-adic Lie groups via a map $P \mapsto l_p(z_P)$, where $\ell_p(T) \in \mathbb{Q}_p[[T]]$ is a certain p-adically convergent power series.

See also Theorem 491 and the notes for Section 25.9 for uniform bounds for points of finite order.

§ 25.5. Theorem 470 is due to Mordell, *Proc Camb. Philos. Soc.,* 21 (1922), 179–92. It was generalized by Weil (*Acta Math.* 52 (1928), 281–315) to number fields and to

† MathSciNet lists almost 2000 papers whose title includes the phrase 'elliptic curve'.

abelian varieties (higher dimensional analogues of elliptic curves), and thus is known as the Mordell–Weil theorem. Theorem 475, or more generally the finiteness of the quotient $E(\mathbb{Q})/mE(\mathbb{Q})$ for all $m \geqslant 1$, is called the 'weak' Mordell–Weil theorem. The structure theorem for finitely generated abelian groups is well-known and may be found in any basic algebra text.

It is conjectured that there are elliptic curves for which $E(\mathbb{Q})$ has arbitrarily high rank. The largest known example is a curve of rank at least 28 that was discovered by Elkies in May 2006. (See Elkies survey article arxiv.org/abs/0709.2908).

Somewhat surprisingly, there is still no proven algorithm for computing the group of rational points on an elliptic curve. All known proofs of Theorem 475 are ineffective in the sense that they do not provide an algorithm for constructing a suitable set of points $Q_1,\ldots,Q_k$ covering all of the congruence classes in the finite quotient group $E(\mathbb{Q})/2E(\mathbb{Q})$. If such points are known, then the remainder of the proof of Theorem 470 is effective, since the constants in Theorem 476 may easily be made effective. There is also an algorithm, due to Manin (*Russian Math. Surveys*, (6) 26 (1971), 7–78), that is effective conditional on various standard, but very deep, conjectures. In practice, there are powerful computer programs, such as Cremona's mwrank (www.maths.nott.ac.uk/personal/jec/mwrank/), that are usually able to compute generators for $E(\mathbb{Q})$ if the coefficients of E are not too large.

Theorem 476 suggests that the height function $h : E(\mathbb{Q}) \to [0, \infty)$ resembles a quadratic form. Néron (*Ann. of Math.* (2) 82 (1965), 249–331) and Tate (unpublished) proved that the limit $\hat{h}(P) = \lim_{n\to\infty} n^{-2}h(nP)$ exists, differs from h by $O(1)$, and is a quadratic form on $E(\mathbb{Q})$ whose extension to $E(\mathbb{Q}) \otimes \mathbb{R}$ is nondegenerate. The function $\hat{h}$, which is called the *canonical* (or *Néron–Tate*) *height*, has many applications. For example, Néron (*op. cit.*) showed that $\#\{P \in E(\mathbb{Q}) : h(P) \leqslant T\} \sim C_E.T^{1/2 \operatorname{rank} E(\mathbb{Q})}$ as $T \to \infty$.

§ 25.6. Theorem 477 is due to Hasse, *Vorläufige Mitteilung, Nachr. Ges. Wiss. Göttingen I, Math.-Phys. Kl. Fachgr. I Math.* 42 (1933), 253–62. A vast generalization to varieties of arbitrary dimension was proposed by Weil (*Bull. Amer. Math. Soc.* 55 (1949), 497–508) and proven by Deligne (*IHES Publ. Math.* 43 (1974), 273–307).

It is an interesting computational problem to compute $\#E(\mathbb{F}_p)$ when p is large. The first polynomial time algorithm is due to Schoof (*Math. Comp.* 44 (1985), 483–94), who also used it to give the first polynomial time algorithm for computing square roots in $\mathbb{F}_p$. A more practical version, although not provably polynomial time, was devised by Elkies and Atkins and is now known as the SEA algorithm (*J. Théor Nombres Bordeaux,* 7 (1995), 219–54). Satoh (*J. Ramanujan Math. Soc.* 15 (2000), 247–70) used cohomological ideas to give a faster algorithm to count $\#E(\mathbb{F}_q)$ when q is a large power of a small prime. Such point counting algorithms have applications to cryptography.

Given two points P and Q in $E(\mathbb{F}_p)$ such that Q is a multiple of P, the problem of determining an integer m with $Q = mP$ is called the *elliptic curve discrete logarithm problem* (ECLDP). The fastest known algorithms for solving the ECDLP are collision algorithms that take $O(\sqrt{p})$ steps. These exponential-time algorithms may be contrasted with the subexponential index calculus, which solves the analogous problem for $\mathbb{F}_p^*$ in $O\left(e^{c(\log p)^{1/3}(\log\log p)^{2/3}}\right)$ steps. The lack of an efficient algorithm to solve the ECDLP led Koblitz (*Math. Comp.* 48 (1977), 203–9) and V. Miller (*Lecture Notes in Comput. Sci.* (Springer), 218 (1986), 417–26) independently to suggest the use of elliptic curves for the construction of public key cryptographic protocols. Thus in addition to any purely intrinsic mathematical interest that the ECDLP might inspire, the existence or nonexistence of faster algorithms to solve the ECDLP is of great practical and finanical importance.

§ 25.7. Theorem 478 is due to V.A. Lebesgue (1869) and Theorem 479 is due to Fermat.

Theorem 483 is due to Siegel (*J. London Math. Soc.* 1 (1926), 66–68 and *Collected Works,* Springer, 1966, 209–66), who gave two different proofs, neither of which provided an effective bound for the size of the solutions. This was remedied by Baker (*J. London Math. Soc.* (1968) 43, 1–9), whose estimates for linear forms in logarithms (*Mathematika* 13 (1966), 204–16; 14 (1967), 102–7; 14 (1967), 220–8) provide effective Diophantine approximation estimates that can be used to prove effective bounds for integer points on elliptic curves. Building on work of Vojta (*Ann. of Math.* 133 (1991), 509–48), Faltings (*Ann. of Math.* 133 (1991), 549–76) generalized Siegel's theorem by proving that an affine subvariety of an abelian variety has only finitely many integral points.

It is trivial to produce Weierstrass equations (25.1.3) having arbitrarily many integer solutions by clearing the denominators of rational solutions. Using this method, Silverman (*J. London Math. Soc.* 28 (1983), 1–7) showed that if there exists an elliptic curve E whose group of rational points $E(\mathbb{Q})$ has rank r, then there exist infinitely many Weierstrass equations (25.1.3) having $\gg (\log \max\{|A|, |B|\})^{r/(r+2)}$ integer solutions.

Lang (*Elliptic Curves: Diophantine Analysis,* Springer, 1978, page 140) conjectured that the number of integer points on a minimal Weierstrass equation should be bounded by a quantity depending only on the rank of the group of rational points. This conjecture was proven for elliptic curves with integral j-invariant by Silverman (*J. Reine Angew. Math.* 378 (1987), 60–100) and, conditional on the *abc*-conjecture of Masser and Oesterlé (see notes to ch. XIII), for all elliptic curves by Hindry and Silverman (*Invent. Math.* 93 (1988), 419–50).

§ 25.8. The quantity a_p defined by (25.8.1) is called the *trace of Frobenius,* because it is the trace of the p-power Frobenius map in the Galois group $\mathrm{Gal}(\bar{\mathbb{Q}}/\mathbb{Q})$ acting as a linear map on the group of points of l-power order in E, where l is any prime other than p.

A conjecture of Sato and Tate (independently) describes the variation of a_p, and thus of $\#E(\mathbb{F}_p)$, as p varies. Theorem 477 says that there is an angle $0 \leqslant \theta_p \leqslant \frac{\pi}{2}$ such that $\cos\theta_p = a_p/2\sqrt{p}$. The Sato–Tate conjecture asserts that for $0 \leqslant \alpha < \beta \leqslant \frac{\pi}{2}$, the density of $\{p{:}\alpha \leqslant \theta_p \leqslant \beta\}$ within the set of primes is $\frac{2}{\pi}\int_\alpha^\beta \sin^2(t)\,dt$. Taylor (*IHES publ. Math.* submitted 2006), building on earlier joint work with Clozel and M. Harris (*IHES Publ. Math.* submitted 2006) and with M. Harris and Sheppard-Barron (*Ann. of Math.* to appear), has proven the Sato–Tate conjecture for elliptic curves whose j-invariant is not an integer.

Theorem 487 was proven by Deuring (*Nachr. Akad. Wiss. Göttingen. Math.-Phys. Kl. Math.-Phys.-Chem. Abt.* (1953), 85–94) for elliptic curves with complex multiplication, by Wiles (*Ann. of Math.* 141 (1995), 443–551), with assistance from Taylor (*Ann. of Math.* 141 (1995), 553–72), for semistable eliptic curves (roughly, curves given by an equation (25.1.3) with $\gcd(A, B) = 1$), and in full generality by Breuil, B. Conrad, Diamond, and Taylor, *J. Amer. Math. Soc.* 14 (2001), 843–939. See § 25.10 and its notes for the connections with Fermat's last theorem.

The conjecture that $\mathrm{ord}_{s=1} L(E, s) = \mathrm{rank}\ E(\mathbb{Q})$, and a refined version describing the leading Taylor coefficient of $L(E, s)$ at $s = 1$, were proposed by Birch and Swinnerton-Dyer (*J. Reine Angew. Math.* 218 (1965), 79–108). An early partial result of Coates and Wiles (*Invent. Math.* 39 (1997), 223–51) showed that if E has complex multiplication and if $L(E, 1) \neq 0$, then $E(\mathbb{Q})$ is finite. Theorem 488 is an amalgamation of work of Gross and Zagier (*Invent. Math.* 84 (1986), 225–320) and Kolyvagin (*Izv. Akad. Nauk SSSR Ser. Mat.* 52 (1988), 522–40, 670–1), combined with Wiles' et al. proof of the Modularity Conjecture (essentially Theorem 487). The conjecture of Birch and Swinnerton-Dyer is one of the seven Millennium Problems proposed by the Clay Mathematics Institute (www.claymath.org/millennium/). Gross and Zagier (*op. cit.*) further show that if $L(E, 1) = 0$ and $L'(E, 1) \neq 0$, then $L'(E, 1) = r\Omega\hat{h}(P)$, where $r \in \mathbb{Q}$, Ω is the value of an elliptic integral, and $\hat{h}(P)$ is the canonical height of a point $P \in E(\mathbb{Q})$ constructed using a method due to Heegner.

A weak form of the Birch–Swinnerton-Dyer conjecture implies that every integer $m \equiv 5, 6, 7 \pmod 8$ is a congruent number. Assuming the same weak form of the Birch–Swinnerton-Dyer conjecture, Tunnell (*Invent. Math.* 72 (1983), 323–34) proved that if m is a squarefree odd integer and if the number of integer solutions to $2x^2 + y^2 + 8z^2 = m$ is twice the number of integer solutions to $2x^2 + y^2 + 32z^2 = m$, then m is a congruent number. He also showed that the converse holds unconditionally, and that similar results hold for squarefree even integers.

§ 25.9. The analytic theory of modular curves and modular functions was extensively studied starting in the 19th century (see, e.g., Kiepert, *Math. Ann.* 32 (1888), 1–135 and 37 (1890), 368–98) and continues to the present day. We have taken a purely algebraic approach, but the reader should be aware that in doing so, we have missed out on much of the theory.

The history of Theorem 491 is quite interesting. Beppo Levi (*Atti Accad. Sci. Torino* 42 (1906), 739–64 and 43 (1908), 99–120, 413–34, 672–81) computed equations of various modular curves $X_1(N)$ and proved that $X_1(N)$ has no nontrivial rational points for $N = 14$, 16, and 20, thereby showing that no elliptic curve can have a rational point of these orders. Prime values of N are more difficult, with $N = 11$ being handled by Billing and Mahler (*J. London Math. Soc.* 15 (1940), 32–43), $N = 17$ by Ogg (*Invent. Math.* 12 (1971), 105–11), and $N = 13$ by Mazur and Tate (*Invent. Math.* 22 (1973), 41–9). Mazur then proved the general result (Theorem 491) in *IHES Publ. Math.* 47 (1978), 33–186.

Mazur's theorem was extended to quadratic number fields by Kamienny (*Invent. Math.* 109 (1992), 221–9), to number fields of degree at most 8 by Kamienny and Mazur, and to number fields of degree at most 14 by Abramovich. Merel (*Invent. Math.* 124 (1996), 437–49) then proved uniform boundedness for all number fields. Merel's theorem states that a point of finite order in $E(k)$ has order bounded by a constant depending only on the degree of the number field k.

§ 25.10. After earlier work by Frey, Hellegouarch, Kubert, and others relating Fermat curves and modular curves, Frey (*Ann. Univ. Sarav. Ser. Math.* 1 (1986), iv+40) suggested that the $E_{a,b,c}$ curves should not be modular. Serre (*Duke Math. J.* 54 (1987), 179–230) formulated a conjecture on modular representations that implies Frey's conjecture. Ribet (*Invent. Math.* 100 (1990), 431–76) then proved Serre's conjecture, thereby showing that $E_{a,b,c}$ is not modular.

Despite their strikingly different statements, Theorem 487 on the analytic continuation of L-series and Theorem 493 on the modularity of elliptic curves are closely related to one another via the theory of modular forms. Work of Eichler (*Arch. Math.* 5 (1954), 355–66), Shimura (*J. Math. Soc. Japan* 10 (1958), 1–28), and Weil (*Math. Ann.* 168 (1967), 149–56) shows that, up to some technical conditions, the two theorems are equivalent. Thus the history of the proof of Theorem 487, which is described in the notes to § 25.8, is equally the history of the proof of Theorem 493.

For a brief, but technical, overview of the proof of Fermat's last theorem. see Stevens, *Modular forms and Fermat's last theorem*, Springer, 1997, 1 15. And for the enterprising reader, the remaining 550+ pages of this instructional conference proceedings provide further details of the many pieces that fit sungly together to form a proof of this famous 350-year-old problem.

APPENDIX

1. Another formula for p_n. We can use Theorem 80 to write down a formula for $\pi(n)$ and so one for p_n. These formulae do not suffer from the disadvantage of those described in § 22.3. In theory, they could be used to calculate $\pi(n)$ and p_n, but at the cost of much heavier calculation than the Sieve of Eratosthenes; indeed the calculation is prohibitive except for fairly small n. It follows from Theorem 80 that

$$(j-2)! \equiv a (\operatorname{mod} j), \qquad (j \geqslant 5)$$

where $a = 1$ or 0, according as j is a prime or composite. Hence we have

$$\pi(n) = 2 + \sum_{j=5}^{n} \left\{ (j-2)! - j \left[\frac{(j-2)!}{j} \right] \right\} \quad (n \geqslant 5),$$

while $\pi(1) = 0$, $\pi(2) = 1$, and $\pi(3) = \pi(4) = 2$.

We now write

$$f(x, x) = 0, \quad f(x, y) = \tfrac{1}{2} \left\{ 1 + \frac{x-y}{|x-y|} \right\} \quad (x \neq y),$$

so that $f(x,y) = 1$ or 0 according as $x > y$ or $x \leqslant y$. Then $f(n, \pi(j)) = 0$ or 1 according as $n \leqslant \pi(j)$ or $n > \pi(j)$, i.e. as $j \geqslant p_n$ or $j < p_n$. But $p_n < 2^n$ by Theorem 418. Hence

$$1 + \sum_{j=1}^{2^n} f(n, \pi(j)) = 1 + \sum_{j=1}^{p_n - 1} 1 = p_n.$$

This is our formula for p_n.

There is a considerable literature on formulae for primes of various kinds. See, for example, Dudley (*American Math. Monthly* 76 (1969), 23–28), Golomb (*ibid.* 81 (1974), 752–4) and Gandhi's review of the latter paper (*Math. Rev.* 50 (1975), 963), which give further references.

2. A generalization of Theorem 22. Theorem 22 can be generalized to a larger number of variables. Thus suppose that $P_i(x_1, \ldots, x_k)$ and $Q_i(x_1, \ldots, x_k)$ are polynomials with integer coefficients, that $a_1, \ldots, a_m$ are positive integers and that

$$F = F(x_1, \ldots, x_k) = \sum_{i=1}^{m} P_i(x_1, \ldots, x_k) a_i^{Q_i(x_1, \ldots, x_k)}.$$

If F takes only prime values for all possible non-negative values of $x_1, \ldots, x_k$, then F must be a constant. On the other hand, Davis, Matijasevic, Putnam, and Robinson have shown how to construct a polynomial $R(x_1, \ldots, x_k)$, all of whose positive values are prime for non-negative integral values of $x_1, \ldots, x_k$ and for which the range of these positive values is precisely the primes, but all of whose negative values are composite. With $k = 42$, the degree of R need be no more than 5. The least value so far found for k is 10, when the degree of R is 15905. See Matijasevic, *Zapiski naučn, Sem. Leningrad. Otd. mat. Inst. Steklov* 68 (1977), 62–82 (Russian, English summary) for this last result and Jones, Sato, Wada, and Wiens, *American Math. Monthly* 83 (1976), 449–65 for an account of this whole topic and full references.

3. Unsolved problems concerning primes. Apart from the correction of a trivial error, the unsolved problems listed in § 2.8 are the same as those listed in the first edition (1938) of this book. None of these conjectures has been proved or disproved in the intervening 70 years. But there have been substantial advances towards their proof and we describe some of them here.

Goldbach enunciated his 'theorem' (mentioned in § 2.8) that every even $n > 3$ is the sum of two primes in a letter to Euler in 1742. Vinogradov proved in 1937 that every sufficiently large odd number is the sum of three primes. Estermann, *Introduction,* gives Vinogradov's proof. Let $E(x)$ be the number of even integers less than x which are not the sum of two primes. Estermann, van der Corput, and Chudakov proved that $E(x) = o(x)$ and Montgomery and Vaughan (*Acta Arith.* 27 (1975), 353–70) improved this to $E(x) = O(x^{1-\delta})$ for a suitable $\delta > 0$. See this last paper for references. Ramaré (*Ann. Scuola Norm. Sup. Pisa Cl. Sci.* (4) 22 (1995), 645–706) has shown that every positive integer is a sum of at most 6 primes. As of 2007, it has been verified that the Goldbach hypothesis is true for $n \leqslant 5 \times 10^{17}$ (Oliveira e Silva, see http://www.ieeta.pt/tos/goldbach.html).

Let us write P_2 to denote any number that is a prime or the product of two primes. Chen has proved that every sufficiently large even number is a sum of a prime and a P_2 (see Ross, *J. London Math. Soc.* (2) 10 (1975), 500–506 for the simplest proof) and also that there are infinitely many primes p such that $p + 2$ is a P_2. There is a P_2 between n^2 and $(n + 1)^2$ (Chen, *Sci Sinica* 18 (1975), 611–27) and there is a prime between n—n^θ and n, where $\theta = 0.525$ (Baker, Harman, and Pintz, *Proc. London Math. Soc.* (3) 83 (2001), 532–562). All the results mentioned in this paragraph have been found by the modern sieve method; see Halberstam and Roth, ch. 4 for an elementary exposition and Halberstam and Richert for a fuller treatment.

Friedlander and Iwaniec (*Ann. of Math.* (2) 148 (1998), 945–1040) have shown that there are infinitely many primes of the form $a^2 + b^4$. Similarly Heath-Brown (*Acta Math.* 186 (2001), 1–84) has shown that there are infinitely many primes of the shape $a^3 + 2b^3$. This latter result has been extended to arbitrary binary cubic forms by Heath-Brown and Moroz (*Proc. London Math. Soc.* (3) 84 (2002), 257–288). Results of this type give the sparsest polynomial sequences currently known to contain infinitely many primes. It would be very interesting to have a similar result for primes of the shape $4a^3 + 27b^2$, since this would show that there are infinitely many cubic polynomials with integer coefficients and prime discriminant. It would also resolve the open conjecture that there are infinitely many non-isomorphic elliptic curves defined over the rationals and having prime conductor.

It follows from the Prime Number Theorem that for numbers around x the average gap between consecutive primes is asymptotically $\log x$. However it is known that gaps which are much smaller, and much larger, can occur. On the one hand, Goldston, Pintz, and Yıldırım, (in work still to appear, as of 2007) have shown that

$$\liminf_{n\to\infty} \frac{p_{n+1} - p_n}{\log p_n} = 0,$$

and even that

$$\liminf_{n\to\infty} \frac{p_{n+1} - p_n}{(\log p_n)^{1/2}(\log\log p_n)^2} < \infty.$$

In the other direction Pintz (*J. Number Theory* 63 (1997), 286–301) has proved that there are infinitely many primes p_n for which

$$p_{n+1} - p_n \geqslant 2(e^\gamma + o(1)) \log p_n \frac{(\log\log p_n)(\log\log\log\log p_n)}{(\log\log\log p_n)^2}$$

(where γ is Euler's constant).

One of the most remarkable recent results on primes is due to Green and Tao (*Annals of Math.* to appear), and states that the primes contain arbitrarily long arithmetic progressions. The longest such progression currently known (2007) has length 23, and consists of the primes

$$56211383760397 + 44546738095860k \quad (k = 0, 2, \ldots, 22),$$

found by Frind, Underwood, and Jobling.

A LIST OF BOOKS

THIS list contains only (*a*) the books which we quote most frequently and (*b*) those which are most likely to be useful to a reader who wishes to study the subject more seriously. Those marked with an asterisk are elementary. Books in this list are usually referred to by the author's name alone ('Ingham' or 'Pólya and Szegő') or by a short title ('Dickson, *History*' or 'Landau, *Vorlesungen*'). Other books mentioned in the text are given their full titles.

W. Ahrens.* *Mathematische Unterhaltungen und Spiele* (2nd edition, Leipzig, Teubner, 1910).

G. E. Andrews. *The theory of partitions* (London, Addison-Wesley, 1976).

G. E. Andrews and B. Berndt. *Ramanujan's lost notebook*, Part I (New York, Springer, 2005).

G. E. Andrews and K. Eriksson. *Integer partitions*, (Cambridge University Press, 2004).

P. Bachmann. 1. *Zahlentheorie* (Leipzig, Teubner, 1872–1923). (i) *Die Elemente der Zahlentheorie* (1892). (ii) *Die analytische Zahlentheorie* (1894). (iii) *Die Lehre von der Kreisteilung und ihre Beziehungen zur Zahlentheorie* (1872). (iv) *Die Arithmetik der quadratischen Formen* (part 1, 1898; part 2, 1923). (v) *Allgemeine Arithmetik der Zahlkörper* (1905).

2. *Niedere Zahlentheorie* (Leipzig, Teubner; part 1, 1902; part 2, 1910).

3. *Grundlehren der neueren Zahlentheorie* (2nd edition, Berlin, de Gruyter, 1921).

A. Baker. *Transcendental number theory* (Cambridge University Press, 1975).

W. W. Rouse Ball.* *Mathematical recreations and essays* (11th edition, revised by H. S. M. Coxeter, London, Macmillan, 1939).

R. Bellman. *Analytic number theory: an introduction* (Reading Mass., Benjamin Cummings, 1980).

R. D. Carmichael. 1*. *Theory of numbers* (*Mathematical monographs,* no. 13, New York, Wiley, 1914).

2*. *Diophantine analysis* (*Mathematical monographs,* no 16, New York, Wiley, 1915).

J. W. S. Cassels. 1. *An introduction to Diophantine approximation* (*Cambridge Tracts in Mathematics,* no. 45, 1957).

2. *An introduction to the geometry of numbers* (*Berlin*, *Springer*, 1959).

J. W. S. Cassels. *Lectures on elliptic curves*, (Cambridge University Press, 1991).

G. Cornell, J. H. Silverman, G. Stevens, eds. *Modular forms and Fermat's last theorem*, (New York, Springer, 1997).

H. Davenport.* *The higher arithmetic* (London, Hutchinson, 1952).

L. E. Dickson. 1*. *Introduction to the theory of numbers* (Chicago University Press, 1929: *Introduction*).

2. *Studies in the theory of numbers* (Chicago University Press, 1930: *Studies*).

3. *History of the theory of numbers* (Carnegie Institution; vol. i, 1919; vol. ii, 1920; vol. iii, 1923: *History*).

P. G. Lejeune Dirichlet. *Vorlesungen über Zahlentheorie,* herausgegeben von R. Dedekind (4th edition, Braunschweig, Vieweg, 1894).

T. Estermann. *Introduction to modern prime number theory* (*Cambridge Tracts in Mathematics,* No. 41, 1952).

C. F. Gauss. *Disquisitiones arithmeticae* (Leipzig, Fleischer, 1801; reprinted in vol. i of Gauss's *Werke: D.A.*).

H. Halberstam and H.-E. Richert. *Sieve methods* (*L.M.S. Monographs,* no. 4, London, Academic Press, 1974).

H. Halberstam and K. F. Roth. *Sequences* (Oxford University Press, 1966).

G. H. Hardy. *Ramanujan* (Cambridge University Press, 1940).

H. Hasse. 1. *Number theory* (Berlin, Akademie-Verlag, 1977).

2. *Number theory,* translated and edited by H. G. Zimmer (Berlin, Springer, 1978).

E. Hecke. *Vorlesungen über die Theorie der algebraischen Zahlen* (Leipzig, Akademische Verlagsgesellschaft, 1923).

D. Hilbert. *Bericht über die Theorie der algebraischen Zahlkörper* (*Jahresbericht der Deutschen Mathematiker-Vereinigung,* iv, 1897: reprinted in vol. i of Hilbert's *Gesammelte Abhandlungen*).

A. E. Ingham. *The distribution of prime numbers* (*Cambridge Tracts in Mathematics,* no. 30, Cambridge University Press, 1932).

H. W. E. Jung. *Einführung in die Theorie der quadratischen Zahlkörper* (Leipzig, Jänicke, 1936).

A. W. Knapp. *Elliptic curves*, (Princeton University Press, 1992).

N. Koblitz. *Introduction to elliptic curves and modular forms*, (New York, Springer, 1993).

J. F. Koksma. *Diophantische Approximationen* (*Ergebnisse der Mathematik,* Band iv, Heft 4, Berlin, Springer, 1937).

L. Kuipers and H. Niederreiter. *Uniform distribution of sequences* (New York, Wiley, 1974).

E. Landau. 1. *Handbuch der Lehre von der Verteilung der Primzahlen* (2 vols., paged consecutively, Leipzig, Teubner, 1909: *Handbuch*).

2. *Vorlesungen über Zahlentheorie* (3 vols., Leipzig, Hirzel, 1927: *Vorlesungen*).

3. *Einführung in die elementare und analytische Theorie der algebraischen Zahlen um der Ideale* (2nd edition, Leipzig, Teubner, 1927: *Algebraische Zahlen*).

4. *Über einige neuere Forschritte der additiven Zahlentheorie* (*Cambridge Tracts in Mathematics,* no. 35, Cambridge University Press, 1937).

S. Lang. *Elliptic curves: Diophantine analysis*, (Berlin, Springer, 1978).

S. Lang. *Elliptic functions*, (New York, Springer, 1987).

C. G. Lekkerkerker. *Geometry of numbers* (Amsterdam, North-Holland, 1969).

W. J. LeVeque (ed.) *Reviews in number theory* (Providence R. I., A.M.S. 1974).

P. A. MacMahon. *Combinatory analysis* (Cambridge University Press, vol. i, 1915; vol. ii, 1916).

H. Minkowski. 1. *Geometrie der Zahlen* (Leipzig, Teubner, 1910).

2. *Diophantische Approximationen* (Leipzig, Teubner, 1927).

L. J. Mordell. *Diophantine equations* (London, Academic Press, 1969).

T. Nagell.* *Introduction to number theory* (New York, Wiley, 1951).

I. Niven. *Irrational Numbers* (Carus Math. Monographs, no. 11, Math. Assoc. of America, 1956).

C. D. Olds.* *Continued fractions* (New York, Random House, 1963).

K. Ono. *The web of modularity* (CBMS, No.102, American Mathematical Society, 2004).

O. Ore.* *Number theory and its history* (New York, McGraw-Hill, 1948).

O. Perron. 1. *Irrationalzahlen* (Berlin, de Gruyter, 1910).

2. *Die Lehre von den Kettenbrüchen* (Leipzig, Teubner, 1929).

G. Pólya and G. Szegő. *Problems and theorems in analysis* ii (reprinted Berlin, Springer, 1976). (References are to the numbers of problems and solutions in Part VIII).

K. Prachar. *Primzahlverteilung* (Berlin, Springer, 1957).

H. Rademacher und O. Toeplitz.* *Von Zahlen und Figuren* (2nd edition, Berlin, Springer, 1933).

C. A. Rogers. *Packing and covering* (Cambridge Tracts in Math. No. 54, 1964).

A. Scholz.* *Einführung in die Zahlentheorie* (Sammlung Göschen Band 1131, Berlin, de Gruyter, 1945).

D. Shanks.* *Solved and unsolved problems in number theory* (Washington D.C., Spartan Books, 1962).

J. H. Silverman. *The arithmetic of elliptic curves*, (New York, Springer, 1986).

J. H. Silverman. *Advanced topics in the arithmetic of elliptic curves*, (New York, Springer, 1994),

J. H. Silverman and J. Tate. *Rational points on elliptic curves*. (New York, Springer, 1992).

H. J. S. Smith. *Report on the theory of numbers* (*Reports of the British Association,* 1859–1865: reprinted in vol. i of Smith's *Collected mathematical papers*).

J. Sommer. *Vorlesungen über Zahlentheorie* (Leipzig, Tuebner, 1907).

H. M. Stark. *An introduction to number theory* (Chicago, Markham, 1970).

J. V. Uspensky and M. A. Heaslet. *Elementary number theory* (New York, Macmillan, 1939).

R. C. Vaughan. *The Hardy-Littlewood method* (Cambridge Tracts in Math. No. 80, 1981).

I. M. Vinogradov. 1. *The method of trigonometrical sums in the theory of numbers,* translated, revised, and annotated by K. F. Roth and Anne Davenport (London and New York, Interscience Publishers, 1954).

2. *An introduction to the theory of numbers,* translated by Helen Popova (London and New York, Pergamon Press, 1955).

INDEX OF SPECIAL SYMBOLS AND WORDS

THE references give the section and page where the definition of the symbol in question is to be found. We include all symbols which occur frequently in standard senses, but not symbols which, like $S(m,n)$ in § 5.6, are used only in particular sections.

Symbols in the list are sometimes also used temporarily for other purposes, as is γ in § 3.11 and elsewhere.

General analytical symbols

Symbols of divisibility, congruence, etc.

Special numbers and functions

Words

We add references to the definitions of a small number of words and phrases which a reader may find difficulty in tracing because they do not occur in the headings of sections.

INDEX OF NAMES

GENERAL INDEX

Note: References to footnotes are denoted by (f.n.) after the page number. Some symbols which have specialized meanings, or which are easily confused, are included at the beginning of this index.